REAL AND COMPLEX ANALYSIS

REAL AND COMPLEX ANALYSIS

Third Edition

Walter Rudin

Professor of Mathematics
University of Wisconsin, Madison

Boston, Massachusetts   Burr Ridge, Illinois
Dubuque, Iowa   Madison, Wisconsin   New York, New York
San Francisco, California   St. Louis, Missouri

WCB/McGraw-Hill
A Division of The McGraw·Hill Companies

This book was set in Times Roman.
The editor was Peter R. Devine; the production supervisor was Diane Renda;
the cover was designed by Laura Stover.
Project supervision was done by Santype International Limited.
Halliday Lithograph Corporation was printer and binder.

REAL AND COMPLEX ANALYSIS

Copyright © 1987, 1974, 1966 by McGraw-Hill, Inc. All rights reserved. Printed in the United States of America. Except as permitted under the United States Copyright Act of 1976, no part of this publication may be reproduced or distributed in any form or by any means, or stored in a data base or retrieval system, without the prior written permission of the publisher.

14 15 16 17 18 19 BKM BKM 9 0 9 8 7 6 5 4 3 2 1 0

ISBN 0-07-054234-1

Library of Congress Cataloging-in-Publication Data

Rudin, Walter, 1921–
    Real and complex analysis.

    Bibliography: p.
    Includes index.
    1. Mathematical analysis. I. Title.
QA300.R82   1987   515   86–7
ISBN 0-07-054234-1

ABOUT THE AUTHOR

Walter Rudin is the author of three textbooks, *Principles of Mathematical Analysis*, *Real and Complex Analysis*, and *Functional Analysis*, whose widespread use is illustrated by the fact that they have been translated into a total of 13 languages. He wrote the first of these while he was a C.L.E. Moore Instructor at M.I.T., just two years after receiving his Ph.D. at Duke University in 1949. Later he taught at the University of Rochester, and is now a Vilas Research Professor at the University of Wisconsin–Madison, where he has been since 1959.

He has spent leaves at Yale University, at the University of California in La Jolla, and at the University of Hawaii.

His research has dealt mainly with harmonic analysis and with complex variables. He has written three research monographs on these topics, *Fourier Analysis on Groups*, *Function Theory in Polydiscs*, and *Function Theory in the Unit Ball of* C^n.

CONTENTS

	Preface	xiii
	Prologue: The Exponential Function	1
Chapter 1	**Abstract Integration**	5
	Set-theoretic notations and terminology	6
	The concept of measurability	8
	Simple functions	15
	Elementary properties of measures	16
	Arithmetic in $[0, \infty]$	18
	Integration of positive functions	19
	Integration of complex functions	24
	The role played by sets of measure zero	27
	Exercises	31
Chapter 2	**Positive Borel Measures**	33
	Vector spaces	33
	Topological preliminaries	35
	The Riesz representation theorem	40
	Regularity properties of Borel measures	47
	Lebesgue measure	49
	Continuity properties of measurable functions	55
	Exercises	57
Chapter 3	L^p**-Spaces**	61
	Convex functions and inequalities	61
	The L^p-spaces	65
	Approximation by continuous functions	69
	Exercises	71

Chapter 4 Elementary Hilbert Space Theory — 76

- Inner products and linear functionals — 76
- Orthonormal sets — 82
- Trigonometric series — 88
- Exercises — 92

Chapter 5 Examples of Banach Space Techniques — 95

- Banach spaces — 95
- Consequences of Baire's theorem — 97
- Fourier series of continuous functions — 100
- Fourier coefficients of L^1-functions — 103
- The Hahn-Banach theorem — 104
- An abstract approach to the Poisson integral — 108
- Exercises — 112

Chapter 6 Complex Measures — 116

- Total variation — 116
- Absolute continuity — 120
- Consequences of the Radon-Nikodym theorem — 124
- Bounded linear functionals on L^p — 126
- The Riesz representation theorem — 129
- Exercises — 132

Chapter 7 Differentiation — 135

- Derivatives of measures — 135
- The fundamental theorem of Calculus — 144
- Differentiable transformations — 150
- Exercises — 156

Chapter 8 Integration on Product Spaces — 160

- Measurability on cartesian products — 160
- Product measures — 163
- The Fubini theorem — 164
- Completion of product measures — 167
- Convolutions — 170
- Distribution functions — 172
- Exercises — 174

Chapter 9 Fourier Transforms — 178

- Formal properties — 178
- The inversion theorem — 180
- The Plancherel theorem — 185
- The Banach algebra L^1 — 190
- Exercises — 193

Chapter 10 Elementary Properties of Holomorphic
Functions 196
    Complex differentiation 196
    Integration over paths 200
    The local Cauchy theorem 204
    The power series representation 208
    The open mapping theorem 214
    The global Cauchy theorem 217
    The calculus of residues 224
    Exercises 227

Chapter 11 Harmonic Functions 231
    The Cauchy-Riemann equations 231
    The Poisson integral 233
    The mean value property 237
    Boundary behavior of Poisson integrals 239
    Representation theorems 245
    Exercises 249

Chapter 12 The Maximum Modulus Principle 253
    Introduction 253
    The Schwarz lemma 254
    The Phragmen-Lindelöf method 256
    An interpolation theorem 260
    A converse of the maximum modulus theorem 262
    Exercises 264

Chapter 13 Approximation by Rational Functions 266
    Preparation 266
    Runge's theorem 270
    The Mittag-Leffler theorem 273
    Simply connected regions 274
    Exercises 276

Chapter 14 Conformal Mapping 278
    Preservation of angles 278
    Linear fractional transformations 279
    Normal families 281
    The Riemann mapping theorem 282
    The class $\mathscr{S}$ 285
    Continuity at the boundary 289
    Conformal mapping of an annulus 291
    Exercises 293

Chapter 15   Zeros of Holomorphic Functions   298

    Infinite products   298
    The Weierstrass factorization theorem   301
    An interpolation problem   304
    Jensen's formula   307
    Blaschke products   310
    The Müntz-Szasz theorem   312
    Exercises   315

Chapter 16   Analytic Continuation   319

    Regular points and singular points   319
    Continuation along curves   323
    The monodromy theorem   326
    Construction of a modular function   328
    The Picard theorem   331
    Exercises   332

Chapter 17 H^p-Spaces   335

    Subharmonic functions   335
    The spaces H^p and N 337
    The theorem of F. and M. Riesz   341
    Factorization theorems   342
    The shift operator   346
    Conjugate functions   350
    Exercises   352

Chapter 18   Elementary Theory of Banach Algebras   356

    Introduction   356
    The invertible elements   357
    Ideals and homomorphisms   362
    Applications   365
    Exercises   369

Chapter 19   Holomorphic Fourier Transforms   371

    Introduction   371
    Two theorems of Paley and Wiener   372
    Quasi-analytic classes   377
    The Denjoy-Carleman theorem   380
    Exercises   383

Chapter 20   Uniform Approximation by Polynomials   386

    Introduction   386
    Some lemmas   387
    Mergelyan's theorem   390
    Exercises   394

Appendix: Hausdorff's Maximality Theorem   395

Notes and Comments   397

Bibliography   405

List of Special Symbols   407

Index   409

PREFACE

This book contains a first-year graduate course in which the basic techniques and theorems of analysis are presented in such a way that the intimate connections between its various branches are strongly emphasized. The traditionally separate subjects of "real analysis" and "complex analysis" are thus united; some of the basic ideas from functional analysis are also included.

Here are some examples of the way in which these connections are demonstrated and exploited. The Riesz representation theorem and the Hahn-Banach theorem allow one to "guess" the Poisson integral formula. They team up in the proof of Runge's theorem. They combine with Blaschke's theorem on the zeros of bounded holomorphic functions to give a proof of the Müntz-Szasz theorem, which concerns approximation on an interval. The fact that L^2 is a Hilbert space is used in the proof of the Radon-Nikodym theorem, which leads to the theorem about differentiation of indefinite integrals, which in turn yields the existence of radial limits of bounded harmonic functions. The theorems of Plancherel and Cauchy combined give a theorem of Paley and Wiener which, in turn, is used in the Denjoy-Carleman theorem about infinitely differentiable functions on the real line. The maximum modulus theorem gives information about linear transformations on L^p-spaces.

Since most of the results presented here are quite classical (the novelty lies in the arrangement, and some of the proofs are new), I have not attempted to document the source of every item. References are gathered at the end, in Notes and Comments. They are not always to the original sources, but more often to more recent works where further references can be found. In no case does the absence of a reference imply any claim to originality on my part.

The prerequisite for this book is a good course in advanced calculus (set-theoretic manipulations, metric spaces, uniform continuity, and uniform convergence). The first seven chapters of my earlier book "Principles of Mathematical Analysis" furnish sufficient preparation.

Experience with the first edition shows that first-year graduate students can study the first 15 chapters in two semesters, plus some topics from 1 or 2 of the remaining 5. These latter are quite independent of each other. The first 15 should be taken up in the order in which they are presented, except for Chapter 9, which can be postponed.

The most important difference between this third edition and the previous ones is the entirely new chapter on differentiation. The basic facts about differentiation are now derived from the existence of Lebesgue points, which in turn is an easy consequence of the so-called "weak type" inequality that is satisfied by the maximal functions of measures on euclidean spaces. This approach yields strong theorems with minimal effort. Even more important is that it familiarizes students with maximal functions, since these have become increasingly useful in several areas of analysis.

One of these is the study of the boundary behavior of Poisson integrals. A related one concerns H^p-spaces. Accordingly, large parts of Chapters 11 and 17 were rewritten and, I hope, simplified in the process.

I have also made several smaller changes in order to improve certain details: For example, parts of Chapter 4 have been simplified; the notions of equicontinuity and weak convergence are presented in more detail; the boundary behavior of conformal maps is studied by means of Lindelöf's theorem about asymptotic values of bounded holomorphic functions in a disc.

Over the last 20 years, numerous students and colleagues have offered comments and criticisms concerning the content of this book. I sincerely appreciated all of these, and have tried to follow some of them. As regards the present edition, my thanks go to Richard Rochberg for some useful last-minute suggestions, and I especially thank Robert Burckel for the meticulous care with which he examined the entire manuscript.

Walter Rudin

PROLOGUE
THE EXPONENTIAL FUNCTION

This is the most important function in mathematics. It is defined, for every complex number z, by the formula

$$\exp(z) = \sum_{n=0}^{\infty} \frac{z^n}{n!}. \tag{1}$$

The series (1) converges absolutely for every z and converges uniformly on every bounded subset of the complex plane. Thus exp is a continuous function. The absolute convergence of (1) shows that the computation

$$\sum_{k=0}^{\infty} \frac{a^k}{k!} \sum_{m=0}^{\infty} \frac{b^m}{m!} = \sum_{n=0}^{\infty} \frac{1}{n!} \sum_{k=0}^{n} \frac{n!}{k!(n-k)!} a^k b^{n-k} = \sum_{n=0}^{\infty} \frac{(a+b)^n}{n!}$$

is correct. It gives the important addition formula

$$\exp(a)\exp(b) = \exp(a+b), \tag{2}$$

valid for all complex numbers a and b.

We define the number e to be $\exp(1)$, and shall usually replace $\exp(z)$ by the customary shorter expression e^z. Note that $e^0 = \exp(0) = 1$, by (1).

Theorem

(a) *For every complex z we have $e^z \neq 0$.*
(b) *exp is its own derivative*: $\exp'(z) = \exp(z)$.

1

(c) *The restriction of* exp *to the real axis is a monotonically increasing positive function, and*
$$e^x \to \infty \text{ as } x \to \infty, \qquad e^x \to 0 \text{ as } x \to -\infty.$$

(d) *There exists a positive number* π *such that* $e^{\pi i/2} = i$ *and such that* $e^z = 1$ *if and only if* $z/(2\pi i)$ *is an integer.*

(e) exp *is a periodic function, with period* $2\pi i$.

(f) *The mapping* $t \to e^{it}$ *maps the real axis onto the unit circle.*

(g) *If* w *is a complex number and* $w \neq 0$, *then* $w = e^z$ *for some* z.

PROOF By (2), $e^z \cdot e^{-z} = e^{z-z} = e^0 = 1$. This implies (a). Next,

$$\exp'(z) = \lim_{h \to 0} \frac{\exp(z+h) - \exp(z)}{h} = \exp(z) \lim_{h \to 0} \frac{\exp(h) - 1}{h} = \exp(z).$$

The first of the above equalities is a matter of definition, the second follows from (2), and the third from (1), and (b) is proved.

That exp is monotonically increasing on the positive real axis, and that $e^x \to \infty$ as $x \to \infty$, is clear from (1). The other assertions of (c) are consequences of $e^x \cdot e^{-x} = 1$.

For any real number t, (1) shows that e^{-it} is the complex conjugate of e^{it}. Thus

$$|e^{it}|^2 = e^{it} \cdot \overline{e^{it}} = e^{it} \cdot e^{-it} = e^{it-it} = e^0 = 1,$$

or

$$|e^{it}| = 1 \qquad (t \text{ real}). \tag{3}$$

In other words, if t is real, e^{it} lies on the unit circle. We define $\cos t$, $\sin t$ to be the real and imaginary parts of e^{it}:

$$\cos t = \text{Re}\,[e^{it}], \qquad \sin t = \text{Im}\,[e^{it}] \qquad (t \text{ real}). \tag{4}$$

If we differentiate both sides of Euler's identity

$$e^{it} = \cos t + i \sin t, \tag{5}$$

which is equivalent to (4), and if we apply (b), we obtain

$$\cos' t + i \sin' t = i e^{it} = -\sin t + i \cos t,$$

so that

$$\cos' = -\sin, \qquad \sin' = \cos. \tag{6}$$

The power series (1) yields the representation

$$\cos t = 1 - \frac{t^2}{2!} + \frac{t^4}{4!} - \frac{t^6}{6!} + \cdots. \tag{7}$$

Take $t = 2$. The terms of the series (7) then decrease in absolute value (except for the first one) and their signs alternate. Hence cos 2 is less than the sum of the first three terms of (7), with $t = 2$; thus $\cos 2 < -\frac{1}{3}$. Since $\cos 0 = 1$ and cos is a continuous real function on the real axis, we conclude that there is a smallest positive number t_0 for which $\cos t_0 = 0$. We define

$$\pi = 2t_0. \tag{8}$$

It follows from (3) and (5) that $\sin t_0 = \pm 1$. Since

$$\sin'(t) = \cos t > 0$$

on the segment $(0, t_0)$ and since $\sin 0 = 0$, we have $\sin t_0 > 0$, hence $\sin t_0 = 1$, and therefore

$$e^{\pi i/2} = i. \tag{9}$$

It follows that $e^{\pi i} = i^2 = -1$, $e^{2\pi i} = (-1)^2 = 1$, and then $e^{2\pi i n} = 1$ for every integer n. Also, (e) follows immediately:

$$e^{z + 2\pi i} = e^z e^{2\pi i} = e^z. \tag{10}$$

If $z = x + iy$, x and y real, then $e^z = e^x e^{iy}$; hence $|e^z| = e^x$. If $e^z = 1$, we therefore must have $e^x = 1$, so that $x = 0$; to prove that $y/2\pi$ must be an integer, it is enough to show that $e^{iy} \neq 1$ if $0 < y < 2\pi$, by (10).

Suppose $0 < y < 2\pi$, and

$$e^{iy/4} = u + iv \quad (u \text{ and } v \text{ real}). \tag{11}$$

Since $0 < y/4 < \pi/2$, we have $u > 0$ and $v > 0$. Also

$$e^{iy} = (u + iv)^4 = u^4 - 6u^2 v^2 + v^4 + 4iuv(u^2 - v^2). \tag{12}$$

The right side of (12) is real only if $u^2 = v^2$; since $u^2 + v^2 = 1$, this happens only when $u^2 = v^2 = \frac{1}{2}$, and then (12) shows that

$$e^{iy} = -1 \neq 1.$$

This completes the proof of (d).

We already know that $t \to e^{it}$ maps the real axis *into* the unit circle. To prove (f), fix w so that $|w| = 1$; we shall show that $w = e^{it}$ for some real t. Write $w = u + iv$, u and v real, and suppose first that $u \geq 0$ and $v \geq 0$. Since $u \leq 1$, the definition of π shows that there exists a t, $0 \leq t \leq \pi/2$, such that $\cos t = u$; then $\sin^2 t = 1 - u^2 = v^2$, and since $\sin t \geq 0$ if $0 \leq t \leq \pi/2$, we have $\sin t = v$. Thus $w = e^{it}$.

If $u < 0$ and $v \geq 0$, the preceding conditions are satisfied by $-iw$. Hence $-iw = e^{it}$ for some real t, and $w = e^{i(t + \pi/2)}$. Finally, if $v < 0$, the preceding two cases show that $-w = e^{it}$ for some real t, hence $w = e^{i(t + \pi)}$. This completes the proof of (f).

If $w \neq 0$, put $\alpha = w/|w|$. Then $w = |w|\alpha$. By (c), there is a real x such that $|w| = e^x$. Since $|\alpha| = 1$, (f) shows that $\alpha = e^{iy}$ for some real y. Hence $w = e^{x+iy}$. This proves (g) and completes the theorem. ////

We shall encounter the integral of $(1 + x^2)^{-1}$ over the real line. To evaluate it, put $\varphi(t) = \sin t/\cos t$ in $(-\pi/2, \pi/2)$. By (6), $\varphi' = 1 + \varphi^2$. Hence φ is a monotonically increasing mapping of $(-\pi/2, \pi/2)$ onto $(-\infty, \infty)$, and we obtain

$$\int_{-\infty}^{\infty} \frac{dx}{1+x^2} = \int_{-\pi/2}^{\pi/2} \frac{\varphi'(t)\,dt}{1+\varphi^2(t)} = \int_{-\pi/2}^{\pi/2} dt = \pi.$$

CHAPTER
ONE

ABSTRACT INTEGRATION

Toward the end of the nineteenth century it became clear to many mathematicians that the Riemann integral (about which one learns in calculus courses) should be replaced by some other type of integral, more general and more flexible, better suited for dealing with limit processes. Among the attempts made in this direction, the most notable ones were due to Jordan, Borel, W. H. Young, and Lebesgue. It was Lebesgue's construction which turned out to be the most successful.

In brief outline, here is the main idea: The Riemann integral of a function f over an interval $[a, b]$ can be approximated by sums of the form

$$\sum_{i=1}^{n} f(t_i)m(E_i)$$

where $E_1, \ldots, E_n$ are disjoint intervals whose union is $[a, b]$, $m(E_i)$ denotes the length of E_i, and $t_i \in E_i$ for $i = 1, \ldots, n$. Lebesgue discovered that a completely satisfactory theory of integration results if the sets E_i in the above sum are allowed to belong to a larger class of subsets of the line, the so-called "measurable sets," and if the class of functions under consideration is enlarged to what he called "measurable functions." The crucial set-theoretic properties involved are the following: The union and the intersection of any countable family of measurable sets are measurable; so is the complement of every measurable set; and, most important, the notion of "length" (now called "measure") can be extended to them in such a way that

$$m(E_1 \cup E_2 \cup E_3 \cup \cdots) = m(E_1) + m(E_2) + m(E_3) + \cdots$$

for every countable collection $\{E_i\}$ of pairwise disjoint measurable sets. This property of m is called *countable additivity*.

The passage from Riemann's theory of integration to that of Lebesgue is a process of completion (in a sense which will appear more precisely later). It is of the same fundamental importance in analysis as is the construction of the real number system from the rationals.

The above-mentioned measure m is of course intimately related to the geometry of the real line. In this chapter we shall present an abstract (axiomatic) version of the Lebesgue integral, relative to *any* countably additive measure on *any* set. (The precise definitions follow.) This abstract theory is not in any way more difficult than the special case of the real line; it shows that a large part of integration theory is independent of any geometry (or topology) of the underlying space; and, of course, it gives us a tool of much wider applicability. The existence of a large class of measures, among them that of Lebesgue, will be established in Chap. 2.

Set-Theoretic Notations and Terminology

1.1 Some sets can be described by listing their members. Thus $\{x_1, \ldots, x_n\}$ is the set whose members are $x_1, \ldots, x_n$; and $\{x\}$ is the set whose only member is x. More often, sets are described by properties. We write

$$\{x: P\}$$

for the set of all elements x which have the property P. The symbol $\emptyset$ denotes the empty set. The words *collection, family,* and *class* will be used synonymously with *set*.

We write $x \in A$ if x is a member of the set A; otherwise $x \notin A$. If B is a subset of A, i.e., if $x \in B$ implies $x \in A$, we write $B \subset A$. If $B \subset A$ and $A \subset B$, then $A = B$. If $B \subset A$ and $A \neq B$, B is a *proper* subset of A. Note that $\emptyset \subset A$ for every set A.

$A \cup B$ and $A \cap B$ are the union and intersection of A and B, respectively. If $\{A_\alpha\}$ is a collection of sets, where α runs through some index set I, we write

$$\bigcup_{\alpha \in I} A_\alpha \quad \text{and} \quad \bigcap_{\alpha \in I} A_\alpha$$

for the union and intersection of $\{A_\alpha\}$:

$$\bigcup_{\alpha \in I} A_\alpha = \{x: x \in A_\alpha \text{ for at least one } \alpha \in I\}$$

$$\bigcap_{\alpha \in I} A_\alpha = \{x: x \in A_\alpha \text{ for every } \alpha \in I\}.$$

If I is the set of all positive integers, the customary notations are

$$\bigcup_{n=1}^{\infty} A_n \quad \text{and} \quad \bigcap_{n=1}^{\infty} A_n.$$

If no two members of $\{A_\alpha\}$ have an element in common, then $\{A_\alpha\}$ is a *disjoint collection* of sets.

We write $A - B = \{x: x \in A, x \notin B\}$, and denote the complement of A by A^c whenever it is clear from the context with respect to which larger set the complement is taken.

The *cartesian product* $A_1 \times \cdots \times A_n$ of the sets $A_1, \ldots, A_n$ is the set of all ordered n-tuples $(a_1, \ldots, a_n)$ where $a_i \in A_i$ for $i = 1, \ldots, n$.

The *real line* (or real number system) is R^1, and

$$R^k = R^1 \times \cdots \times R^1 \quad (k \text{ factors}).$$

The *extended real number system* is R^1 with two symbols, ∞ and $-\infty$, adjoined, and with the obvious ordering. If $-\infty \leq a \leq b \leq \infty$, the *interval* $[a, b]$ and the *segment* (a, b) are defined to be

$$[a, b] = \{x: a \leq x \leq b\}, \quad (a, b) = \{x: a < x < b\}.$$

We also write

$$[a, b) = \{x: a \leq x < b\}, \quad (a, b] = \{x: a < x \leq b\}.$$

If $E \subset [-\infty, \infty]$ and $E \neq \emptyset$, the least upper bound (supremum) and greatest lower bound (infimum) of E exist in $[-\infty, \infty]$ and are denoted by sup E and inf E.

Sometimes (but only when sup $E \in E$) we write max E for sup E.

The symbol

$$f: X \to Y$$

means that f is a *function* (or *mapping* or *transformation*) of the set X into the set Y; i.e., f assigns to each $x \in X$ an element $f(x) \in Y$. If $A \subset X$ and $B \subset Y$, the *image* of A and the *inverse image* (or *pre-image*) of B are

$$f(A) = \{y: y = f(x) \text{ for some } x \in A\},$$
$$f^{-1}(B) = \{x: f(x) \in B\}.$$

Note that $f^{-1}(B)$ may be empty even when $B \neq \emptyset$.

The *domain* of f is X. The *range* of f is $f(X)$.

If $f(X) = Y$, f is said to map X *onto* Y.

We write $f^{-1}(y)$, instead of $f^{-1}(\{y\})$, for every $y \in Y$. If $f^{-1}(y)$ consists of at most one point, for each $y \in Y$, f is said to be *one-to-one*. If f is one-to-one, then f^{-1} is a function with domain $f(X)$ and range X.

If $f: X \to [-\infty, \infty]$ and $E \subset X$, it is customary to write $\sup_{x \in E} f(x)$ rather than sup $f(E)$.

If $f: X \to Y$ and $g: Y \to Z$, the *composite function* $g \circ f: X \to Z$ is defined by the formula

$$(g \circ f)(x) = g(f(x)) \quad (x \in X).$$

If the range of f lies in the real line (or in the complex plane), then f is said to be a *real function* (or a *complex function*). For a complex function f, the statement "$f \geq 0$" means that all values $f(x)$ of f are nonnegative real numbers.

The Concept of Measurability

The class of measurable functions plays a fundamental role in integration theory. It has some basic properties in common with another most important class of functions, namely, the continuous ones. It is helpful to keep these similarities in mind. Our presentation is therefore organized in such a way that the analogies between the concepts *topological space*, *open set*, and *continuous function*, on the one hand, and *measurable space*, *measurable set*, and *measurable function*, on the other, are strongly emphasized. It seems that the relations between these concepts emerge most clearly when the setting is quite abstract, and this (rather than a desire for mere generality) motivates our approach to the subject.

1.2 Definition

(a) A collection τ of subsets of a set X is said to be a *topology in X* if τ has the following three properties:

   (i) $\emptyset \in \tau$ and $X \in \tau$.
   (ii) If $V_i \in \tau$ for $i = 1, \ldots, n$, then $V_1 \cap V_2 \cap \cdots \cap V_n \in \tau$.
   (iii) If $\{V_\alpha\}$ is an arbitrary collection of members of τ (finite, countable, or uncountable), then $\bigcup_\alpha V_\alpha \in \tau$.

(b) If τ is a topology in X, then X is called a *topological space*, and the members of τ are called the *open sets* in X.

(c) If X and Y are topological spaces and if f is a mapping of X into Y, then f is said to be *continuous* provided that $f^{-1}(V)$ is an open set in X for every open set V in Y.

1.3 Definition

(a) A collection $\mathfrak{M}$ of subsets of a set X is said to be a *σ-algebra in X* if $\mathfrak{M}$ has the following properties:

   (i) $X \in \mathfrak{M}$.
   (ii) If $A \in \mathfrak{M}$, then $A^c \in \mathfrak{M}$, where A^c is the complement of A relative to X.
   (iii) If $A = \bigcup_{n=1}^\infty A_n$ and if $A_n \in \mathfrak{M}$ for $n = 1, 2, 3, \ldots$, then $A \in \mathfrak{M}$.

(b) If $\mathfrak{M}$ is a σ-algebra in X, then X is called a *measurable space*, and the members of $\mathfrak{M}$ are called the *measurable sets* in X.

(c) If X is a measurable space, Y is a topological space, and f is a mapping of X into Y, then f is said to be *measurable* provided that $f^{-1}(V)$ is a measurable set in X for every open set V in Y.

It would perhaps be more satisfactory to apply the term "measurable space" to the ordered pair $(X, \mathfrak{M})$, rather than to X. After all, X is a set, and X has not been changed in any way by the fact that we now also have a σ-algebra of its subsets in mind. Similarly, a topological space is an ordered pair (X, τ). But if this sort of thing were systematically done in all mathematics, the terminology would become awfully cumbersome. We shall discuss this again at somewhat greater length in Sec. 1.21.

1.4 Comments on Definition 1.2 The most familiar topological spaces are the *metric spaces*. We shall assume some familiarity with metric spaces but shall give the basic definitions, for the sake of completeness.

A *metric space* is a set X in which a *distance function* (or *metric*) ρ is defined, with the following properties:

(a) $0 \leq \rho(x, y) < \infty$ for all x and $y \in X$.
(b) $\rho(x, y) = 0$ if and only if $x = y$.
(c) $\rho(x, y) = \rho(y, x)$ for all x and $y \in X$.
(d) $\rho(x, y) \leq \rho(x, z) + \rho(z, y)$ for all x, y, and $z \in X$.

Property (d) is called the *triangle inequality*.

If $x \in X$ and $r \geq 0$, the *open ball* with center at x and radius r is the set $\{y \in X: \rho(x, y) < r\}$.

If X is a metric space and if τ is the collection of all sets $E \subset X$ which are arbitrary unions of open balls, then τ is a topology in X. This is not hard to verify; the intersection property depends on the fact that if $x \in B_1 \cap B_2$, where B_1 and B_2 are open balls, then x is the center of an open ball $B \subset B_1 \cap B_2$. We leave this as an exercise.

For instance, in the real line R^1 a set is open if and only if it is a union of open segments (a, b). In the plane R^2, the open sets are those which are unions of open circular discs.

Another topological space, which we shall encounter frequently, is the extended real line $[-\infty, \infty]$; its topology is defined by declaring the following sets to be open: $(a, b), [-\infty, a), (a, \infty]$, and any union of segments of this type.

The definition of continuity given in Sec. 1.2(c) is a global one. Frequently it is desirable to define continuity locally: A mapping f of X into Y is said to be *continuous at the point* $x_0 \in X$ if to every neighborhood V of $f(x_0)$ there corresponds a neighborhood W of x_0 such that $f(W) \subset V$.

(A *neighborhood* of a point x is, by definition, an open set which contains x.)

When X and Y are metric spaces, this local definition is of course the same as the usual epsilon-delta definition, and is equivalent to the requirement that $\lim f(x_n) = f(x_0)$ in Y whenever $\lim x_n = x_0$ in X.

The following easy proposition relates the local and global definitions of continuity in the expected manner:

1.5 Proposition *Let X and Y be topological spaces. A mapping f of X into Y is continuous if and only if f is continuous at every point of X.*

PROOF If f is continuous and $x_0 \in X$, then $f^{-1}(V)$ is a neighborhood of x_0, for every neighborhood V of $f(x_0)$. Since $f(f^{-1}(V)) \subset V$, it follows that f is continuous at x_0.

If f is continuous at every point of X and if V is open in Y, every point $x \in f^{-1}(V)$ has a neighborhood W_x such that $f(W_x) \subset V$. Therefore $W_x \subset f^{-1}(V)$. It follows that $f^{-1}(V)$ is the union of the open sets W_x, so $f^{-1}(V)$ is itself open. Thus f is continuous. ////

1.6 Comments on Definition 1.3 Let $\mathfrak{M}$ be a σ-algebra in a set X. Referring to Properties (i) to (iii) of Definition 1.3(a), we immediately derive the following facts.

(a) Since $\emptyset = X^c$, (i) and (ii) imply that $\emptyset \in \mathfrak{M}$.
(b) Taking $A_{n+1} = A_{n+2} = \cdots = \emptyset$ in (iii), we see that $A_1 \cup A_2 \cup \cdots \cup A_n \in \mathfrak{M}$ if $A_i \in \mathfrak{M}$ for $i = 1, \ldots, n$.
(c) Since

$$\bigcap_{n=1}^{\infty} A_n = \left(\bigcup_{n=1}^{\infty} A_n^c \right)^c,$$

$\mathfrak{M}$ is closed under the formation of countable (and also finite) intersections.
(d) Since $A - B = B^c \cap A$, we have $A - B \in \mathfrak{M}$ if $A \in \mathfrak{M}$ and $B \in \mathfrak{M}$.

The prefix σ refers to the fact that (iii) is required to hold for all *countable* unions of members of $\mathfrak{M}$. If (iii) is required for finite unions only, then $\mathfrak{M}$ is called an *algebra* of sets.

1.7 Theorem *Let Y and Z be topological spaces, and let $g: Y \to Z$ be continuous.*

(a) *If X is a topological space, if $f: X \to Y$ is continuous, and if $h = g \circ f$, then $h: X \to Z$ is continuous.*
(b) *If X is a measurable space, if $f: X \to Y$ is measurable, and if $h = g \circ f$, then $h: X \to Z$ is measurable.*

Stated informally, continuous functions of continuous functions are continuous; continuous functions of measurable functions are measurable.

PROOF If V is open in Z, then $g^{-1}(V)$ is open in Y, and

$$h^{-1}(V) = f^{-1}(g^{-1}(V)).$$

If f is continuous, it follows that $h^{-1}(V)$ is open, proving (a).
If f is measurable, it follows that $h^{-1}(V)$ is measurable, proving (b). ////

1.8 Theorem *Let u and v be real measurable functions on a measurable space X, let Φ be a continuous mapping of the plane into a topological space Y, and define*

$$h(x) = \Phi(u(x), v(x))$$

for $x \in X$. Then $h: X \to Y$ is measurable.

PROOF Put $f(x) = (u(x), v(x))$. Then f maps X into the plane. Since $h = \Phi \circ f$, Theorem 1.7 shows that it is enough to prove the measurability of f.

If R is any open rectangle in the plane, with sides parallel to the axes, then R is the cartesian product of two segments I_1 and I_2, and

$$f^{-1}(R) = u^{-1}(I_1) \cap v^{-1}(I_2),$$

which is measurable, by our assumption on u and v. Every open set V in the plane is a countable union of such rectangles R_i, and since

$$f^{-1}(V) = f^{-1}\left(\bigcup_{i=1}^{\infty} R_i\right) = \bigcup_{i=1}^{\infty} f^{-1}(R_i),$$

$f^{-1}(V)$ is measurable. ////

1.9 Let X be a measurable space. The following propositions are corollaries of Theorems 1.7 and 1.8:

(a) *If $f = u + iv$, where u and v are real measurable functions on X, then f is a complex measurable function on X.*

This follows from Theorem 1.8, with $\Phi(z) = z$.

(b) *If $f = u + iv$ is a complex measurable function on X, then u, v, and $|f|$ are real measurable functions on X.*

This follows from Theorem 1.7, with $g(z) = \text{Re}(z)$, $\text{Im}(z)$, and $|z|$.

(c) *If f and g are complex measurable functions on X, then so are $f + g$ and fg.*

For real f and g this follows from Theorem 1.8, with

$$\Phi(s, t) = s + t$$

and $\Phi(s, t) = st$. The complex case then follows from (a) and (b).

(d) *If E is a measurable set in X and if*

$$\chi_E(x) = \begin{cases} 1 & \text{if } x \in E \\ 0 & \text{if } x \notin E \end{cases}$$

then χ_E is a measurable function.

This is obvious. We call χ_E the *characteristic function* of the set E. The letter χ will be reserved for characteristic functions throughout this book.

(e) *If f is a complex measurable function on X, there is a complex measurable function α on X such that $|\alpha| = 1$ and $f = \alpha|f|$.*

PROOF Let $E = \{x : f(x) = 0\}$, let Y be the complex plane with the origin removed, define $\varphi(z) = z/|z|$ for $z \in Y$, and put

$$\alpha(x) = \varphi(f(x) + \chi_E(x)) \qquad (x \in X).$$

If $x \in E$, $\alpha(x) = 1$; if $x \notin E$, $\alpha(x) = f(x)/|f(x)|$. Since φ is continuous on Y and since E is measurable (why?), the measurability of α follows from (c), (d), and Theorem 1.7.   ////

We now show that σ-algebras exist in great profusion.

1.10 Theorem *If $\mathscr{F}$ is any collection of subsets of X, there exists a smallest σ-algebra $\mathfrak{M}^*$ in X such that $\mathscr{F} \subset \mathfrak{M}^*$.*

This $\mathfrak{M}^*$ is sometimes called the σ-algebra *generated* by $\mathscr{F}$.

PROOF Let Ω be the family of all σ-algebras $\mathfrak{M}$ in X which contain $\mathscr{F}$. Since the collection of all subsets of X is such a σ-algebra, Ω is not empty. Let $\mathfrak{M}^*$ be the intersection of all $\mathfrak{M} \in \Omega$. It is clear that $\mathscr{F} \subset \mathfrak{M}^*$ and that $\mathfrak{M}^*$ lies in every σ-algebra in X which contains $\mathscr{F}$. To complete the proof, we have to show that $\mathfrak{M}^*$ is itself a σ-algebra.

If $A_n \in \mathfrak{M}^*$ for $n = 1, 2, 3, \ldots$, and if $\mathfrak{M} \in \Omega$, then $A_n \in \mathfrak{M}$, so $\bigcup A_n \in \mathfrak{M}$, since $\mathfrak{M}$ is a σ-algebra. Since $\bigcup A_n \in \mathfrak{M}$ for *every* $\mathfrak{M} \in \Omega$, we conclude that $\bigcup A_n \in \mathfrak{M}^*$. The other two defining properties of a σ-algebra are verified in the same manner.   ////

1.11 Borel Sets Let X be a topological space. By Theorem 1.10, there exists a smallest σ-algebra $\mathscr{B}$ in X such that every open set in X belongs to $\mathscr{B}$. The members of $\mathscr{B}$ are called the *Borel sets* of X.

In particular, closed sets are Borel sets (being, by definition, the complements of open sets), and so are all countable unions of closed sets and all countable intersections of open sets. These last two are called F_σ's and G_δ's, respectively, and play a considerable role. The notation is due to Hausdorff. The letters F and G were used for closed and open sets, respectively, and σ refers to union (*Summe*), δ to intersection (*Durchschnitt*). For example, every half-open interval $[a, b)$ is a G_δ and an F_σ in R^1.

Since $\mathscr{B}$ is a σ-algebra, we may now regard X as a measurable space, with the Borel sets playing the role of the measurable sets; more concisely, we consider the measurable space $(X, \mathscr{B})$. If $f \colon X \to Y$ is a continuous mapping of X, where Y is any topological space, then it is evident from the definitions that $f^{-1}(V) \in \mathscr{B}$ for every open set V in Y. In other words, *every continuous mapping of X is Borel measurable*.

Borel measurable mappings are often called *Borel mappings*, or *Borel functions*.

1.12 Theorem *Suppose $\mathfrak{M}$ is a σ-algebra in X, and Y is a topological space. Let f map X into Y.*

(a) *If Ω is the collection of all sets $E \subset Y$ such that $f^{-1}(E) \in \mathfrak{M}$, then Ω is a σ-algebra in Y.*

(b) *If f is measurable and E is a Borel set in Y, then $f^{-1}(E) \in \mathfrak{M}$.*

(c) *If $Y = [-\infty, \infty]$ and $f^{-1}((\alpha, \infty]) \in \mathfrak{M}$ for every real α, then f is measurable.*

(d) *If f is measurable, if Z is a topological space, if $g: Y \to Z$ is a Borel mapping, and if $h = g \circ f$, then $h: X \to Z$ is measurable.*

Part (c) is a frequently used criterion for the measurability of real-valued functions. (See also Exercise 3.) Note that (d) generalizes Theorem 1.7(b).

PROOF (a) follows from the relations

$$f^{-1}(Y) = X,$$
$$f^{-1}(Y - A) = X - f^{-1}(A),$$

and $\quad f^{-1}(A_1 \cup A_2 \cup \cdots) = f^{-1}(A_1) \cup f^{-1}(A_2) \cup \cdots.$

To prove (b), let Ω be as in (a); the measurability of f implies that Ω contains all open sets in Y, and since Ω is a σ-algebra, Ω contains all Borel sets in Y.

To prove (c), let Ω be the collection of all $E \subset [-\infty, \infty]$ such that $f^{-1}(E) \in \mathfrak{M}$. Choose a real number α, and choose $\alpha_n < \alpha$ so that $\alpha_n \to \alpha$ as $n \to \infty$. Since $(\alpha_n, \infty] \in \Omega$ for each n, since

$$[-\infty, \alpha) = \bigcup_{n=1}^{\infty} [-\infty, \alpha_n] = \bigcup_{n=1}^{\infty} (\alpha_n, \infty]^c,$$

and since (a) shows that Ω is a σ-algebra, we see that $[-\infty, \alpha) \in \Omega$. The same is then true of

$$(\alpha, \beta) = [-\infty, \beta) \cap (\alpha, \infty].$$

Since every open set in $[-\infty, \infty]$ is a countable union of segments of the above types, Ω contains every open set. Thus f is measurable.

To prove (d), let $V \subset Z$ be open. Then $g^{-1}(V)$ is a Borel set of Y, and since

$$h^{-1}(V) = f^{-1}(g^{-1}(V)),$$

(b) shows that $h^{-1}(V) \in \mathfrak{M}$. ////

1.13 Definition Let $\{a_n\}$ be a sequence in $[-\infty, \infty]$, and put

$$b_k = \sup \{a_k, a_{k+1}, a_{k+2}, \ldots\} \qquad (k = 1, 2, 3, \ldots) \qquad (1)$$

and
$$\beta = \inf \{b_1, b_2, b_3, \ldots\}. \tag{2}$$

We call β the *upper limit* of $\{a_n\}$, and write
$$\beta = \limsup_{n \to \infty} a_n. \tag{3}$$

The following properties are easily verified: First, $b_1 \geq b_2 \geq b_3 \geq \cdots$, so that $b_k \to \beta$ as $k \to \infty$; secondly, there is a subsequence $\{a_{n_i}\}$ of $\{a_n\}$ such that $a_{n_i} \to \beta$ as $i \to \infty$, and β is the largest number with this property.

The *lower limit* is defined analogously: simply interchange sup and inf in (1) and (2). Note that
$$\liminf_{n \to \infty} a_n = -\limsup_{n \to \infty} (-a_n). \tag{4}$$

If $\{a_n\}$ converges, then evidently
$$\limsup_{n \to \infty} a_n = \liminf_{n \to \infty} a_n = \lim_{n \to \infty} a_n. \tag{5}$$

Suppose $\{f_n\}$ is a sequence of extended-real functions on a set X. Then $\sup_n f_n$ and $\limsup_{n \to \infty} f_n$ are the functions defined on X by
$$\left(\sup_n f_n\right)(x) = \sup_n (f_n(x)), \tag{6}$$

$$\left(\limsup_{n \to \infty} f_n\right)(x) = \limsup_{n \to \infty} (f_n(x)). \tag{7}$$

If
$$f(x) = \lim_{n \to \infty} f_n(x), \tag{8}$$

the limit being assumed to exist at every $x \in X$, then we call f the *pointwise limit* of the sequence $\{f_n\}$.

1.14 Theorem *If $f_n: X \to [-\infty, \infty]$ is measurable, for $n = 1, 2, 3, \ldots$, and*
$$g = \sup_{n \geq 1} f_n, \qquad h = \limsup_{n \to \infty} f_n,$$

then g and h are measurable.

PROOF $g^{-1}((\alpha, \infty]) = \bigcup_{n=1}^{\infty} f_n^{-1}((\alpha, \infty])$. Hence Theorem 1.12(c) implies that g is measurable. The same result holds of course with inf in place of sup, and since
$$h = \inf_{k \geq 1} \left\{\sup_{i \geq k} f_i\right\},$$

it follows that h is measurable. ////

Corollaries

(a) *The limit of every pointwise convergent sequence of complex measurable functions is measurable.*

(b) *If f and g are measurable (with range in $[-\infty, \infty]$), then so are* max $\{f, g\}$ *and* min $\{f, g\}$. *In particular, this is true of the functions*

$$f^+ = \max\{f, 0\} \quad \text{and} \quad f^- = -\min\{f, 0\}.$$

1.15 The above functions f^+ and f^- are called the *positive and negative parts* of f. We have $|f| = f^+ + f^-$ and $f = f^+ - f^-$, a standard representation of f as a difference of two nonnegative functions, with a certain minimality property:

Proposition *If $f = g - h$, $g \geq 0$, and $h \geq 0$, then $f^+ \leq g$ and $f^- \leq h$.*

PROOF $f \leq g$ and $0 \leq g$ clearly implies max $\{f, 0\} \leq g$. ////

Simple Functions

1.16 Definition A complex function s on a measurable space X whose range consists of only finitely many points will be called a *simple function*. Among these are the nonnegative simple functions, whose range is a finite subset of $[0, \infty)$. Note that we explicitly exclude ∞ from the values of a simple function.

If $\alpha_1, \ldots, \alpha_n$ are the distinct values of a simple function s, and if we set $A_i = \{x: s(x) = \alpha_i\}$, then clearly

$$s = \sum_{i=1}^{n} \alpha_i \chi_{A_i},$$

where χ_{A_i} is the characteristic function of A_i, as defined in Sec. 1.9(d).

It is also clear that s is measurable if and only if each of the sets A_i is measurable.

1.17 Theorem *Let $f: X \to [0, \infty]$ be measurable. There exist simple measurable functions s_n on X such that*

(a) $0 \leq s_1 \leq s_2 \leq \cdots \leq f$,
(b) $s_n(x) \to f(x)$ as $n \to \infty$, for every $x \in X$.

PROOF Put $\delta_n = 2^{-n}$. To each positive integer n and each real number t corresponds a unique integer $k = k_n(t)$ that satisfies $k\delta_n \leq t < (k+1)\delta_n$. Define

$$\varphi_n(t) = \begin{cases} k_n(t)\delta_n & \text{if } 0 \leq t < n \\ n & \text{if } n \leq t \leq \infty. \end{cases} \quad (1)$$

Each φ_n is then a Borel function on $[0, \infty]$,

$$t - \delta_n < \varphi_n(t) \leq t \quad \text{if } 0 \leq t \leq n, \tag{2}$$

$0 \leq \varphi_1 \leq \varphi_2 \leq \cdots \leq t$, and $\varphi_n(t) \to t$ as $n \to \infty$, for every $t \in [0, \infty]$. It follows that the functions

$$s_n = \varphi_n \circ f \tag{3}$$

satisfy (a) and (b); they are measurable, by Theorem 1.12(d). ////

Elementary Properties of Measures

1.18 Definition

(a) A *positive measure* is a function μ, defined on a σ-algebra $\mathfrak{M}$, whose range is in $[0, \infty]$ and which is *countably additive*. This means that if $\{A_i\}$ is a *disjoint* countable collection of members of $\mathfrak{M}$, then

$$\mu\left(\bigcup_{i=1}^{\infty} A_i\right) = \sum_{i=1}^{\infty} \mu(A_i). \tag{1}$$

To avoid trivialities, we shall also assume that $\mu(A) < \infty$ for at least one $A \in \mathfrak{M}$.

(b) A *measure space* is a measurable space which has a positive measure defined on the σ-algebra of its measurable sets.

(c) A *complex measure* is a complex-valued countably additive function defined on a σ-algebra.

Note: What we have called a *positive measure* is frequently just called a *measure*; we add the word "positive" for emphasis. If $\mu(E) = 0$ for every $E \in \mathfrak{M}$, then μ is a positive measure, by our definition. The value ∞ is admissible for a positive measure; but when we talk of a *complex* measure μ, it is understood that $\mu(E)$ is a complex number, for every $E \in \mathfrak{M}$. The *real measures* form a subclass of the complex ones, of course.

1.19 Theorem *Let μ be a positive measure on a σ-algebra $\mathfrak{M}$. Then*

(a) $\mu(\varnothing) = 0$.
(b) $\mu(A_1 \cup \cdots \cup A_n) = \mu(A_1) + \cdots + \mu(A_n)$ *if $A_1, \ldots, A_n$ are pairwise disjoint members of $\mathfrak{M}$.*
(c) $A \subset B$ *implies* $\mu(A) \leq \mu(B)$ *if $A \in \mathfrak{M}$, $B \in \mathfrak{M}$.*
(d) $\mu(A_n) \to \mu(A)$ *as $n \to \infty$ if $A = \bigcup_{n=1}^{\infty} A_n$, $A_n \in \mathfrak{M}$, and*

$$A_1 \subset A_2 \subset A_3 \subset \cdots.$$

(e) $\mu(A_n) \to \mu(A)$ *as $n \to \infty$ if $A = \bigcap_{n=1}^{\infty} A_n$, $A_n \in \mathfrak{M}$,*

$$A_1 \supset A_2 \supset A_3 \supset \cdots,$$

and $\mu(A_1)$ is finite.

As the proof will show, these properties, with the exception of (c), also hold for complex measures; (b) is called *finite additivity*; (c) is called *monotonicity*.

PROOF

(a) Take $A \in \mathfrak{M}$ so that $\mu(A) < \infty$, and take $A_1 = A$ and $A_2 = A_3 = \cdots = \emptyset$ in 1.18(1).
(b) Take $A_{n+1} = A_{n+2} = \cdots = \emptyset$ in 1.18(1).
(c) Since $B = A \cup (B - A)$ and $A \cap (B - A) = \emptyset$, we see that (b) implies $\mu(B) = \mu(A) + \mu(B - A) \geq \mu(A)$.
(d) Put $B_1 = A_1$, and put $B_n = A_n - A_{n-1}$ for $n = 2, 3, 4, \ldots$. Then $B_n \in \mathfrak{M}$, $B_i \cap B_j = \emptyset$ if $i \neq j$, $A_n = B_1 \cup \cdots \cup B_n$, and $A = \bigcup_{i=1}^{\infty} B_i$. Hence

$$\mu(A_n) = \sum_{i=1}^{n} \mu(B_i) \quad \text{and} \quad \mu(A) = \sum_{i=1}^{\infty} \mu(B_i).$$

Now (d) follows, by the definition of the sum of an infinite series.
(e) Put $C_n = A_1 - A_n$. Then $C_1 \subset C_2 \subset C_3 \subset \cdots$,

$$\mu(C_n) = \mu(A_1) - \mu(A_n),$$

$A_1 - A = \bigcup C_n$, and so (d) shows that

$$\mu(A_1) - \mu(A) = \mu(A_1 - A) = \lim_{n \to \infty} \mu(C_n) = \mu(A_1) - \lim_{n \to \infty} \mu(A_n).$$

This implies (e). ////

1.20 Examples The construction of interesting measure spaces requires some labor, as we shall see. However, a few simple-minded examples can be given immediately:

(a) For any $E \subset X$, where X is any set, define $\mu(E) = \infty$ if E is an infinite set, and let $\mu(E)$ be the number of points in E if E is finite. This μ is called the *counting measure* on X.
(b) Fix $x_0 \in X$, define $\mu(E) = 1$ if $x_0 \in E$ and $\mu(E) = 0$ if $x_0 \notin E$, for any $E \subset X$. This μ may be called the *unit mass* concentrated at x_0.
(c) Let μ be the counting measure on the set $\{1, 2, 3, \ldots\}$, let $A_n = \{n, n + 1, n + 2, \ldots\}$. Then $\bigcap A_n = \emptyset$ but $\mu(A_n) = \infty$ for $n = 1, 2, 3, \ldots$. This shows that the hypothesis

$$\mu(A_1) < \infty$$

is not superfluous in Theorem 1.19(e).

1.21 A Comment on Terminology One frequently sees measure spaces referred to as "ordered triples" $(X, \mathfrak{M}, \mu)$ where X is a set, $\mathfrak{M}$ is a σ-algebra in X, and μ is a measure defined on $\mathfrak{M}$. Similarly, measurable spaces are "ordered pairs" $(X, \mathfrak{M})$.

This is logically all right, and often convenient, though somewhat redundant. For instance, in $(X, \mathfrak{M})$ the set X is merely the largest member of $\mathfrak{M}$, so if we know $\mathfrak{M}$ we also know X. Similarly, every measure has a σ-algebra for its domain, by definition, so if we know a measure μ we also know the σ-algebra $\mathfrak{M}$ on which μ is defined and we know the set X in which $\mathfrak{M}$ is a σ-algebra.

It is therefore perfectly legitimate to use expressions like "Let μ be a measure" or, if we wish to emphasize the σ-algebra or the set in question, to say "Let μ be a measure on $\mathfrak{M}$" or "Let μ be a measure on X."

What is logically rather meaningless but customary (and we shall often follow mathematical custom rather than logic) is to say "Let X be a measure space"; the emphasis should not be on the set, but on the measure. Of course, when this wording is used, it is tacitly understood that there is a measure defined on some σ-algebra in X and that it is this measure which is really under discussion.

Similarly, a topological space is an ordered pair (X, τ), where τ is a topology in the set X, and the significant data are contained in τ, not in X, but "the topological space X" is what one talks about.

This sort of tacit convention is used throughout mathematics. Most mathematical systems are sets with some class of distinguished subsets or some binary operations or some relations (which are required to have certain properties), and one can list these and then describe the system as an ordered pair, triple, etc., depending on what is needed. For instance, the real line may be described as a quadruple $(R^1, +, \cdot, <)$, where $+$, $\cdot$, and $<$ satisfy the axioms of a complete archimedean ordered field. But it is a safe bet that very few mathematicians think of the real field as an ordered quadruple.

Arithmetic in $[0, \infty]$

1.22 Throughout integration theory, one inevitably encounters ∞. One reason is that one wants to be able to integrate over sets of infinite measure; after all, the real line has infinite length. Another reason is that even if one is primarily interested in real-valued functions, the lim sup of a sequence of positive real functions or the sum of a sequence of positive real functions may well be ∞ at some points, and much of the elegance of theorems like 1.26 and 1.27 would be lost if one had to make some special provisions whenever this occurs.

Let us define $a + \infty = \infty + a = \infty$ if $0 \le a \le \infty$, and

$$a \cdot \infty = \infty \cdot a = \begin{cases} \infty & \text{if } 0 < a \le \infty \\ 0 & \text{if } a = 0; \end{cases}$$

sums and products of real numbers are of course defined in the usual way.

It may seem strange to define $0 \cdot \infty = 0$. However, one verifies without difficulty that with this definition the *commutative, associative, and distributive laws* hold in $[0, \infty]$ *without any restriction*.

The cancellation laws have to be treated with some care: $a + b = a + c$ implies $b = c$ only when $a < \infty$, and $ab = ac$ implies $b = c$ only when $0 < a < \infty$. Observe that the following useful proposition holds:

If $0 \le a_1 \le a_2 \le \cdots$, $0 \le b_1 \le b_2 \le \cdots$, $a_n \to a$, and $b_n \to b$, then $a_n b_n \to ab$.

If we combine this with Theorems 1.17 and 1.14, we see that *sums and products of measurable functions into* $[0, \infty]$ *are measurable.*

Integration of Positive Functions

In this section, $\mathfrak{M}$ will be a σ-algebra in a set X and μ will be a positive measure on $\mathfrak{M}$.

1.23 Definition If $s: X \to [0, \infty)$ is a measurable simple function, of the form

$$s = \sum_{i=1}^{n} \alpha_i \chi_{A_i}, \tag{1}$$

where $\alpha_1, \ldots, \alpha_n$ are the distinct values of s (compare Definition 1.16), and if $E \in \mathfrak{M}$, we define

$$\int_E s \, d\mu = \sum_{i=1}^{n} \alpha_i \mu(A_i \cap E). \tag{2}$$

The convention $0 \cdot \infty = 0$ is used here; it may happen that $\alpha_i = 0$ for some i and that $\mu(A_i \cap E) = \infty$.

If $f: X \to [0, \infty]$ is measurable, and $E \in \mathfrak{M}$, we define

$$\int_E f \, d\mu = \sup \int_E s \, d\mu, \tag{3}$$

the supremum being taken over all simple measurable functions s such that $0 \le s \le f$.

The left member of (3) is called the *Lebesgue integral* of f over E, with respect to the measure μ. It is a number in $[0, \infty]$.

Observe that we apparently have two definitions for $\int_E f \, d\mu$ if f is simple, namely, (2) and (3). However, these assign the same value to the integral, since f is, in this case, the largest of the functions s which occur on the right of (3).

1.24 The following propositions are immediate consequences of the definitions. The functions and sets occurring in them are assumed to be measurable:

(a) If $0 \le f \le g$, then $\int_E f \, d\mu \le \int_E g \, d\mu$.
(b) If $A \subset B$ and $f \ge 0$, then $\int_A f \, d\mu \le \int_B f \, d\mu$.

(c) If $f \geq 0$ and c is a constant, $0 \leq c < \infty$, then

$$\int_E cf \, d\mu = c \int_E f \, d\mu.$$

(d) If $f(x) = 0$ for all $x \in E$, then $\int_E f \, d\mu = 0$, even if $\mu(E) = \infty$.
(e) If $\mu(E) = 0$, then $\int_E f \, d\mu = 0$, even if $f(x) = \infty$ for every $x \in E$.
(f) If $f \geq 0$, then $\int_E f \, d\mu = \int_X \chi_E f \, d\mu$.

This last result shows that we could have restricted our definition of integration to integrals over all of X, without losing any generality. If we wanted to integrate over subsets, we could then use (f) as the definition. It is purely a matter of taste which definition is preferred.

One may also remark here that every measurable subset E of a measure space X is again a measure space, in a perfectly natural way: The new measurable sets are simply those measurable subsets of X which lie in E, and the measure is unchanged, except that its domain is restricted. This shows again that as soon as we have integration defined over every measure space, we automatically have it defined over every measurable subset of every measure space.

1.25 Proposition Let s and t be nonnegative measurable simple functions on X. For $E \in \mathfrak{M}$, define

$$\varphi(E) = \int_E s \, d\mu. \tag{1}$$

Then φ is a measure on $\mathfrak{M}$. Also

$$\int_X (s + t) \, d\mu = \int_X s \, d\mu + \int_X t \, d\mu. \tag{2}$$

(This proposition contains provisional forms of Theorems 1.27 and 1.29.)

PROOF If s is as in Definition 1.23, and if $E_1, E_2, \ldots$ are disjoint members of $\mathfrak{M}$ whose union is E, the countable additivity of μ shows that

$$\varphi(E) = \sum_{i=1}^n \alpha_i \mu(A_i \cap E) = \sum_{i=1}^n \alpha_i \sum_{r=1}^\infty \mu(A_i \cap E_r)$$

$$= \sum_{r=1}^\infty \sum_{i=1}^n \alpha_i \mu(A_i \cap E_r) = \sum_{r=1}^\infty \varphi(E_r).$$

Also, $\varphi(\emptyset) = 0$, so that φ is not identically ∞.

Next, let s be as before, let $\beta_1, \ldots, \beta_m$ be the distinct values of t, and let $B_j = \{x: t(x) = \beta_j\}$. If $E_{ij} = A_i \cap B_j$, then

$$\int_{E_{ij}} (s + t)\, d\mu = (\alpha_i + \beta_j)\mu(E_{ij})$$

and

$$\int_{E_{ij}} s\, d\mu + \int_{E_{ij}} t\, d\mu = \alpha_i \mu(E_{ij}) + \beta_j \mu(E_{ij}).$$

Thus (2) holds with E_{ij} in place of X. Since X is the disjoint union of the sets E_{ij} ($1 \le i \le n$, $1 \le j \le m$), the first half of our proposition implies that (2) holds. ////

We now come to the interesting part of the theory. One of its most remarkable features is the ease with which it handles limit operations.

1.26 Lebesgue's Monotone Convergence Theorem *Let $\{f_n\}$ be a sequence of measurable functions on X, and suppose that*

(a) $0 \le f_1(x) \le f_2(x) \le \cdots \le \infty$ *for every $x \in X$,*
(b) $f_n(x) \to f(x)$ *as $n \to \infty$, for every $x \in X$.*

Then f is measurable, and

$$\int_X f_n\, d\mu \to \int_X f\, d\mu \qquad \text{as } n \to \infty.$$

PROOF Since $\int f_n \le \int f_{n+1}$, there exists an $\alpha \in [0, \infty]$ such that

$$\int_X f_n\, d\mu \to \alpha \qquad \text{as } n \to \infty. \tag{1}$$

By Theorem 1.14, f is measurable. Since $f_n \le f$, we have $\int f_n \le \int f$ for every n, so (1) implies

$$\alpha \le \int_X f\, d\mu. \tag{2}$$

Let s be any simple measurable function such that $0 \le s \le f$, let c be a constant, $0 < c < 1$, and define

$$E_n = \{x: f_n(x) \ge cs(x)\} \qquad (n = 1, 2, 3, \ldots). \tag{3}$$

Each E_n is measurable, $E_1 \subset E_2 \subset E_3 \subset \cdots$, and $X = \bigcup E_n$. To see this equality, consider some $x \in X$. If $f(x) = 0$, then $x \in E_1$; if $f(x) > 0$, then $cs(x) < f(x)$, since $c < 1$; hence $x \in E_n$ for some n. Also

$$\int_X f_n\, d\mu \ge \int_{E_n} f_n\, d\mu \ge c \int_{E_n} s\, d\mu \qquad (n = 1, 2, 3, \ldots). \tag{4}$$

Let $n \to \infty$, applying Proposition 1.25 and Theorem 1.19(d) to the last integral in (4). The result is

$$\alpha \geq c \int_X s \, d\mu. \tag{5}$$

Since (5) holds for every $c < 1$, we have

$$\alpha \geq \int_X s \, d\mu \tag{6}$$

for every simple measurable s satisfying $0 \leq s \leq f$, so that

$$\alpha \geq \int_X f \, d\mu. \tag{7}$$

The theorem follows from (1), (2), and (7). ////

1.27 Theorem *If $f_n: X \to [0, \infty]$ is measurable, for $n = 1, 2, 3, \ldots$, and*

$$f(x) = \sum_{n=1}^{\infty} f_n(x) \qquad (x \in X), \tag{1}$$

then

$$\int_X f \, d\mu = \sum_{n=1}^{\infty} \int_X f_n \, d\mu. \tag{2}$$

PROOF First, there are sequences $\{s_i'\}$, $\{s_i''\}$ of simple measurable functions such that $s_i' \to f_1$ and $s_i'' \to f_2$, as in Theorem 1.17. If $s_i = s_i' + s_i''$, then $s_i \to f_1 + f_2$, and the monotone convergence theorem, combined with Proposition 1.25, shows that

$$\int_X (f_1 + f_2) \, d\mu = \int_X f_1 \, d\mu + \int_X f_2 \, d\mu. \tag{3}$$

Next, put $g_N = f_1 + \cdots + f_N$. The sequence $\{g_N\}$ converges monotonically to f, and if we apply induction to (3) we see that

$$\int_X g_N \, d\mu = \sum_{n=1}^{N} \int_X f_n \, d\mu. \tag{4}$$

Applying the monotone convergence theorem once more, we obtain (2), and the proof is complete. ////

If we let μ be the counting measure on a countable set, Theorem 1.27 is a statement about double series of nonnegative real numbers (which can of course be proved by more elementary means):

Corollary *If $a_{ij} \geq 0$ for i and $j = 1, 2, 3, \ldots$, then*

$$\sum_{i=1}^{\infty} \sum_{j=1}^{\infty} a_{ij} = \sum_{j=1}^{\infty} \sum_{i=1}^{\infty} a_{ij}.$$

1.28 Fatou's Lemma *If $f_n : X \to [0, \infty]$ is measurable, for each positive integer n, then*

$$\int_X \left(\liminf_{n \to \infty} f_n \right) d\mu \leq \liminf_{n \to \infty} \int_X f_n \, d\mu. \tag{1}$$

Strict inequality can occur in (1); see Exercise 8.

PROOF Put

$$g_k(x) = \inf_{i \geq k} f_i(x) \qquad (k = 1, 2, 3, \ldots; x \in X). \tag{2}$$

Then $g_k \leq f_k$, so that

$$\int_X g_k \, d\mu \leq \int_X f_k \, d\mu \qquad (k = 1, 2, 3, \ldots). \tag{3}$$

Also, $0 \leq g_1 \leq g_2 \leq \cdots$, each g_k is measurable, by Theorem 1.14, and $g_k(x) \to \liminf f_n(x)$ as $k \to \infty$, by Definition 1.13. The monotone convergence theorem shows therefore that the left side of (3) tends to the left side of (1), as $k \to \infty$. Hence (1) follows from (3). ////

1.29 Theorem *Suppose $f : X \to [0, \infty]$ is measurable, and*

$$\varphi(E) = \int_E f \, d\mu \qquad (E \in \mathfrak{M}). \tag{1}$$

Then φ is a measure on $\mathfrak{M}$, and

$$\int_X g \, d\varphi = \int_X gf \, d\mu \tag{2}$$

for every measurable g on X with range in $[0, \infty]$.

PROOF Let $E_1, E_2, E_3, \ldots$ be disjoint members of $\mathfrak{M}$ whose union is E. Observe that

$$\chi_E f = \sum_{j=1}^{\infty} \chi_{E_j} f \tag{3}$$

and that

$$\varphi(E) = \int_X \chi_E f \, d\mu, \qquad \varphi(E_j) = \int_X \chi_{E_j} f \, d\mu. \tag{4}$$

It now follows from Theorem 1.27 that

$$\varphi(E) = \sum_{j=1}^{\infty} \varphi(E_j). \tag{5}$$

Since $\varphi(\emptyset) = 0$, (5) proves that φ is a measure.

Next, (1) shows that (2) holds whenever $g = \chi_E$ for some $E \in \mathfrak{M}$. Hence (2) holds for every simple measurable function g, and the general case follows from the monotone convergence theorem.   ////

Remark The second assertion of Theorem 1.29 is sometimes written in the form

$$d\varphi = f\, d\mu. \tag{6}$$

We assign no independent meaning to the symbols $d\varphi$ and $d\mu$; (6) merely means that (2) holds for every measurable $g \geq 0$.

Theorem 1.29 has a very important converse, the Radon-Nikodym theorem, which will be proved in Chap. 6.

Integration of Complex Functions

As before, μ will in this section be a positive measure on an arbitrary measurable space X.

1.30 Definition We define $L^1(\mu)$ to be the collection of all complex measurable functions f on X for which

$$\int_X |f|\, d\mu < \infty.$$

Note that the measurability of f implies that of $|f|$, as we saw in Proposition 1.9(b); hence the above integral is defined.

The members of $L^1(\mu)$ are called *Lebesgue integrable* functions (with respect to μ) or *summable functions*. The significance of the exponent 1 will become clear in Chap. 3.

1.31 Definition If $f = u + iv$, where u and v are real measurable functions on X, and if $f \in L^1(\mu)$, we define

$$\int_E f\, d\mu = \int_E u^+\, d\mu - \int_E u^-\, d\mu + i\int_E v^+\, d\mu - i\int_E v^-\, d\mu \tag{1}$$

for every measurable set E.

Here u^+ and u^- are the positive and negative parts of u, as defined in Sec. 1.15; v^+ and v^- are similarly obtained from v. These four functions are measurable, real, and nonnegative; hence the four integrals on the right of (1) exist, by Definition 1.23. Furthermore, we have $u^+ \leq |u| \leq |f|$, etc., so that each of these four integrals is finite. Thus (1) defines the integral on the left as a complex number.

Occasionally it is desirable to define the integral of a measurable function f with range in $[-\infty, \infty]$ to be

$$\int_E f \, d\mu = \int_E f^+ \, d\mu - \int_E f^- \, d\mu, \tag{2}$$

provided that at least one of the integrals on the right of (2) is finite. The left side of (2) is then a number in $[-\infty, \infty]$.

1.32 Theorem *Suppose f and $g \in L^1(\mu)$ and α and β are complex numbers. Then $\alpha f + \beta g \in L^1(\mu)$, and*

$$\int_X (\alpha f + \beta g) \, d\mu = \alpha \int_X f \, d\mu + \beta \int_X g \, d\mu. \tag{1}$$

PROOF The measurability of $\alpha f + \beta g$ follows from Proposition 1.9(c). By Sec. 1.24 and Theorem 1.27,

$$\int_X |\alpha f + \beta g| \, d\mu \leq \int_X (|\alpha| |f| + |\beta| |g|) \, d\mu$$

$$= |\alpha| \int_X |f| \, d\mu + |\beta| \int_X |g| \, d\mu < \infty.$$

Thus $\alpha f + \beta g \in L^1(\mu)$.

To prove (1), it is clearly sufficient to prove

$$\int_X (f + g) \, d\mu = \int_X f \, d\mu + \int_X g \, d\mu \tag{2}$$

and

$$\int_X (\alpha f) \, d\mu = \alpha \int_X f \, d\mu, \tag{3}$$

and the general case of (2) will follow if we prove (2) for real f and g in $L^1(\mu)$.

Assuming this, and setting $h = f + g$, we have

$$h^+ - h^- = f^+ - f^- + g^+ - g^-$$

or

$$h^+ + f^- + g^- = f^+ + g^+ + h^-. \tag{4}$$

By Theorem 1.27,

$$\int h^+ + \int f^- + \int g^- = \int f^+ + \int g^+ + \int h^-, \tag{5}$$

and since each of these integrals is finite, we may transpose and obtain (2).

That (3) holds if $\alpha \geq 0$ follows from Proposition 1.24(c). It is easy to verify that (3) holds if $\alpha = -1$, using relations like $(-u)^+ = u^-$. The case $\alpha = i$ is also easy: If $f = u + iv$, then

$$\int (if) = \int (iu - v) = \int (-v) + i \int u = -\int v + i \int u = i\left(\int u + i \int v\right)$$
$$= i \int f.$$

Combining these cases with (2), we obtain (3) for any complex α. ////

1.33 Theorem *If $f \in L^1(\mu)$, then*

$$\left| \int_X f \, d\mu \right| \leq \int_X |f| \, d\mu.$$

PROOF Put $z = \int_X f \, d\mu$. Since z is a complex number, there is a complex number α, with $|\alpha| = 1$, such that $\alpha z = |z|$. Let u be the real part of αf. Then $u \leq |\alpha f| = |f|$. Hence

$$\left| \int_X f \, d\mu \right| = \alpha \int_X f \, d\mu = \int_X \alpha f \, d\mu = \int_X u \, d\mu \leq \int_X |f| \, d\mu.$$

The third of the above equalities holds since the preceding ones show that $\int \alpha f \, d\mu$ is real. ////

We conclude this section with another important convergence theorem.

1.34 Lebesgue's Dominated Convergence Theorem *Suppose $\{f_n\}$ is a sequence of complex measurable functions on X such that*

$$f(x) = \lim_{n \to \infty} f_n(x) \qquad (1)$$

exists for every $x \in X$. If there is a function $g \in L^1(\mu)$ such that

$$|f_n(x)| \leq g(x) \qquad (n = 1, 2, 3, \ldots; x \in X), \qquad (2)$$

then $f \in L^1(\mu)$,

$$\lim_{n \to \infty} \int_X |f_n - f| \, d\mu = 0, \qquad (3)$$

and

$$\lim_{n \to \infty} \int_X f_n \, d\mu = \int_X f \, d\mu. \qquad (4)$$

PROOF Since $|f| \le g$ and f is measurable, $f \in L^1(\mu)$. Since $|f_n - f| \le 2g$, Fatou's lemma applies to the functions $2g - |f_n - f|$ and yields

$$\int_X 2g \, d\mu \le \liminf_{n \to \infty} \int_X (2g - |f_n - f|) \, d\mu$$

$$= \int_X 2g \, d\mu + \liminf_{n \to \infty} \left(- \int_X |f_n - f| \, d\mu \right)$$

$$= \int_X 2g \, d\mu - \limsup_{n \to \infty} \int_X |f_n - f| \, d\mu.$$

Since $\int 2g \, d\mu$ is finite, we may subtract it and obtain

$$\limsup_{n \to \infty} \int_X |f_n - f| \, d\mu \le 0. \tag{5}$$

If a sequence of nonnegative real numbers fails to converge to 0, then its upper limit is positive. Thus (5) implies (3). By Theorem 1.33, applied to $f_n - f$, (3) implies (4).   ////

The Role Played by Sets of Measure Zero

1.35 Definition Let P be a property which a point x may or may not have. For instance, P might be the property "$f(x) > 0$" if f is a given function, or it might be "$\{f_n(x)\}$ converges" if $\{f_n\}$ is a given sequence of functions.

If μ is a measure on a σ-algebra $\mathfrak{M}$ and if $E \in \mathfrak{M}$, the statement "P holds almost everywhere on E" (abbreviated to "P holds a.e. on E") means that there exists an $N \in \mathfrak{M}$ such that $\mu(N) = 0$, $N \subset E$, and P holds at every point of $E - N$. This concept of a.e. depends of course very strongly on the given measure, and we shall write "a.e. $[\mu]$" whenever clarity requires that the measure be indicated.

For example, if f and g are measurable functions and if

$$\mu(\{x : f(x) \ne g(x)\}) = 0, \tag{1}$$

we say that $f = g$ a.e. $[\mu]$ on X, and we may write $f \sim g$. This is easily seen to be an equivalence relation. The transitivity ($f \sim g$ and $g \sim h$ implies $f \sim h$) is a consequence of the fact that the union of two sets of measure 0 has measure 0.

Note that if $f \sim g$, then, for every $E \in \mathfrak{M}$,

$$\int_E f \, d\mu = \int_E g \, d\mu. \tag{2}$$

To see this, let N be the set which appears in (1); then E is the union of the disjoint sets $E - N$ and $E \cap N$; on $E - N$, $f = g$, and $\mu(E \cap N) = 0$.

Thus, generally speaking, sets of measure 0 are negligible in integration. It ought to be true that every subset of a negligible set is negligible. But it may happen that some set $N \in \mathfrak{M}$ with $\mu(N) = 0$ has a subset E which is not a member of $\mathfrak{M}$. Of course we can *define* $\mu(E) = 0$ in this case. But will this extension of μ still be a measure, i.e., will it still be defined on a σ-algebra? It is a pleasant fact that the answer is affirmative:

1.36 Theorem *Let $(X, \mathfrak{M}, \mu)$ be a measure space, let $\mathfrak{M}^*$ be the collection of all $E \subset X$ for which there exist sets A and $B \in \mathfrak{M}$ such that $A \subset E \subset B$ and $\mu(B - A) = 0$, and define $\mu(E) = \mu(A)$ in this situation. Then $\mathfrak{M}^*$ is a σ-algebra, and μ is a measure on $\mathfrak{M}^*$.*

This extended measure μ is called *complete*, since all subsets of sets of measure 0 are now measurable; the σ-algebra $\mathfrak{M}^*$ is called the μ-*completion* of $\mathfrak{M}$. The theorem says that every measure can be completed, so, whenever it is convenient, we may assume that any given measure is complete; this just gives us more measurable sets, hence more measurable functions. Most measures that one meets in the ordinary course of events are already complete, but there are exceptions; one of these will occur in the proof of Fubini's theorem in Chap. 8.

PROOF We begin by checking that μ is well defined for every $E \in \mathfrak{M}^*$. Suppose $A \subset E \subset B$, $A_1 \subset E \subset B_1$, and $\mu(B - A) = \mu(B_1 - A_1) = 0$. (The letters A and B will denote members of $\mathfrak{M}$ throughout this proof.) Since

$$A - A_1 \subset E - A_1 \subset B_1 - A_1$$

we have $\mu(A - A_1) = 0$, hence $\mu(A) = \mu(A \cap A_1)$. For the same reason, $\mu(A_1) = \mu(A_1 \cap A)$. We conclude that indeed $\mu(A_1) = \mu(A)$.

Next, let us verify that $\mathfrak{M}^*$ has the three defining properties of a σ-algebra.

(i) $X \in \mathfrak{M}^*$, because $X \in \mathfrak{M}$ and $\mathfrak{M} \subset \mathfrak{M}^*$.
(ii) If $A \subset E \subset B$ then $B^c \subset E^c \subset A^c$. Thus $E \in \mathfrak{M}^*$ implies $E^c \in \mathfrak{M}^*$, because $A^c - B^c = A^c \cap B = B - A$.
(iii) If $A_i \subset E_i \subset B_i$, $E = \bigcup E_i$, $A = \bigcup A_i$, $B = \bigcup B_i$, then $A \subset E \subset B$ and

$$B - A = \bigcup_1^\infty (B_i - A) \subset \bigcup_1^\infty (B_i - A_i).$$

Since countable unions of sets of measure zero have measure zero, it follows that $E \in \mathfrak{M}^*$ if $E_i \in \mathfrak{M}^*$ for $i = 1, 2, 3, \ldots$.

Finally, if the sets E_i are disjoint in step (iii), the same is true of the sets A_i, and we conclude that

$$\mu(E) = \mu(A) = \sum_1^\infty \mu(A_i) = \sum_1^\infty \mu(E_i).$$

This proves that μ is countably additive on $\mathfrak{M}^*$. ////

1.37 The fact that functions which are equal a.e. are indistinguishable as far as integration is concerned suggests that our definition of measurable function might profitably be enlarged. Let us call a function f defined on a set $E \in \mathfrak{M}$ *measurable on* X if $\mu(E^c) = 0$ and if $f^{-1}(V) \cap E$ is measurable for every open set V. If we define $f(x) = 0$ for $x \in E^c$, we obtain a measurable function on X, in the old sense. If our measure happens to be complete, we can define f on E^c in a perfectly arbitrary manner, and we still get a measurable function. The integral of f over any set $A \in \mathfrak{M}$ is independent of the definition of f on E^c; therefore this definition need not even be specified at all.

There are many situations where this occurs naturally. For instance, a function f on the real line may be differentiable only almost everywhere (with respect to Lebesgue measure), but under certain conditions it is still true that f is the integral of its derivative; this will be discussed in Chap. 7. Or a sequence $\{f_n\}$ of measurable functions on X may converge only almost everywhere; with our new definition of measurability, the limit is still a measurable function on X, and we do not have to cut down to the set on which convergence actually occurs.

To illustrate, let us state a corollary of Lebesgue's dominated convergence theorem in a form in which exceptional sets of measure zero are admitted:

1.38 Theorem *Suppose $\{f_n\}$ is a sequence of complex measurable functions defined a.e. on X such that*

$$\sum_{n=1}^{\infty} \int_X |f_n|\, d\mu < \infty. \qquad (1)$$

Then the series

$$f(x) = \sum_{n=1}^{\infty} f_n(x) \qquad (2)$$

converges for almost all x, $f \in L^1(\mu)$, and

$$\int_X f\, d\mu = \sum_{n=1}^{\infty} \int_X f_n\, d\mu. \qquad (3)$$

PROOF Let S_n be the set on which f_n is defined, so that $\mu(S_n^c) = 0$. Put $\varphi(x) = \sum |f_n(x)|$, for $x \in S = \bigcap S_n$. Then $\mu(S^c) = 0$. By (1) and Theorem 1.27,

$$\int_S \varphi\, d\mu < \infty. \qquad (4)$$

If $E = \{x \in S : \varphi(x) < \infty\}$, it follows from (4) that $\mu(E^c) = 0$. The series (2) converges absolutely for every $x \in E$, and if $f(x)$ is defined by (2) for $x \in E$, then $|f(x)| \le \varphi(x)$ on E, so that $f \in L^1(\mu)$ on E, by (4). If $g_n = f_1 + \cdots + f_n$, then $|g_n| \le \varphi$, $g_n(x) \to f(x)$ for all $x \in E$, and Theorem 1.34 gives (3) with E in place of X. This is equivalent to (3), since $\mu(E^c) = 0$. ////

Note that even if the f_n were defined at *every* point of X, (1) would only imply that (2) converges *almost everywhere*. Here are some other situations in which we can draw conclusions only almost everywhere:

1.39 Theorem

(a) Suppose $f: X \to [0, \infty]$ is measurable, $E \in \mathfrak{M}$, and $\int_E f \, d\mu = 0$. Then $f = 0$ a.e. on E.
(b) Suppose $f \in L^1(\mu)$ and $\int_E f \, d\mu = 0$ for every $E \in \mathfrak{M}$. Then $f = 0$ a.e. on X.
(c) Suppose $f \in L^1(\mu)$ and

$$\left| \int_X f \, d\mu \right| = \int_X |f| \, d\mu.$$

Then there is a constant α such that $\alpha f = |f|$ a.e. on X.

Note that (c) describes the condition under which equality holds in Theorem 1.33.

PROOF

(a) If $A_n = \{x \in E : f(x) > 1/n\}$, $n = 1, 2, 3, \ldots$, then

$$\frac{1}{n} \mu(A_n) \le \int_{A_n} f \, d\mu \le \int_E f \, d\mu = 0,$$

so that $\mu(A_n) = 0$. Since $\{x \in E : f(x) > 0\} = \bigcup A_n$, (a) follows.

(b) Put $f = u + iv$, let $E = \{x : u(x) \ge 0\}$. The real part of $\int_E f \, d\mu$ is then $\int_E u^+ \, d\mu$. Hence $\int_E u^+ \, d\mu = 0$, and (a) implies that $u^+ = 0$ a.e. We conclude similarly that

$$u^- = v^+ = v^- = 0 \quad \text{a.e.}$$

(c) Examine the proof of Theorem 1.33. Our present assumption implies that the last inequality in the proof of Theorem 1.33 must actually be an equality. Hence $\int (|f| - u) \, d\mu = 0$. Since $|f| - u \ge 0$, (a) shows that $|f| = u$ a.e. This says that the real part of αf is equal to $|\alpha f|$ a.e., hence $\alpha f = |\alpha f| = |f|$ a.e., which is the desired conclusion. ////

1.40 Theorem
Suppose $\mu(X) < \infty$, $f \in L^1(\mu)$, S is a closed set in the complex plane, and the averages

$$A_E(f) = \frac{1}{\mu(E)} \int_E f \, d\mu$$

lie in S for every $E \in \mathfrak{M}$ with $\mu(E) > 0$. Then $f(x) \in S$ for almost all $x \in X$.

PROOF Let Δ be a closed circular disc (with center at α and radius $r > 0$, say) in the complement of S. Since S^c is the union of countably many such discs, it is enough to prove that $\mu(E) = 0$, where $E = f^{-1}(\Delta)$.

If we had $\mu(E) > 0$, then

$$|A_E(f) - \alpha| = \frac{1}{\mu(E)} \left| \int_E (f - \alpha) \, d\mu \right| \leq \frac{1}{\mu(E)} \int_E |f - \alpha| \, d\mu \leq r,$$

which is impossible, since $A_E(f) \in S$. Hence $\mu(E) = 0$. ////

1.41 Theorem *Let $\{E_k\}$ be a sequence of measurable sets in X, such that*

$$\sum_{k=1}^{\infty} \mu(E_k) < \infty. \tag{1}$$

Then almost all $x \in X$ lie in at most finitely many of the sets E_k.

PROOF If A is the set of all x which lie in infinitely many E_k, we have to prove that $\mu(A) = 0$. Put

$$g(x) = \sum_{k=1}^{\infty} \chi_{E_k}(x) \qquad (x \in X). \tag{2}$$

For each x, each term in this series is either 0 or 1. Hence $x \in A$ if and only if $g(x) = \infty$. By Theorem 1.27, the integral of g over X is equal to the sum in (1). Thus $g \in L^1(\mu)$, and so $g(x) < \infty$ a.e. ////

Exercises

1 Does there exist an infinite σ-algebra which has only countably many members?

2 Prove an analogue of Theorem 1.8 for n functions.

3 Prove that if f is a real function on a measurable space X such that $\{x: f(x) \geq r\}$ is measurable for every rational r, then f is measurable.

4 Let $\{a_n\}$ and $\{b_n\}$ be sequences in $[-\infty, \infty]$, and prove the following assertions:

(a) $$\limsup_{n \to \infty} (-a_n) = -\liminf_{n \to \infty} a_n.$$

(b) $$\limsup_{n \to \infty} (a_n + b_n) \leq \limsup_{n \to \infty} a_n + \limsup_{n \to \infty} b_n$$

provided none of the sums is of the form $\infty - \infty$.

(c) If $a_n \leq b_n$ for all n, then

$$\liminf_{n \to \infty} a_n \leq \liminf_{n \to \infty} b_n.$$

Show by an example that strict inequality can hold in (b).

5 (a) Suppose $f: X \to [-\infty, \infty]$ and $g: X \to [-\infty, \infty]$ are measurable. Prove that the sets
$$\{x: f(x) < g(x)\}, \{x: f(x) = g(x)\}$$
are measurable.

(b) Prove that the set of points at which a sequence of measurable real-valued functions converges (to a finite limit) is measurable.

6 Let X be an uncountable set, let $\mathfrak{M}$ be the collection of all sets $E \subset X$ such that either E or E^c is at most countable, and define $\mu(E) = 0$ in the first case, $\mu(E) = 1$ in the second. Prove that $\mathfrak{M}$ is a σ-algebra in X and that μ is a measure on $\mathfrak{M}$. Describe the corresponding measurable functions and their integrals.

7 Suppose $f_n: X \to [0, \infty]$ is measurable for $n = 1, 2, 3, \ldots$, $f_1 \geq f_2 \geq f_3 \geq \cdots \geq 0$, $f_n(x) \to f(x)$ as $n \to \infty$, for every $x \in X$, and $f_1 \in L^1(\mu)$. Prove that then
$$\lim_{n \to \infty} \int_X f_n \, d\mu = \int_X f \, d\mu$$
and show that this conclusion does *not* follow if the condition "$f_1 \in L^1(\mu)$" is omitted.

8 Put $f_n = \chi_E$ if n is odd, $f_n = 1 - \chi_E$ if n is even. What is the relevance of this example to Fatou's lemma?

9 Suppose μ is a positive measure on X, $f: X \to [0, \infty]$ is measurable, $\int_X f \, d\mu = c$, where $0 < c < \infty$, and α is a constant. Prove that
$$\lim_{n \to \infty} \int_X n \log [1 + (f/n)^\alpha] \, d\mu = \begin{cases} \infty & \text{if } 0 < \alpha < 1, \\ c & \text{if } \alpha = 1, \\ 0 & \text{if } 1 < \alpha < \infty. \end{cases}$$

Hint: If $\alpha \geq 1$, prove that the integrands are dominated by αf. If $\alpha < 1$, Fatou's lemma can be applied.

10 Suppose $\mu(X) < \infty$, $\{f_n\}$ is a sequence of bounded complex measurable functions on X, and $f_n \to f$ uniformly on X. Prove that
$$\lim_{n \to \infty} \int_X f_n \, d\mu = \int_X f \, d\mu,$$
and show that the hypothesis "$\mu(X) < \infty$" cannot be omitted.

11 Show that
$$A = \bigcap_{n=1}^{\infty} \bigcup_{k=n}^{\infty} E_k$$
in Theorem 1.41, and hence prove the theorem without any reference to integration.

12 Suppose $f \in L^1(\mu)$. Prove that to each $\epsilon > 0$ there exists a $\delta > 0$ such that $\int_E |f| \, d\mu < \epsilon$ whenever $\mu(E) < \delta$.

13 Show that proposition 1.24(c) is also true when $c = \infty$.

CHAPTER
TWO
POSITIVE BOREL MEASURES

Vector Spaces

2.1 Definition A *complex vector space* (or a vector space over the complex field) is a set V, whose elements are called *vectors* and in which two operations, called *addition* and *scalar multiplication*, are defined, with the following familiar algebraic properties:

To every pair of vectors x and y there corresponds a vector $x + y$, in such a way that $x + y = y + x$ and $x + (y + z) = (x + y) + z$; V contains a unique vector 0 (the *zero vector* or *origin* of V) such that $x + 0 = x$ for every $x \in V$; and to each $x \in V$ there corresponds a unique vector $-x$ such that $x + (-x) = 0$.

To each pair (α, x), where $x \in V$ and α is a scalar (in this context, the word *scalar* means *complex number*), there is associated a vector $\alpha x \in V$, in such a way that $1x = x$, $\alpha(\beta x) = (\alpha\beta)x$, and such that the two distributive laws

$$\alpha(x + y) = \alpha x + \alpha y, (\alpha + \beta)x = \alpha x + \beta x \tag{1}$$

hold.

A *linear transformation* of a vector space V into a vector space V_1 is a mapping Λ of V into V_1 such that

$$\Lambda(\alpha x + \beta y) = \alpha \Lambda x + \beta \Lambda y \tag{2}$$

for all x and $y \in V$ and for all scalars α and β. In the special case in which V_1 is the field of scalars (this is the simplest example of a vector space, except for the trivial one consisting of 0 alone), Λ is called a *linear functional*. A linear functional is thus a complex function on V which satisfies (2).

Note that one often writes Λx, rather than $\Lambda(x)$, if Λ is linear.

The preceding definitions can of course be made equally well with any field whatsoever in place of the complex field. Unless the contrary is explicitly stated, however, all vector spaces occurring in this book will be complex, with one notable exception: the euclidean spaces R^k are vector spaces over the *real* field.

2.2 Integration as a Linear Functional Analysis is full of vector spaces and linear transformations, and there is an especially close relationship between integration on the one hand and linear functionals on the other.

For instance, Theorem 1.32 shows that $L^1(\mu)$ is a vector space, for any positive measure μ, and that the mapping

$$f \to \int_X f \, d\mu \tag{1}$$

is a linear functional on $L^1(\mu)$. Similarly, if g is any bounded measurable function, the mapping

$$f \to \int_X fg \, d\mu \tag{2}$$

is a linear functional on $L^1(\mu)$; we shall see in Chap. 6 that the functionals (2) are, in a sense, the only interesting ones on $L^1(\mu)$.

For another example, let C be the set of all continuous complex functions on the unit interval $I = [0, 1]$. The sum of the two continuous functions is continuous, and so is any scalar multiple of a continuous function. Hence C is a vector space, and if

$$\Lambda f = \int_0^1 f(x) \, dx \qquad (f \in C), \tag{3}$$

the integral being the ordinary Riemann integral, then Λ is clearly a linear functional on C; Λ has an additional interesting property: it is a *positive linear functional*. This means that $\Lambda f \geq 0$ whenever $f \geq 0$.

One of the tasks which is still ahead of us is the construction of the Lebesgue measure. The construction can be based on the linear functional (3), by the following observation: Consider a segment $(a, b) \subset I$ and consider the class of all $f \in C$ such that $0 \leq f \leq 1$ on I and $f(x) = 0$ for all x not in (a, b). We have $\Lambda f < b - a$ for all such f, but we can choose f so that Λf is as close to $b - a$ as desired. Thus the length (or measure) of (a, b) is intimately related to the values of the functional Λ.

The preceding observation, when looked at from a more general point of view, leads to a remarkable and extremely important theorem of F. Riesz:

To every positive linear functional Λ on C corresponds a finite positive Borel measure μ on I such that

$$\Lambda f = \int_I f \, d\mu \qquad (f \in C). \tag{4}$$

[The converse is obvious: if μ is a finite positive Borel measure on I and if Λ is defined by (4), then Λ is a positive linear functional on C.]

It is clearly of interest to replace the bounded interval I by R^1. We can do this by restricting attention to those continuous functions on R^1 which vanish outside some bounded interval. (These functions are Riemann integrable, for instance.) Next, functions of several variables occur frequently in analysis. Thus we ought to move from R^1 to R^n. It turns out that the proof of the Riesz theorem still goes through, with hardly any changes. Moreover, it turns out that the euclidean properties of R^n (coordinates, orthogonality, etc.) play no role in the proof; in fact, if one thinks of them too much they just get in the way. Essential to the proof are certain *topological* properties of R^n. (Naturally. We are now dealing with *continuous* functions.) The crucial property is that of *local compactness*: Each point of R^n has a neighborhood whose closure is compact.

We shall therefore establish the Riesz theorem in a very general setting (Theorem 2.14). The existence of Lebesgue measure follows then as a special case. Those who wish to concentrate on a more concrete situation may skip lightly over the following section on topological preliminaries (Urysohn's lemma is the item of greatest interest there; see Exercise 3) and may replace locally compact Hausdorff spaces by locally compact metric spaces, or even by euclidean spaces, without missing any of the principal ideas.

It should also be mentioned that there are situations, especially in probability theory, where measures occur naturally on spaces without topology, or on topological spaces that are not locally compact. An example is the so-called Wiener measure which assigns numbers to certain sets of continuous functions and which is a basic tool in the study of Brownian motion. These topics will not be discussed in this book.

Topological Preliminaries

2.3 Definitions Let X be a topological space, as defined in Sec. 1.2.

(a) A set $E \subset X$ is *closed* if its complement E^c is open. (Hence $\emptyset$ and X are closed, finite unions of closed sets are closed, and arbitrary intersections of closed sets are closed.)

(b) The *closure* $\bar{E}$ of a set $E \subset X$ is the smallest closed set in X which contains E. (The following argument proves the existence of $\bar{E}$: The collection Ω of all closed subsets of X which contain E is not empty, since $X \in \Omega$; let $\bar{E}$ be the intersection of all members of Ω.)

(c) A set $K \subset X$ is *compact* if every open cover of K contains a finite subcover. More explicitly, the requirement is that if $\{V_\alpha\}$ is a collection of open sets whose union contains K, then the union of some finite subcollection of $\{V_\alpha\}$ also contains K.

In particular, if X is itself compact, then X is called a *compact space*.

(d) A *neighborhood* of a point $p \in X$ is any open subset of X which contains p. (The use of this term is not quite standardized; some use "neighborhood of p" for any set which contains an open set containing p.)

(e) X is a *Hausdorff space* if the following is true: If $p \in X$, $q \in X$, and $p \neq q$, then p has a neighborhood U and q has a neighborhood V such that $U \cap V = \emptyset$.

(f) X is *locally compact* if every point of X has a neighborhood whose closure is compact.

Obviously, every compact space is locally compact.

We recall the Heine-Borel theorem: *The compact subsets of a euclidean space R^n are precisely those that are closed and bounded* ([26],† Theorem 2.41). From this it follows easily that R^n is a locally compact Hausdorff space. Also, every metric space is a Hausdorff space.

2.4 Theorem *Suppose K is compact and F is closed, in a topological space X. If $F \subset K$, then F is compact.*

PROOF If $\{V_\alpha\}$ is an open cover of F and $W = F^c$, then $W \cup \bigcup_\alpha V_\alpha$ covers X; hence there is a finite collection $\{V_{\alpha_i}\}$ such that

$$K \subset W \cup V_{\alpha_1} \cup \cdots \cup V_{\alpha_n}.$$

Then $F \subset V_{\alpha_1} \cup \cdots \cup V_{\alpha_n}$. ////

Corollary *If $A \subset B$ and if B has compact closure, so does A.*

2.5 Theorem *Suppose X is a Hausdorff space, $K \subset X$, K is compact, and $p \in K^c$. Then there are open sets U and W such that $p \in U$, $K \subset W$, and $U \cap W = \emptyset$.*

PROOF If $q \in K$, the Hausdorff separation axiom implies the existence of disjoint open sets U_q and V_q, such that $p \in U_q$ and $q \in V_q$. Since K is compact, there are points $q_1, \ldots, q_n \in K$ such that

$$K \subset V_{q_1} \cup \cdots \cup V_{q_n}.$$

Our requirements are then satisfied by the sets

$$U = U_{q_1} \cap \cdots \cap U_{q_n} \quad \text{and} \quad W = V_{q_1} \cup \cdots \cup V_{q_n}. \quad ////$$

Corollaries

(a) Compact subsets of Hausdorff spaces are closed.
(b) If F is closed and K is compact in a Hausdorff space, then $F \cap K$ is compact.

Corollary (b) follows from (a) and Theorem 2.4.

† Numbers in brackets refer to the Bibliography.

2.6 Theorem *If $\{K_\alpha\}$ is a collection of compact subsets of a Hausdorff space and if $\bigcap_\alpha K_\alpha = \varnothing$, then some finite subcollection of $\{K_\alpha\}$ also has empty intersection.*

PROOF Put $V_\alpha = K_\alpha^c$. Fix a member K_1 of $\{K_\alpha\}$. Since no point of K_1 belongs to every K_α, $\{V_\alpha\}$ is an open cover of K_1. Hence $K_1 \subset V_{\alpha_1} \cup \cdots \cup V_{\alpha_n}$ for some finite collection $\{V_{\alpha_i}\}$. This implies that

$$K_1 \cap K_{\alpha_1} \cap \cdots \cap K_{\alpha_n} = \varnothing. \qquad ////$$

2.7 Theorem *Suppose U is open in a locally compact Hausdorff space X, $K \subset U$, and K is compact. Then there is an open set V with compact closure such that*

$$K \subset V \subset \bar{V} \subset U.$$

PROOF Since every point of K has a neighborhood with compact closure, and since K is covered by the union of finitely many of these neighborhoods, K lies in an open set G with compact closure. If $U = X$, take $V = G$.

Otherwise, let C be the complement of U. Theorem 2.5 shows that to each $p \in C$ there corresponds an open set W_p such that $K \subset W_p$ and $p \notin \overline{W_p}$. Hence $\{C \cap \bar{G} \cap \overline{W_p}\}$, where p ranges over C, is a collection of compact sets with empty intersection. By Theorem 2.6 there are points $p_1, \ldots, p_n \in C$ such that

$$C \cap \bar{G} \cap \overline{W_{p_1}} \cap \cdots \cap \overline{W_{p_n}} = \varnothing.$$

The set

$$V = G \cap W_{p_1} \cap \cdots \cap W_{p_n}$$

then has the required properties, since

$$\bar{V} \subset \bar{G} \cap \overline{W_{p_1}} \cap \cdots \cap \overline{W_{p_n}}. \qquad ////$$

2.8 Definition Let f be a real (or extended-real) function on a topological space. If

$$\{x : f(x) > \alpha\}$$

is open for every real α, f is said to be *lower semicontinuous*. If

$$\{x : f(x) < \alpha\}$$

is open for every real α, f is said to be *upper semicontinuous*.

A real function is obviously continuous if and only if it is both upper and lower semicontinuous.

The simplest examples of semicontinuity are furnished by characteristic functions:

(a) Characteristic functions of *open* sets are *lower* semicontinuous.
(b) Characteristic functions of *closed* sets are *upper* semicontinuous.

The following property is an almost immediate consequence of the definitions:

(c) *The supremum of any collection of lower semicontinuous functions is lower semicontinuous. The infimum of any collection of upper semicontinuous functions is upper semicontinuous.*

2.9 Definition The *support* of a complex function f on a topological space X is the closure of the set

$$\{x : f(x) \neq 0\}.$$

The collection of all continuous complex functions on X whose support is compact is denoted by $C_c(X)$.

Observe that $C_c(X)$ *is a vector space*. This is due to two facts:

(a) The support of $f + g$ lies in the union of the support of f and the support of g, and any finite union of compact sets is compact.
(b) The sum of two continuous complex functions is continuous, as are scalar multiples of continuous functions.

(Statement and proof of Theorem 1.8 hold verbatim if "measurable function" is replaced by "continuous function," "measurable space" by "topological space"; take $\Phi(s, t) = s + t$, or $\Phi(s, t) = st$, to prove that sums and products of continuous functions are continuous.)

2.10 Theorem *Let X and Y be topological spaces, and let $f: X \to Y$ be continuous. If K is a compact subset of X, then $f(K)$ is compact.*

PROOF If $\{V_\alpha\}$ is an open cover of $f(K)$, then $\{f^{-1}(V_\alpha)\}$ is an open cover of K, hence $K \subset f^{-1}(V_{\alpha_1}) \cup \cdots \cup f^{-1}(V_{\alpha_n})$ for some $\alpha_1, \ldots, \alpha_n$, and therefore $f(K) \subset V_{\alpha_1} \cup \cdots \cup V_{\alpha_n}$. ////

Corollary *The range of any $f \in C_c(X)$ is a compact subset of the complex plane.*

In fact, if K is the support of $f \in C_c(X)$, then $f(X) \subset f(K) \cup \{0\}$. If X is not compact, then $0 \in f(X)$, but 0 need not lie in $f(K)$, as is seen by easy examples.

2.11 Notation In this chapter the following conventions will be used. The notation

$$K \prec f \tag{1}$$

will mean that K is a compact subset of X, that $f \in C_c(X)$, that $0 \leq f(x) \leq 1$ for all $x \in X$, and that $f(x) = 1$ for all $x \in K$. The notation

$$f \prec V \qquad (2)$$

will mean that V is open, that $f \in C_c(X)$, $0 \leq f \leq 1$, and that the support of f lies in V. The notation

$$K \prec f \prec V \qquad (3)$$

will be used to indicate that both (1) and (2) hold.

2.12 Urysohn's Lemma *Suppose X is a locally compact Hausdorff space, V is open in X, $K \subset V$, and K is compact. Then there exists an $f \in C_c(X)$, such that*

$$K \prec f \prec V. \qquad (1)$$

In terms of characteristic functions, the conclusion asserts the existence of a *continuous function* f which satisfies the inequalities $\chi_K \leq f \leq \chi_V$. Note that it is easy to find *semicontinuous* functions which do this; examples are χ_K and χ_V.

PROOF Put $r_1 = 0$, $r_2 = 1$, and let $r_3, r_4, r_5, \ldots$ be an enumeration of the rationals in $(0, 1)$. By Theorem 2.7, we can find open sets V_0 and then V_1 such that $\bar{V}_0$ is compact and

$$K \subset V_1 \subset \bar{V}_1 \subset V_0 \subset \bar{V}_0 \subset V. \qquad (2)$$

Suppose $n \geq 2$ and $V_{r_1}, \ldots, V_{r_n}$ have been chosen in such a manner that $r_i < r_j$ implies $\bar{V}_{r_j} \subset V_{r_i}$. Then one of the numbers $r_1, \ldots, r_n$, say r_i, will be the largest one which is smaller than r_{n+1}, and another, say r_j, will be the smallest one larger than r_{n+1}. Using Theorem 2.7 again, we can find $V_{r_{n+1}}$ so that

$$\bar{V}_{r_j} \subset V_{r_{n+1}} \subset \bar{V}_{r_{n+1}} \subset V_{r_i}.$$

Continuing, we obtain a collection $\{V_r\}$ of open sets, one for every rational $r \in [0, 1]$, with the following properties: $K \subset V_1$, $\bar{V}_0 \subset V$, each $\bar{V}_r$ is compact, and

$$s > r \text{ implies } \bar{V}_s \subset V_r. \qquad (3)$$

Define

$$f_r(x) = \begin{cases} r & \text{if } x \in V_r, \\ 0 & \text{otherwise,} \end{cases} \qquad g_s(x) = \begin{cases} 1 & \text{if } x \in \bar{V}_s, \\ s & \text{otherwise,} \end{cases} \qquad (4)$$

and

$$f = \sup_r f_r, \qquad g = \inf_s g_s. \qquad (5)$$

The remarks following Definition 2.8 show that f is lower semicontinuous and that g is upper semicontinuous. It is clear that $0 \leq f \leq 1$, that

$f(x) = 1$ if $x \in K$, and that f has its support in $\bar{V}_0$. The proof will be completed by showing that $f = g$.

The inequality $f_r(x) > g_s(x)$ is possible only if $r > s$, $x \in V_r$, and $x \notin \bar{V}_s$. But $r > s$ implies $V_r \subset V_s$. Hence $f_r \leq g_s$ for all r and s, so $f \leq g$.

Suppose $f(x) < g(x)$ for some x. Then there are rationals r and s such that $f(x) < r < s < g(x)$. Since $f(x) < r$, we have $x \notin V_r$; since $g(x) > s$, we have $x \in \bar{V}_s$. By (3), this is a contradiction. Hence $f = g$. ////

2.13 Theorem *Suppose $V_1, \ldots, V_n$ are open subsets of a locally compact Hausdorff space X, K is compact, and*

$$K \subset V_1 \cup \cdots \cup V_n.$$

Then there exist functions $h_i \prec V_i$ ($i = 1, \ldots, n$) such that

$$h_1(x) + \cdots + h_n(x) = 1 \qquad (x \in K). \tag{1}$$

Because of (1), the collection $\{h_1, \ldots, h_n\}$ is called a *partition of unity on K*, subordinate to the cover $\{V_1, \ldots, V_n\}$.

PROOF By Theorem 2.7, each $x \in K$ has a neighborhood W_x with compact closure $\bar{W}_x \subset V_i$ for some i (depending on x). There are points $x_1, \ldots, x_m$ such that $W_{x_1} \cup \cdots \cup W_{x_m} \supset K$. If $1 \leq i \leq n$, let H_i be the union of those W_{x_j} which lie in V_i. By Urysohn's lemma, there are functions g_i such that $H_i \prec g_i \prec V_i$. Define

$$\begin{aligned} h_1 &= g_1 \\ h_2 &= (1 - g_1)g_2 \\ &\cdots\cdots\cdots \\ h_n &= (1 - g_1)(1 - g_2) \cdots (1 - g_{n-1})g_n. \end{aligned} \tag{2}$$

Then $h_i \prec V_i$. It is easily verified, by induction, that

$$h_1 + h_2 + \cdots + h_n = 1 - (1 - g_1)(1 - g_2) \cdots (1 - g_n). \tag{3}$$

Since $K \subset H_1 \cup \cdots \cup H_n$, at least one $g_i(x) = 1$ at each point $x \in K$; hence (3) shows that (1) holds. ////

The Riesz Representation Theorem

2.14 Theorem *Let X be a locally compact Hausdorff space, and let Λ be a positive linear functional on $C_c(X)$. Then there exists a σ-algebra $\mathfrak{M}$ in X which contains all Borel sets in X, and there exists a unique positive measure μ on $\mathfrak{M}$ which represents Λ in the sense that*

(a) $\Lambda f = \int_X f \, d\mu$ for every $f \in C_c(X)$,

and which has the following additional properties:

(b) $\mu(K) < \infty$ for every compact set $K \subset X$.
(c) For every $E \in \mathfrak{M}$, we have
$$\mu(E) = \inf \{\mu(V): E \subset V, V \text{ open}\}.$$

(d) The relation
$$\mu(E) = \sup \{\mu(K): K \subset E, K \text{ compact}\}$$
holds for every open set E, and for every $E \in \mathfrak{M}$ with $\mu(E) < \infty$.
(e) If $E \in \mathfrak{M}$, $A \subset E$, and $\mu(E) = 0$, then $A \in \mathfrak{M}$.

For the sake of clarity, let us be more explicit about the meaning of the word "positive" in the hypothesis: Λ is assumed to be a linear functional on the complex vector space $C_c(X)$, with the additional property that Λf is a nonnegative real number for every f whose range consists of nonnegative real numbers. Briefly, if $f(X) \subset [0, \infty)$ then $\Lambda f \in [0, \infty)$.

Property (a) is of course the one of greatest interest. After we define $\mathfrak{M}$ and μ, (b) to (d) will be established in the course of proving that $\mathfrak{M}$ is a σ-algebra and that μ is countably additive. We shall see later (Theorem 2.18) that in "reasonable" spaces X every Borel measure which satisfies (b) also satisfies (c) and (d) and that (d) actually holds for every $E \in \mathfrak{M}$, in those cases. Property (e) merely says that $(X, \mathfrak{M}, \mu)$ is a complete measure space, in the sense of Theorem 1.36.

Throughout the proof of this theorem, the letter K will stand for a compact subset of X, and V will denote an open set in X.

Let us begin by proving the uniqueness of μ. If μ satisfies (c) and (d), it is clear that μ is determined on $\mathfrak{M}$ by its values on compact sets. Hence it suffices to prove that $\mu_1(K) = \mu_2(K)$ for all K, whenever μ_1 and μ_2 are measures for which the theorem holds. So, fix K and $\epsilon > 0$. By (b) and (c), there exists a $V \supset K$ with $\mu_2(V) < \mu_2(K) + \epsilon$; by Urysohn's lemma, there exists an f so that $K \prec f \prec V$; hence

$$\mu_1(K) = \int_X \chi_K \, d\mu_1 \le \int_X f \, d\mu_1 = \Lambda f = \int_X f \, d\mu_2$$
$$\le \int_X \chi_V \, d\mu_2 = \mu_2(V) < \mu_2(K) + \epsilon.$$

Thus $\mu_1(K) \le \mu_2(K)$. If we interchange the roles of μ_1 and μ_2, the opposite inequality is obtained, and the uniqueness of μ is proved.

Incidentally, the above computation shows that (a) forces (b).

Construction of μ and $\mathfrak{M}$

For every open set V in X, define

$$\mu(V) = \sup \{\Lambda f: f \prec V\}. \tag{1}$$

If $V_1 \subset V_2$, it is clear that (1) implies $\mu(V_1) \leq \mu(V_2)$. Hence

$$\mu(E) = \inf \{\mu(V): E \subset V, V \text{ open}\}, \quad (2)$$

if E is an open set, and it is consistent with (1) to *define* $\mu(E)$ by (2), for *every* $E \subset X$.

Note that although we have defined $\mu(E)$ for every $E \subset X$, the countable additivity of μ will be proved only on a certain σ-algebra $\mathfrak{M}$ in X.

Let $\mathfrak{M}_F$ be the class of all $E \subset X$ which satisfy two conditions: $\mu(E) < \infty$, and

$$\mu(E) = \sup \{\mu(K): K \subset E, K \text{ compact}\}. \quad (3)$$

Finally, let $\mathfrak{M}$ be the class of all $E \subset X$ such that $E \cap K \in \mathfrak{M}_F$ for every compact K.

Proof that μ and $\mathfrak{M}$ have the required properties

It is evident that μ is *monotone*, i.e., that $\mu(A) \leq \mu(B)$ if $A \subset B$ and that $\mu(E) = 0$ implies $E \in \mathfrak{M}_F$ and $E \in \mathfrak{M}$. Thus (e) holds, and so does (c), by definition.

Since the proof of the other assertions is rather long, it will be convenient to divide it into several steps.

Observe that the positivity of Λ implies that Λ is *monotone*: $f \leq g$ implies $\Lambda f \leq \Lambda g$. This is clear, since $\Lambda g = \Lambda f + \Lambda(g - f)$ and $g - f \geq 0$. This monotonicity will be used in Steps II and X.

STEP I *If $E_1, E_2, E_3, \ldots$ are arbitrary subsets of X, then*

$$\mu\left(\bigcup_{i=1}^{\infty} E_i\right) \leq \sum_{i=1}^{\infty} \mu(E_i). \quad (4)$$

PROOF We first show that

$$\mu(V_1 \cup V_2) \leq \mu(V_1) + \mu(V_2), \quad (5)$$

if V_1 and V_2 are open. Choose $g \prec V_1 \cup V_2$. By Theorem 2.13 there are functions h_1 and h_2 such that $h_i \prec V_i$ and $h_1(x) + h_2(x) = 1$ for all x in the support of g. Hence $h_i g \prec V_i$, $g = h_1 g + h_2 g$, and so

$$\Lambda g = \Lambda(h_1 g) + \Lambda(h_2 g) \leq \mu(V_1) + \mu(V_2). \quad (6)$$

Since (6) holds for *every* $g \prec V_1 \cup V_2$, (5) follows.

If $\mu(E_i) = \infty$ for some i, then (4) is trivially true. Suppose therefore that $\mu(E_i) < \infty$ for every i. Choose $\epsilon > 0$. By (2) there are open sets $V_i \supset E_i$ such that

$$\mu(V_i) < \mu(E_i) + 2^{-i}\epsilon \quad (i = 1, 2, 3, \ldots).$$

Put $V = \bigcup_1^\infty V_i$, and choose $f \prec V$. Since f has compact support, we see that $f \prec V_1 \cup \cdots \cup V_n$ for some n. Applying induction to (5), we therefore obtain

$$\Lambda f \leq \mu(V_1 \cup \cdots \cup V_n) \leq \mu(V_1) + \cdots + \mu(V_n) \leq \sum_{i=1}^\infty \mu(E_i) + \epsilon.$$

Since this holds for every $f \prec V$, and since $\bigcup E_i \subset V$, it follows that

$$\mu\left(\bigcup_{i=1}^\infty E_i\right) \leq \mu(V) \leq \sum_{i=1}^\infty \mu(E_i) + \epsilon,$$

which proves (4), since ϵ was arbitrary. ////

STEP II *If K is compact, then $K \in \mathfrak{M}_F$ and*

$$\mu(K) = \inf \{\Lambda f : K \prec f\}. \tag{7}$$

This implies assertion (b) of the theorem.

PROOF If $K \prec f$ and $0 < \alpha < 1$, let $V_\alpha = \{x : f(x) > \alpha\}$. Then $K \subset V_\alpha$, and $\alpha g \leq f$ whenever $g \prec V_\alpha$. Hence

$$\mu(K) \leq \mu(V_\alpha) = \sup \{\Lambda g : g \prec V_\alpha\} \leq \alpha^{-1} \Lambda f.$$

Let $\alpha \to 1$, to conclude that

$$\mu(K) \leq \Lambda f. \tag{8}$$

Thus $\mu(K) < \infty$. Since K evidently satisfies (3), $K \in \mathfrak{M}_F$.

If $\epsilon > 0$, there exists $V \supset K$ with $\mu(V) < \mu(K) + \epsilon$. By Urysohn's lemma, $K \prec f \prec V$ for some f. Thus

$$\Lambda f \leq \mu(V) < \mu(K) + \epsilon,$$

which, combined with (8), gives (7). ////

STEP III *Every open set satisfies (3). Hence $\mathfrak{M}_F$ contains every open set V with $\mu(V) < \infty$.*

PROOF Let α be a real number such that $\alpha < \mu(V)$. There exists an $f \prec V$ with $\alpha < \Lambda f$. If W is *any* open set which contains the support K of f, then $f \prec W$, hence $\Lambda f \leq \mu(W)$. Thus $\Lambda f \leq \mu(K)$. This exhibits a compact $K \subset V$ with $\alpha < \mu(K)$, so that (3) holds for V. ////

STEP IV *Suppose $E = \bigcup_{i=1}^\infty E_i$, where $E_1, E_2, E_3, \ldots$ are pairwise disjoint members of $\mathfrak{M}_F$. Then*

$$\mu(E) = \sum_{i=1}^\infty \mu(E_i). \tag{9}$$

If, in addition, $\mu(E) < \infty$, then also $E \in \mathfrak{M}_F$.

PROOF We first show that

$$\mu(K_1 \cup K_2) = \mu(K_1) + \mu(K_2) \tag{10}$$

if K_1 and K_2 are disjoint compact sets. Choose $\epsilon > 0$. By Urysohn's lemma, there exists $f \in C_c(X)$ such that $f(x) = 1$ on K_1, $f(x) = 0$ on K_2, and $0 \le f \le 1$. By Step II there exists g such that

$$K_1 \cup K_2 \prec g \quad \text{and} \quad \Lambda g < \mu(K_1 \cup K_2) + \epsilon.$$

Note that $K_1 \prec fg$ and $K_2 \prec (1-f)g$. Since Λ is linear, it follows from (8) that

$$\mu(K_1) + \mu(K_2) \le \Lambda(fg) + \Lambda(g - fg) = \Lambda g < \mu(K_1 \cup K_2) + \epsilon.$$

Since ϵ was arbitrary, (10) follows now from Step I.

If $\mu(E) = \infty$, (9) follows from Step I. Assume therefore that $\mu(E) < \infty$, and choose $\epsilon > 0$. Since $E_i \in \mathfrak{M}_F$, there are compact sets $H_i \subset E_i$ with

$$\mu(H_i) > \mu(E_i) - 2^{-i}\epsilon \quad (i = 1, 2, 3, \ldots). \tag{11}$$

Putting $K_n = H_1 \cup \cdots \cup H_n$ and using induction on (10), we obtain

$$\mu(E) \ge \mu(K_n) = \sum_{i=1}^{n} \mu(H_i) > \sum_{i=1}^{n} \mu(E_i) - \epsilon. \tag{12}$$

Since (12) holds for every n and every $\epsilon > 0$, the left side of (9) is not smaller than the right side, and so (9) follows from Step I.

But if $\mu(E) < \infty$ and $\epsilon > 0$, (9) shows that

$$\mu(E) \le \sum_{i=1}^{N} \mu(E_i) + \epsilon \tag{13}$$

for some N. By (12), it follows that $\mu(E) \le \mu(K_N) + 2\epsilon$, and this shows that E satisfies (3); hence $E \in \mathfrak{M}_F$. ////

STEP V *If $E \in \mathfrak{M}_F$ and $\epsilon > 0$, there is a compact K and an open V such that $K \subset E \subset V$ and $\mu(V - K) < \epsilon$.*

PROOF Our definitions show that there exist $K \subset E$ and $V \supset E$ so that

$$\mu(V) - \frac{\epsilon}{2} < \mu(E) < \mu(K) + \frac{\epsilon}{2}.$$

Since $V - K$ is open, $V - K \in \mathfrak{M}_F$, by Step III. Hence Step IV implies that

$$\mu(K) + \mu(V - K) = \mu(V) < \mu(K) + \epsilon. \quad ////$$

STEP VI *If $A \in \mathfrak{M}_F$ and $B \in \mathfrak{M}_F$, then $A - B$, $A \cup B$, and $A \cap B$ belong to $\mathfrak{M}_F$.*

PROOF If $\epsilon > 0$, Step V shows that there are sets K_i and V_i such that $K_1 \subset A \subset V_1$, $K_2 \subset B \subset V_2$, and $\mu(V_i - K_i) < \epsilon$, for $i = 1, 2$. Since

$$A - B \subset V_1 - K_2 \subset (V_1 - K_1) \cup (K_1 - V_2) \cup (V_2 - K_2),$$

Step I shows that

$$\mu(A - B) \leq \epsilon + \mu(K_1 - V_2) + \epsilon. \tag{14}$$

Since $K_1 - V_2$ is a compact subset of $A - B$, (14) shows that $A - B$ satisfies (3), so that $A - B \in \mathfrak{M}_F$.

Since $A \cup B = (A - B) \cup B$, an application of Step IV shows that $A \cup B \in \mathfrak{M}_F$. Since $A \cap B = A - (A - B)$, we also have $A \cap B \in \mathfrak{M}_F$. ////

STEP VII $\mathfrak{M}$ *is a σ-algebra in X which contains all Borel sets.*

PROOF Let K be an arbitrary compact set in X.

If $A \in \mathfrak{M}$, then $A^c \cap K = K - (A \cap K)$, so that $A^c \cap K$ is a difference of two members of $\mathfrak{M}_F$. Hence $A^c \cap K \in \mathfrak{M}_F$, and we conclude: $A \in \mathfrak{M}$ implies $A^c \in \mathfrak{M}$.

Next, suppose $A = \bigcup_1^\infty A_i$, where each $A_i \in \mathfrak{M}$. Put $B_1 = A_1 \cap K$, and

$$B_n = (A_n \cap K) - (B_1 \cup \cdots \cup B_{n-1}) \qquad (n = 2, 3, 4, \ldots). \tag{15}$$

Then $\{B_n\}$ is a disjoint sequence of members of $\mathfrak{M}_F$, by Step VI, and $A \cap K = \bigcup_1^\infty B_n$. It follows from Step IV that $A \cap K \in \mathfrak{M}_F$. Hence $A \in \mathfrak{M}$.

Finally, if C is closed, then $C \cap K$ is compact, hence $C \cap K \in \mathfrak{M}_F$, so $C \in \mathfrak{M}$. In particular, $X \in \mathfrak{M}$.

We have thus proved that $\mathfrak{M}$ is a σ-algebra in X which contains all closed subsets of X. Hence $\mathfrak{M}$ contains all Borel sets in X. ////

STEP VIII $\mathfrak{M}_F$ *consists of precisely those sets $E \in \mathfrak{M}$ for which $\mu(E) < \infty$.*

This implies assertion (d) of the theorem.

PROOF If $E \in \mathfrak{M}_F$, Steps II and VI imply that $E \cap K \in \mathfrak{M}_F$ for every compact K, hence $E \in \mathfrak{M}$.

Conversely, suppose $E \in \mathfrak{M}$ and $\mu(E) < \infty$, and choose $\epsilon > 0$. There is an open set $V \supset E$ with $\mu(V) < \infty$; by III and V, there is a compact $K \subset V$ with $\mu(V - K) < \epsilon$. Since $E \cap K \in \mathfrak{M}_F$, there is a compact set $H \subset E \cap K$ with

$$\mu(E \cap K) < \mu(H) + \epsilon.$$

Since $E \subset (E \cap K) \cup (V - K)$, it follows that

$$\mu(E) \leq \mu(E \cap K) + \mu(V - K) < \mu(H) + 2\epsilon,$$

which implies that $E \in \mathfrak{M}_F$. ////

STEP IX μ *is a measure on $\mathfrak{M}$.*

46 REAL AND COMPLEX ANALYSIS

PROOF The countable additivity of μ on $\mathfrak{M}$ follows immediately from Steps IV and VIII. ////

STEP X *For every $f \in C_c(X)$, $\Lambda f = \int_X f \, d\mu$.*

This proves (a), and completes the theorem.

PROOF Clearly, it is enough to prove this for real f. Also, it is enough to prove the *inequality*

$$\Lambda f \leq \int_X f \, d\mu \tag{16}$$

for every real $f \in C_c(X)$. For once (16) is established, the linearity of Λ shows that

$$-\Lambda f = \Lambda(-f) \leq \int_X (-f) \, d\mu = -\int_X f \, d\mu,$$

which, together with (16), shows that equality holds in (16).

Let K be the support of a real $f \in C_c(X)$, let $[a, b]$ be an interval which contains the range of f (note the Corollary to Theorem 2.10), choose $\epsilon > 0$, and choose y_i, for $i = 0, 1, \ldots, n$, so that $y_i - y_{i-1} < \epsilon$ and

$$y_0 < a < y_1 < \cdots < y_n = b. \tag{17}$$

Put

$$E_i = \{x : y_{i-1} < f(x) \leq y_i\} \cap K \qquad (i = 1, \ldots, n). \tag{18}$$

Since f is continuous, f is Borel measurable, and the sets E_i are therefore disjoint Borel sets whose union is K. There are open sets $V_i \supset E_i$ such that

$$\mu(V_i) < \mu(E_i) + \frac{\epsilon}{n} \qquad (i = 1, \ldots, n) \tag{19}$$

and such that $f(x) < y_i + \epsilon$ for all $x \in V_i$. By Theorem 2.13, there are functions $h_i \prec V_i$ such that $\sum h_i = 1$ on K. Hence $f = \sum h_i f$, and Step II shows that

$$\mu(K) \leq \Lambda(\sum h_i) = \sum \Lambda h_i.$$

Since $h_i f \leq (y_i + \epsilon)h_i$, and since $y_i - \epsilon < f(x)$ on E_i, we have

$$\Lambda f = \sum_{i=1}^{n} \Lambda(h_i f) \leq \sum_{i=1}^{n} (y_i + \epsilon)\Lambda h_i$$

$$= \sum_{i=1}^{n} (|a| + y_i + \epsilon)\Lambda h_i - |a| \sum_{i=1}^{n} \Lambda h_i$$

$$\leq \sum_{i=1}^{n} (|a| + y_i + \epsilon)[\mu(E_i) + \epsilon/n] - |a|\mu(K)$$

$$= \sum_{i=1}^{n} (y_i - \epsilon)\mu(E_i) + 2\epsilon\mu(K) + \frac{\epsilon}{n}\sum_{i=1}^{n}(|a| + y_i + \epsilon)$$

$$\leq \int_X f\, d\mu + \epsilon[2\mu(K) + |a| + b + \epsilon].$$

Since ϵ was arbitrary, (16) is established, and the proof of the theorem is complete. ////

Regularity Properties of Borel Measures

2.15 Definition A measure μ defined on the σ-algebra of all Borel sets in a locally compact Hausdorff space X is called a *Borel measure on* X. If μ is positive, a Borel set $E \subset X$ is *outer regular* or *inner regular*, respectively, if E has property (c) or (d) of Theorem 2.14. If every Borel set in X is both outer and inner regular, μ is called *regular*.

In our proof of the Riesz theorem, outer regularity of every set E was built into the construction, but inner regularity was proved only for the open sets and for those $E \in \mathfrak{M}$ for which $\mu(E) < \infty$. It turns out that this flaw is in the nature of things. One cannot prove regularity of μ under the hypothesis of Theorem 2.14; an example is described in Exercise 17.

However, a slight strengthening of the hypotheses does give us a regular measure. Theorem 2.17 shows this. And if we specialize a little more, Theorem 2.18 shows that all regularity problems neatly disappear.

2.16 Definition A set E in a topological space is called σ-*compact* if E is a countable union of compact sets.

A set E in a measure space (with measure μ) is said to have σ-*finite measure* if E is a countable union of sets E_i with $\mu(E_i) < \infty$.

For example, in the situation described in Theorem 2.14, every σ-compact set has σ-finite measure. Also, it is easy to see that if $E \in \mathfrak{M}$ and E has σ-finite measure, then E is inner regular.

2.17 Theorem *Suppose X is a locally compact, σ-compact Hausdorff space. If $\mathfrak{M}$ and μ are as described in the statement of Theorem 2.14, then $\mathfrak{M}$ and μ have the following properties:*

(a) *If $E \in \mathfrak{M}$ and $\epsilon > 0$, there is a closed set F and an open set V such that $F \subset E \subset V$ and $\mu(V - F) < \epsilon$.*
(b) *μ is a regular Borel measure on X.*
(c) *If $E \in \mathfrak{M}$, there are sets A and B such that A is an F_σ, B is a G_δ, $A \subset E \subset B$, and $\mu(B - A) = 0$.*

As a corollary of (c) we see that every $E \in \mathfrak{M}$ is the union of an F_σ and a set of measure 0.

PROOF Let $X = K_1 \cup K_2 \cup K_3 \cup \cdots$, where each K_n is compact. If $E \in \mathfrak{M}$ and $\epsilon > 0$, then $\mu(K_n \cap E) < \infty$, and there are open sets $V_n \supset K_n \cap E$ such that

$$\mu(V_n - (K_n \cap E)) < \frac{\epsilon}{2^{n+1}} \qquad (n = 1, 2, 3, \ldots). \tag{1}$$

If $V = \bigcup V_n$, then $V - E \subset \bigcup (V_n - (K_n \cap E))$, so that

$$\mu(V - E) < \frac{\epsilon}{2}.$$

Apply this to E^c in place of E: There is an open set $W \supset E^c$ such that $\mu(W - E^c) < \epsilon/2$. If $F = W^c$, then $F \subset E$, and $E - F = W - E^c$. Now (a) follows.

Every closed set $F \subset X$ is σ-compact, because $F = \bigcup (F \cap K_n)$. Hence (a) implies that every set $E \in \mathfrak{M}$ is inner regular. This proves (b).

If we apply (a) with $\epsilon = 1/j$ ($j = 1, 2, 3, \ldots$), we obtain closed sets F_j and open sets V_j such that $F_j \subset E \subset V_j$ and $\mu(V_j - F_j) < 1/j$. Put $A = \bigcup F_j$ and $B = \bigcap V_j$. Then $A \subset E \subset B$, A is an F_σ, B is a G_δ, and $\mu(B - A) = 0$ since $B - A \subset V_j - F_j$ for $j = 1, 2, 3, \ldots$. This proves (c). ////

2.18 Theorem *Let X be a locally compact Hausdorff space in which every open set is σ-compact. Let λ be any positive Borel measure on X such that $\lambda(K) < \infty$ for every compact set K. Then λ is regular.*

Note that every euclidean space R^k satisfies the present hypothesis, since every open set in R^k is a countable union of closed balls.

PROOF Put $\Lambda f = \int_X f \, d\lambda$, for $f \in C_c(X)$. Since $\lambda(K) < \infty$ for every compact K, Λ is a positive linear functional on $C_c(X)$, and there is a regular measure μ, satisfying the conclusions of Theorem 2.17, such that

$$\int_X f \, d\lambda = \int_X f \, d\mu \qquad (f \in C_c(X)). \tag{1}$$

We will show that $\lambda = \mu$.

Let V be open in X. Then $V = \bigcup K_i$, where K_i is compact, $i = 1, 2, 3, \ldots$. By Urysohn's lemma we can choose f_i so that $K_i \prec f_i \prec V$. Let $g_n = \max(f_1, \ldots, f_n)$. Then $g_n \in C_c(X)$ and $g_n(x)$ increases to $\chi_V(x)$ at every point $x \in X$. Hence (1) and the monotone convergence theorem imply

$$\lambda(V) = \lim_{n \to \infty} \int_X g_n \, d\lambda = \lim_{n \to \infty} \int_X g_n \, d\mu = \mu(V). \tag{2}$$

Now let E be a Borel set in X, and choose $\epsilon > 0$. Since μ satisfies Theorem 2.17, there is a closed set F and an open set V such that $F \subset E \subset V$ and $\mu(V - F) < \epsilon$. Hence $\mu(V) \leq \mu(F) + \epsilon \leq \mu(E) + \epsilon$.

Since $V - F$ is open, (2) shows that $\lambda(V - F) < \epsilon$, hence $\lambda(V) \leq \lambda(E) + \epsilon$. Consequently

$$\lambda(E) \leq \lambda(V) = \mu(V) \leq \mu(E) + \epsilon$$

and

$$\mu(E) \leq \mu(V) = \lambda(V) \leq \lambda(E) + \epsilon,$$

so that $|\lambda(E) - \mu(E)| < \epsilon$ for every $\epsilon > 0$. Hence $\lambda(E) = \mu(E)$. ////

In Exercise 18 a compact Hausdorff space is described in which the complement of a certain point fails to be σ-compact and in which the conclusion of the preceding theorem is not true.

Lebesgue Measure

2.19 Euclidean Spaces Euclidean k-dimensional space R^k is the set of all points $x = (\xi_1, \ldots, \xi_k)$ whose coordinates ξ_i are real numbers, with the following algebraic and topological structure:

If $x = (\xi_1, \ldots, \xi_k)$, $y = (\eta_1, \ldots, \eta_k)$, and α is a real number, $x + y$ and αx are defined by

$$x + y = (\xi_1 + \eta_1, \ldots, \xi_k + \eta_k), \qquad \alpha x = (\alpha \xi_1, \ldots, \alpha \xi_k). \tag{1}$$

This makes R^k into a real vector space. If $x \cdot y = \sum \xi_i \eta_i$ and $|x| = (x \cdot x)^{1/2}$, the Schwarz inequality $|x \cdot y| \leq |x| |y|$ leads to the triangle inequality

$$|x - y| \leq |x - z| + |z - y|; \tag{2}$$

hence we obtain a metric by setting $\rho(x, y) = |x - y|$. We assume that these facts are familiar and shall prove them in greater generality in Chap. 4.

If $E \subset R^k$ and $x \in R^k$, the *translate of E by x* is the set

$$E + x = \{y + x : y \in E\}. \tag{3}$$

A set of the form

$$W = \{x : \alpha_i < \xi_i < \beta_i, 1 \leq i \leq k\}, \tag{4}$$

or any set obtained by replacing any or all of the $<$ signs in (4) by $\leq$, is called a *k-cell*; its volume is defined to be

$$\text{vol}(W) = \prod_{i=1}^{k} (\beta_i - \alpha_i). \tag{5}$$

If $a \in R^k$ and $\delta > 0$, we shall call the set

$$Q(a; \delta) = \{x : \alpha_i \leq \xi_i < \alpha_i + \delta, 1 \leq i \leq k\} \tag{6}$$

the *δ-box with corner at a*. Here $a = (\alpha_1, \ldots, \alpha_k)$.

For $n = 1, 2, 3, \ldots$, we let P_n be the set of all $x \in R^k$ whose coordinates are integral multiples of 2^{-n}, and we let Ω_n be the collection of all 2^{-n} boxes with corners at points of P_n. We shall need the following four properties of $\{\Omega_n\}$. The first three are obvious by inspection.

(a) If n is fixed, each $x \in R^k$ lies in one and only one member of Ω_n.
(b) If $Q' \in \Omega_n$, $Q'' \in \Omega_r$, and $r < n$, then either $Q' \subset Q''$ or $Q' \cap Q'' = \emptyset$.
(c) If $Q \in \Omega_r$, then $\text{vol}(Q) = 2^{-rk}$; and if $n > r$, the set P_n has exactly $2^{(n-r)k}$ points in Q.
(d) Every nonempty open set in R^k is a countable union of disjoint boxes belonging to $\Omega_1 \cup \Omega_2 \cup \Omega_3 \cup \cdots$.

PROOF OF (d) If V is open, every $x \in V$ lies in an open ball which lies in V; hence $x \in Q \subset V$ for some Q belonging to some Ω_n. In other words, V is the union of all boxes which lie in V and which belong to some Ω_n. From this collection of boxes, select those which belong to Ω_1, and remove those in Ω_2, $\Omega_3, \ldots$ which lie in any of the selected boxes. From the remaining collection, select those boxes of Ω_2 which lie in V, and remove those in $\Omega_3, \Omega_4, \ldots$ which lie in any of the selected boxes. If we proceed in this way, (a) and (b) show that (d) holds. ////

2.20 Theorem *There exists a positive complete measure m defined on a σ-algebra $\mathfrak{M}$ in R^k, with the following properties:*

(a) $m(W) = \text{vol}(W)$ *for every k-cell W*.
(b) *$\mathfrak{M}$ contains all Borel sets in R^k; more precisely, $E \in \mathfrak{M}$ if and only if there are sets A and $B \subset R^k$ such that $A \subset E \subset B$, A is an F_σ, B is a G_δ, and $m(B - A) = 0$. Also, m is regular.*

(c) m is translation-invariant, i.e.,

$$m(E + x) = m(E)$$

for every $E \in \mathfrak{M}$ and every $x \in R^k$.

(d) If μ is any positive translation-invariant Borel measure on R^k such that $\mu(K) < \infty$ for every compact set K, then there is a constant c such that $\mu(E) = cm(E)$ for all Borel sets $E \subset R^k$.

(e) To every linear transformation T of R^k into R^k corresponds a real number $\Delta(T)$ such that

$$m(T(E)) = \Delta(T)m(E)$$

for every $E \in \mathfrak{M}$. In particular, $m(T(E)) = m(E)$ when T is a rotation.

The members of $\mathfrak{M}$ are the *Lebesgue measurable* sets in R^k; m is the *Lebesgue measure* on R^k. When clarity requires it, we shall write m_k in place of m.

PROOF If f is any complex function on R^k, with compact support, define

$$\Lambda_n f = 2^{-nk} \sum_{x \in P_n} f(x) \qquad (n = 1, 2, 3, \ldots), \tag{1}$$

where P_n is as in Sec. 2.19.

Now suppose $f \in C_c(R^k)$, f is real, W is an open k-cell which contains the support of f, and $\epsilon > 0$. The uniform continuity of f ([26], Theorem 4.19) shows that there is an integer N and that there are functions g and h with support in W, such that (i) g and h are constant on each box belonging to Ω_N, (ii) $g \le f \le h$, and (iii) $h - g < \epsilon$. If $n > N$, Property 2.19(c) shows that

$$\Lambda_N g = \Lambda_n g \le \Lambda_n f \le \Lambda_n h = \Lambda_N h. \tag{2}$$

Thus the upper and lower limits of $\{\Lambda_n f\}$ differ by at most ϵ vol (W), and since ϵ was arbitrary, we have proved the existence of

$$\Lambda f = \lim_{n \to \infty} \Lambda_n f \qquad (f \in C_c(R^k)). \tag{3}$$

It is immediate that Λ is a positive linear functional on $C_c(R^k)$. (In fact, Λf is precisely the Riemann integral of f over R^k. We went through the preceding construction in order not to have to rely on any theorems about Riemann integrals in several variables.)

We define m and $\mathfrak{M}$ to be the measure and σ-algebra associated with this Λ as in Theorem 2.14.

Since Theorem 2.14 gives us a complete measure and since R^k is σ-compact, Theorem 2.17 implies assertion (b) of Theorem 2.20.

To prove (a), let W be the open cell 2.19(4), let E_r be the union of those boxes belonging to Ω_r whose closures lie in W, choose f_r so that $\bar{E}_r \prec f_r \prec W$, and put $g_r = \max\{f_1, \ldots, f_r\}$. Our construction of Λ shows that

$$\text{vol}(E_r) \le \Lambda f_r \le \Lambda g_r \le \text{vol } W. \tag{4}$$

As $r \to \infty$, $\text{vol}(E_r) \to \text{vol}(W)$, and

$$\Lambda g_r = \int g_r \, dm \to m(W) \tag{5}$$

by the monotone convergence theorem, since $g_r(x) \to \chi_W(x)$ for all $x \in R^k$. Thus $m(W) = \text{vol}(W)$ for every open cell W, and since every k-cell is the intersection of a decreasing sequence of open k-cells, we obtain (a).

The proofs of (c), (d), and (e) will use the following observation: If λ is a positive Borel measure on R^k and $\lambda(E) = m(E)$ for all boxes E, then the same equality holds for all open sets E, by property 2.19(d), and therefore for all Borel sets E, since λ and m are regular (Theorem 2.18).

To prove (c), fix $x \in R^k$ and define $\lambda(E) = m(E + x)$. It is clear that λ is then a measure; by (a), $\lambda(E) = m(E)$ for all boxes, hence $m(E + x) = m(E)$ for all Borel sets E. The same equality holds for every $E \in \mathfrak{M}$, because of (b).

Suppose next that μ satisfies the hypotheses of (d). Let Q_0 be a 1-box, put $c = \mu(Q_0)$. Since Q_0 is the union of 2^{nk} disjoint 2^{-n} boxes that are translates of each other, we have

$$2^{nk}\mu(Q) = \mu(Q_0) = cm(Q_0) = c \cdot 2^{nk}m(Q)$$

for every 2^{-n}-box Q. Property 2.19(d) implies now that $\mu(E) = cm(E)$ for all open sets $E \subset R^k$. This proves (d).

To prove (e), let $T: R^k \to R^k$ be linear. If the range of T is a subspace Y of lower dimension, then $m(Y) = 0$ and the desired conclusion holds with $\Delta(T) = 0$. In the other case, elementary linear algebra tells us that T is a one-to-one map of R^k onto R^k whose inverse is also linear. Thus T is a homeomorphism of R^k onto R^k, so that $T(E)$ is a Borel set for every Borel set E, and we can therefore define a positive Borel measure μ on R^k by

$$\mu(E) = m(T(E)).$$

The linearity of T, combined with the translation-invariance of m, gives

$$\mu(E + x) = m(T(E + x)) = m(T(E) + Tx) = m(T(E)) = \mu(E).$$

Thus μ is translation-invariant, and the first assertion of (e) follows from (d), first for Borel sets E, then for all $E \in \mathfrak{M}$ by (b).

To find $\Delta(T)$, we merely need to know $m(T(E))/m(E)$ for one set E with $0 < m(E) < \infty$. If T is a rotation, let E be the unit ball of R^k; then $T(E) = E$, and $\Delta(T) = 1$. ////

2.21 Remarks If m is the Lebesgue measure on R^k, it is customary to write $L^1(R^k)$ in place of $L^1(m)$. If E is a Lebesgue measurable subset of R^k, and if m is restricted to the measurable subsets of E, a new measure space is obtained in an obvious fashion. The phrase "$f \in L^1$ on E" or "$f \in L^1(E)$" is used to indicate that f is integrable on this measure space.

If $k = 1$, if I is any of the sets (a, b), $(a, b]$, $[a, b)$, $[a, b]$, and if $f \in L^1(I)$, it is customary to write

$$\int_a^b f(x)\, dx \quad \text{in place of} \quad \int_I f\, dm.$$

Since the Lebesgue measure of any single point is 0, it makes no difference over which of these four sets the integral is extended.

Everything learned about integration in elementary Calculus courses is still useful in the present context, for *if f is a continuous complex function on $[a, b]$, then the Riemann integral of f and the Lebesgue integral of f over $[a, b]$ coincide.* This is obvious from our construction if $f(a) = f(b) = 0$ and if $f(x)$ is defined to be 0 for $x < a$ and for $x > b$. The general case follows without difficulty. Actually the same thing is true for every Riemann integrable f on $[a, b]$. Since we shall have no occasion to discuss Riemann integrable functions in the sequel, we omit the proof and refer to Theorem 11.33 of [26].

Two natural questions may have occurred to some readers by now: Is every Lebesgue measurable set a Borel set? Is every subset of R^k Lebesgue measurable? The answer is negative in both cases, even when $k = 1$.

The first question can be settled by a cardinality argument which we sketch briefly. Let c be the cardinality of the continuum (the real line or, equivalently, the collection of all sets of integers). We know that R^k has a *countable* base (open balls with rational radii and with centers in some countable dense subset of R^k), and that $\mathcal{B}_k$ (the collection of all Borel sets of R^k) is the σ-algebra generated by this base. It follows from this (we omit the proof) that $\mathcal{B}_k$ has cardinality c. On the other hand, there exist Cantor sets $E \subset R^1$ with $m(E) = 0$. (Exercise 5.) The completeness of m implies that each of the 2^c subsets of E is Lebesgue measurable. Since $2^c > c$, most subsets of E are not Borel sets.

The following theorem answers the second question.

2.22 Theorem *If $A \subset R^1$ and every subset of A is Lebesgue measurable then $m(A) = 0$.*

Corollary *Every set of positive measure has nonmeasurable subsets.*

PROOF We shall use the fact that R^1 is a group, relative to addition. Let Q be the subgroup that consists of the rational numbers, and let E be a set that contains exactly one point from each coset of Q in R^1. (The assertion that there is such a set is a direct application of the axiom of choice.) Then E has the following two properties.

(a) $(E + r) \cap (E + s) = \emptyset$ if $r \in Q$, $s \in Q$, $r \neq s$.
(b) Every $x \in R^1$ lies in $E + r$ for some $r \in Q$.

To prove (a), suppose $x \in (E + r) \cap (E + s)$. Then $x = y + r = z + s$ for some $y \in E$, $z \in E$, $y \neq z$. But $y - z = s - r \in Q$, so that y and z lie in the same coset of Q, a contradiction.

To prove (b), let y be the point of E that lies in the same coset as x, put $r = x - y$.

Fix $t \in Q$, for the moment, and put $A_t = A \cap (E + t)$. By hypothesis, A_t is measurable. Let $K \subset A_t$ be compact, let H be the union of the translates $K + r$, where r ranges over $Q \cap [0, 1]$. Then H is bounded, hence $m(H) < \infty$. Since $K \subset E + t$, (a) shows that the sets $K + r$ are pairwise disjoint. Thus $m(H) = \sum_r m(K + r)$. But $m(K + r) = m(K)$. It follows that $m(K) = 0$. This holds for every compact $K \subset A_t$. Hence $m(A_t) = 0$.

Finally, (b) shows that $A = \bigcup A_t$, where t ranges over Q. Since Q is countable, we conclude that $m(A) = 0$. ////

2.23 Determinants The scale factors $\Delta(T)$ that occur in Theorem 2.20(e) can be interpreted algebraically by means of determinants.

Let $\{e_1, \ldots, e_k\}$ be the standard basis for R^k: the ith coordinate of e_j is 1 if $i = j$, 0 if $i \neq j$. If $T: R^k \to R^k$ is linear and

$$Te_j = \sum_{i=1}^{k} \alpha_{ij} e_i \qquad (1 \leq j \leq k) \tag{1}$$

then det T is, by definition, the determinant of the matrix $[T]$ that has α_{ij} in row i and column j.

We claim that

$$\Delta(T) = |\det T|. \tag{2}$$

If $T = T_1 T_2$, it is clear that $\Delta(T) = \Delta(T_1)\Delta(T_2)$. The multiplication theorem for determinants shows therefore that if (2) holds for T_1 and T_2, then (2) also holds for T. Since every linear operator on R^k is a product of finitely many linear operators of the following three types, it suffices to establish (2) for each of these:

(I) $\{Te_1, \ldots, Te_k\}$ is a permutation of $\{e_1, \ldots, e_k\}$.
(II) $Te_1 = \alpha e_1$, $Te_i = e_i$ for $i = 2, \ldots, k$.
(III) $Te_1 = e_1 + e_2$, $Te_i = e_i$ for $i = 2, \ldots, k$.

Let Q be the cube consisting of all $x = (\xi_1, \ldots, \xi_k)$ with $0 \leq \xi_i < 1$ for $i = 1, \ldots, k$.

If T is of type (I), then $[T]$ has exactly one 1 in each row and each column and has 0 in all other places. So det $T = \pm 1$. Also, $T(Q) = Q$. So $\Delta(T) = 1 = |\det T|$.

If T is of type (II), then clearly $\Delta(T) = |\alpha| = |\det T|$.

If T is of type (III), then $\det T = 1$ and $T(Q)$ is the set of all points $\sum \xi_i e_i$ whose coordinates satisfy

$$\xi_1 \le \xi_2 < \xi_1 + 1, \qquad 0 \le \xi_i < 1 \quad \text{if} \quad i \ne 2. \tag{3}$$

If S_1 is the set of points in $T(Q)$ that have $\xi_2 < 1$ and if S_2 is the rest of $T(Q)$, then

$$S_1 \cup (S_2 - e_2) = Q, \tag{4}$$

and $S_1 \cap (S_2 - e_2)$ is empty. Hence $\Delta(T) = m(S_1 \cup S_2) = m(S_1) + m(S_2 - e_2) = m(Q) = 1$, so that we again have $\Delta(T) = |\det T|$.

Continuity Properties of Measurable Functions

Since the continuous functions played such a prominent role in our construction of Borel measures, and of Lebesgue measure in particular, it seems reasonable to expect that there are some interesting relations between continuous functions and measurable functions. In this section we shall give two theorems of this kind.

We shall assume, in both of them, *that μ is a measure on a locally compact Hausdorff space X which has the properties stated in Theorem 2.14.* In particular, μ could be Lebesgue measure on some R^k.

2.24 Lusin's Theorem *Suppose f is a complex measurable function on X, $\mu(A) < \infty$, $f(x) = 0$ if $x \notin A$, and $\epsilon > 0$. Then there exists a $g \in C_c(X)$ such that*

$$\mu(\{x: f(x) \ne g(x)\}) < \epsilon. \tag{1}$$

Furthermore, we may arrange it so that

$$\sup_{x \in X} |g(x)| \le \sup_{x \in X} |f(x)|. \tag{2}$$

PROOF Assume first that $0 \le f < 1$ and that A is compact. Attach a sequence $\{s_n\}$ to f, as in the proof of Theorem 1.17, and put $t_1 = s_1$ and $t_n = s_n - s_{n-1}$ for $n = 2, 3, 4, \ldots$. Then $2^n t_n$ is the characteristic function of a set $T_n \subset A$, and

$$f(x) = \sum_{n=1}^{\infty} t_n(x) \qquad (x \in X). \tag{3}$$

Fix an open set V such that $A \subset V$ and $\bar{V}$ is compact. There are compact sets K_n and open sets V_n such that $K_n \subset T_n \subset V_n \subset V$ and $\mu(V_n - K_n) < 2^{-n}\epsilon$. By Urysohn's lemma, there are functions h_n such that $K_n \prec h_n \prec V_n$. Define

$$g(x) = \sum_{n=1}^{\infty} 2^{-n} h_n(x) \qquad (x \in X). \tag{4}$$

This series converges uniformly on X, so g is continuous. Also, the support of g lies in $\bar{V}$. Since $2^{-n} h_n(x) = t_n(x)$ except in $V_n - K_n$, we have $g(x) = f(x)$

except in $\bigcup (V_n - K_n)$, and this latter set has measure less than ϵ. Thus (1) holds if A is compact and $0 \le f \le 1$.

It follows that (1) holds if A is compact and f is a bounded measurable function. The compactness of A is easily removed, for if $\mu(A) < \infty$ then A contains a compact set K with $\mu(A - K)$ smaller than any preassigned positive number. Next, if f is a complex measurable function and if $B_n = \{x : |f(x)| > n\}$, then $\bigcap B_n = \emptyset$, so $\mu(B_n) \to 0$, by Theorem 1.19(e). Since f coincides with the bounded function $(1 - \chi_{B_n}) \cdot f$ except on B_n, (1) follows in the general case.

Finally, let $R = \sup \{|f(x)| : x \in X\}$, and define $\varphi(z) = z$ if $|z| \le R$, $\varphi(z) = Rz/|z|$ if $|z| > R$. Then φ is a continuous mapping of the complex plane onto the disc of radius R. If g satisfies (1) and $g_1 = \varphi \circ g$, then g_1 satisfies (1) and (2).   ////

Corollary *Assume that the hypotheses of Lusin's theorem are satisfied and that $|f| \le 1$. Then there is a sequence $\{g_n\}$ such that $g_n \in C_c(X)$, $|g_n| \le 1$, and*

$$f(x) = \lim_{n \to \infty} g_n(x) \quad \text{a.e.} \tag{5}$$

PROOF The theorem implies that to each n there corresponds a $g_n \in C_c(X)$, with $|g_n| \le 1$, such that $\mu(E_n) \le 2^{-n}$, where E_n is the set of all x at which $f(x) \ne g_n(x)$. For almost every x it is then true that x lies in at most finitely many of the sets E_n (Theorem 1.41). For any such x, it follows that $f(x) = g_n(x)$ for all large enough n. This gives (5).   ////

2.25 The Vitali-Carathéodory Theorem *Suppose $f \in L^1(\mu)$, f is real-valued, and $\epsilon > 0$. Then there exist functions u and v on X such that $u \le f \le v$, u is upper semicontinuous and bounded above, v is lower semicontinuous and bounded below, and*

$$\int_X (v - u) \, d\mu < \epsilon. \tag{1}$$

PROOF Assume first that $f \ge 0$ and that f is not identically 0. Since f is the pointwise limit of an increasing sequence of simple functions s_n, f is the sum of the simple functions $t_n = s_n - s_{n-1}$ (taking $s_0 = 0$), and since t_n is a linear combination of characteristic functions, we see that there are measurable sets E_i (not necessarily disjoint) and constants $c_i > 0$ such that

$$f(x) = \sum_{i=1}^{\infty} c_i \chi_{E_i}(x) \quad (x \in X). \tag{2}$$

Since
$$\int_X f\,d\mu = \sum_{i=1}^{\infty} c_i \mu(E_i), \qquad (3)$$

the series in (3) converges. There are compact sets K_i and open sets V_i such that $K_i \subset E_i \subset V_i$ and

$$c_i \mu(V_i - K_i) < 2^{-i-1}\epsilon \qquad (i = 1, 2, 3, \ldots). \qquad (4)$$

Put

$$v = \sum_{i=1}^{\infty} c_i \chi_{V_i}, \qquad u = \sum_{i=1}^{N} c_i \chi_{K_i}, \qquad (5)$$

where N is chosen so that

$$\sum_{N+1}^{\infty} c_i \mu(E_i) < \frac{\epsilon}{2}. \qquad (6)$$

Then v is lower semicontinuous, u is upper semicontinuous, $u \leq f \leq v$, and

$$v - u = \sum_{i=1}^{N} c_i(\chi_{V_i} - \chi_{K_i}) + \sum_{N+1}^{\infty} c_i \chi_{V_i}$$
$$\leq \sum_{i=1}^{\infty} c_i(\chi_{V_i} - \chi_{K_i}) + \sum_{N+1}^{\infty} c_i \chi_{E_i}$$

so that (4) and (6) imply (1).

In the general case, write $f = f^+ - f^-$, attach u_1 and v_1 to f^+, attach u_2 and v_2 to f^-, as above, and put $u = u_1 - v_2$, $v = v_1 - u_2$. Since $-v_2$ is upper semicontinuous and since the sum of two upper semicontinuous functions is upper semicontinuous (similarly for lower semicontinuous; we leave the proof of this as an exercise), u and v have the desired properties. ////

Exercises

1 Let $\{f_n\}$ be a sequence of real nonnegative functions on R^1, and consider the following four statements:

(a) If f_1 and f_2 are upper semicontinuous, then $f_1 + f_2$ is upper semicontinuous.
(b) If f_1 and f_2 are lower semicontinuous, then $f_1 + f_2$ is lower semicontinuous.
(c) If each f_n is upper semicontinuous, then $\sum_1^{\infty} f_n$ is upper semicontinuous.
(d) If each f_n is lower semicontinuous, then $\sum_1^{\infty} f_n$ is lower semicontinuous.

Show that three of these are true and that one is false. What happens if the word "nonnegative" is omitted? Is the truth of the statements affected if R^1 is replaced by a general topological space?

2 Let f be an arbitrary complex function on R^1, and define

$$\varphi(x, \delta) = \sup\{|f(s) - f(t)| : s, t \in (x - \delta, x + \delta)\},$$
$$\varphi(x) = \inf\{\varphi(x, \delta) : \delta > 0\}.$$

Prove that φ is upper semicontinuous, that f is continuous at a point x if and only if $\varphi(x) = 0$, and hence that the set of points of continuity of an arbitrary complex function is a G_δ.

Formulate and prove an analogous statement for general topological spaces in place of R^1.

3 Let X be a metric space, with metric ρ. For any nonempty $E \subset X$, define

$$\rho_E(x) = \inf \{\rho(x, y): y \in E\}.$$

Show that ρ_E is a uniformly continuous function on X. If A and B are disjoint nonempty closed subsets of X, examine the relevance of the function

$$f(x) = \frac{\rho_A(x)}{\rho_A(x) + \rho_B(x)}$$

to Urysohn's lemma.

4 Examine the proof of the Riesz theorem and prove the following two statements:

(a) If $E_1 \subset V_1$ and $E_2 \subset V_2$, where V_1 and V_2 are disjoint open sets, then $\mu(E_1 \cup E_2) = \mu(E_1) + \mu(E_2)$, even if E_1 and E_2 are not in $\mathfrak{M}$.

(b) If $E \in \mathfrak{M}_F$, then $E = N \cup K_1 \cup K_2 \cup \cdots$, where $\{K_i\}$ is a disjoint countable collection of compact sets and $\mu(N) = 0$.

In Exercises 5 to 8, m stands for Lebesgue measure on R^1.

5 Let E be Cantor's familiar "middle thirds" set. Show that $m(E) = 0$, even though E and R^1 have the same cardinality.

6 Construct a totally disconnected compact set $K \subset R^1$ such that $m(K) > 0$. (K is to have no connected subset consisting of more than one point.)

If v is lower semicontinuous and $v \leq \chi_K$, show that actually $v \leq 0$. Hence χ_K cannot be approximated from below by lower semicontinuous functions, in the sense of the Vitali-Carathéodory theorem.

7 If $0 < \epsilon < 1$, construct an open set $E \subset [0, 1]$ which is dense in $[0, 1]$, such that $m(E) = \epsilon$. (To say that A is dense in B means that the closure of A contains B.)

8 Construct a Borel set $E \subset R^1$ such that

$$0 < m(E \cap I) < m(I)$$

for every nonempty segment I. Is it possible to have $m(E) < \infty$ for such a set?

9 Construct a sequence of continuous function f_n on $[0, 1]$ such that $0 \leq f_n \leq 1$, such that

$$\lim_{n \to \infty} \int_0^1 f_n(x) \, dx = 0,$$

but such that the sequence $\{f_n(x)\}$ converges for no $x \in [0, 1]$.

10 If $\{f_n\}$ is a sequence of continuous functions on $[0, 1]$ such that $0 \leq f_n \leq 1$ and such that $f_n(x) \to 0$ as $n \to \infty$, for every $x \in [0, 1]$, then

$$\lim_{n \to \infty} \int_0^1 f_n(x) \, dx = 0.$$

Try to prove this without using any measure theory or any theorems about Lebesgue integration. (This is to impress you with the power of the Lebesgue integral. A nice proof was given by W. F. Eberlein in *Communications on Pure and Applied Mathematics*, vol. X, pp. 357–360, 1957.)

11 Let μ be a regular Borel measure on a compact Hausdorff space X; assume $\mu(X) = 1$. Prove that there is a compact set $K \subset X$ (the *carrier* or *support* of μ) such that $\mu(K) = 1$ but $\mu(H) < 1$ for every proper compact subset H of K. *Hint*: Let K be the intersection of all compact K_α with $\mu(K_\alpha) = 1$; show that every open set V which contains K also contains some K_α. Regularity of μ is needed; compare Exercise 18. Show that K^c is the largest open set in X whose measure is 0.

12 Show that every compact subset of R^1 is the support of a Borel measure.

13 Is it true that every compact subset of R^1 is the support of a continuous function? If not, can you describe the class of all compact sets in R^1 which are supports of continuous functions? Is your description valid in other topological spaces?

14 Let f be a real-valued Lebesgue measurable function on R^k. Prove that there exist Borel functions g and h such that $g(x) = h(x)$ a.e. $[m]$, and $g(x) \leq f(x) \leq h(x)$ for every $x \in R^k$.

15 It is easy to guess the limits of

$$\int_0^n \left(1 - \frac{x}{n}\right)^n e^{x/2} \, dx \quad \text{and} \quad \int_0^n \left(1 + \frac{x}{n}\right)^n e^{-2x} \, dx,$$

as $n \to \infty$. Prove that your guesses are correct.

16 Why is $m(Y) = 0$ in the proof of Theorem 2.20(e)?

17 Define the distance between points (x_1, y_1) and (x_2, y_2) in the plane to be

$$|y_1 - y_2| \quad \text{if } x_1 = x_2, \qquad 1 + |y_1 - y_2| \quad \text{if } x_1 \neq x_2.$$

Show that this is indeed a metric, and that the resulting metric space X is locally compact.

If $f \in C_c(X)$, let $x_1, \ldots, x_n$ be those values of x for which $f(x, y) \neq 0$ for at least one y (there are only finitely many such x!), and define

$$\Lambda f = \sum_{j=1}^n \int_{-\infty}^\infty f(x_j, y) \, dy.$$

Let μ be the measure associated with this Λ by Theorem 2.14. If E is the x-axis, show that $\mu(E) = \infty$ although $\mu(K) = 0$ for every compact $K \subset E$.

18 This exercise requires more set-theoretic skill than the preceding ones. Let X be a well-ordered uncountable set which has a last element ω_1, such that every predecessor of ω_1 has at most countably many predecessors. ("Construction": Take any well-ordered set which has elements with uncountably many predecessors, and let ω_1 be the first of these; ω_1 is called the first uncountable ordinal.) For $\alpha \in X$, let $P_\alpha [S_\alpha]$ be the set of all predecessors (successors) of α, and call a subset of X open if it is a P_α or an S_β or a $P_\alpha \cap S_\beta$ or a union of such sets. Prove that X is then a compact Hausdorff space. (*Hint*: No well-ordered set contains an infinite decreasing sequence.)

Prove that the complement of the point ω_1 is an open set which is not σ-compact.

Prove that to every $f \in C(X)$ there corresponds an $\alpha \neq \omega_1$ such that f is constant on S_α.

Prove that the intersection of every countable collection $\{K_n\}$ of uncountable compact subsets of X is uncountable. (*Hint*: Consider limits of increasing countable sequences in X which intersect each K_n in infinitely many points.)

Let $\mathfrak{M}$ be the collection of all $E \subset X$ such that either $E \cup \{\omega_1\}$ or $E^c \cup \{\omega_1\}$ contains an uncountable compact set; in the first case, define $\lambda(E) = 1$; in the second case, define $\lambda(E) = 0$. Prove that $\mathfrak{M}$ is a σ-algebra which contains all Borel sets in X, that λ is a measure on $\mathfrak{M}$ which is *not* regular (every neighborhood of ω_1 has measure 1), and that

$$f(\omega_1) = \int_X f \, d\lambda$$

for every $f \in C(X)$. Describe the regular μ which Theorem 2.14 associates with this linear functional.

19 Go through the proof of Theorem 2.14, assuming X to be compact (or even compact metric) rather than just locally compact, and see what simplifications you can find.

20 Find continuous functions $f_n: [0, 1] \to [0, \infty)$ such that $f_n(x) \to 0$ for all $x \in [0, 1]$ as $n \to \infty$, $\int_0^1 f_n(x) \, dx \to 0$, but $\sup_n f_n$ is not in L^1. (This shows that the conclusion of the dominated convergence theorem may hold even when part of its hypothesis is violated.)

21 If X is compact and $f: X \to (-\infty, \infty)$ is upper semicontinuous, prove that f attains its maximum at some point of X.

22 Suppose that X is a metric space, with metric d, and that $f: X \to [0, \infty]$ is lower semicontinuous, $f(p) < \infty$ for at least one $p \in X$. For $n = 1, 2, 3, \ldots$, $x \in X$, define

$$g_n(x) = \inf \{f(p) + nd(x, p): p \in X\}$$

and prove that

(i) $|g_n(x) - g_n(y)| \le nd(x, y)$,
(ii) $0 \le g_1 \le g_2 \le \cdots \le f$,
(iii) $g_n(x) \to f(x)$ as $n \to \infty$, for all $x \in X$.

Thus f is the pointwise limit of an increasing sequence of continuous functions. (Note that the converse is almost trivial.)

23 Suppose V is open in R^k and μ is a finite positive Borel measure on R^k. Is the function that sends x to $\mu(V + x)$ necessarily continuous? lower semicontinuous? upper semicontinuous?

24 A *step function* is, by definition, a finite linear combination of characteristic functions of bounded intervals in R^1. Assume $f \in L^1(R^1)$, and prove that there is a sequence $\{g_n\}$ of step functions so that

$$\lim_{n \to \infty} \int_{-\infty}^{\infty} |f(x) - g_n(x)| \, dx = 0.$$

25 (i) Find the smallest constant c such that

$$\log(1 + e^t) < c + t \qquad (0 < t < \infty).$$

(ii) Does

$$\lim_{n \to \infty} \frac{1}{n} \int_0^1 \log\{1 + e^{nf(x)}\} \, dx$$

exist for every real $f \in L^1$? If it exists, what is it?

CHAPTER
THREE

L^p-SPACES

Convex Functions and Inequalities

Many of the most common inequalities in analysis have their origin in the notion of convexity.

3.1 Definition A real function φ defined on a segment (a, b), where $-\infty \leq a < b \leq \infty$, is called *convex* if the inequality

$$\varphi((1 - \lambda)x + \lambda y) \leq (1 - \lambda)\varphi(x) + \lambda\varphi(y) \tag{1}$$

holds whenever $a < x < b$, $a < y < b$, and $0 \leq \lambda \leq 1$.

Graphically, the condition is that if $x < t < y$, then the point $(t, \varphi(t))$ should lie below or on the line connecting the points $(x, \varphi(x))$ and $(y, \varphi(y))$ in the plane. Also, (1) is equivalent to the requirement that

$$\frac{\varphi(t) - \varphi(s)}{t - s} \leq \frac{\varphi(u) - \varphi(t)}{u - t} \tag{2}$$

whenever $a < s < t < u < b$.

The mean value theorem for differentiation, combined with (2), shows immediately that a real differentiable function φ is convex in (a, b) if and only if $a < s < t < b$ implies $\varphi'(s) \leq \varphi'(t)$, i.e., if and only if the derivative φ' is a monotonically increasing function.

For example, the exponential function is convex on $(-\infty, \infty)$.

3.2 Theorem *If φ is convex on (a, b) then φ is continuous on (a, b).*

PROOF The idea of the proof is most easily conveyed in geometric language. Those who may worry that this is not "rigorous" are invited to transcribe it in terms of epsilons and deltas.

Suppose $a < s < x < y < t < b$. Write S for the point $(s, \varphi(s))$ in the plane, and deal similarly with x, y, and t. Then X is on or below the line SY, hence Y is on or above the line through S and X; also, Y is on or below XT. As $y \to x$, it follows that $Y \to X$, i.e., $\varphi(y) \to \varphi(x)$. Left-hand limits are handled in the same manner, and the continuity of φ follows. ////

Note that this theorem depends on the fact that we are working on an *open* segment. For instance, if $\varphi(x) = 0$ on $[0, 1)$ and $\varphi(1) = 1$, then φ satisfies 3.1(1) on $[0, 1]$ without being continuous.

3.3 Theorem (Jensen's Inequality) *Let μ be a positive measure on a σ-algebra $\mathfrak{M}$ in a set Ω, so that $\mu(\Omega) = 1$. If f is a real function in $L^1(\mu)$, if $a < f(x) < b$ for all $x \in \Omega$, and if φ is convex on (a, b), then*

$$\varphi\left(\int_\Omega f \, d\mu\right) \leq \int_\Omega (\varphi \circ f) \, d\mu. \tag{1}$$

Note: The cases $a = -\infty$ and $b = \infty$ are not excluded. It may happen that $\varphi \circ f$ is not in $L^1(\mu)$; in that case, as the proof will show, the integral of $\varphi \circ f$ exists in the extended sense described in Sec. 1.31, and its value is $+\infty$.

PROOF Put $t = \int_\Omega f \, d\mu$. Then $a < t < b$. If β is the supremum of the quotients on the left of 3.1(2), where $a < s < t$, then β is no larger than any of the quotients on the right of 3.1(2), for any $u \in (t, b)$. It follows that

$$\varphi(s) \geq \varphi(t) + \beta(s - t) \qquad (a < s < b). \tag{2}$$

Hence

$$\varphi(f(x)) - \varphi(t) - \beta(f(x) - t) \geq 0 \tag{3}$$

for every $x \in \Omega$. Since φ is continuous, $\varphi \circ f$ is measurable. If we integrate both sides of (3) with respect to μ, (1) follows from our choice of t and the assumption $\mu(\Omega) = 1$. ////

To give an example, take $\varphi(x) = e^x$. Then (1) becomes

$$\exp\left\{\int_\Omega f \, d\mu\right\} \leq \int_\Omega e^f \, d\mu. \tag{4}$$

If Ω is a finite set, consisting of points $p_1, \ldots, p_n$, say, and if

$$\mu(\{p_i\}) = 1/n, \qquad f(p_i) = x_i,$$

(4) becomes

$$\exp\left\{\frac{1}{n}(x_1 + \cdots + x_n)\right\} \le \frac{1}{n}(e^{x_1} + \cdots + e^{x_n}), \tag{5}$$

for real x_i. Putting $y_i = e^{x_i}$, we obtain the familiar inequality between the arithmetic and geometric means of n positive numbers:

$$(y_1 y_2 \cdots y_n)^{1/n} \le \frac{1}{n}(y_1 + y_2 + \cdots + y_n). \tag{6}$$

Going back from this to (4), it should become clear why the left and right sides of

$$\exp\left\{\int_\Omega \log g\, d\mu\right\} \le \int_\Omega g\, d\mu \tag{7}$$

are often called the geometric and arithmetic means, respectively, of the positive function g.

If we take $\mu(\{p_i\}) = \alpha_i > 0$, where $\sum \alpha_i = 1$, then we obtain

$$y_1^{\alpha_1} y_2^{\alpha_2} \cdots y_n^{\alpha_n} \le \alpha_1 y_1 + \alpha_2 y_2 + \cdots + \alpha_n y_n \tag{8}$$

in place of (6). These are just a few samples of what is contained in Theorem 3.3.

For a converse, see Exercise 20.

3.4 Definition If p and q are positive real numbers such that $p + q = pq$, or equivalently

$$\frac{1}{p} + \frac{1}{q} = 1, \tag{1}$$

then we call p and q a pair of *conjugate exponents*. It is clear that (1) implies $1 < p < \infty$ and $1 < q < \infty$. An important special case is $p = q = 2$.

As $p \to 1$, (1) forces $q \to \infty$. Consequently 1 and ∞ are also regarded as a pair of conjugate exponents. Many analysts denote the exponent conjugate to p by p', often without saying so explicitly.

3.5 Theorem *Let p and q be conjugate exponents, $1 < p < \infty$. Let X be a measure space, with measure μ. Let f and g be measurable functions on X, with range in $[0, \infty]$. Then*

$$\int_X fg\, d\mu \le \left\{\int_X f^p\, d\mu\right\}^{1/p} \left\{\int_X g^q\, d\mu\right\}^{1/q} \tag{1}$$

and

$$\left\{\int_X (f+g)^p\, d\mu\right\}^{1/p} \le \left\{\int_X f^p\, d\mu\right\}^{1/p} + \left\{\int_X g^p\, d\mu\right\}^{1/p}. \tag{2}$$

The inequality (1) is Hölder's; (2) is Minkowski's. If $p = q = 2$, (1) is known as the Schwarz inequality.

PROOF Let A and B be the two factors on the right of (1). If $A = 0$, then $f = 0$ a.e. (by Theorem 1.39); hence $fg = 0$ a.e., so (1) holds. If $A > 0$ and $B = \infty$, (1) is again trivial. So we need consider only the case $0 < A < \infty$, $0 < B < \infty$. Put

$$F = \frac{f}{A}, \qquad G = \frac{g}{B}. \tag{3}$$

This gives

$$\int_X F^p \, d\mu = \int_X G^q \, d\mu = 1. \tag{4}$$

If $x \in X$ is such that $0 < F(x) < \infty$ and $0 < G(x) < \infty$, there are real numbers s and t such that $F(x) = e^{s/p}$, $G(x) = e^{t/q}$. Since $1/p + 1/q = 1$, the convexity of the exponential function implies that

$$e^{s/p + t/q} \leq p^{-1} e^s + q^{-1} e^t. \tag{5}$$

It follows that

$$F(x) G(x) \leq p^{-1} F(x)^p + q^{-1} G(x)^q \tag{6}$$

for every $x \in X$. Integration of (6) yields

$$\int_X FG \, d\mu \leq p^{-1} + q^{-1} = 1, \tag{7}$$

by (4); inserting (3) into (7), we obtain (1).

Note that (6) could also have been obtained as a special case of the inequality 3.3(8).

To prove (2), we write

$$(f + g)^p = f \cdot (f + g)^{p-1} + g \cdot (f + g)^{p-1}. \tag{8}$$

Hölder's inequality gives

$$\int f \cdot (f + g)^{p-1} \leq \left\{ \int f^p \right\}^{1/p} \left\{ \int (f + g)^{(p-1)q} \right\}^{1/q}. \tag{9}$$

Let (9') be the inequality (9) with f and g interchanged. Since $(p - 1)q = p$, addition of (9) and (9') gives

$$\int (f + g)^p \leq \left\{ \int (f + g)^p \right\}^{1/q} \left[\left\{ \int f^p \right\}^{1/p} + \left\{ \int g^p \right\}^{1/p} \right]. \tag{10}$$

Clearly, it is enough to prove (2) in the case that the left side is greater than 0 and the right side is less than ∞. The convexity of the function t^p for $0 < t < \infty$ shows that

$$\left(\frac{f + g}{2} \right)^p \leq \frac{1}{2} (f^p + g^p).$$

Hence the left side of (2) is less than ∞, and (2) follows from (10) if we divide by the first factor on the right of (10), bearing in mind that $1 - 1/q = 1/p$. This completes the proof. ////

It is sometimes useful to know the conditions under which equality can hold in an inequality. In many cases this information may be obtained by examining the proof of the inequality.

For instance, equality holds in (7) if and only if equality holds in (6) for almost every x. In (5), equality holds if and only if $s = t$. Hence "$F^p = G^q$ a.e." is a necessary and sufficient condition for equality in (7), if (4) is assumed. In terms of the original functions f and g, the following result is then obtained:

Assuming $A < \infty$ and $B < \infty$, equality holds in (1) if and only if there are constants α and β, not both 0, such that $\alpha f^p = \beta g^q$ a.e.

We leave the analogous discussion of equality in (2) as an exercise.

The L^p-spaces

In this section, X will be an arbitrary measure space with a positive measure μ.

3.6 Definition If $0 < p < \infty$ and if f is a complex measurable function on X, define

$$\|f\|_p = \left\{ \int_X |f|^p \, d\mu \right\}^{1/p} \tag{1}$$

and let $L^p(\mu)$ consist of all f for which

$$\|f\|_p < \infty. \tag{2}$$

We call $\|f\|_p$ the L^p-norm of f.

If μ is Lebesgue measure on R^k, we write $L^p(R^k)$ instead of $L^p(\mu)$, as in Sec. 2.21. If μ is the counting measure on a set A, it is customary to denote the corresponding L^p-space by $\ell^p(A)$, or simply by ℓ^p, if A is countable. An element of ℓ^p may be regarded as a complex sequence $x = \{\xi_n\}$, and

$$\|x\|_p = \left\{ \sum_{n=1}^{\infty} |\xi_n|^p \right\}^{1/p}.$$

3.7 Definition Suppose $g: X \to [0, \infty]$ is measurable. Let S be the set of all real α such that

$$\mu(g^{-1}((\alpha, \infty])) = 0. \tag{1}$$

If $S = \emptyset$, put $\beta = \infty$. If $S \neq \emptyset$, put $\beta = \inf S$. Since

$$g^{-1}((\beta, \infty]) = \bigcup_{n=1}^{\infty} g^{-1}\left(\left(\beta + \frac{1}{n}, \infty\right]\right), \tag{2}$$

and since the union of a countable collection of sets of measure 0 has measure 0, we see that $\beta \in S$. We call β the *essential supremum* of g.

If f is a complex measurable function on X, we define $\|f\|_\infty$ to be the essential supremum of $|f|$, and we let $L^\infty(\mu)$ consist of all f for which $\|f\|_\infty < \infty$. The members of $L^\infty(\mu)$ are sometimes called *essentially bounded* measurable functions on X.

It follows from this definition that the inequality $|f(x)| \leq \lambda$ holds for almost all x if and only if $\lambda \geq \|f\|_\infty$.

As in Definition 3.6, $L^\infty(R^k)$ denotes the class of all essentially bounded (with respect to Lebesgue measure) functions on R^k, and $\ell^\infty(A)$ is the class of all bounded functions on A. (Here bounded means the same as essentially bounded, since every nonempty set has positive measure!)

3.8 Theorem *If p and q are conjugate exponents, $1 \leq p \leq \infty$, and if $f \in L^p(\mu)$ and $g \in L^q(\mu)$, then $fg \in L^1(\mu)$, and*

$$\|fg\|_1 \leq \|f\|_p \|g\|_q. \tag{1}$$

PROOF For $1 < p < \infty$, (1) is simply Hölder's inequality, applied to $|f|$ and $|g|$. If $p = \infty$, note that

$$|f(x)g(x)| \leq \|f\|_\infty |g(x)| \tag{2}$$

for almost all x; integrating (2), we obtain (1). If $p = 1$, then $q = \infty$, and the same argument applies. ////

3.9 Theorem *Suppose $1 \leq p \leq \infty$, and $f \in L^p(\mu)$, $g \in L^p(\mu)$. Then $f + g \in L^p(\mu)$, and*

$$\|f + g\|_p \leq \|f\|_p + \|g\|_p. \tag{1}$$

PROOF For $1 < p < \infty$, this follows from Minkowski's inequality, since

$$\int_X |f + g|^p \, d\mu \leq \int_X (|f| + |g|)^p \, d\mu.$$

For $p = 1$ or $p = \infty$, (1) is a trivial consequence of the inequality $|f + g| \leq |f| + |g|$. ////

3.10 Remarks Fix p, $1 \leq p \leq \infty$. If $f \in L^p(\mu)$ and α is a complex number, it is clear that $\alpha f \in L^p(\mu)$. In fact,

$$\|\alpha f\|_p = |\alpha| \|f\|_p. \tag{1}$$

In conjunction with Theorem 3.9, this shows that $L^p(\mu)$ is a *complex vector space*.

Suppose f, g, and h are in $L^p(\mu)$. Replacing f by $f - g$ and g by $g - h$ in Theorem 3.9, we obtain

$$\|f - h\|_p \leq \|f - g\|_p + \|g - h\|_p. \tag{2}$$

This suggests that a metric may be introduced in $L^p(\mu)$ by defining the distance between f and g to be $\|f - g\|_p$. Call this distance $d(f, g)$ for the moment. Then $0 \leq d(f, g) < \infty$, $d(f, f) = 0$, $d(f, g) = d(g, f)$, and (2) shows that the triangle inequality $d(f, h) \leq d(f, g) + d(g, h)$ is satisfied. The only other property which d should have to define a metric space is that $d(f, g) = 0$ should imply that $f = g$. In our present situation this need not be so; we have $d(f, g) = 0$ precisely when $f(x) = g(x)$ for almost all x.

Let us write $f \sim g$ if and only if $d(f, g) = 0$. It is clear that this is an equivalence relation in $L^p(\mu)$ which partitions $L^p(\mu)$ into equivalence classes; each class consists of all functions which are equivalent to a given one. If F and G are two equivalence classes, choose $f \in F$ and $g \in G$, and define $d(F, G) = d(f, g)$; note that $f \sim f_1$ and $g \sim g_1$ implies

$$d(f, g) = d(f_1, g_1),$$

so that $d(F, G)$ is well defined.

With this definition, the set of equivalence classes is now a metric space. Note that it is also a vector space, since $f \sim f_1$ and $g \sim g_1$ implies $f + g \sim f_1 + g_1$ and $\alpha f \sim \alpha f_1$.

When $L^p(\mu)$ is regarded as a metric space, then the space which is really under consideration is therefore *not a space whose elements are functions, but a space whose elements are equivalence classes of functions*. For the sake of simplicity of language, it is, however, customary to relegate this distinction to the status of a tacit understanding and to continue to speak of $L^p(\mu)$ as a space of functions. We shall follow this custom.

If $\{f_n\}$ is a sequence in $L^p(\mu)$, if $f \in L^p(\mu)$, and if $\lim_{n \to \infty} \|f_n - f\|_p = 0$, we say that $\{f_n\}$ *converges to f in $L^p(\mu)$* (or that $\{f_n\}$ converges to f in the mean of order p, or that $\{f_n\}$ is L^p-convergent to f). If to every $\epsilon > 0$ there corresponds an integer N such that $\|f_n - f_m\|_p < \epsilon$ as soon as $n > N$ and $m > N$, we call $\{f_n\}$ a *Cauchy sequence* in $L^p(\mu)$. These definitions are exactly as in any metric space.

It is a very important fact that $L^p(\mu)$ is a *complete* metric space, i.e., that every Cauchy sequence in $L^p(\mu)$ converges to an element of $L^p(\mu)$:

3.11 Theorem *$L^p(\mu)$ is a complete metric space, for $1 \leq p \leq \infty$ and for every positive measure μ.*

PROOF Assume first that $1 \leq p < \infty$. Let $\{f_n\}$ be a Cauchy sequence in $L^p(\mu)$. There is a subsequence $\{f_{n_i}\}$, $n_1 < n_2 < \cdots$, such that

$$\|f_{n_{i+1}} - f_{n_i}\|_p < 2^{-i} \quad (i = 1, 2, 3, \ldots). \tag{1}$$

Put

$$g_k = \sum_{i=1}^{k} |f_{n_{i+1}} - f_{n_i}|, \qquad g = \sum_{i=1}^{\infty} |f_{n_{i+1}} - f_{n_i}|. \qquad (2)$$

Since (1) holds, the Minkowski inequality shows that $\|g_k\|_p < 1$ for $k = 1$, 2, 3, ... Hence an application of Fatou's lemma to $\{g_k^p\}$ gives $\|g\|_p \le 1$. In particular, $g(x) < \infty$ a.e., so that the series

$$f_{n_1}(x) + \sum_{i=1}^{\infty} (f_{n_{i+1}}(x) - f_{n_i}(x)) \qquad (3)$$

converges absolutely for almost every $x \in X$. Denote the sum of (3) by $f(x)$, for those x at which (3) converges; put $f(x) = 0$ on the remaining set of measure zero. Since

$$f_{n_1} + \sum_{i=1}^{k-1} (f_{n_{i+1}} - f_{n_i}) = f_{n_k}, \qquad (4)$$

we see that

$$f(x) = \lim_{i \to \infty} f_{n_i}(x) \qquad \text{a.e.} \qquad (5)$$

Having found a function f which is the pointwise limit a.e. of $\{f_{n_i}\}$, we now have to prove that this f is the L^p-limit of $\{f_n\}$. Choose $\epsilon > 0$. There exists an N such that $\|f_n - f_m\|_p < \epsilon$ if $n > N$ and $m > N$. For every $m > N$, Fatou's lemma shows therefore that

$$\int_X |f - f_m|^p \, d\mu \le \liminf_{i \to \infty} \int_X |f_{n_i} - f_m|^p \, d\mu \le \epsilon^p. \qquad (6)$$

We conclude from (6) that $f - f_m \in L^p(\mu)$, hence that $f \in L^p(\mu)$ [since $f = (f - f_m) + f_m$], and finally that $\|f - f_m\|_p \to 0$ as $m \to \infty$. This completes the proof for the case $1 \le p < \infty$.

In $L^\infty(\mu)$ the proof is much easier. Suppose $\{f_n\}$ is a Cauchy sequence in $L^\infty(\mu)$, let A_k and $B_{m,n}$ be the sets where $|f_k(x)| > \|f_k\|_\infty$ and where $|f_n(x) - f_m(x)| > \|f_n - f_m\|_\infty$, and let E be the union of these sets, for k, m, $n = 1, 2, 3, \ldots$. Then $\mu(E) = 0$, and on the complement of E the sequence $\{f_n\}$ converges uniformly to a bounded function f. Define $f(x) = 0$ for $x \in E$. Then $f \in L^\infty(\mu)$, and $\|f_n - f\|_\infty \to 0$ as $n \to \infty$. ////

The preceding proof contains a result which is interesting enough to be stated separately:

3.12 Theorem *If $1 \le p \le \infty$ and if $\{f_n\}$ is a Cauchy sequence in $L^p(\mu)$, with limit f, then $\{f_n\}$ has a subsequence which converges pointwise almost everywhere to $f(x)$.*

The simple functions play an interesting role in $L^p(\mu)$:

3.13 Theorem *Let S be the class of all complex, measurable, simple functions on X such that*

$$\mu(\{x: s(x) \neq 0\}) < \infty. \tag{1}$$

If $1 \leq p < \infty$, then S is dense in $L^p(\mu)$.

PROOF First, it is clear that $S \subset L^p(\mu)$. Suppose $f \geq 0$, $f \in L^p(\mu)$, and let $\{s_n\}$ be as in Theorem 1.17. Since $0 \leq s_n \leq f$, we have $s_n \in L^p(\mu)$, hence $s_n \in S$. Since $|f - s_n|^p \leq f^p$, the dominated convergence theorem shows that $\|f - s_n\|_p \to 0$ as $n \to \infty$. Thus f is in the L^p-closure of S. The general case (f complex) follows from this. ////

Approximation by Continuous Functions

So far we have considered $L^p(\mu)$ on any measure space. Now let X be a locally compact Hausdorff space, and let μ be a measure on a σ-algebra $\mathfrak{M}$ in X, with the properties stated in Theorem 2.14. For example, X might be R^k, and μ might be Lebesgue measure on R^k.

Under these circumstances, we have the following analogue of Theorem 3.13:

3.14 Theorem *For $1 \leq p < \infty$, $C_c(X)$ is dense in $L^p(\mu)$.*

PROOF Define S as in Theorem 3.13. If $s \in S$ and $\epsilon > 0$, there exists a $g \in C_c(X)$ such that $g(x) = s(x)$ except on a set of measure $< \epsilon$, and $|g| \leq \|s\|_\infty$ (Lusin's theorem). Hence

$$\|g - s\|_p \leq 2\epsilon^{1/p}\|s\|_\infty. \tag{1}$$

Since S is dense in $L^p(\mu)$, this completes the proof. ////

3.15 Remarks Let us discuss the relations between the spaces $L^p(R^k)$ (the L^p-spaces in which the underlying measure is Lebesgue measure on R^k) and the space $C_c(R^k)$ in some detail. We consider a fixed dimension k.

For every $p \in [1, \infty]$ we have a metric on $C_c(R^k)$; the distance between f and g is $\|f - g\|_p$. Note that this is a genuine metric, and that we do not have to pass to equivalence classes. The point is that if two continuous functions on R^k are not identical, then they differ on some nonempty open set V, and $m(V) > 0$, since V contains a k-cell. Thus if two members of $C_c(R^k)$ are equal a.e., they are equal. It is also of interest to note that in $C_c(R^k)$ the essential supremum is the same as the actual supremum: for $f \in C_c(R^k)$

$$\|f\|_\infty = \sup_{x \in R^k} |f(x)|. \tag{1}$$

If $1 \le p < \infty$, Theorem 3.14 says that $C_c(R^k)$ is dense in $L^p(R^k)$, and Theorem 3.11 shows that $L^p(R^k)$ is complete. Thus *$L^p(R^k)$ is the completion of the metric space which is obtained by endowing $C_c(R^k)$ with the L^p-metric.*

The cases $p = 1$ and $p = 2$ are the ones of greatest interest. Let us state once more, in different words, what the preceding result says if $p = 1$ and $k = 1$; the statement shows that the Lebesgue integral is indeed the "right" generalization of the Riemann integral:

If the distance between two continuous functions f and g, with compact supports in R^1, is defined to be

$$\int_{-\infty}^{\infty} |f(t) - g(t)|\, dt, \qquad (2)$$

the completion of the resulting metric space consists precisely of the Lebesgue integrable functions on R^1, provided we identify any two that are equal almost everywhere.

Of course, *every* metric space S has a completion S^* whose elements may be viewed abstractly as equivalence classes of Cauchy sequences in S (see [26], p. 82). The important point in the present situation is that the various L^p-completions of $C_c(R^k)$ again turn out to be spaces of functions on R^k.

The case $p = \infty$ differs from the cases $p < \infty$. *The L^∞-completion of $C_c(R^k)$ is not $L^\infty(R^k)$, but is $C_0(R^k)$, the space of all continuous functions on R^k which "vanish at infinity,"* a concept which will be defined in Sec. 3.16. Since (1) shows that the L^∞-norm coincides with the supremum norm on $C_c(R^k)$, the above assertion about $C_0(R^k)$ is a special case of Theorem 3.17.

3.16 Definition A complex function f on a locally compact Hausdorff space X is said to *vanish at infinity* if to every $\epsilon > 0$ there exists a compact set $K \subset X$ such that $|f(x)| < \epsilon$ for all x not in K.

The class of all continuous f on X which vanish at infinity is called $C_0(X)$.

It is clear that $C_c(X) \subset C_0(X)$, and that the two classes coincide if X is compact. In that case we write $C(X)$ for either of them.

3.17 Theorem *If X is a locally compact Hausdorff space, then $C_0(X)$ is the completion of $C_c(X)$, relative to the metric defined by the supremum norm*

$$\|f\| = \sup_{x \in X} |f(x)|. \qquad (1)$$

PROOF An elementary verification shows that $C_0(X)$ satisfies the axioms of a metric space if the distance between f and g is taken to be $\|f - g\|$. We have to show that (a) $C_c(X)$ is dense in $C_0(X)$ and (b) $C_0(X)$ is a complete metric space.

Given $f \in C_0(X)$ and $\epsilon > 0$, there is a compact set K so that $|f(x)| < \epsilon$ outside K. Urysohn's lemma gives us a function $g \in C_c(X)$ such that $0 \le g \le 1$ and $g(x) = 1$ on K. Put $h = fg$. Then $h \in C_c(X)$ and $\|f - h\| < \epsilon$. This proves (a).

To prove (b), let $\{f_n\}$ be a Cauchy sequence in $C_0(X)$, i.e., assume that $\{f_n\}$ converges uniformly. Then its pointwise limit function f is continuous. Given $\epsilon > 0$, there exists an n so that $\|f_n - f\| < \epsilon/2$ and there is a compact set K so that $|f_n(x)| < \epsilon/2$ outside K. Hence $|f(x)| < \epsilon$ outside K, and we have proved that f vanishes at infinity. Thus $C_0(X)$ is complete. ////

Exercises

1 Prove that the supremum of any collection of convex functions on (a, b) is convex on (a, b) (if it is finite) and that pointwise limits of sequences of convex functions are convex. What can you say about upper and lower limits of sequences of convex functions?

2 If φ is convex on (a, b) and if ψ is convex and nondecreasing on the range of φ, prove that $\psi \circ \varphi$ is convex on (a, b). For $\varphi > 0$, show that the convexity of $\log \varphi$ implies the convexity of φ, but not vice versa.

3 Assume that φ is a continuous real function on (a, b) such that

$$\varphi\left(\frac{x+y}{2}\right) \le \frac{1}{2}\varphi(x) + \frac{1}{2}\varphi(y)$$

for all x and $y \in (a, b)$. Prove that φ is convex. (The conclusion does *not* follow if continuity is omitted from the hypotheses.)

4 Suppose f is a complex measurable function on X, μ is a positive measure on X, and

$$\varphi(p) = \int_X |f|^p \, d\mu = \|f\|_p^p \qquad (0 < p < \infty).$$

Let $E = \{p: \varphi(p) < \infty\}$. Assume $\|f\|_\infty > 0$.

(a) If $r < p < s$, $r \in E$, and $s \in E$, prove that $p \in E$.

(b) Prove that $\log \varphi$ is convex in the interior of E and that φ is continuous on E.

(c) By (a), E is connected. Is E necessarily open? Closed? Can E consist of a single point? Can E be any connected subset of $(0, \infty)$?

(d) If $r < p < s$, prove that $\|f\|_p \le \max(\|f\|_r, \|f\|_s)$. Show that this implies the inclusion $L^r(\mu) \cap L^s(\mu) \subset L^p(\mu)$.

(e) Assume that $\|f\|_r < \infty$ for some $r < \infty$ and prove that

$$\|f\|_p \to \|f\|_\infty \qquad \text{as } p \to \infty.$$

5 Assume, in addition to the hypotheses of Exercise 4, that

$$\mu(X) = 1.$$

(a) Prove that $\|f\|_r \le \|f\|_s$ if $0 < r < s \le \infty$.

(b) Under what conditions does it happen that $0 < r < s \le \infty$ and $\|f\|_r = \|f\|_s < \infty$?

(c) Prove that $L^r(\mu) \supset L^s(\mu)$ if $0 < r < s$. Under what conditions do these two spaces contain the same functions?

(d) Assume that $\|f\|_r < \infty$ for some $r > 0$, and prove that

$$\lim_{p \to 0} \|f\|_p = \exp\left\{\int_X \log |f| \, d\mu\right\}$$

if $\exp\{-\infty\}$ is defined to be 0.

6 Let m be Lebesgue measure on $[0, 1]$, and define $\|f\|_p$ with respect to m. Find all functions Φ on $[0, \infty)$ such that the relation

$$\Phi(\lim_{p \to 0} \|f\|_p) = \int_0^1 (\Phi \circ f)\, dm$$

holds for every bounded, measurable, positive f. Show first that

$$c\Phi(x) + (1 - c)\Phi(1) = \Phi(x^c) \qquad (x > 0,\ 0 \le c \le 1).$$

Compare with Exercise 5(d).

7 For some measures, the relation $r < s$ implies $L^r(\mu) \subset L^s(\mu)$; for others, the inclusion is reversed; and there are some for which $L^r(\mu)$ does not contain $L^s(\mu)$ if $r \ne s$. Give examples of these situations, and find conditions on μ under which these situations will occur.

8 If g is a positive function on $(0, 1)$ such that $g(x) \to \infty$ as $x \to 0$, then there is a convex function h on $(0, 1)$ such that $h \le g$ and $h(x) \to \infty$ as $x \to 0$. True or false? Is the problem changed if $(0, 1)$ is replaced by $(0, \infty)$ and $x \to 0$ is replaced by $x \to \infty$?

9 Suppose f is Lebesgue measurable on $(0, 1)$, and not essentially bounded. By Exercise 4(e), $\|f\|_p \to \infty$ as $p \to \infty$. Can $\|f\|_p$ tend to ∞ arbitrarily slowly? More precisely, is it true that to every positive function Φ on $(0, \infty)$ such that $\Phi(p) \to \infty$ as $p \to \infty$ one can find an f such that $\|f\|_p \to \infty$ as $p \to \infty$, but $\|f\|_p \le \Phi(p)$ for all sufficiently large p?

10 Suppose $f_n \in L^p(\mu)$, for $n = 1, 2, 3, \ldots$, and $\|f_n - f\|_p \to 0$ and $f_n \to g$ a.e., as $n \to \infty$. What relation exists between f and g?

11 Suppose $\mu(\Omega) = 1$, and suppose f and g are positive measurable functions on Ω such that $fg \ge 1$. Prove that

$$\int_\Omega f\, d\mu \cdot \int_\Omega g\, d\mu \ge 1.$$

12 Suppose $\mu(\Omega) = 1$ and $h: \Omega \to [0, \infty]$ is measurable. If

$$A = \int_\Omega h\, d\mu,$$

prove that

$$\sqrt{1 + A^2} \le \int_\Omega \sqrt{1 + h^2}\, d\mu \le 1 + A.$$

If μ is Lebesgue measure on $[0, 1]$ and if h is continuous, $h = f'$, the above inequalities have a simple geometric interpretation. From this, conjecture (for general Ω) under what conditions on h equality can hold in either of the above inequalities, and prove your conjecture.

13 Under what conditions on f and g does equality hold in the conclusions of Theorems 3.8 and 3.9? You may have to treat the cases $p = 1$ and $p = \infty$ separately.

14 Suppose $1 < p < \infty$, $f \in L^p = L^p((0, \infty))$, relative to Lebesgue measure, and

$$F(x) = \frac{1}{x} \int_0^x f(t)\, dt \qquad (0 < x < \infty).$$

(a) Prove Hardy's inequality

$$\|F\|_p \le \frac{p}{p - 1} \|f\|_p$$

which shows that the mapping $f \to F$ carries L^p into L^p.

(b) Prove that equality holds only if $f = 0$ a.e.
(c) Prove that the constant $p/(p - 1)$ cannot be replaced by a smaller one.
(d) If $f > 0$ and $f \in L^1$, prove that $F \notin L^1$.

Suggestions: (a) Assume first that $f \geq 0$ and $f \in C_c((0, \infty))$. Integration by parts gives

$$\int_0^\infty F^p(x)\, dx = -p \int_0^\infty F^{p-1}(x) x F'(x)\, dx.$$

Note that $xF' = f - F$, and apply Hölder's inequality to $\int F^{p-1} f$. Then derive the general case. (c) Take $f(x) = x^{-1/p}$ on $[1, A]$, $f(x) = 0$ elsewhere, for large A. See also Exercise 14, Chap. 8.

15 Suppose $\{a_n\}$ is a sequence of positive numbers. Prove that

$$\sum_{N=1}^\infty \left(\frac{1}{N} \sum_{n=1}^N a_n\right)^p \leq \left(\frac{p}{p-1}\right)^p \sum_{n=1}^\infty a_n^p$$

if $1 < p < \infty$. *Hint*: If $a_n \geq a_{n+1}$, the result can be made to follow from Exercise 14. This special case implies the general one.

16 Prove Egoroff's theorem: If $\mu(X) < \infty$, if $\{f_n\}$ is a sequence of complex measurable functions which converges pointwise at every point of X, and if $\epsilon > 0$, there is a measurable set $E \subset X$, with $\mu(X - E) < \epsilon$, such that $\{f_n\}$ converges uniformly on E.

(The conclusion is that by redefining the f_n on a set of arbitrarily small measure we can convert a pointwise convergent sequence to a uniformly convergent one; note the similarity with Lusin's theorem.)

Hint: Put

$$S(n, k) = \bigcap_{i, j > n} \left\{x : |f_i(x) - f_j(x)| < \frac{1}{k}\right\},$$

show that $\mu(S(n, k)) \to \mu(X)$ as $n \to \infty$, for each k, and hence that there is a suitably increasing sequence $\{n_k\}$ such that $E = \bigcap S(n_k, k)$ has the desired property.

Show that the theorem does not extend to σ-finite spaces.

Show that the theorem does extend, with essentially the same proof, to the situation in which the sequence $\{f_n\}$ is replaced by a family $\{f_t\}$, where t ranges over the positive reals; the assumptions are now that, for all $x \in X$,

(i) $\lim_{t \to \infty} f_t(x) = f(x)$ and

(ii) $t \to f_t(x)$ is continuous.

17 (a) If $0 < p < \infty$, put $\gamma_p = \max(1, 2^{p-1})$, and show that

$$|\alpha - \beta|^p \leq \gamma_p(|\alpha|^p + |\beta|^p)$$

for arbitrary complex numbers α and β.

(b) Suppose μ is a positive measure on X, $0 < p < \infty$, $f \in L^p(\mu)$, $f_n \in L^p(\mu)$, $f_n(x) \to f(x)$ a.e., and $\|f_n\|_p \to \|f\|_p$ as $n \to \infty$. Show that then $\lim \|f - f_n\|_p = 0$, by completing the two proofs that are sketched below.

  (i) By Egoroff's theorem, $X = A \cup B$ in such a way that $\int_A |f|^p < \epsilon$, $\mu(B) < \infty$, and $f_n \to f$ uniformly on B. Fatou's lemma, applied to $\int_B |f_n|^p$, leads to

  $$\limsup \int_A |f_n|^p\, d\mu \leq \epsilon.$$

  (ii) Put $h_n = \gamma_p(|f|^p + |f_n|^p) - |f - f_n|^p$, and use Fatou's lemma as in the proof of Theorem 1.34.

(c) Show that the conclusion of (b) is false if the hypothesis $\|f_n\|_p \to \|f\|_p$ is omitted, even if $\mu(X) < \infty$.

18 Let μ be a positive measure on X. A sequence $\{f_n\}$ of complex measurable functions on X is said to *converge in measure* to the measurable function f if to every $\epsilon > 0$ there corresponds an N such that

$$\mu(\{x: |f_n(x) - f(x)| > \epsilon\}) < \epsilon$$

for all $n > N$. (This notion is of importance in probability theory.) Assume $\mu(X) < \infty$ and prove the following statements:
  (a) If $f_n(x) \to f(x)$ a.e., then $f_n \to f$ in measure.
  (b) If $f_n \in L^p(\mu)$ and $\|f_n - f\|_p \to 0$, then $f_n \to f$ in measure; here $1 \le p \le \infty$.
  (c) If $f_n \to f$ in measure, then $\{f_n\}$ has a subsequence which converges to f a.e.

Investigate the converses of (a) and (b). What happens to (a), (b), and (c) if $\mu(X) = \infty$, for instance, if μ is Lebesgue measure on R^1?

19 Define the *essential range* of a function $f \in L^\infty(\mu)$ to be the set R_f consisting of all complex numbers w such that

$$\mu(\{x: |f(x) - w| < \epsilon\}) > 0$$

for every $\epsilon > 0$. Prove that R_f is compact. What relation exists between the set R_f and the number $\|f\|_\infty$?

Let A_f be the set of all averages

$$\frac{1}{\mu(E)} \int_E f \, d\mu$$

where $E \in \mathfrak{M}$ and $\mu(E) > 0$. What relations exist between A_f and R_f? Is A_f always closed? Are there measures μ such that A_f is convex for every $f \in L^\infty(\mu)$? Are there measures μ such that A_f fails to be convex for some $f \in L^\infty(\mu)$?

How are these results affected if $L^\infty(\mu)$ is replaced by $L^1(\mu)$, for instance?

20 Suppose φ is a real function on R^1 such that

$$\varphi\left(\int_0^1 f(x) \, dx\right) \le \int_0^1 \varphi(f) \, dx$$

for every real bounded measurable f. Prove that φ is then convex.

21 Call a metric space Y a *completion* of a metric space X if X is dense in Y and Y is complete. In Sec. 3.15 reference was made to "the" completion of a metric space. State and prove a uniqueness theorem which justifies this terminology.

22 Suppose X is a metric space in which every Cauchy sequence has a convergent subsequence. Does it follow that X is complete? (See the proof of Theorem 3.11.)

23 Suppose μ is a positive measure on X, $\mu(X) < \infty$, $f \in L^\infty(\mu)$, $\|f\|_\infty > 0$, and

$$\alpha_n = \int_X |f|^n \, d\mu \qquad (n = 1, 2, 3, \ldots).$$

Prove that

$$\lim_{n \to \infty} \frac{\alpha_{n+1}}{\alpha_n} = \|f\|_\infty.$$

24 Suppose μ is a positive measure, $f \in L^p(\mu)$, $g \in L^p(\mu)$.
  (a) If $0 < p < 1$, prove that

$$\int \left| |f|^p - |g|^p \right| \, d\mu \le \int |f - g|^p \, d\mu$$

that $\Delta(f, g) = \int |f - g|^p \, d\mu$ defines a metric on $L^p(\mu)$, and that the resulting metric space is complete.

(b) If $1 \leq p < \infty$ and $\|f\|_p \leq R$, $\|g\|_p \leq R$, prove that

$$\int \left| |f|^p - |g|^p \right| d\mu \leq 2pR^{p-1} \|f - g\|_p.$$

Hint: Prove first, for $x \geq 0$, $y \geq 0$, that

$$|x^p - y^p| \leq \begin{cases} |x - y|^p & \text{if } 0 < p < 1, \\ p|x - y|(x^{p-1} + y^{p-1}) & \text{if } 1 \leq p < \infty. \end{cases}$$

Note that (a) and (b) establish the continuity of the mapping $f \to |f|^p$ that carries $L^p(\mu)$ into $L^1(\mu)$.

25 Suppose μ is a positive measure on X and $f: X \to (0, \infty)$ satisfies $\int_X f \, d\mu = 1$. Prove, for every $E \subset X$ with $0 < \mu(E) < \infty$, that

$$\int_E (\log f) \, d\mu \leq \mu(E) \log \frac{1}{\mu(E)}$$

and, when $0 < p < 1$,

$$\int_E f^p \, d\mu \leq \mu(E)^{1-p}.$$

26 If f is a positive measurable function on $[0, 1]$, which is larger,

$$\int_0^1 f(x) \log f(x) \, dx \quad \text{or} \quad \int_0^1 f(s) \, ds \int_0^1 \log f(t) \, dt?$$

CHAPTER FOUR

ELEMENTARY HILBERT SPACE THEORY

Inner Products and Linear Functionals

4.1 Definition A complex vector space H is called an *inner product space* (or *unitary space*) if to each ordered pair of vectors x and $y \in H$ there is associated a complex number (x, y), the so-called "inner product" (or "scalar product") of x and y, such that the following rules hold:

(a) $(y, x) = \overline{(x, y)}$. (The bar denotes complex conjugation.)
(b) $(x + y, z) = (x, z) + (y, z)$ if x, y, and $z \in H$.
(c) $(\alpha x, y) = \alpha(x, y)$ if x and $y \in H$ and α is a scalar.
(d) $(x, x) \geq 0$ for all $x \in H$.
(e) $(x, x) = 0$ only if $x = 0$.

Let us list some immediate consequences of these axioms:

(c) implies that $(0, y) = 0$ for all $y \in H$.
(b) and (c) may be combined into the statement: *For every $y \in H$, the mapping $x \to (x, y)$ is a linear functional on H.*
(a) and (c) show that $(x, \alpha y) = \bar{\alpha}(x, y)$.
(a) and (b) imply the second distributive law:

$$(z, x + y) = (z, x) + (z, y).$$

By (d), we may define $\|x\|$, the *norm* of the vector $x \in H$, to be the nonnegative square root of (x, x). Thus

(f) $$\|x\|^2 = (x, x).$$

4.2 The Schwarz Inequality
The properties 4.1 (a) to (d) imply that

$$|(x, y)| \leq \|x\| \, \|y\|$$

for all x and y ∈ H.

PROOF Put $A = \|x\|^2$, $B = |(x, y)|$, and $C = \|y\|^2$. There is a complex number α such that $|\alpha| = 1$ and $\alpha(y, x) = B$. For any real r, we then have

$$(x - r\alpha y, x - r\alpha y) = (x, x) - r\alpha(y, x) - r\bar{\alpha}(x, y) + r^2(y, y). \tag{1}$$

The expression on the left is real and not negative. Hence

$$A - 2Br + Cr^2 \geq 0 \tag{2}$$

for every real r. If $C = 0$, we must have $B = 0$; otherwise (2) is false for large positive r. If $C > 0$, take $r = B/C$ in (2), and obtain $B^2 \leq AC$. ////

4.3 The Triangle Inequality *For x and y ∈ H, we have*

$$\|x + y\| \leq \|x\| + \|y\|.$$

PROOF By the Schwarz inequality,

$$\|x + y\|^2 = (x + y, x + y) = (x, x) + (x, y) + (y, x) + (y, y)$$

$$\leq \|x\|^2 + 2\|x\| \, \|y\| + \|y\|^2 = (\|x\| + \|y\|)^2.$$ ////

4.4 Definition
It follows from the triangle inequality that

$$\|x - z\| \leq \|x - y\| + \|y - z\| \qquad (x, y, z \in H). \tag{1}$$

If we define the distance between x and y to be $\|x - y\|$, all the axioms for a metric space are satisfied; here, for the first time, we use part (e) of Definition 4.1.

Thus H is now a metric space. If this metric space is *complete*, i.e., if every Cauchy sequence converges in H, then H is called a *Hilbert space*.

Throughout the rest of this chapter, the letter H will denote a Hilbert space.

4.5 Examples

(a) For any fixed n, the set C^n of all n-tuples

$$x = (\xi_1, \ldots, \xi_n),$$

where $\xi_1, \ldots, \xi_n$ are complex numbers, is a Hilbert space if addition and scalar multiplication are defined componentwise, as usual, and if

$$(x, y) = \sum_{j=1}^{n} \xi_j \bar{\eta}_j \qquad (y = (\eta_1, \ldots, \eta_n)).$$

(b) If μ is any positive measure, $L^2(\mu)$ is a Hilbert space, with inner product

$$(f, g) = \int_X f\bar{g}\, d\mu.$$

The integrand on the right is in $L^1(\mu)$, by Theorem 3.8, so that (f, g) is well defined. Note that

$$\|f\| = (f,f)^{1/2} = \left\{\int_X |f|^2\, d\mu\right\}^{1/2} = \|f\|_2.$$

The completeness of $L^2(\mu)$ (Theorem 3.11) shows that $L^2(\mu)$ is indeed a Hilbert space. [We recall that $L^2(\mu)$ should be regarded as a space of *equivalence classes* of functions; compare the discussion in Sec. 3.10.]

For $H = L^2(\mu)$, the inequalities 4.2 and 4.3 turn out to be special cases of the inequalities of Hölder and Minkowski.

Note that Example (a) is a special case of (b). What is the measure in (a)?

(c) The vector space of all continuous complex functions on [0, 1] is an inner product space if

$$(f, g) = \int_0^1 f(t)\overline{g(t)}\, dt$$

but is not a Hilbert space.

4.6 Theorem *For any fixed $y \in H$, the mappings*

$$x \to (x, y), \quad x \to (y, x), \quad x \to \|x\|$$

are continuous functions on H.

PROOF The Schwarz inequality implies that

$$|(x_1, y) - (x_2, y)| = |(x_1 - x_2, y)| \le \|x_1 - x_2\|\, \|y\|,$$

which proves that $x \to (x, y)$ is, in fact, uniformly continuous, and the same is true for $x \to (y, x)$. The triangle inequality $\|x_1\| \le \|x_1 - x_2\| + \|x_2\|$ yields

$$\|x_1\| - \|x_2\| \le \|x_1 - x_2\|,$$

and if we interchange x_1 and x_2 we see that

$$|\,\|x_1\| - \|x_2\|\,| \le \|x_1 - x_2\|$$

for all x_1 and $x_2 \in H$. Thus $x \to \|x\|$ is also uniformly continuous. ////

4.7 Subspaces A subset M of a vector space V is called a *subspace* of V if M is itself a vector space, relative to the addition and scalar multiplication which are defined in V. A necessary and sufficient condition for a set $M \subset V$ to be a subspace is that $x + y \in M$ and $\alpha x \in M$ whenever x and $y \in M$ and α is a scalar.

In the vector space context, the word "subspace" will always have this meaning. Sometimes, for emphasis, we may use the term "linear subspace" in place of subspace.

For example, if V is the vector space of all complex functions on a set S, the set of all bounded complex functions on S is a subspace of V, but the set of all $f \in V$ with $|f(x)| \leq 1$ for all $x \in S$ is not. The real vector space R^3 has the following subspaces, and no others: (a) R^3, (b) all planes through the origin 0, (c) all straight lines through 0, and (d) $\{0\}$.

A *closed subspace* of H is a subspace that is a closed set relative to the topology induced by the metric of H.

Note that *if M is a subspace of H, so is its closure $\bar{M}$*. To see this, pick x and y in $\bar{M}$ and let α be a scalar. There are sequences $\{x_n\}$ and $\{y_n\}$ in M that converge to x and y, respectively. It is then easy to verify that $x_n + y_n$ and αx_n converge to $x + y$ and αx, respectively. Thus $x + y \in \bar{M}$ and $\alpha x \in \bar{M}$.

4.8 Convex Sets A set E in a vector space V is said to be *convex* if it has the following geometric property: Whenever $x \in E$, $y \in E$, and $0 < t < 1$, the point

$$z_t = (1 - t)x + ty$$

also lies in E. As t runs from 0 to 1, one may visualize z_t as describing a straight line segment in V, from x to y. Convexity requires that E contain the segments between any two of its points.

It is clear that every subspace of V is convex.

Also, if E is convex, so is each of its translates

$$E + x = \{y + x : y \in E\}.$$

4.9 Orthogonality If $(x, y) = 0$ for some x and $y \in H$, we say that x is orthogonal to y, and sometimes write $x \perp y$. Since $(x, y) = 0$ implies $(y, x) = 0$, the relation $\perp$ is symmetric.

Let $x^\perp$ denote the set of all $y \in H$ which are orthogonal to x; and if M is a subspace of H, let $M^\perp$ be the set of all $y \in H$ which are orthogonal to every $x \in M$.

Note that $x^\perp$ is a subspace of H, since $x \perp y$ and $x \perp y'$ implies $x \perp (y + y')$ and $x \perp \alpha y$. Also, $x^\perp$ is precisely the set of points where the continuous function $y \to (x, y)$ is 0. Hence $x^\perp$ is a *closed* subspace of H. Since

$$M^\perp = \bigcap_{x \in M} x^\perp,$$

$M^\perp$ is an intersection of closed subspaces, and it follows that $M^\perp$ *is a closed subspace of H*.

4.10 Theorem *Every nonempty, closed, convex set E in a Hilbert space H contains a unique element of smallest norm.*

In other words, there is one and only one $x_0 \in E$ such that $\|x_0\| \le \|x\|$ for every $x \in E$.

PROOF An easy computation, using only the properties listed in Definition 4.1, establishes the identity

$$\|x + y\|^2 + \|x - y\|^2 = 2\|x\|^2 + 2\|y\|^2 \qquad (x \text{ and } y \in H). \tag{1}$$

This is known as the *parallelogram law*: If we interpret $\|x\|$ to be the length of the vector x, (1) says that the sum of the squares of the diagonals of a parallelogram is equal to the sum of the squares of its sides, a familiar proposition in plane geometry.

Let $\delta = \inf\{\|x\|: x \in E\}$. For any x and $y \in E$, we apply (1) to $\tfrac{1}{2}x$ and $\tfrac{1}{2}y$ and obtain

$$\tfrac{1}{4}\|x - y\|^2 = \tfrac{1}{2}\|x\|^2 + \tfrac{1}{2}\|y\|^2 - \left\|\frac{x+y}{2}\right\|^2. \tag{2}$$

Since E is convex, $(x + y)/2 \in E$. Hence

$$\|x - y\|^2 \le 2\|x\|^2 + 2\|y\|^2 - 4\delta^2 \qquad (x \text{ and } y \in E). \tag{3}$$

If also $\|x\| = \|y\| = \delta$, then (3) implies $x = y$, and we have proved the uniqueness assertion of the theorem.

The definition of δ shows that there is a sequence $\{y_n\}$ in E so that $\|y_n\| \to \delta$ as $n \to \infty$. Replace x and y in (3) by y_n and y_m. Then, as $n \to \infty$ and $m \to \infty$, the right side of (3) will tend to 0. This shows that $\{y_n\}$ is a Cauchy sequence. Since H is complete, there exists an $x_0 \in H$ so that $y_n \to x_0$, i.e., $\|y_n - x_0\| \to 0$, as $n \to \infty$. Since $y_n \in E$ and E is closed, $x_0 \in E$. Since the norm is a continuous function on H (Theorem 4.6), it follows that

$$\|x_0\| = \lim_{n \to \infty} \|y_n\| = \delta. \qquad //// $$

4.11 Theorem *Let M be a closed subspace of a Hilbert space H.*
(a) *Every $x \in H$ has then a unique decomposition*

$$x = Px + Qx$$

into a sum of $Px \in M$ and $Qx \in M^\perp$.
(b) *Px and Qx are the nearest points to x in M and in $M^\perp$, respectively.*
(c) *The mappings $P: H \to M$ and $Q: H \to M^\perp$ are linear.*
(d) *$\|x\|^2 = \|Px\|^2 + \|Qx\|^2$.*

Corollary *If $M \ne H$, then there exists $y \in H$, $y \ne 0$, such that $y \perp M$.*

P and Q are called the *orthogonal projections* of H onto M and $M^\perp$.

PROOF As regards the uniqueness in (a), suppose that $x' + y' = x'' + y''$ for some vectors x', x'' in M and y', y'' in $M^\perp$. Then

$$x' - x'' = y'' - y'.$$

Since $x' - x'' \in M$, $y'' - y' \in M^\perp$, and $M \cap M^\perp = \{0\}$ [an immediate consequence of the fact that $(x, x) = 0$ implies $x = 0$], we have $x'' = x'$, $y'' = y'$.

To prove the existence of the decomposition, note that the set

$$x + M = \{x + y : y \in M\}$$

is closed and convex. Define Qx to be the element of smallest norm in $x + M$; this exists, by Theorem 4.10. Define $Px = x - Qx$.

Since $Qx \in x + M$, it is clear that $Px \in M$. Thus P maps H into M.

To prove that Q maps H into $M^\perp$ we show that $(Qx, y) = 0$ for all $y \in M$. Assume $\|y\| = 1$, without loss of generality, and put $z = Qx$. The minimizing property of Qx shows that

$$(z, z) = \|z\|^2 \leq \|z - \alpha y\|^2 = (z - \alpha y, z - \alpha y)$$

for every scalar α. This simplifies to

$$0 \leq -\alpha(y, z) - \bar{\alpha}(z, y) + \alpha\bar{\alpha}.$$

With $\alpha = (z, y)$, this gives $0 \leq -|(z, y)|^2$, so that $(z, y) = 0$. Thus $Qx \in M^\perp$.

We have already seen that $Px \in M$. If $y \in M$, it follows that

$$\|x - y\|^2 = \|Qx + (Px - y)\|^2 = \|Qx\|^2 + \|Px - y\|^2$$

which is obviously minimized when $y = Px$.

We have now proved (a) and (b). If we apply (a) to x, to y, and to $\alpha x + \beta y$, we obtain

$$P(\alpha x + \beta y) - \alpha Px - \beta Py = \alpha Qx + \beta Qy - Q(\alpha x + \beta y).$$

The left side is in M, the right side in $M^\perp$. Hence both are 0, so P and Q are linear.

Since $Px \perp Qx$, (d) follows from (a).

To prove the corollary, take $x \in H$, $x \notin M$, and put $y = Qx$. Since $Px \in M$, $x \neq Px$, hence $y = x - Px \neq 0$. ////

We have already observed that $x \to (x, y)$ is, for each $y \in H$, a continuous linear functional on H. It is a very important fact that *all* continuous linear functionals on H are of this type.

4.12 Theorem *If L is a continuous linear functional on H, then there is a unique $y \in H$ such that*

$$Lx = (x, y) \qquad (x \in H). \tag{1}$$

PROOF If $Lx = 0$ for all x, take $y = 0$. Otherwise, define

$$M = \{x: Lx = 0\}. \tag{2}$$

The linearity of L shows that M is a subspace. The continuity of L shows that M is closed. Since $Lx \neq 0$ for some $x \in H$, Theorem 4.11 shows that $M^\perp$ does not consist of 0 alone.

Hence there exists $z \in M^\perp$, with $\|z\| = 1$. Put

$$u = (Lx)z - (Lz)x. \tag{3}$$

Since $Lu = (Lx)(Lz) - (Lz)(Lx) = 0$, we have $u \in M$. Thus $(u, z) = 0$. This gives

$$Lx = (Lx)(z, z) = (Lz)(x, z). \tag{4}$$

Thus (1) holds with $y = \alpha z$, where $\bar{\alpha} = Lz$.

The uniqueness of y is easily proved, for if $(x, y) = (x, y')$ for all $x \in H$, set $z = y - y'$; then $(x, z) = 0$ for all $x \in H$; in particular, $(z, z) = 0$, hence $z = 0$.
////

Orthonormal Sets

4.13 Definitions If V is a vector space, if $x_1, \ldots, x_k \in V$, and if $c_1, \ldots, c_k$ are scalars, then $c_1 x_1 + \cdots + c_k x_k$ is called a *linear combination* of $x_1, \ldots, x_k$. The set $\{x_1, \ldots, x_k\}$ is called *independent* if $c_1 x_1 + \cdots + c_k x_k = 0$ implies that $c_1 = \cdots = c_k = 0$. A set $S \subset V$ is independent if every finite subset of S is independent. The set $[S]$ of all linear combinations of all finite subsets of S (also called the set of all *finite linear combinations* of members of S) is clearly a vector space; $[S]$ is the smallest subspace of V which contains S; $[S]$ is called the *span* of S, or the space spanned by S.

A set of vectors u_α in a Hilbert space H, where α runs through some index set A, is called *orthonormal* if it satisfies the orthogonality relations $(u_\alpha, u_\beta) = 0$ for all $\alpha \neq \beta$, $\alpha \in A$, and $\beta \in A$, and if it is normalized so that $\|u_\alpha\| = 1$ for each $\alpha \in A$. In other words, $\{u_\alpha\}$ is orthonormal provided that

$$(u_\alpha, u_\beta) = \begin{cases} 1 & \text{if } \alpha = \beta, \\ 0 & \text{if } \alpha \neq \beta. \end{cases} \tag{1}$$

If $\{u_\alpha : \alpha \in A\}$ is orthonormal, we associate with each $x \in H$ a complex function $\hat{x}$ on the index set A, defined by

$$\hat{x}(\alpha) = (x, u_\alpha) \qquad (\alpha \in A). \tag{2}$$

One sometimes calls the numbers $\hat{x}(\alpha)$ the *Fourier coefficients* of x, relative to the set $\{u_\alpha\}$.

We begin with some simple facts about *finite* orthonormal sets.

4.14 Theorem *Suppose that $\{u_\alpha : \alpha \in A\}$ is an orthonormal set in H and that F is a finite subset of A. Let M_F be the span of $\{u_\alpha : \alpha \in F\}$.*

(a) If φ is a complex function on A that is 0 outside F, then there is a vector $y \in M_F$, namely

$$y = \sum_{\alpha \in F} \varphi(\alpha) u_\alpha \tag{1}$$

that has $\hat{y}(\alpha) = \varphi(\alpha)$ for every $\alpha \in A$. Also,

$$\|y\|^2 = \sum_{\alpha \in F} |\varphi(\alpha)|^2. \tag{2}$$

(b) If $x \in H$ and

$$s_F(x) = \sum_{\alpha \in F} \hat{x}(\alpha) u_\alpha \tag{3}$$

then

$$\|x - s_F(x)\| < \|x - s\| \tag{4}$$

for every $s \in M_F$, except for $s = s_F(x)$, and

$$\sum_{\alpha \in F} |\hat{x}(\alpha)|^2 \leq \|x\|^2. \tag{5}$$

PROOF Part (a) is an immediate consequence of the orthogonality relations 4.13(1).

In the proof of (b), let us write s_F in place of $s_F(x)$, and note that $\hat{s}_F(\alpha) = \hat{x}(\alpha)$ for all $\alpha \in F$. This says that $(x - s_F) \perp u_\alpha$ if $\alpha \in F$, hence $(x - s_F) \perp (s_F - s)$ for every $s \in M_F$, and therefore

$$\|x - s\|^2 = \|(x - s_F) + (s_F - s)\|^2 = \|x - s_F\|^2 + \|s_F - s\|^2. \tag{6}$$

This gives (4). With $s = 0$, (6) gives $\|s_F\|^2 \leq \|x\|^2$, which is the same as (5), because of (2). ////

The inequality (4) states that the "partial sum" $s_F(x)$ of the "Fourier series" $\sum \hat{x}(\alpha) u_\alpha$ of x is the unique best approximation to x in M_F, relative to the metric defined by the Hilbert space norm.

4.15 We want to drop the finiteness condition that appears in Theorem 4.14 (thus obtaining Theorems 4.17 and 4.18) without even restricting ourselves to sets that are necessarily countable. For this reason it seems advisable to clarify the meaning of the symbol $\sum_{\alpha \in A} \varphi(\alpha)$ when α ranges over an arbitrary set A.

Assume $0 \leq \varphi(\alpha) \leq \infty$ for each $\alpha \in A$. Then

$$\sum_{\alpha \in A} \varphi(\alpha) \tag{1}$$

denotes the supremum of the set of all finite sums $\varphi(\alpha_1) + \cdots + \varphi(\alpha_n)$, where $\alpha_1, \ldots, \alpha_n$ are distinct members of A.

A moment's consideration will show that *the sum* (1) *is thus precisely the Lebesgue integral of φ relative to the counting measure μ on A.*

In this context one usually writes $\ell^p(A)$ for $L^p(\mu)$. A complex function φ with domain A is thus in $\ell^2(A)$ if and only if

$$\sum_{\alpha \in A} |\varphi(\alpha)|^2 < \infty. \tag{2}$$

Example 4.5(b) shows that $\ell^2(A)$ is a Hilbert space, with inner product

$$(\varphi, \psi) = \sum_{\alpha \in A} \varphi(\alpha)\overline{\psi(\alpha)}. \tag{3}$$

Here, again, the sum over A stands for the integral of $\varphi\bar{\psi}$ with respect to the counting measure; note that $\varphi\bar{\psi} \in \ell^1(A)$ because φ and ψ are in $\ell^2(A)$.

Theorem 3.13 shows that *the functions φ that are zero except on some finite subset of A are dense in $\ell^2(A)$.*

Moreover, if $\varphi \in \ell^2(A)$, then $\{\alpha \in A : \varphi(\alpha) \neq 0\}$ is at most countable. For if A_n is the set of all α where $|\varphi(\alpha)| > 1/n$, then the number of elements of A_n is at most

$$\sum_{\alpha \in A_n} |n\varphi(\alpha)|^2 \leq n^2 \sum_{\alpha \in A} |\varphi(\alpha)|^2.$$

Each A_n ($n = 1, 2, 3, \ldots$) is thus a finite set.

The following lemma about complete metric spaces will make it easy to pass from finite orthonormal sets to infinite ones.

4.16 Lemma *Suppose that*

(a) *X and Y are metric spaces, X is complete,*
(b) *$f: X \to Y$ is continuous,*
(c) *X has a dense subset X_0 on which f is an isometry, and*
(d) *$f(X_0)$ is dense in Y.*

Then f is an isometry of X onto Y.

The most important part of the conclusion is that f maps X onto all of Y.

Recall that an isometry is simply a mapping that preserves distances. Thus, by assumption, the distance between $f(x_1)$ and $f(x_2)$ in Y is equal to that between x_1 and x_2 in X, for all points x_1, x_2 in X_0.

PROOF The fact that f is an isometry on X is an immediate consequence of the continuity of f, since X_0 is dense in X.

Pick $y \in Y$. Since $f(X_0)$ is dense in Y, there is a sequence $\{x_n\}$ in X_0 such that $f(x_n) \to y$ as $n \to \infty$. Thus $\{f(x_n)\}$ is a Cauchy sequence in Y. Since f is an isometry on X_0, it follows that $\{x_n\}$ is also a Cauchy sequence. The completeness of X implies now that $\{x_n\}$ converges to some $x \in X$, and the continuity of f shows that $f(x) = \lim f(x_n) = y$. ////

4.17 Theorem Let $\{u_\alpha : \alpha \in A\}$ be an orthonormal set in H, and let P be the space of all finite linear combinations of the vectors u_α.

The inequality

$$\sum_{\alpha \in A} |\hat{x}(\alpha)|^2 \leq \|x\|^2 \tag{1}$$

holds then for every $x \in H$, and $x \to \hat{x}$ is a continuous linear mapping of H onto $\ell^2(A)$ whose restriction to the closure $\bar{P}$ of P is an isometry of $\bar{P}$ onto $\ell^2(A)$.

PROOF Since the inequality 4.14(5) holds for every finite set $F \subset A$, we have (1), the so-called *Bessel inequality*.

Define f on H by $f(x) = \hat{x}$. Then (1) shows explicitly that f maps H into $\ell^2(A)$. The linearity of f is obvious. If we apply (1) to $x - y$ we see that

$$\|f(y) - f(x)\|_2 = \|\hat{y} - \hat{x}\|_2 \leq \|y - x\|.$$

Thus f is continuous. Theorem 4.14(a) shows that f is an isometry of P onto the dense subspace of $\ell^2(A)$ consisting of those functions whose support is a finite set $F \subset A$. The theorem follows therefore from Lemma 4.16, applied with $X = \bar{P}$, $X_0 = P$, $Y = \ell^2(A)$; note that $\bar{P}$, being a closed subset of the complete metric space H, is itself complete. ////

The fact that the mapping $x \to \hat{x}$ carries H onto $\ell^2(A)$ is known as the *Riesz-Fischer theorem*.

4.18 Theorem Let $\{u_\alpha : \alpha \in A\}$ be an orthonormal set in H. Each of the following four conditions on $\{u_\alpha\}$ implies the other three:

(i) $\{u_\alpha\}$ is a maximal orthonormal set in H.
(ii) The set P of all finite linear combinations of members of $\{u_\alpha\}$ is dense in H.
(iii) The equality

$$\sum_{\alpha \in A} |\hat{x}(\alpha)|^2 = \|x\|^2$$

holds for every $x \in H$.
(iv) The equality

$$\sum_{\alpha \in A} \hat{x}(\alpha)\overline{\hat{y}(\alpha)} = (x, y)$$

holds for all $x \in H$ and $y \in H$.

The last formula is known as *Parseval's identity*. Observe that $\hat{x}$ and $\hat{y}$ are in $\ell^2(A)$, hence $\hat{x}\hat{y}$ is in $\ell^1(A)$, so that the sum in (iv) is well defined. Of course, (iii) is the special case $x = y$ of (iv).

Maximal orthonormal sets are often called *complete orthonormal sets* or *orthonormal bases*.

PROOF To say that $\{u_\alpha\}$ is maximal means simply that no vector of H can be adjoined to $\{u_\alpha\}$ in such a way that the resulting set is still orthonormal. This happens precisely when there is no $x \neq 0$ in H that is orthogonal to every u_α.

We shall prove that (i)$\to$(ii)$\to$(iii)$\to$(iv)$\to$(i).

If P is not dense in H, then its closure $\bar{P}$ is not all of H, and the corollary to Theorem 4.11 implies that $P^\perp$ contains a nonzero vector. Thus $\{u_\alpha\}$ is not maximal when P is not dense, and (i) implies (ii).

If (ii) holds, so does (iii), by Theorem 4.17.

The implication (iii)$\to$(iv) follows from the easily proved Hilbert space identity (sometimes called the "polarization identity")

$$4(x, y) = \|x + y\|^2 - \|x - y\|^2 + i\|x + iy\|^2 - i\|x - iy\|^2$$

which expresses the inner product (x, y) in terms of norms and which is equally valid with $\hat{x}, \hat{y}$ in place of x, y, simply because $\ell^2(A)$ is also a Hilbert space. (See Exercise 19 for other identities of this type.) Note that the sums in (iii) and (iv) are $\|\hat{x}\|_2^2$ and $(\hat{x}, \hat{y})$, respectively.

Finally, if (i) is false, there exists $u \neq 0$ in H so that $(u, u_\alpha) = 0$ for all $\alpha \in A$. If $x = y = u$, then $(x, y) = \|u\|^2 > 0$ but $\hat{x}(\alpha) = 0$ for all $\alpha \in A$, hence (iv) fails. Thus (iv) implies (i), and the proof is complete.   ////

4.19 Isomorphisms Speaking informally, two algebraic systems of the same nature are said to be *isomorphic* if there is a one-to-one mapping of one onto the other which preserves all relevant properties. For instance, we may ask whether two groups are isomorphic or whether two fields are isomorphic. Two vector spaces are isomorphic if there is a one-to-one *linear* mapping of one onto the other. The linear mappings are the ones which preserve the relevant concepts in a vector space, namely, addition and scalar multiplication.

In the same way, two Hilbert spaces H_1 and H_2 are isomorphic if there is a one-to-one linear mapping Λ of H_1 onto H_2 which also preserves inner products: $(\Lambda x, \Lambda y) = (x, y)$ for all x and $y \in H_1$. Such a Λ is an isomorphism (or, more specifically, a *Hilbert space isomorphism*) of H_1 onto H_2. Using this terminology, Theorems 4.17 and 4.18 yield the following statement:

If $\{u_\alpha : \alpha \in A\}$ is a maximal orthonormal set in a Hilbert space H, and if $\hat{x}(\alpha) = (x, u_\alpha)$, then the mapping $x \to \hat{x}$ is a Hilbert space isomorphism of H onto $\ell^2(A)$.

One can prove (we shall omit this) that $\ell^2(A)$ and $\ell^2(B)$ are isomorphic if and only if the sets A and B have the same cardinal number. But we shall prove that every nontrivial Hilbert space (this means that the space does not consist of 0 alone) is isomorphic to some $\ell^2(A)$, by proving that every such space contains a maximal orthonormal set (Theorem 4.22). The proof will depend on a property of partially ordered sets which is equivalent to the axiom of choice.

4.20 Partially Ordered Sets A set $\mathscr{P}$ is said to be *partially ordered* by a binary relation $\leq$ if

(a) $a \leq b$ and $b \leq c$ implies $a \leq c$.
(b) $a \leq a$ for every $a \in \mathscr{P}$.
(c) $a \leq b$ and $b \leq a$ implies $a = b$.

A subset $\mathscr{Q}$ of a partially ordered set $\mathscr{P}$ is said to be *totally ordered* (or *linearly ordered*) if every pair $a, b \in \mathscr{Q}$ satisfies either $a \leq b$ or $b \leq a$.

For example, every collection of subsets of a given set is partially ordered by the inclusion relation $\subset$.

To give a more specific example, let $\mathscr{P}$ be the collection of all open subsets of the plane, partially ordered by set inclusion, and let $\mathscr{Q}$ be the collection of all open circular discs with center at the origin. Then $\mathscr{Q} \subset \mathscr{P}$, $\mathscr{Q}$ is totally ordered by $\subset$, and $\mathscr{Q}$ is a *maximal* totally ordered subset of $\mathscr{P}$. This means that if any member of $\mathscr{P}$ not in $\mathscr{Q}$ is adjoined to $\mathscr{Q}$, the resulting collection of sets is no longer totally ordered by $\subset$.

4.21 The Hausdorff Maximality Theorem *Every nonempty partially ordered set contains a maximal totally ordered subset.*

This is a consequence of the axiom of choice and is, in fact, equivalent to it; another (very similar) form of it is known as Zorn's lemma. We give the proof in the Appendix.

If now H is a nontrivial Hilbert space, then there exists a $u \in H$ with $\|u\| = 1$, so that there is a nonempty orthonormal set in H. The existence of a maximal orthonormal set is therefore a consequence of the following theorem:

4.22 Theorem *Every orthonormal set B in a Hilbert space H is contained in a maximal orthonormal set in H.*

PROOF Let $\mathscr{P}$ be the class of all orthonormal sets in H which contain the given set B. Partially order $\mathscr{P}$ by set inclusion. Since $B \in \mathscr{P}$, $\mathscr{P} \neq \varnothing$. Hence $\mathscr{P}$ contains a maximal totally ordered class Ω. Let S be the union of all members of Ω. It is clear that $B \subset S$. We claim that S is a maximal orthonormal set:

If u_1 and $u_2 \in S$, then $u_1 \in A_1$ and $u_2 \in A_2$ for some A_1 and $A_2 \in \Omega$. Since Ω is total ordered, $A_1 \subset A_2$ (or $A_2 \subset A_1$), so that $u_1 \in A_2$ and $u_2 \in A_2$. Since A_2 is orthonormal, $(u_1, u_2) = 0$ if $u_1 \neq u_2$, $(u_1, u_2) = 1$ if $u_1 = u_2$. Thus S is an orthonormal set.

Suppose S is not maximal. Then S is a proper subset of an orthonormal set S^*. Clearly, $S^* \notin \Omega$, and S^* contains every member of Ω. Hence we may adjoin S^* to Ω and still have a total order. This contradicts the maximality of Ω. ////

Trigonometric Series

4.23 Definitions Let T be the unit circle in the complex plane, i.e., the set of all complex numbers of absolute value 1. If F is any function on T and if f is defined on R^1 by

$$f(t) = F(e^{it}), \tag{1}$$

then f is a periodic function of period 2π. This means that $f(t + 2\pi) = f(t)$ for all real t. Conversely, if f is a function on R^1, with period 2π, then there is a function F on T such that (1) holds. Thus we may identify functions on T with 2π-periodic functions on R^1; and, for simplicity of notation, we shall sometimes write $f(t)$ rather than $f(e^{it})$, even if we think of f as being defined on T.

With these conventions in mind, we define $L^p(T)$, for $1 \leq p < \infty$, to be the class of all complex, Lebesgue measurable, 2π-periodic functions on R^1 for which the norm

$$\|f\|_p = \left\{\frac{1}{2\pi} \int_{-\pi}^{\pi} |f(t)|^p \, dt\right\}^{1/p} \tag{2}$$

is finite.

In other words, we are looking at $L^p(\mu)$, where μ is Lebesgue measure on $[0, 2\pi]$ (or on T), divided by 2π. $L^\infty(T)$ will be the class of all 2π-periodic members of $L^\infty(R^1)$, with the essential supremum norm, and $C(T)$ consists of all continuous complex functions on T (or, equivalently, of all continuous, complex, 2π-periodic functions on R^1), with norm

$$\|f\|_\infty = \sup_t |f(t)|, \tag{3}$$

The factor $1/(2\pi)$ in (2) simplifies the formalism we are about to develop. For instance, the L^p-norm of the constant function 1 is 1.

A *trigonometric polynomial* is a finite sum of the form

$$f(t) = a_0 + \sum_{n=1}^{N} (a_n \cos nt + b_n \sin nt) \qquad (t \in R^1) \tag{4}$$

where $a_0, a_1, \ldots, a_N$ and $b_1, \ldots, b_N$ are complex numbers. On account of the Euler identities, (4) can also be written in the form

$$f(t) = \sum_{n=-N}^{N} c_n e^{int} \tag{5}$$

which is more convenient for most purposes. It is clear that every trigonometric polynomial has period 2π.

We shall denote the set of all integers (positive, zero, and negative) by Z, and put

$$u_n(t) = e^{int} \qquad (n \in Z). \tag{6}$$

If we define the inner product in $L^2(T)$ by

$$(f, g) = \frac{1}{2\pi} \int_{-\pi}^{\pi} f(t)\overline{g(t)}\, dt \tag{7}$$

[note that this is in agreement with (2)], an easy computation shows that

$$(u_n, u_m) = \frac{1}{2\pi} \int_{-\pi}^{\pi} e^{i(n-m)t}\, dt = \begin{cases} 1 & \text{if } n = m, \\ 0 & \text{if } n \neq m. \end{cases} \tag{8}$$

Thus $\{u_n : n \in Z\}$ is an orthonormal set in $L^2(T)$, usually called the *trigonometric system*. We shall now prove that this system is maximal, and shall then derive concrete versions of the abstract theorems previously obtained in the Hilbert space context.

4.24 The Completeness of the Trigonometric System Theorem 4.18 shows that the maximality (or completeness) of the trigonometric system will be proved as soon as we can show that the set of all trigonometric polynomials is dense in $L^2(T)$. Since $C(T)$ is dense in $L^2(T)$, by Theorem 3.14 (note that T is compact), it suffices to show that to every $f \in C(T)$ and to every $\epsilon > 0$ there is a trigonometric polynomial P such that $\|f - P\|_2 < \epsilon$. Since $\|g\|_2 \leq \|g\|_\infty$ for every $g \in C(T)$, the estimate $\|f - P\|_2 < \epsilon$ will follow from $\|f - P\|_\infty < \epsilon$, and it is this estimate which we shall prove.

Suppose we had trigonometric polynomials $Q_1, Q_2, Q_3, \ldots$, with the following properties:

(a) $$Q_k(t) \geq 0 \text{ for } t \in R^1.$$

(b) $$\frac{1}{2\pi} \int_{-\pi}^{\pi} Q_k(t)\, dt = 1.$$

(c) If $\eta_k(\delta) = \sup \{Q_k(t) : \delta \leq |t| \leq \pi\}$, then

$$\lim_{k \to \infty} \eta_k(\delta) = 0$$

for every $\delta > 0$.

Another way of stating (c) is to say: for every $\delta > 0$, $Q_k(t) \to 0$ uniformly on $[-\pi, -\delta] \cup [\delta, \pi]$.

To each $f \in C(T)$ we associate the functions P_k defined by

$$P_k(t) = \frac{1}{2\pi} \int_{-\pi}^{\pi} f(t - s)Q_k(s)\, ds \qquad (k = 1, 2, 3, \ldots). \tag{1}$$

If we replace s by $-s$ (using Theorem 2.20(e)) and then by $s - t$, the periodicity of f and Q_k shows that the value of the integral is not affected. Hence

$$P_k(t) = \frac{1}{2\pi} \int_{-\pi}^{\pi} f(s) Q_k(t - s)\, ds \qquad (k = 1, 2, 3, \ldots). \tag{2}$$

Since each Q_k is a trigonometric polynomial, Q_k is of the form

$$Q_k(t) = \sum_{n=-N_k}^{N_k} a_{n,k} e^{int}, \tag{3}$$

and if we replace t by $t - s$ in (3) and substitute the result in (2), we see that each P_k is a trigonometric polynomial.

Let $\epsilon > 0$ be given. Since f is uniformly continuous on T, there exists a $\delta > 0$ such that $|f(t) - f(s)| < \epsilon$ whenever $|t - s| < \delta$. By (b), we have

$$P_k(t) - f(t) = \frac{1}{2\pi} \int_{-\pi}^{\pi} \{f(t - s) - f(t)\} Q_k(s)\, ds,$$

and (a) implies, for all t, that

$$|P_k(t) - f(t)| \le \frac{1}{2\pi} \int_{-\pi}^{\pi} |f(t - s) - f(t)|\, Q_k(s)\, ds = A_1 + A_2,$$

where A_1 is the integral over $[-\delta, \delta]$ and A_2 is the integral over $[-\pi, -\delta] \cup [\delta, \pi]$. In A_1, the integrand is less than $\epsilon Q_k(s)$, so $A_1 < \epsilon$, by (b). In A_2, we have $Q_k(s) \le \eta_k(\delta)$, hence

$$A_2 \le 2\|f\|_\infty \cdot \eta_k(\delta) < \epsilon \tag{4}$$

for sufficiently large k, by (c). Since these estimates are independent of t, we have proved that

$$\lim_{k \to \infty} \|f - P_k\|_\infty = 0. \tag{5}$$

It remains to construct the Q_k. This can be done in many ways. Here is a simple one. Put

$$Q_k(t) = c_k \left\{ \frac{1 + \cos t}{2} \right\}^k, \tag{6}$$

where c_k is chosen so that (b) holds. Since (a) is clear, we only need to show (c). Since Q_k is even, (b) shows that

$$1 = \frac{c_k}{\pi} \int_0^{\pi} \left\{ \frac{1 + \cos t}{2} \right\}^k dt > \frac{c_k}{\pi} \int_0^{\pi} \left\{ \frac{1 + \cos t}{2} \right\}^k \sin t\, dt = \frac{2c_k}{\pi(k + 1)}.$$

Since Q_k is decreasing on $[0, \pi]$, it follows that

$$Q_k(t) \le Q_k(\delta) \le \frac{\pi(k+1)}{2} \left(\frac{1 + \cos \delta}{2} \right)^k \qquad (0 < \delta \le |t| \le \pi). \tag{7}$$

This implies (c), since $1 + \cos \delta < 2$ if $0 < \delta \leq \pi$.

We have proved the following important result:

4.25 Theorem *If $f \in C(T)$ and $\epsilon > 0$, there is a trigonometric polynomial P such that*

$$|f(t) - P(t)| < \epsilon$$

for every real t.

A more precise result was proved by Fejér (1904): *The arithmetic means of the partial sums of the Fourier series of any $f \in C(T)$ converge uniformly to f.* For a proof (quite similar to the above) see Theorem 3.1 of [45], or p. 89 of [36], vol. I.

4.26 Fourier Series For any $f \in L^1(T)$, we define the *Fourier coefficients* of f by the formula

$$\hat{f}(n) = \frac{1}{2\pi} \int_{-\pi}^{\pi} f(t)e^{-int}\, dt \qquad (n \in Z), \tag{1}$$

where, we recall, Z is the set of all integers. We thus associate with each $f \in L^1(T)$ a function $\hat{f}$ on Z. The *Fourier series* of f is

$$\sum_{-\infty}^{\infty} \hat{f}(n)e^{int}, \tag{2}$$

and its *partial sums* are

$$s_N(t) = \sum_{-N}^{N} \hat{f}(n)e^{int} \qquad (N = 0, 1, 2, \ldots). \tag{3}$$

Since $L^2(T) \subset L^1(T)$, (1) can be applied to every $f \in L^2(T)$. Comparing the definitions made in Secs. 4.23 and 4.13, we can now restate Theorems 4.17 and 4.18 in concrete terms:

The *Riesz-Fischer theorem* asserts that if $\{c_n\}$ is a sequence of complex numbers such that

$$\sum_{n=-\infty}^{\infty} |c_n|^2 < \infty, \tag{4}$$

then there exists an $f \in L^2(T)$ such that

$$c_n = \frac{1}{2\pi} \int_{-\pi}^{\pi} f(t)e^{-int}\, dt \qquad (n \in Z). \tag{5}$$

The *Parseval theorem* asserts that

$$\sum_{n=-\infty}^{\infty} \hat{f}(n)\overline{\hat{g}(n)} = \frac{1}{2\pi} \int_{-\pi}^{\pi} f(t)\overline{g(t)}\, dt \tag{6}$$

whenever $f \in L^2(T)$ and $g \in L^2(T)$; the series on the left of (6) converges absolutely; and if s_N is as in (3), then

$$\lim_{N \to \infty} \|f - s_N\|_2 = 0, \tag{7}$$

since a special case of (6) yields

$$\|f - s_N\|_2^2 = \sum_{|n| > N} |\hat{f}(n)|^2. \tag{8}$$

Note that (7) says that every $f \in L^2(T)$ is the L^2-limit of the partial sums of its Fourier series; i.e., the Fourier series of f converges to f, in the L^2-sense. Pointwise convergence presents a more delicate problem, as we shall see in Chap. 5.

The Riesz-Fischer theorem and the Parseval theorem may be summarized by saying that the mapping $f \leftrightarrow \hat{f}$ is a Hilbert space isomorphism of $L^2(T)$ onto $\ell^2(Z)$.

The theory of Fourier series in other function spaces, for instance in $L^1(T)$, is much more difficult than in $L^2(T)$, and we shall touch only a few aspects of it.

Observe that the crucial ingredient in the proof of the Riesz-Fischer theorem is the fact that L^2 is complete. This is so well recognized that the name "Riesz-Fischer theorem" is sometimes given to the theorem which asserts the completeness of L^2, or even of any L^p.

Exercises

In this set of exercises, H always denotes a Hilbert space.

1 If M is a closed subspace of H, prove that $M = (M^\perp)^\perp$. Is there a similar true statement for subspaces M which are not necessarily closed?

2 Let $\{x_n : n = 1, 2, 3, \ldots\}$ be a linearly independent set of vectors in H. Show that the following construction yields an orthonormal set $\{u_n\}$ such that $\{x_1, \ldots, x_N\}$ and $\{u_1, \ldots, u_N\}$ have the same span for all N.

Put $u_1 = x_1/\|x_1\|$. Having $u_1, \ldots, u_{n-1}$ define

$$v_n = x_n - \sum_{i=1}^{n-1} (x_n, u_i) u_i, \qquad u_n = v_n/\|v_n\|.$$

Note that this leads to a proof of the existence of a maximal orthonormal set in separable Hilbert spaces which makes no appeal to the Hausdorff maximality principle. (A space is *separable* if it contains a countable dense subset.)

3 Show that $L^p(T)$ is separable if $1 \leq p < \infty$, but that $L^\infty(T)$ is not separable.

4 Show that H is separable if and only if H contains a maximal orthonormal system which is at most countable.

5 If $M = \{x : Lx = 0\}$, where L is a continuous linear functional on H, prove that $M^\perp$ is a vector space of dimension 1 (unless $M = H$).

6 Let $\{u_n\}$ ($n = 1, 2, 3, \ldots$) be an orthonormal set in H. Show that this gives an example of a closed and bounded set which is not compact. Let Q be the set of all $x \in H$ of the form

$$x = \sum_1^\infty c_n u_n \qquad \left(\text{where } |c_n| \leq \frac{1}{n}\right).$$

Prove that Q is compact. (Q is called the Hilbert cube.)

More generally, let $\{\delta_n\}$ be a sequence of positive numbers, and let S be the set of all $x \in H$ of the form

$$x = \sum_1^\infty c_n u_n \quad \text{(where } |c_n| \le \delta_n \text{)}.$$

Prove that S is compact if and only if $\sum_1^\infty \delta_n^2 < \infty$.

Prove that H is not locally compact.

7 Suppose $\{a_n\}$ is a sequence of positive numbers such that $\sum a_n b_n < \infty$ whenever $b_n \ge 0$ and $\sum b_n^2 < \infty$. Prove that $\sum a_n^2 < \infty$.

Suggestion: If $\sum a_n^2 = \infty$ then there are disjoint sets E_k ($k = 1, 2, 3, \ldots$) so that

$$\sum_{n \in E_k} a_n^2 > 1.$$

Define b_n so that $b_n = c_k a_n$ for $n \in E_k$. For suitably chosen c_k, $\sum a_n b_n = \infty$ although $\sum b_n^2 < \infty$.

8 If H_1 and H_2 are two Hilbert spaces, prove that one of them is isomorphic to a subspace of the other. (Note that every closed subspace of a Hilbert space is a Hilbert space.)

9 If $A \subset [0, 2\pi]$ and A is measurable, prove that

$$\lim_{n \to \infty} \int_A \cos nx \, dx = \lim_{n \to \infty} \int_A \sin nx \, dx = 0.$$

10 Let $n_1 < n_2 < n_3 < \cdots$ be positive integers, and let E be the set of all $x \in [0, 2\pi]$ at which $\{\sin n_k x\}$ converges. Prove that $m(E) = 0$. *Hint:* $2 \sin^2 \alpha = 1 - \cos 2\alpha$, so $\sin n_k x \to \pm 1/\sqrt{2}$ a.e. on E, by Exercise 9.

11 Find a nonempty closed set E in $L^2(T)$ that contains no element of smallest norm.

12 The constants c_k in Sec. 4.24 were shown to be such that $k^{-1} c_k$ is bounded. Estimate the relevant integral more precisely and show that

$$0 < \lim_{k \to \infty} k^{-1/2} c_k < \infty.$$

13 Suppose f is a continuous function on R^1, with period 1. Prove that

$$\lim_{N \to \infty} \frac{1}{N} \sum_{n=1}^N f(n\alpha) = \int_0^1 f(t) \, dt$$

for every irrational real number α. *Hint:* Do it first for

$$f(t) = \exp(2\pi i k t), \quad k = 0, \pm 1, \pm 2, \ldots.$$

14 Compute

$$\min_{a, b, c} \int_{-1}^1 |x^3 - a - bx - cx^2|^2 \, dx$$

and find

$$\max \int_{-1}^1 x^3 g(x) \, dx,$$

where g is subject to the restrictions

$$\int_{-1}^1 g(x) \, dx = \int_{-1}^1 x g(x) \, dx = \int_{-1}^1 x^2 g(x) \, dx = 0; \quad \int_{-1}^1 |g(x)|^2 \, dx = 1.$$

15 Compute
$$\min_{a,b,c} \int_0^\infty |x^3 - a - bx - cx^2|^2 e^{-x}\, dx.$$
State and solve the corresponding maximum problem, as in Exercise 14.

16 If $x_0 \in H$ and M is a closed linear subspace of H, prove that
$$\min\{\|x - x_0\| : x \in M\} = \max\{|(x_0, y)| : y \in M^\perp, \|y\| = 1\}.$$

17 Show that there is a continuous one-to-one mapping γ of $[0, 1]$ into H such that $\gamma(b) - \gamma(a)$ is orthogonal to $\gamma(d) - \gamma(c)$ whenever $0 \le a \le b \le c \le d \le 1$. ($\gamma$ may be called a "curve with orthogonal increments.") *Hint*: Take $H = L^2$, and consider characteristic functions of certain subsets of $[0, 1]$.

18 Define $u_s(t) = e^{ist}$ for all $s \in R^1$, $t \in R^1$. Let X be the complex vector space consisting of all finite linear combinations of these functions u_s. If $f \in X$ and $g \in X$, show that
$$(f, g) = \lim_{A \to \infty} \frac{1}{2A} \int_{-A}^{A} f(t)\overline{g(t)}\, dt$$
exists. Show that this inner product makes X into a unitary space whose completion is a nonseparable Hilbert space H. Show also that $\{u_s : s \in R^1\}$ is a maximal orthonormal set in H.

19 Fix a positive integer N, put $\omega = e^{2\pi i/N}$, prove the orthogonality relations
$$\frac{1}{N} \sum_{n=1}^N \omega^{nk} = \begin{cases} 1 & \text{if } k = 0 \\ 0 & \text{if } 1 \le k \le N-1 \end{cases}$$
and use them to derive the identities
$$(x, y) = \frac{1}{N} \sum_{n=1}^N \|x + \omega^n y\|^2 \omega^n$$
that hold in every inner product space if $N \ge 3$. Show also that
$$(x, y) = \frac{1}{2\pi} \int_{-\pi}^{\pi} \|x + e^{i\theta} y\|^2 e^{i\theta}\, d\theta.$$

CHAPTER
FIVE

EXAMPLES OF BANACH SPACE TECHNIQUES

Banach Spaces

5.1 In the preceding chapter we saw how certain analytic facts about trigonometric series can be made to emerge from essentially goemetric considerations about general Hilbert spaces, involving the notions of convexity, subspaces, orthogonality, and completeness. There are many problems in analysis that can be attacked with greater ease when they are placed within a suitably chosen abstract framework. The theory of Hilbert spaces is not always suitable since orthogonality is something rather special. The class of all Banach spaces affords greater variety. In this chapter we shall develop some of the basic properties of Banach spaces and illustrate them by applications to concrete problems.

5.2 Definition A complex vector space X is said to be a *normed linear space* if to each $x \in X$ there is associated a nonnegative real number $\|x\|$, called the *norm* of x, such that

(a) $\|x + y\| \le \|x\| + \|y\|$ for all x and $y \in X$,
(b) $\|\alpha x\| = |\alpha| \|x\|$ if $x \in X$ and α is a scalar,
(c) $\|x\| = 0$ implies $x = 0$.

By (a), the triangle inequality

$$\|x - y\| \le \|x - z\| + \|z - y\| \qquad (x, y, z \in X)$$

holds. Combined with (b) (take $\alpha = 0$, $\alpha = -1$) and (c) this shows that every normed linear space may be regarded as a metric space, the distance between x and y being $\|x - y\|$.

A *Banach space* is a normed linear space which is *complete* in the metric defined by its norm.

For instance, every Hilbert space is a Banach space, so is every $L^p(\mu)$ normed by $\|f\|_p$ (provided we identify functions which are equal a.e.) if $1 \leq p \leq \infty$, and so is $C_0(X)$ with the supremum norm. The simplest Banach space is of course the complex field itself, normed by $\|x\| = |x|$.

One can equally well discuss *real* Banach spaces; the definition is exactly the same, except that all scalars are assumed to be real.

5.3 Definition Consider a linear transformation Λ from a normed linear space X into a normed linear space Y, and define its *norm* by

$$\|\Lambda\| = \sup \{\|\Lambda x\| : x \in X, \|x\| \leq 1\}. \tag{1}$$

If $\|\Lambda\| < \infty$, then Λ is called a *bounded linear transformation*.

In (1), $\|x\|$ is the norm of x in X, $\|\Lambda x\|$ is the norm of Λx in Y; it will frequently happen that several norms occur together, and the context will make it clear which is which.

Observe that we could restrict ourselves to *unit vectors* x in (1), i.e., to x with $\|x\| = 1$, without changing the supremum, since

$$\|\Lambda(\alpha x)\| = \|\alpha \Lambda x\| = |\alpha| \|\Lambda x\|. \tag{2}$$

Observe also that $\|\Lambda\|$ is the smallest number such that the inequality

$$\|\Lambda x\| \leq \|\Lambda\| \|x\| \tag{3}$$

holds for *every* $x \in X$.

The following geometric picture is helpful: Λ maps the *closed unit ball* in X, i.e., the set

$$\{x \in X : \|x\| \leq 1\}, \tag{4}$$

into the closed ball in Y with center at 0 and radius $\|\Lambda\|$.

An important special case is obtained by taking the complex field for Y; in that case we talk about *bounded linear functionals*.

5.4 Theorem *For a linear transformation Λ of a normed linear space X into a normed linear space Y, each of the following three conditions implies the other two:*

(a) Λ *is bounded*.
(b) Λ *is continuous*.
(c) Λ *is continuous at one point of X.*

PROOF Since $\|\Lambda(x_1 - x_2)\| \leq \|\Lambda\| \|x_1 - x_2\|$, it is clear that (a) implies (b), and (b) implies (c) trivially. Suppose Λ is continuous at x_0. To each $\epsilon > 0$ one can then find a $\delta > 0$ so that $\|x - x_0\| < \delta$ implies $\|\Lambda x - \Lambda x_0\| < \epsilon$. In other words, $\|x\| < \delta$ implies

$$\|\Lambda(x_0 + x) - \Lambda x_0\| < \epsilon.$$

But then the linearity of Λ shows that $\|\Lambda x\| < \epsilon$. Hence $\|\Lambda\| \le \epsilon/\delta$, and (c) implies (a). ////

Consequences of Baire's Theorem

5.5 The manner in which the completeness of Banach spaces is frequently exploited depends on the following theorem about complete metric spaces, which also has many applications in other parts of mathematics. It implies two of the three most important theorems which make Banach spaces useful tools in analysis, the *Banach-Steinhaus theorem* and the *open mapping theorem*. The third is the *Hahn-Banach extension theorem*, in which completeness plays no role.

5.6 Baire's Theorem *If X is a complete metric space, the intersection of every countable collection of dense open subsets of X is dense in X.*

In particular (except in the trivial case $X = \emptyset$), the intersection is not empty. This is often the principal significance of the theorem.

PROOF Suppose $V_1, V_2, V_3, \ldots$ are dense and open in X. Let W be any open set in X. We have to show that $\bigcap V_n$ has a point in W if $W \ne \emptyset$.

Let ρ be the metric of X; let us write

$$S(x, r) = \{y \in X : \rho(x, y) < r\} \tag{1}$$

and let $\bar{S}(x, r)$ be the closure of $S(x, r)$. [*Note*: There exist situations in which $\bar{S}(x, r)$ does *not* contain all y with $\rho(x, y) \le r$!]

Since V_1 is dense, $W \cap V_1$ is a nonempty open set, and we can therefore find x_1 and r_1 such that

$$\bar{S}(x_1, r_1) \subset W \cap V_1 \quad \text{and} \quad 0 < r_1 < 1. \tag{2}$$

If $n \ge 2$ and x_{n-1} and r_{n-1} are chosen, the denseness of V_n shows that $V_n \cap S(x_{n-1}, r_{n-1})$ is not empty, and we can therefore find x_n and r_n such that

$$\bar{S}(x_n, r_n) \subset V_n \cap S(x_{n-1}, r_{n-1}) \quad \text{and} \quad 0 < r_n < \frac{1}{n}. \tag{3}$$

By induction, this process produces a sequence $\{x_n\}$ in X. If $i > n$ and $j > n$, the construction shows that x_i and x_j both lie in $S(x_n, r_n)$, so that $\rho(x_i, x_j) < 2r_n < 2/n$, and hence $\{x_n\}$ is a Cauchy sequence. Since X is complete, there is a point $x \in X$ such that $x = \lim_{n \to \infty} x_n$.

Since x_i lies in the closed set $\bar{S}(x_n, r_n)$ if $i > n$, it follows that x lies in each $\bar{S}(x_n, r_n)$, and (3) shows that x lies in each V_n. By (2), $x \in W$. This completes the proof. ////

Corollary *In a complete metric space, the intersection of any countable collection of dense G_δ's is again a dense G_δ.*

This follows from the theorem, since every G_δ is the intersection of a countable collection of open sets, and since the union of countably many countable sets is countable.

5.7 Baire's theorem is sometimes called the *category theorem*, for the following reason.

Call a set $E \subset X$ *nowhere dense* if its closure $\bar{E}$ contains no nonempty open subset of X. Any countable union of nowhere dense sets is called a set of the *first category*; all other subsets of X are of the *second category* (Baire's terminology). Theorem 5.6 is equivalent to the statement that *no complete metric space is of the first category*. To see this, just take complements in the statement of Theorem 5.6.

5.8 The Banach-Steinhaus Theorem *Suppose X is a Banach space, Y is a normed linear space, and $\{\Lambda_\alpha\}$ is a collection of bounded linear transformations of X into Y, where α ranges over some index set A. Then either there exists an $M < \infty$ such that*

$$\|\Lambda_\alpha\| \leq M \tag{1}$$

for every $\alpha \in A$, or

$$\sup_{\alpha \in A} \|\Lambda_\alpha x\| = \infty \tag{2}$$

for all x belonging to some dense G_δ in X.

In geometric terminology, the alternatives are as follows: *Either* there is a ball B in Y (with radius M and center at 0) such that every Λ_α maps the unit ball of X into B, *or* there exist $x \in X$ (in fact, a whole dense G_δ of them) such that no ball in Y contains $\Lambda_\alpha x$ for all α.

The theorem is sometimes referred to as the *uniform boundedness principle*.

PROOF Put

$$\varphi(x) = \sup_{\alpha \in A} \|\Lambda_\alpha x\| \qquad (x \in X) \tag{3}$$

and let

$$V_n = \{x \colon \varphi(x) > n\} \qquad (n = 1, 2, 3, \ldots). \tag{4}$$

Since each Λ_α is continuous and since the norm of Y is a continuous function on Y (an immediate consequence of the triangle inequality, as in the proof of Theorem 4.6), each function $x \to \|\Lambda_\alpha x\|$ is continuous on X. Hence φ is lower semicontinuous, and each V_n is open.

If one of these sets, say V_N, fails to be dense in X, then there exist an $x_0 \in X$ and an $r > 0$ such that $\|x\| \leq r$ implies $x_0 + x \notin V_N$; this means that $\varphi(x_0 + x) \leq N$, or

$$\|\Lambda_\alpha(x_0 + x)\| \leq N \tag{5}$$

for all $\alpha \in A$ and all x with $\|x\| \leq r$. Since $x = (x_0 + x) - x_0$, we then have

$$\|\Lambda_\alpha x\| \leq \|\Lambda_\alpha(x_0 + x)\| + \|\Lambda_\alpha x_0\| \leq 2N, \tag{6}$$

and it follows that (1) holds with $M = 2N/r$.

The other possibility is that every V_n is dense in X. In that case, $\bigcap V_n$ is a dense G_δ in X, by Baire's theorem; and since $\varphi(x) = \infty$ for every $x \in \bigcap V_n$, the proof is complete. ////

5.9 The Open Mapping Theorem *Let U and V be the open unit balls of the Banach spaces X and Y. To every bounded linear transformation Λ of X onto Y there corresponds a $\delta > 0$ so that*

$$\Lambda(U) \supset \delta V. \tag{1}$$

Note the word "onto" in the hypothesis. The symbol δV stands for the set $\{\delta y : y \in V\}$, i.e., the set of all $y \in Y$ with $\|y\| < \delta$.

It follows from (1) and the linearity of Λ that the image of every open ball in X, with center at x_0, say, contains an open ball in Y with center at Λx_0. Hence the image of every open set is open. This explains the name of the theorem.

Here is another way of stating (1): *To every y with $\|y\| < \delta$ there corresponds an x with $\|x\| < 1$ so that $\Lambda x = y$.*

PROOF Given $y \in Y$, there exists an $x \in X$ such that $\Lambda x = y$; if $\|x\| < k$, it follows that $y \in \Lambda(kU)$. Hence Y is the union of the sets $\Lambda(kU)$, for $k = 1, 2, 3, \ldots$. Since Y is complete, the Baire theorem implies that there is a nonempty open set W in the closure of some $\Lambda(kU)$. This means that every point of W is the limit of a sequence $\{\Lambda x_i\}$, where $x_i \in kU$; from now on, k and W are fixed.

Choose $y_0 \in W$, and choose $\eta > 0$ so that $y_0 + y \in W$ if $\|y\| < \eta$. For any such y there are sequences $\{x_i'\}, \{x_i''\}$ in kU such that

$$\Lambda x_i' \to y_0, \quad \Lambda x_i'' \to y_0 + y \quad (i \to \infty). \tag{2}$$

Setting $x_i = x_i'' - x_i'$, we have $\|x_i\| < 2k$ and $\Lambda x_i \to y$. Since this holds for every y with $\|y\| < \eta$, the linearity of Λ shows that the following is true, if $\delta = \eta/2k$:

To each $y \in Y$ and to each $\epsilon > 0$ there corresponds an $x \in X$ such that

$$\|x\| \leq \delta^{-1}\|y\| \quad \text{and} \quad \|y - \Lambda x\| < \epsilon. \tag{3}$$

This is almost the desired conclusion, as stated just before the start of the proof, except that there we had $\epsilon = 0$.

Fix $y \in \delta V$, and fix $\epsilon > 0$. By (3) there exists an x_1 with $\|x_1\| < 1$ and

$$\|y - \Lambda x_1\| < \tfrac{1}{2}\delta\epsilon. \tag{4}$$

Suppose $x_1, \ldots, x_n$ are chosen so that

$$\|y - \Lambda x_1 - \cdots - \Lambda x_n\| < 2^{-n}\delta\epsilon. \tag{5}$$

Use (3), with y replaced by the vector on the left side of (5), to obtain an x_{n+1} so that (5) holds with $n + 1$ in place of n, and

$$\|x_{n+1}\| < 2^{-n}\epsilon \quad (n = 1, 2, 3, \ldots). \tag{6}$$

If we set $s_n = x_1 + \cdots + x_n$, (6) shows that $\{s_n\}$ is a Cauchy sequence in X. Since X is complete, there exists an $x \in X$ so that $s_n \to x$. The inequality $\|x_1\| < 1$, together with (6), shows that $\|x\| < 1 + \epsilon$. Since Λ is continuous, $\Lambda s_n \to \Lambda x$. By (5), $\Lambda s_n \to y$. Hence $\Lambda x = y$.

We have now proved that

$$\Lambda((1 + \epsilon)U) \supset \delta V, \tag{7}$$

or

$$\Lambda(U) \supset (1 + \epsilon)^{-1}\delta V, \tag{8}$$

for every $\epsilon > 0$. The union of the sets on the right of (8), taken over all $\epsilon > 0$, is δV. This proves (1). ////

5.10 Theorem *If X and Y are Banach spaces and if Λ is a bounded linear transformation of X onto Y which is also one-to-one, then there is a $\delta > 0$ such that*

$$\|\Lambda x\| \geq \delta \|x\| \quad (x \in X). \tag{1}$$

In other words, Λ^{-1} is a bounded linear transformation of Y onto X.

PROOF If δ is chosen as in the statement of Theorem 5.9, the conclusion of that theorem, combined with the fact that Λ is now one-to-one, shows that $\|\Lambda x\| < \delta$ implies $\|x\| < 1$. Hence $\|x\| \geq 1$ implies $\|\Lambda x\| \geq \delta$, and (1) is proved.

The transformation Λ^{-1} is defined on Y by the requirement that $\Lambda^{-1} y = x$ if $y = \Lambda x$. A trivial verification shows that Λ^{-1} is linear, and (1) implies that $\|\Lambda^{-1}\| \leq 1/\delta$. ////

Fourier Series of Continuous Functions

5.11 A Convergence Problem *Is it true for every $f \in C(T)$ that the Fourier series of f converges to $f(x)$ at every point x?*

Let us recall that the nth partial sum of the Fourier series of f at the point x is given by

$$s_n(f; x) = \frac{1}{2\pi} \int_{-\pi}^{\pi} f(t) D_n(x - t) \, dt \qquad (n = 0, 1, 2, \ldots), \tag{1}$$

where
$$D_n(t) = \sum_{k=-n}^{n} e^{ikt}. \tag{2}$$

This follows directly from formulas 4.26(1) and 4.26(3).

The problem is to determine whether

$$\lim_{n \to \infty} s_n(f; x) = f(x) \tag{3}$$

for every $f \in C(T)$ and for every real x. We observed in Sec. 4.26 that the partial sums do converge to f in the L^2-norm, and Theorem 3.12 implies therefore that each $f \in L^2(T)$ [hence also each $f \in C(T)$] is the pointwise limit a.e. of some subsequence of the full sequence of the partial sums. But this does not answer the present question.

We shall see that the Banach-Steinhaus theorem answers the question *negatively*. Put

$$s^*(f; x) = \sup_n |s_n(f; x)|. \tag{4}$$

To begin with, take $x = 0$, and define

$$\Lambda_n f = s_n(f; 0) \qquad (f \in C(T); n = 1, 2, 3, \ldots). \tag{5}$$

We know that $C(T)$ is a Banach space, relative to the supremum norm $\|f\|_\infty$. It follows from (1) that each Λ_n is a bounded linear functional on $C(T)$, of norm

$$\|\Lambda_n\| \le \frac{1}{2\pi} \int_{-\pi}^{\pi} |D_n(t)| \, dt = \|D_n\|_1. \tag{6}$$

We claim that

$$\|\Lambda_n\| \to \infty \qquad \text{as } n \to \infty. \tag{7}$$

This will be proved by showing that equality holds in (6) and that

$$\|D_n\|_1 \to \infty \qquad \text{as } n \to \infty. \tag{8}$$

Multiply (2) by $e^{it/2}$ and by $e^{-it/2}$ and subtract one of the resulting two equations from the other, to obtain

$$D_n(t) = \frac{\sin(n + \tfrac{1}{2})t}{\sin(t/2)}. \tag{9}$$

Since $|\sin x| \leq |x|$ for all real x, (9) shows that

$$\|D_n\|_1 > \frac{2}{\pi} \int_0^\pi \left|\sin\left(n + \frac{1}{2}\right)t\right| \frac{dt}{t} = \frac{2}{\pi} \int_0^{(n+1/2)\pi} |\sin t| \frac{dt}{t}$$

$$> \frac{2}{\pi} \sum_{k=1}^n \frac{1}{k\pi} \int_{(k-1)\pi}^{k\pi} |\sin t| \, dt = \frac{4}{\pi^2} \sum_{k=1}^n \frac{1}{k} \to \infty,$$

which proves (8).

Next, fix n, and put $g(t) = 1$ if $D_n(t) \geq 0$, $g(t) = -1$ if $D_n(t) < 0$. There exist $f_j \in C(T)$ such that $-1 \leq f_j \leq 1$ and $f_j(t) \to g(t)$ for every t, as $j \to \infty$. By the dominated convergence theorem,

$$\lim_{j \to \infty} \Lambda_n(f_j) = \lim_{j \to \infty} \frac{1}{2\pi} \int_{-\pi}^\pi f_j(-t) D_n(t) \, dt = \frac{1}{2\pi} \int_{-\pi}^\pi g(-t) D_n(t) \, dt = \|D_n\|_1.$$

Thus equality holds in (6), and we have proved (7).

Since (7) holds, the Banach-Steinhaus theorem asserts now that $s^*(f; 0) = \infty$ for every f in some dense G_δ-set in $C(T)$.

We chose $x = 0$ just for convenience. It is clear that the same result holds for every other x:

To each real number x there corresponds a set $E_x \subset C(T)$ which is a dense G_δ in $C(T)$, such that $s^*(f; x) = \infty$ for every $f \in E_x$.

In particular, the Fourier series of each $f \in E_x$ diverges at x, and we have a negative answer to our question. (Exercise 22 shows the answer is positive if mere continuity is replaced by a somewhat stronger smoothness assumption.)

It is interesting to observe that the above result can be strengthened by another application of Baire's theorem. Let us take countably many points x_i, and let E be the intersection of the corresponding sets

$$E_{x_i} \subset C(T).$$

By Baire's theorem, E is a dense G_δ in $C(T)$. Every $f \in E$ has

$$s^*(f; x_i) = \infty$$

at every point x_i.

For each f, $s^*(f; x)$ is a lower semicontinuous function of x, since (4) exhibits it as the supremum of a collection of continuous functions. Hence $\{x : s^*(f; x) = \infty\}$ is a G_δ in R^1, for each f. If the above points x_i are taken so that their union is dense in $(-\pi, \pi)$, we obtain the following result:

5.12 Theorem *There is a set $E \subset C(T)$ which is a dense G_δ in $C(T)$ and which has the following property: For each $f \in E$, the set*

$$Q_f = \{x : s^*(f; x) = \infty\}$$

is a dense G_δ in R^1.

This gains in interest if we realize that E, as well as each Q_f, is an *uncountable* set:

5.13 Theorem *In a complete metric space X which has no isolated points, no countable dense set is a G_δ.*

PROOF Let x_k be the points of a countable dense set E in X. Assume that E is a G_δ. Then $E = \bigcap V_n$, where each V_n is dense and open. Let

$$W_n = V_n - \bigcup_{k=1}^{n} \{x_k\}.$$

Then each W_n is still a dense open set, but $\bigcap W_n = \varnothing$, in contradiction to Baire's theorem. ////

Note: A slight change in the proof of Baire's theorem shows actually that every dense G_δ contains a perfect set if X is as above.

Fourier Coefficients of L^1-functions

5.14 As in Sec. 4.26, we associate to every $f \in L^1(T)$ a function $\hat{f}$ on Z defined by

$$\hat{f}(n) = \frac{1}{2\pi} \int_{-\pi}^{\pi} f(t) e^{-int} \, dt \qquad (n \in Z). \tag{1}$$

It is easy to prove that $\hat{f}(n) \to 0$ as $|n| \to \infty$, for every $f \in L^1$. For we know that $C(T)$ is dense in $L^1(T)$ (Theorem 3.14) and that the trigonometric polynomials are dense in $C(T)$ (Theorem 4.25). If $\epsilon > 0$ and $f \in L^1(T)$, this says that there is a $g \in C(T)$ and a trigonometric polynomial P such that $\|f - g\|_1 < \epsilon$ and $\|g - P\|_\infty < \epsilon$. Since

$$\|g - P\|_1 \leq \|g - P\|_\infty$$

if follows that $\|f - P\|_1 < 2\epsilon$; and if $|n|$ is large enough (depending on P), then

$$|\hat{f}(n)| = \left| \frac{1}{2\pi} \int_{-\pi}^{\pi} \{f(t) - P(t)\} e^{-int} \, dt \right| \leq \|f - P\|_1 < 2\epsilon. \tag{2}$$

Thus $\hat{f}(n) \to 0$ as $n \to \pm\infty$. This is known as the Riemann-Lebesgue lemma.

The question we wish to raise is whether the converse is true. That is to say, if $\{a_n\}$ is a sequence of complex numbers such that $a_n \to 0$ as $n \to \pm\infty$, does it follow that there is an $f \in L^1(T)$ such that $\hat{f}(n) = a_n$ for all $n \in Z$? In other words, is something like the Riesz-Fischer theorem true in this situation?

This can easily be answered (negatively) with the aid of the open mapping theorem.

Let c_0 be the space of all complex functions φ on Z such that $\varphi(n) \to 0$ as $n \to \pm\infty$, with the supremum norm

$$\|\varphi\|_\infty = \sup\{|\varphi(n)|: n \in Z\}. \tag{3}$$

Then c_0 is easily seen to be a Banach space. In fact, if we declare every subset of Z to be open, then Z is a locally compact Hausdorff space, and c_0 is nothing but $C_0(Z)$.

The following theorem contains the answer to our question:

5.15 Theorem *The mapping $f \to \hat{f}$ is a one-to-one bounded linear transformation of $L^1(T)$ into (but not onto) c_0.*

PROOF Define Λ by $\Lambda f = \hat{f}$. It is clear that Λ is linear. We have just proved that Λ maps $L^1(T)$ into c_0, and formula 5.14(1) shows that $|\hat{f}(n)| \le \|f\|_1$, so that $\|\Lambda\| \le 1$. (Actually, $\|\Lambda\| = 1$; to see this, take $f = 1$.) Let us now prove that Λ is one-to-one. Suppose $f \in L^1(T)$ and $\hat{f}(n) = 0$ for every $n \in Z$. Then

$$\int_{-\pi}^{\pi} f(t)g(t)\,dt = 0 \tag{1}$$

if g is any trigonometric polynomial. By Theorem 4.25 and the dominated convergence theorem, (1) holds for every $g \in C(T)$. Apply the dominated convergence theorem once more, in conjunction with the Corollary to Lusin's theorem, to conclude that (1) holds if g is the characteristic function of any measurable set in T. Now Theorem 1.39(b) shows that $f = 0$ a.e.

If the range of Λ were all of c_0, Theorem 5.10 would imply the existence of a $\delta > 0$ such that

$$\|\hat{f}\|_\infty \ge \delta \|f\|_1 \tag{2}$$

for every $f \in L^1(T)$. But if $D_n(t)$ is defined as in Sec. 5.11, then $D_n \in L^1(T)$, $\|\hat{D}_n\|_\infty = 1$ for $n = 1, 2, 3, \ldots$, and $\|D_n\|_1 \to \infty$ as $n \to \infty$. Hence there is no $\delta > 0$ such that the inequalities

$$\|\hat{D}_n\|_\infty \ge \delta \|D_n\|_1 \tag{3}$$

hold for every n.

This completes the proof. ////

The Hahn-Banach Theorem

5.16 Theorem *If M is a subspace of a normed linear space X and if f is a bounded linear functional on M, then f can be extended to a bounded linear functional F on X so that $\|F\| = \|f\|$.*

Note that M need not be closed.

Before we turn to the proof, some comments seem called for. First, to say (in the most general situation) that a function F is an *extension* of f means that the domain of F includes that of f and that $F(x) = f(x)$ for all x in the domain of f. Second, the norms $\|F\|$ and $\|f\|$ are computed relative to the domains of F and f; explicitly,

$$\|f\| = \sup\{|f(x)|: x \in M, \|x\| \leq 1\}, \qquad \|F\| = \sup\{|F(x)|: x \in X, \|x\| \leq 1\}.$$

The third comment concerns the field of scalars. So far everything has been stated for complex scalars, but the complex field could have been replaced by the real field without any changes in statements or proofs. The Hahn-Banach theorem is also true in both cases; nevertheless, it appears to be essentially a "real" theorem. The fact that the complex case was not yet proved when Banach wrote his classical book "Opérations linéaires" may be the main reason that real scalars are the only ones considered in his work.

It will be helpful to introduce some temporary terminology. Recall that V is a complex (real) vector space if $x + y \in V$ for x and $y \in V$, and if $\alpha x \in V$ for all complex (real) numbers α. It follows trivially that *every complex vector space is also a real vector space*. A complex function φ on a complex vector space V is a *complex-linear functional* if

$$\varphi(x + y) = \varphi(x) + \varphi(y) \quad \text{and} \quad \varphi(\alpha x) = \alpha \varphi(x) \tag{1}$$

for all x and $y \in V$ and all *complex* α. A real-valued function φ on a complex (real) vector space V is a *real-linear functional* if (1) holds for all *real* α.

If u is the real part of a complex-linear functional f, i.e., if $u(x)$ is the real part of the complex number $f(x)$ for all $x \in V$, it is easily seen that u is a real-linear functional. The following relations hold between f and u:

5.17 Proposition *Let V be a complex vector space.*

(a) *If u is the real part of a complex-linear functional f on V, then*

$$f(x) = u(x) - iu(ix) \qquad (x \in V). \tag{1}$$

(b) *If u is a real-linear functional on V and if f is defined by (1), then f is a complex-linear functional on V.*

(c) *If V is a normed linear space and f and u are related as in (1), then $\|f\| = \|u\|$.*

PROOF If α and β are real numbers and $z = \alpha + i\beta$, the real part of iz is $-\beta$. This gives the identity

$$z = \operatorname{Re} z - i \operatorname{Re}(iz) \tag{2}$$

for all complex numbers z. Since

$$\operatorname{Re}(if(x)) = \operatorname{Re} f(ix) = u(ix), \tag{3}$$

(1) follows from (2) with $z = f(x)$.

Under the hypotheses (b), it is clear that $f(x+y) = f(x) + f(y)$ and that $f(\alpha x) = \alpha f(x)$ for all real α. But we also have

$$f(ix) = u(ix) - iu(-x) = u(ix) + iu(x) = if(x), \tag{4}$$

which proves that f is complex-linear.

Since $|u(x)| \leq |f(x)|$, we have $\|u\| \leq \|f\|$. On the other hand, to every $x \in V$ there corresponds a complex number α, $|\alpha| = 1$, so that $\alpha f(x) = |f(x)|$. Then

$$|f(x)| = f(\alpha x) = u(\alpha x) \leq \|u\| \cdot \|\alpha x\| = \|u\| \cdot \|x\|, \tag{5}$$

which proves that $\|f\| \leq \|u\|$. ////

5.18 Proof of Theorem 5.16 We first assume that X is a real normed linear space and, consequently, that f is a real-linear bounded functional on M. If $\|f\| = 0$, the desired extension is $F = 0$. Omitting this case, there is no loss of generality in assuming that $\|f\| = 1$.

Choose $x_0 \in X$, $x_0 \notin M$, and let M_1 be the vector space spanned by M and x_0. Then M_1 consists of all vectors of the form $x + \lambda x_0$, where $x \in M$ and λ is a real scalar. If we define $f_1(x + \lambda x_0) = f(x) + \lambda \alpha$, where α is any fixed real number, it is trivial to verify that an extension of f to a linear functional on M_1 is obtained. The problem is to choose α so that the extended functional still has norm 1. This will be the case provided that

$$|f(x) + \lambda \alpha| \leq \|x + \lambda x_0\| \quad (x \in M, \lambda \text{ real}). \tag{1}$$

Replace x by $-\lambda x$ and divide both sides of (1) by $|\lambda|$. The requirement is then that

$$|f(x) - \alpha| \leq \|x - x_0\| \quad (x \in M), \tag{2}$$

i.e., that $A_x \leq \alpha \leq B_x$ for all $x \in M$, where

$$A_x = f(x) - \|x - x_0\| \quad \text{and} \quad B_x = f(x) + \|x - x_0\|. \tag{3}$$

There exists such an α if and only if all the intervals $[A_x, B_x]$ have a common point, i.e., if and only if

$$A_x \leq B_y \tag{4}$$

for all x and $y \in M$. But

$$f(x) - f(y) = f(x - y) \leq \|x - y\| \leq \|x - x_0\| + \|y - x_0\|, \tag{5}$$

and so (4) follows from (3).

We have now proved that there exists a norm-preserving extension f_1 of f on M_1.

Let $\mathscr{P}$ be the collection of all ordered pairs (M', f'), where M' is a subspace of X which contains M and where f' is a real-linear extension of f to M', with $\|f'\| = 1$. Partially order $\mathscr{P}$ by declaring $(M', f') \leq (M'', f'')$ to mean that $M' \subset M''$ and $f''(x) = f'(x)$ for all $x \in M'$. The axioms of a partial order are clearly satisfied, $\mathscr{P}$ is not empty since it contains (M, f), and so the Hausdorff maximality theorem asserts the existence of a maximal totally ordered subcollection Ω of $\mathscr{P}$.

Let Φ be the collection of all M' such that $(M', f') \in \Omega$. Then Φ is totally ordered, by set inclusion, and *therefore* the union $\tilde{M}$ of all members of Φ is a subspace of X. (Note that in general the union of two subspaces is not a subspace. An example is two planes through the origin in R^3.) If $x \in \tilde{M}$, then $x \in M'$ for some $M' \in \Phi$; define $F(x) = f'(x)$, where f' is the function which occurs in the pair $(M', f') \in \Omega$. Our definition of the partial order in Ω shows that it is immaterial which $M' \in \Phi$ we choose to define $F(x)$, as long as M' contains x.

It is now easy to check that F is a linear functional on $\tilde{M}$, with $\|F\| = 1$. If $\tilde{M}$ were a proper subspace X, the first part of the proof would give us a further extension of F, and this would contradict the maximality of Ω. Thus $\tilde{M} = X$, and the proof is complete for the case of real scalars.

If now f is a complex-linear functional on the subspace M of the complex normed linear space X, let u be the real part of f, use the real Hahn-Banach theorem to extend u to a real-linear functional U on X, with $\|U\| = \|u\|$, and define

$$F(x) = U(x) - iU(ix) \qquad (x \in X). \tag{6}$$

By Proposition 5.17, F is a complex-linear extension of f, and

$$\|F\| = \|U\| = \|u\| = \|f\|.$$

This completes the proof. ////

Let us mention two important consequences of the Hahn-Banach theorem:

5.19 Theorem *Let M be a linear subspace of a normed linear space X, and let $x_0 \in X$. Then x_0 is in the closure $\bar{M}$ of M if and only if there is no bounded linear functional f on X such that $f(x) = 0$ for all $x \in M$ but $f(x_0) \neq 0$.*

PROOF If $x_0 \in \bar{M}$, f is a bounded linear functional on X, and $f(x) = 0$ for all $x \in M$, the continuity of f shows that we also have $f(x_0) = 0$.

Conversely, suppose $x_0 \notin \bar{M}$. Then there exists a $\delta > 0$ such that $\|x - x_0\| > \delta$ for all $x \in M$. Let M' be the subspace generated by M and x_0, and define $f(x + \lambda x_0) = \lambda$ if $x \in M$ and λ is a scalar. Since

$$\delta |\lambda| \leq |\lambda| \|x_0 + \lambda^{-1} x\| = \|\lambda x_0 + x\|,$$

we see that f is a linear functional on M' whose norm is at most δ^{-1}. Also $f(x) = 0$ on M, $f(x_0) = 1$. The Hahn-Banach theorem allows us to extend this f from M' to X. ////

5.20 Theorem *If X is a normed linear space and if $x_0 \in X$, $x_0 \neq 0$, there is a bounded linear functional f on X, of norm 1, so that $f(x_0) = \|x_0\|$.*

PROOF Let $M = \{\lambda x_0\}$, and define $f(\lambda x_0) = \lambda \|x_0\|$. Then f is a linear functional of norm 1 on M, and the Hahn-Banach theorem can again be applied.   ////

5.21 Remarks If X is a normed linear space, let X^* be the collection of all bounded linear functionals on X. If addition and scalar multiplication of linear functionals are defined in the obvious manner, it is easy to see that X^* is again a normed linear space. In fact, X^* is a Banach space; this follows from the fact that the field of scalars is a complete metric space. We leave the verification of these properties of X^* as an exercise.

One of the consequences of Theorem 5.20 is that X^* is not the trivial vector space (i.e., X^* consists of more than 0) if X is not trivial. In fact, X^* *separates points on* X. This means that if $x_1 \neq x_2$ in X there exists an $f \in X^*$ such that $f(x_1) \neq f(x_2)$. To prove this, merely take $x_0 = x_2 - x_1$ in Theorem 5.20.

Another consequence is that, for $x \in X$,

$$\|x\| = \sup\{|f(x)| : f \in X^*, \|f\| = 1\}.$$

Hence, for fixed $x \in X$, the mapping $f \to f(x)$ is a bounded linear functional on X^*, of norm $\|x\|$.

This interplay between X and X^* (the so-called "dual space" of X) forms the basis of a large portion of that part of mathematics which is known as *functional analysis*.

An Abstract Approach to the Poisson Integral

5.22 Successful applications of the Hahn-Banach theorem to concrete problems depend of course on a knowledge of the bounded linear functionals on the normed linear space under consideration. So far we have only determined the bounded linear functionals on a Hilbert space (where a much simpler proof of the Hahn-Banach theorem exists; see Exercise 6), and we know the positive linear functionals on $C_c(X)$.

We shall now describe a general situation in which the last-mentioned functionals occur naturally.

Let K be a compact Hausdorff space, let H be a compact subset of K, and let A be a subspace of $C(K)$ such that $1 \in A$ (1 denotes the function which assigns the number 1 to each $x \in K$) and such that

$$\|f\|_K = \|f\|_H \qquad (f \in A). \tag{1}$$

Here we used the notation

$$\|f\|_E = \sup\{|f(x)| : x \in E\}. \tag{2}$$

Because of the example discussed in Sec. 5.23, H is sometimes called a *boundary* of K, corresponding to the space A.

If $f \in A$ and $x \in K$, (1) says that

$$|f(x)| \le \|f\|_H. \tag{3}$$

In particular, if $f(y) = 0$ for every $y \in H$, then $f(x) = 0$ for all $x \in K$. Hence if f_1 and $f_2 \in A$ and $f_1(y) = f_2(y)$ for every $y \in H$, then $f_1 = f_2$; to see this, put $f = f_1 - f_2$.

Let M be the set of all functions on H that are restrictions to H of members of A. It is clear that M is a subspace of $C(H)$. The preceding remark shows that each member of M has a unique extension to a member of A. Thus we have a natural one-to-one correspondence between M and A, which is also norm-preserving, by (1). Hence it will cause no confusion if we use the same letter to designate a member of A and its restriction to H.

Fix a point $x \in K$. The inequality (3) shows that the mapping $f \to f(x)$ is a bounded linear functional on M, of norm 1 [since equality holds in (3) if $f = 1$]. By the Hahn-Banach theorem there is a linear functional Λ on $C(H)$, of norm 1, such that

$$\Lambda f = f(x) \qquad (f \in M). \tag{4}$$

We claim that the properties

$$\Lambda 1 = 1, \qquad \|\Lambda\| = 1 \tag{5}$$

imply that Λ is a *positive* linear functional on $C(H)$.

To prove this, suppose $f \in C(H)$, $0 \le f \le 1$, put $g = 2f - 1$, and put $\Lambda g = \alpha + i\beta$, where α and β are real. Note that $-1 \le g \le 1$, so that $|g + ir|^2 \le 1 + r^2$ for every real constant r. Hence (5) implies that

$$(\beta + r)^2 \le |\alpha + i(\beta + r)|^2 = |\Lambda(g + ir)|^2 \le 1 + r^2. \tag{6}$$

Thus $\beta^2 + 2r\beta \le 1$ for every real r, which forces $\beta = 0$. Since $\|g\|_H \le 1$, we have $|\alpha| \le 1$; hence

$$\Lambda f = \tfrac{1}{2}\Lambda(1 + g) = \tfrac{1}{2}(1 + \alpha) \ge 0. \tag{7}$$

Now Theorem 2.14 can be applied. It shows that there is a regular positive Borel measure μ_x on H such that

$$\Lambda f = \int_H f \, d\mu_x \qquad (f \in C(H)). \tag{8}$$

In particular, we get the representation formula

$$f(x) = \int_H f \, d\mu_x \qquad (f \in A). \tag{9}$$

What we have proved is that *to each $x \in K$ there corresponds a positive measure μ_x on the "boundary" H which "represents" x in the sense that* (9) *holds for every $f \in A$*.

Note that Λ determines μ_x uniquely; but there is no reason to expect the Hahn-Banach extension to be unique. Hence, in general, we cannot say much about the uniqueness of the representing measures. Under special circumstances we do get uniqueness, as we shall see presently.

5.23 To see an example of the preceding situation, let $U = \{z: |z| < 1\}$ be the open unit disc in the complex plane, put $K = \bar{U}$ (the closed unit disc), and take for H the boundary T of U. We claim that every polynomial f, i.e., every function of the form

$$f(z) = \sum_{n=0}^{N} a_n z^n, \tag{1}$$

where $a_0, \ldots, a_N$ are complex numbers, satisfies the relation

$$\|f\|_U = \|f\|_T. \tag{2}$$

(Note that the continuity of f shows that the supremum of $|f|$ over U is the same as that over $\bar{U}$.)

Since $\bar{U}$ is compact, there exists a $z_0 \in \bar{U}$ such that $|f(z_0)| \geq |f(z)|$ for all $z \in \bar{U}$. Assume $z_0 \in U$. Then

$$f(z) = \sum_{n=0}^{N} b_n (z - z_0)^n, \tag{3}$$

and if $0 < r < 1 - |z_0|$, we obtain

$$\sum_{n=0}^{N} |b_n|^2 r^{2n} = \frac{1}{2\pi} \int_{-\pi}^{\pi} |f(z_0 + re^{i\theta})|^2 \, d\theta \leq \frac{1}{2\pi} \int_{-\pi}^{\pi} |f(z_0)|^2 \, d\theta = |b_0|^2,$$

so that $b_1 = b_2 = \cdots = b_N = 0$; i.e., f is constant. Thus $z_0 \in T$ for every nonconstant polynomial f, and this proves (2).

(We have just proved a special case of the *maximum modulus theorem*; we shall see later that this is an important property of all holomorphic functions.)

5.24 The Poisson Integral Let A be any subspace of $C(\bar{U})$ (where $\bar{U}$ is the closed unit disc, as above) such that A contains all polynomials and such that

$$\|f\|_U = \|f\|_T \tag{1}$$

holds for every $f \in A$. We do not exclude the possibility that A consists of precisely the polynomials, but A might be larger.

The general result obtained in Sec. 5.22 applies to A and shows that to each $z \in U$ there corresponds a positive Borel measure μ_z on T such that

$$f(z) = \int_T f \, d\mu_z \qquad (f \in A). \tag{2}$$

(This also holds for $z \in T$, but is then trivial: μ_z is simply the unit mass concentrated at the point z.)

We now fix $z \in U$ and write $z = re^{i\theta}$, $0 \le r < 1$, θ real.
If $u_n(w) = w^n$, then $u_n \in A$ for $n = 0, 1, 2, \ldots$; hence (2) shows that

$$r^n e^{in\theta} = \int_T u_n \, d\mu_z \qquad (n = 0, 1, 2, \ldots). \tag{3}$$

Since $u_{-n} = \bar{u}_n$ on T, (3) leads to

$$\int_T u_n \, d\mu_z = r^{|n|} e^{in\theta} \qquad (n = 0, \pm 1, \pm 2, \ldots). \tag{4}$$

This suggests that we look at the real function

$$P_r(\theta - t) = \sum_{n=-\infty}^{\infty} r^{|n|} e^{in(\theta - t)} \qquad (t \text{ real}), \tag{5}$$

since

$$\frac{1}{2\pi} \int_{-\pi}^{\pi} P_r(\theta - t) e^{int} \, dt = r^{|n|} e^{in\theta} \qquad (n = 0, \pm 1, \pm 2, \ldots). \tag{6}$$

Note that the series (5) is dominated by the convergent geometric series $\sum r^{|n|}$, so that it is legitimate to insert the series into the integral (6) and to integrate term by term, which gives (6). Comparison of (4) and (6) gives

$$\int_T f \, d\mu_z = \frac{1}{2\pi} \int_{-\pi}^{\pi} f(e^{it}) P_r(\theta - t) \, dt \tag{7}$$

for $f = u_n$, hence for every trigonometric polynomial f, and Theorem 4.25 now implies that (7) holds for every $f \in C(T)$. [This shows that μ_z was uniquely determined by (2). Why?]

In particular, (7) holds if $f \in A$, and then (2) gives the representation

$$f(z) = \frac{1}{2\pi} \int_{-\pi}^{\pi} f(e^{it}) P_r(\theta - t) \, dt \qquad (f \in A). \tag{8}$$

The series (5) can be summed explicitly, since it is the real part of

$$1 + 2 \sum_{1}^{\infty} (ze^{-it})^n = \frac{e^{it} + z}{e^{it} - z} = \frac{1 - r^2 + 2ir \sin(\theta - t)}{|1 - ze^{-it}|^2}.$$

Thus

$$P_r(\theta - t) = \frac{1 - r^2}{1 - 2r \cos(\theta - t) + r^2}. \tag{9}$$

This is the so-called "Poisson kernel." Note that $P_r(\theta - t) \ge 0$ if $0 \le r < 1$.
We now summarize what we have proved:

5.25 Theorem *Suppose A is a vector space of continuous complex functions on the closed unit disc $\bar{U}$. If A contains all polynomials, and if*

$$\sup_{z \in U} |f(z)| = \sup_{z \in T} |f(z)| \qquad (1)$$

for every $f \in A$ (where T is the unit circle, the boundary of U), then the Poisson integral representation

$$f(z) = \frac{1}{2\pi} \int_{-\pi}^{\pi} \frac{1 - r^2}{1 - 2r \cos(\theta - t) + r^2} f(e^{it}) \, dt \qquad (z = re^{i\theta}) \qquad (2)$$

is valid for every $f \in A$ and every $z \in U$.

Exercises

1 Let X consist of two points a and b, put $\mu(\{a\}) = \mu(\{b\}) = \frac{1}{2}$, and let $L^p(\mu)$ be the resulting *real* L^p-space. Identify each real function f on X with the point $(f(a), f(b))$ in the plane, and sketch the unit balls of $L^p(\mu)$, for $0 < p \le \infty$. Note that they are convex if and only if $1 \le p \le \infty$. For which p is this unit ball a square? A circle? If $\mu(\{a\}) \ne \mu(b)$, how does the situation differ from the preceding one?

2 Prove that the unit ball (open or closed) is convex in every normed linear space.

3 If $1 < p < \infty$, prove that the unit ball of $L^p(\mu)$ is *strictly convex*; this means that if

$$\|f\|_p = \|g\|_p = 1, \qquad f \ne g, \qquad h = \tfrac{1}{2}(f + g),$$

then $\|h\|_p < 1$. (Geometrically, the surface of the ball contains no straight lines.) Show that this fails in every $L^1(\mu)$, in every $L^\infty(\mu)$, and in every $C(X)$. (Ignore trivialities, such as spaces consisting of only one point.)

4 Let C be the space of all continuous functions on $[0, 1]$, with the supremum norm. Let M consist of all $f \in C$ for which

$$\int_0^{1/2} f(t) \, dt - \int_{1/2}^1 f(t) \, dt = 1.$$

Prove that M is a closed convex subset of C which contains no element of minimal norm.

5 Let M be the set of all $f \in L^1([0, 1])$, relative to Lebesgue measure, such that

$$\int_0^1 f(t) \, dt = 1.$$

Show that M is a closed convex subset of $L^1([0, 1])$ which contains infinitely many elements of minimal norm. (Compare this and Exercise 4 with Theorem 4.10.)

6 Let f be a bounded linear functional on a subspace M of a Hilbert space H. Prove that f has a *unique* norm-preserving extension to a bounded linear functional on H, and that this extension vanishes on $M^\perp$.

7 Construct a bounded linear functional on some subspace of some $L^1(\mu)$ which has two (hence infinitely many) distinct norm-preserving linear extensions to $L^1(\mu)$.

8 Let X be a normed linear space, and let X^* be its dual space, as defined in Sec. 5.21, with the norm

$$\|f\| = \sup \{|f(x)| : \|x\| \le 1\}.$$

(a) Prove that X^* is a Banach space.

(b) Prove that the mapping $f \to f(x)$ is, for each $x \in X$, a bounded linear functional on X^*, of norm $\|x\|$. (This gives a natural imbedding of X in its "second dual" X^{**}, the dual space of X^*.)

(c) Prove that $\{\|x_n\|\}$ is bounded if $\{x_n\}$ is a sequence in X such that $\{f(x_n)\}$ is bounded for every $f \in X^*$.

9 Let c_0, ℓ^1, and ℓ^∞ be the Banach spaces consisting of all complex sequences $x = \{\xi_i\}$, $i = 1, 2, 3, \ldots$, defined as follows:

$$x \in \ell^1 \text{ if and only if } \|x\|_1 = \sum |\xi_i| < \infty.$$

$$x \in \ell^\infty \text{ if and only if } \|x\|_\infty = \sup |\xi_i| < \infty.$$

c_0 is the subspace of ℓ^∞ consisting of all $x \in \ell^\infty$ for which $\xi_i \to 0$ as $i \to \infty$.

Prove the following four statements.

(a) If $y = \{\eta_i\} \in \ell^1$ and $\Lambda x = \sum \xi_i \eta_i$ for every $x \in c_0$, then Λ is a bounded linear functional on c_0, and $\|\Lambda\| = \|y\|_1$. Moreover, every $\Lambda \in (c_0)^*$ is obtained in this way. In brief, $(c_0)^* = \ell^1$.

(More precisely, these two spaces are not equal; the preceding statement exhibits an isometric vector space isomorphism between them.)

(b) In the same sense, $(\ell^1)^* = \ell^\infty$.

(c) Every $y \in \ell^1$ induces a bounded linear functional on ℓ^∞, as in (a). However, this does *not* give all of $(\ell^\infty)^*$, since $(\ell^\infty)^*$ contains nontrivial functionals that vanish on all of c_0.

(d) c_0 and ℓ^1 are separable but ℓ^∞ is not.

10 If $\sum \alpha_i \xi_i$ converges for every sequence $\{\xi_i\}$ such that $\xi_i \to 0$ as $i \to \infty$, prove that $\sum |\alpha_i| < \infty$.

11 For $0 < \alpha \leq 1$, let Lip α denote the space of all complex functions f on $[a, b]$ for which

$$M_f = \sup_{s \neq t} \frac{|f(s) - f(t)|}{|s - t|^\alpha} < \infty.$$

Prove that Lip α is a Banach space, if $\|f\| = |f(a)| + M_f$; also, if

$$\|f\| = M_f + \sup_x |f(x)|.$$

(The members of Lip α are said to satisfy a *Lipschitz condition* of order α.)

12 Let K be a triangle (two-dimensional figure) in the plane, let H be the set consisting of the vertices of K, and let A be the set of all real functions f on K, of the form

$$f(x, y) = \alpha x + \beta y + \gamma \qquad (\alpha, \beta, \text{ and } \gamma \text{ real}).$$

Show that to each $(x_0, y_0) \in K$ there corresponds a unique measure μ on H such that

$$f(x_0, y_0) = \int_H f \, d\mu.$$

(Compare Sec. 5.22.)

Replace K by a square, let H again be the set of its vertices, and let A be as above. Show that to each point of K there still corresponds a measure on H, with the above property, but that uniqueness is now lost.

Can you extrapolate to a more general theorem? (Think of other figures, higher dimensional spaces.)

13 Let $\{f_n\}$ be a sequence of continuous complex functions on a (nonempty) complete metric space X, such that $f(x) = \lim f_n(x)$ exists (as a complex number) for every $x \in X$.

(a) Prove that there is an open set $V \neq \emptyset$ and a number $M < \infty$ such that $|f_n(x)| < M$ for all $x \in V$ and for $n = 1, 2, 3, \ldots$.

(b) If $\epsilon > 0$, prove that there is an open set $V \neq \emptyset$ and an integer N such that $|f(x) - f_n(x)| \leq \epsilon$ if $x \in V$ and $n \geq N$.

Hint for (b): For $N = 1, 2, 3, \ldots$, put

$$A_N = \{x : |f_m(x) - f_n(x)| \leq \epsilon \text{ if } m \geq N \text{ and } n \geq N\}.$$

Since $X = \bigcup A_N$, some A_N has a nonempty interior.

14 Let C be the space of all real continuous functions on $I = [0, 1]$ with the supremum norm. Let X_n be the subset of C consisting of those f for which there exists a $t \in I$ such that $|f(s) - f(t)| \leq n|s - t|$ for all $s \in I$. Fix n and prove that each open set in C contains an open set which does not intersect X_n. (Each $f \in C$ can be uniformly approximated by a zigzag function g with very large slopes, and if $\|g - h\|$ is small, $h \notin X_n$.) Show that this implies the existence of a dense G_δ in C which consists entirely of nowhere differentiable functions.

15 Let $A = (a_{ij})$ be an infinite matrix with complex entries, where $i, j = 0, 1, 2, \ldots$. A associates with each sequence $\{s_j\}$ a sequence $\{\sigma_i\}$, defined by

$$\sigma_i = \sum_{j=0}^{\infty} a_{ij} s_j \quad (i = 1, 2, 3, \ldots),$$

provided that these series converge.

Prove that A transforms every convergent sequence $\{s_j\}$ to a sequence $\{\sigma_i\}$ which converges to the same limit if and only if the following conditions are satisfied:

(a) $\quad \lim\limits_{i \to \infty} a_{ij} = 0 \quad$ for each j.

(b) $\quad \sup_i \sum\limits_{j=0}^{\infty} |a_{ij}| < \infty$.

(c) $\quad \lim\limits_{i \to \infty} \sum\limits_{j=0}^{\infty} a_{ij} = 1$.

The process of passing from $\{s_j\}$ to $\{\sigma_i\}$ is called a *summability method*. Two examples are:

$$a_{ij} = \begin{cases} \dfrac{1}{i+1} & \text{if } 0 \leq j \leq i, \\ 0 & \text{if } i < j, \end{cases}$$

and $\quad a_{ij} = (1 - r_i) r_i^j, \quad 0 < r_i < 1, \quad r_i \to 1$.

Prove that each of these also transforms some divergent sequences $\{s_j\}$ (even some unbounded ones) to convergent sequences $\{\sigma_i\}$.

16 Suppose X and Y are Banach spaces, and suppose Λ is a linear mapping of X into Y, with the following property: For every sequence $\{x_n\}$ in X for which $x = \lim x_n$ and $y = \lim \Lambda x_n$ exist, it is true that $y = \Lambda x$. Prove that Λ is continuous.

This is the so-called "closed graph theorem." *Hint*: Let $X \oplus Y$ be the set of all ordered pairs (x, y), $x \in X$ and $y \in Y$, with addition and scalar multiplication defined componentwise. Prove that $X \oplus Y$ is a Banach space, if $\|(x, y)\| = \|x\| + \|y\|$. The graph G of Λ is the subset of $X \oplus Y$ formed by the pairs $(x, \Lambda x)$, $x \in X$. Note that our hypothesis says that G is closed; hence G is a Banach space. Note that $(x, \Lambda x) \to x$ is continuous, one-to-one, and linear and maps G onto X.

Observe that there exist *nonlinear* mappings (of R^1 onto R^1, for instance) whose graph is closed although they are not continuous: $f(x) = 1/x$ if $x \neq 0$, $f(0) = 0$.

17 If μ is a positive measure, each $f \in L^\infty(\mu)$ defines a multiplication operator M_f on $L^2(\mu)$ into $L^2(\mu)$, such that $M_f(g) = fg$. Prove that $\|M_f\| \leq \|f\|_\infty$. For which measures μ is it true that $\|M_f\| = \|f\|_\infty$ for all $f \in L^\infty(\mu)$? For which $f \in L^\infty(\mu)$ does M_f map $L^2(\mu)$ onto $L^2(\mu)$?

18 Suppose $\{\Lambda_n\}$ is a sequence of bounded linear transformations from a normed linear space X to a Banach space Y, suppose $\|\Lambda_n\| \leq M < \infty$ for all n, and suppose there is a dense set $E \subset X$ such that $\{\Lambda_n x\}$ converges for each $x \in E$. Prove that $\{\Lambda_n x\}$ converges for each $x \in X$.

19 If s_n is the nth partial sum of the Fourier series of a function $f \in C(T)$, prove that $s_n/\log n \to 0$ uniformly, as $n \to \infty$, for each $f \in C(T)$. That is, prove that

$$\lim_{n \to \infty} \frac{\|s_n\|_\infty}{\log n} = 0.$$

On the other hand, if $\lambda_n/\log n \to 0$, prove that there exists an $f \in C(T)$ such that the sequence $\{s_n(f; 0)/\lambda_n\}$ is unbounded. *Hint:* Apply the reasoning of Exercise 18 and that of Sec. 5.11, with a better estimate of $\|D_n\|_1$ than was used there.

20 (a) Does there exist a sequence of continuous positive functions f_n on R^1 such that $\{f_n(x)\}$ is unbounded if and only if x is rational?

(b) Replace "rational" by "irrational" in (a) and answer the resulting question.

(c) Replace "$\{f_n(x)\}$ is unbounded" by "$f_n(x) \to \infty$ as $n \to \infty$" and answer the resulting analogues of (a) and (b).

21 Suppose $E \subset R^1$ is measurable, and $m(E) = 0$. Must there be a translate $E + x$ of E that does not intersect E? Must there be a homeomorphism h of R^1 onto R^1 so that $h(E)$ does not intersect E?

22 Suppose $f \in C(T)$ and $f \in \text{Lip } \alpha$ for some $\alpha > 0$. (See Exercise 11.) Prove that the Fourier series of f converges to $f(x)$, by completing the following outline: It is enough to consider the case $x = 0$, $f(0) = 0$. The difference between the partial sums $s_n(f; 0)$ and the integrals

$$\frac{1}{\pi} \int_{-\pi}^{\pi} f(t) \frac{\sin nt}{t} dt$$

tends to 0 as $n \to \infty$. The function $f(t)/t$ is in $L^1(T)$. Apply the Riemann-Lebesgue lemma. More careful reasoning shows that the convergence is actually uniform on T.

CHAPTER
SIX

COMPLEX MEASURES

Total Variation

6.1 Introduction Let $\mathfrak{M}$ be a σ-algebra in a set X. Call a countable collection $\{E_i\}$ of members of $\mathfrak{M}$ a *partition of* E if $E_i \cap E_j = \emptyset$ whenever $i \neq j$, and if $E = \bigcup E_i$. A *complex measure* μ on $\mathfrak{M}$ is then a complex function on $\mathfrak{M}$ such that

$$\mu(E) = \sum_{i=1}^{\infty} \mu(E_i) \qquad (E \in \mathfrak{M}) \tag{1}$$

for every partition $\{E_i\}$ of E.

Observe that the convergence of the series in (1) is now part of the requirement (unlike for positive measures, where the series could either converge or diverge to ∞). Since the union of the sets E_i is not changed if the subscripts are permuted, every rearrangement of the series (1) must also converge. Hence ([26], Theorem 3.56) the series actually converges absolutely.

Let us consider the problem of finding a positive measure λ which dominates a given complex measure μ on $\mathfrak{M}$, in the sense that $|\mu(E)| \leq \lambda(E)$ for every $E \in \mathfrak{M}$, and let us try to keep λ as small as we can. Every solution to our problem (if there is one at all) must satisfy

$$\lambda(E) = \sum_{i=1}^{\infty} \lambda(E_i) \geq \sum_{1}^{\infty} |\mu(E_i)|, \tag{2}$$

for every partition $\{E_i\}$ of any set $E \in \mathfrak{M}$, so that $\lambda(E)$ is at least equal to the supremum of the sums on the right of (2), taken over all partitions of E. This suggests that we *define a set function* $|\mu|$ *on* $\mathfrak{M}$ *by*

$$|\mu|(E) = \sup \sum_{i=1}^{\infty} |\mu(E_i)| \qquad (E \in \mathfrak{M}), \tag{3}$$

the supremum being taken over all partitions $\{E_i\}$ *of* E.

This notation is perhaps not the best, but it is the customary one. Note that $|\mu|(E) \geq |\mu(E)|$, but that in general $|\mu|(E)$ is not equal to $|\mu(E)|$.

It turns out, as will be proved below, that $|\mu|$ actually *is* a measure, so that our problem does have a solution. The discussion which led to (3) shows then clearly that $|\mu|$ is the minimal solution, in the sense that any other solution λ has the property $\lambda(E) \geq |\mu|(E)$ for all $E \in \mathfrak{M}$.

The set function $|\mu|$ is called the *total variation* of μ, or sometimes, to avoid misunderstanding, the *total variation measure*. The term "total variation of μ" is also frequently used to denote the number $|\mu|(X)$.

If μ is a positive measure, then of course $|\mu| = \mu$.

Besides being a measure, $|\mu|$ has another unexpected property: $|\mu|(X) < \infty$. Since $|\mu(E)| \leq |\mu|(E) \leq |\mu|(X)$, this implies that every complex measure μ on any σ-algebra is bounded: If the range of μ lies in the complex plane, then it actually lies in some disc of finite radius. This property (proved in Theorem 6.4) is sometimes expressed by saying that μ is *of bounded variation*.

6.2 Theorem *The total variation $|\mu|$ of a complex measure μ on $\mathfrak{M}$ is a positive measure on $\mathfrak{M}$.*

PROOF Let $\{E_i\}$ be a partition of $E \in \mathfrak{M}$. Let t_i be real numbers such that $t_i < |\mu|(E_i)$. Then each E_i has a partition $\{A_{ij}\}$ such that

$$\sum_j |\mu(A_{ij})| > t_i \qquad (i = 1, 2, 3, \ldots). \tag{1}$$

Since $\{A_{ij}\}$ $(i, j = 1, 2, 3, \ldots)$ is a partition of E, it follows that

$$\sum_i t_i \leq \sum_{i,j} |\mu(A_{ij})| \leq |\mu|(E). \tag{2}$$

Taking the supremum of the left side of (2), over all admissible choices of $\{t_i\}$, we see that

$$\sum_i |\mu|(E_i) \leq |\mu|(E). \tag{3}$$

To prove the opposite inequality, let $\{A_j\}$ be any partition of E. Then for any fixed j, $\{A_j \cap E_i\}$ is a partition of A_j, and for any fixed i, $\{A_j \cap E_i\}$ is a partition of E_i. Hence

$$\sum_j |\mu(A_j)| = \sum_j \left| \sum_i \mu(A_j \cap E_i) \right|$$
$$\leq \sum_j \sum_i |\mu(A_j \cap E_i)|$$
$$= \sum_i \sum_j |\mu(A_j \cap E_i)| \leq \sum_i |\mu|(E_i). \tag{4}$$

Since (4) holds for every partition $\{A_j\}$ of E, we have

$$|\mu|(E) \leq \sum_i |\mu|(E_i). \tag{5}$$

By (3) and (5), $|\mu|$ is countably additive.

Note that the Corollary to Theorem 1.27 was used in (2) and (4).

That $|\mu|$ is not identically ∞ is a trivial consequence of Theorem 6.4 but can also be seen right now, since $|\mu|(\emptyset) = 0$. ////

6.3 Lemma *If $z_1, \ldots, z_N$ are complex numbers then there is a subset S of $\{1, \ldots, N\}$ for which*

$$\left|\sum_{k \in S} z_k\right| \geq \frac{1}{\pi} \sum_{k=1}^N |z_k|.$$

PROOF Write $z_k = |z_k| e^{i\alpha_k}$. For $-\pi \leq \theta \leq \pi$, let $S(\theta)$ be the set of all k for which $\cos(\alpha_k - \theta) > 0$. Then

$$\left|\sum_{S(\theta)} z_k\right| = \left|\sum_{S(\theta)} e^{-i\theta} z_k\right| \geq \text{Re} \sum_{S(\theta)} e^{-i\theta} z_k = \sum_{k=1}^N |z_k| \cos^+ (\alpha_k - \theta).$$

Choose θ_0 so as to maximize the last sum, and put $S = S(\theta_0)$. This maximum is at least as large as the average of the sum over $[-\pi, \pi]$, and this average is $\pi^{-1} \sum |z_k|$, because

$$\frac{1}{2\pi} \int_{-\pi}^{\pi} \cos^+ (\alpha - \theta) \, d\theta = \frac{1}{\pi}$$

for every α. ////

6.4 Theorem *If μ is a complex measure on X, then*

$$|\mu|(X) < \infty.$$

PROOF Suppose first that some set $E \in \mathfrak{M}$ has $|\mu|(E) = \infty$. Put $t = \pi(1 + |\mu(E)|)$. Since $|\mu|(E) > t$, there is a partition $\{E_i\}$ of E such that

$$\sum_{i=1}^N |\mu(E_i)| > t$$

for some N. Apply Lemma 6.3, with $z_i = \mu(E_i)$, to conclude that there is a set $A \subset E$ (a union of some of the sets E_i) for which

$$|\mu(A)| > t/\pi > 1.$$

Setting $B = E - A$, it follows that

$$|\mu(B)| = |\mu(E) - \mu(A)| \geq |\mu(A)| - |\mu(E)| > \frac{t}{\pi} - |\mu(E)| = 1.$$

We have thus split E into disjoint sets A and B with $|\mu(A)| > 1$ and $|\mu(B)| > 1$. Evidently, at least one of $|\mu|(A)$ and $|\mu|(B)$ is ∞, by Theorem 6.2.

Now if $|\mu|(X) = \infty$, split X into A_1, B_1, as above, with $|\mu(A_1)| > 1$, $|\mu|(B_1) = \infty$. Split B_1 into A_2, B_2, with $|\mu(A_2)| > 1$, $|\mu|(B_2) = \infty$. Continuing in this way, we get a countably infinite disjoint collection $\{A_i\}$, with $|\mu(A_i)| > 1$ for each i. The countable additivity of μ implies that

$$\mu\left(\bigcup_i A_i\right) = \sum_i \mu(A_i).$$

But this series cannot converge, since $\mu(A_i)$ does not tend to 0 as $i \to \infty$. This contradiction shows that $|\mu|(X) < \infty$. ////

6.5 If μ and λ are complex measures on the same σ-algebra $\mathfrak{M}$, we define $\mu + \lambda$ and $c\mu$ by

$$(\mu + \lambda)(E) = \mu(E) + \lambda(E)$$
$$(c\mu)(E) = c\mu(E) \qquad (E \in \mathfrak{M}) \qquad (1)$$

for any scalar c, in the usual manner. It is then trivial to verify that $\mu + \lambda$ and $c\mu$ are complex measures. The collection of all complex measures on $\mathfrak{M}$ is thus a vector space. If we put

$$\|\mu\| = |\mu|(X), \qquad (2)$$

it is easy to verify that all axioms of a normed linear space are satisfied.

6.6 Positive and Negative Variations Let us now specialize and consider a *real* measure μ on a σ-algebra $\mathfrak{M}$. (Such measures are frequently called *signed* measures.) Define $|\mu|$ as before, and define

$$\mu^+ = \tfrac{1}{2}(|\mu| + \mu), \qquad \mu^- = \tfrac{1}{2}(|\mu| - \mu). \qquad (1)$$

Then both μ^+ and μ^- are positive measures on $\mathfrak{M}$, and they are bounded, by Theorem 6.4. Also,

$$\mu = \mu^+ - \mu^-, \qquad |\mu| = \mu^+ + \mu^-. \qquad (2)$$

The measures μ^+ and μ^- are called the *positive* and *negative variations* of μ, respectively. This representation of μ as the difference of the positive measures μ^+ and μ^- is known as the *Jordan decomposition* of μ. Among all representations of μ as a difference of two positive measures, the Jordan decomposition has a certain minimum property which will be established as a corollary to Theorem 6.14.

Absolute Continuity

6.7 Definitions Let μ be a positive measure on a σ-algebra $\mathfrak{M}$, and let λ be an arbitrary measure on $\mathfrak{M}$; λ may be positive or complex. (Recall that a complex measure has its range in the complex plane, but that our usage of the term "positive measure" includes ∞ as an admissible value. Thus the positive measures do not form a subclass of the complex ones.)

We say that λ is *absolutely continuous* with respect to μ, and write

$$\lambda \ll \mu \qquad (1)$$

if $\lambda(E) = 0$ for every $E \in \mathfrak{M}$ for which $\mu(E) = 0$.

If there is a set $A \in \mathfrak{M}$ such that $\lambda(E) = \lambda(A \cap E)$ for every $E \in \mathfrak{M}$, we say that λ is *concentrated on* A. This is equivalent to the hypothesis that $\lambda(E) = 0$ whenever $E \cap A = \emptyset$.

Suppose λ_1 and λ_2 are measures on $\mathfrak{M}$, and suppose there exists a pair of disjoint sets A and B such that λ_1 is concentrated on A and λ_2 is concentrated on B. Then we say that λ_1 and λ_2 are *mutually singular*, and write

$$\lambda_1 \perp \lambda_2. \qquad (2)$$

Here are some elementary properties of these concepts.

6.8 Proposition *Suppose, μ, λ, λ_1, and λ_2 are measures on a σ-algebra $\mathfrak{M}$, and μ is positive.*

(a) *If λ is concentrated on A, so is $|\lambda|$.*
(b) *If $\lambda_1 \perp \lambda_2$, then $|\lambda_1| \perp |\lambda_2|$.*
(c) *If $\lambda_1 \perp \mu$ and $\lambda_2 \perp \mu$, then $\lambda_1 + \lambda_2 \perp \mu$.*
(d) *If $\lambda_1 \ll \mu$ and $\lambda_2 \ll \mu$, then $\lambda_1 + \lambda_2 \ll \mu$.*
(e) *If $\lambda \ll \mu$, then $|\lambda| \ll \mu$.*
(f) *If $\lambda_1 \ll \mu$ and $\lambda_2 \perp \mu$, then $\lambda_1 \perp \lambda_2$.*
(g) *If $\lambda \ll \mu$ and $\lambda \perp \mu$, then $\lambda = 0$.*

PROOF

(a) If $E \cap A = \emptyset$ and $\{E_j\}$ is any partition of E, then $\lambda(E_j) = 0$ for all j. Hence $|\lambda|(E) = 0$.
(b) This follows immediately from (a).
(c) There are disjoint sets A_1 and B_1 such that λ_1 is concentrated on A_1 and μ on B_1, and there are disjoint sets A_2 and B_2 such that λ_2 is concentrated on A_2 and μ on B_2. Hence $\lambda_1 + \lambda_2$ is concentrated on $A = A_1 \cup A_2$, μ is concentrated on $B = B_1 \cap B_2$, and $A \cap B = \emptyset$.
(d) This is obvious.
(e) Suppose $\mu(E) = 0$, and $\{E_j\}$ is a partition of E. Then $\mu(E_j) = 0$; and since $\lambda \ll \mu$, $\lambda(E_j) = 0$ for all j, hence $\sum |\lambda(E_j)| = 0$. This implies $|\lambda|(E) = 0$.

(f) Since $\lambda_2 \perp \mu$, there is a set A with $\mu(A) = 0$ on which λ_2 is concentrated. Since $\lambda_1 \ll \mu$, $\lambda_1(E) = 0$ for every $E \subset A$. So λ_1 is concentrated on the complement of A.

(g) By (f), the hypothesis of (g) implies, that $\lambda \perp \lambda$, and this clearly forces $\lambda = 0$. ////

We come now to the principal theorem about absolute continuity. In fact, it is probably the most important theorem in measure theory. Its statement will involve σ-finite measures. The following lemma describes one of their significant properties.

6.9 Lemma *If μ is a positive σ-finite measure on a σ-algebra $\mathfrak{M}$ in a set X, then there is a function $w \in L^1(\mu)$ such that $0 < w(x) < 1$ for every $x \in X$.*

PROOF To say that μ is σ-finite means that X is the union of countably many sets $E_n \in \mathfrak{M}$ ($n = 1, 2, 3, \ldots$) for which $\mu(E_n)$ is finite. Put $w_n(x) = 0$ if $x \in X - E_n$ and put

$$w_n(x) = 2^{-n}/(1 + \mu(E_n))$$

if $x \in E_n$. Then $w = \sum_1^\infty w_n$ has the required properties. ////

The point of the lemma is that μ can be replaced by a *finite* measure $\tilde{\mu}$ (namely, $d\tilde{\mu} = w\, d\mu$) which, because of the strict positivity of w, has *precisely* the same sets of measure 0 as μ.

6.10 The Theorem of Lebesgue-Radon-Nikodym *Let μ be a positive σ-finite measure on a σ-algebra $\mathfrak{M}$ in a set X, and let λ be a complex measure on $\mathfrak{M}$.*

(a) *There is then a unique pair of complex measures λ_a and λ_s on $\mathfrak{M}$ such that*

$$\lambda = \lambda_a + \lambda_s, \qquad \lambda_a \ll \mu, \qquad \lambda_s \perp \mu. \tag{1}$$

If λ is positive and finite, then so are λ_a and λ_s.

(b) *There is a unique $h \in L^1(\mu)$ such that*

$$\lambda_a(E) = \int_E h\, d\mu \tag{2}$$

for every set $E \in \mathfrak{M}$.

The pair (λ_a, λ_s) is called the *Lebesgue decomposition* of λ relative to μ. The uniqueness of the decomposition is easily seen, for if (λ_a', λ_s') is another pair which satisfies (1), then

$$\lambda_a' - \lambda_a = \lambda_s - \lambda_s', \tag{3}$$

$\lambda_a' - \lambda_a \ll \mu$, and $\lambda_s - \lambda_s' \perp \mu$, hence both sides of (3) are 0; we have used 6.8(c), 6.8(d), and 6.8(g).

The *existence* of the decomposition is the significant part of (a).

Assertion (b) is known as the *Radon-Nikodym* theorem. Again, uniqueness of h is immediate, from Theorem 1.39(b). Also, if h is *any* member of $L^1(\mu)$, the integral in (2) defines a measure on $\mathfrak{M}$ (Theorem 1.29) which is clearly absolutely continuous with respect to μ. The point of the Radon-Nikodym theorem is the converse: *Every $\lambda \ll \mu$* (in which case $\lambda_a = \lambda$) is obtained in this way.

The function h which occurs in (2) is called the *Radon-Nikodym derivative* of λ_a with respect to μ. As noted after Theorem 1.29, we may express (2) in the form $d\lambda_a = h\, d\mu$, or even in the form $h = d\lambda_a/d\mu$.

The idea of the following proof, which yields both (a) and (b) at one stroke, is due to von Neumann.

PROOF Assume first that λ is a positive bounded measure on $\mathfrak{M}$. Associate w to μ as in Lemma 6.9. Then $d\varphi = d\lambda + w\, d\mu$ defines a positive bounded measure φ on $\mathfrak{M}$. The definition of the sum of two measures shows that

$$\int_X f\, d\varphi = \int_X f\, d\lambda + \int_X fw\, d\mu \tag{4}$$

for $f = \chi_E$, hence for simple f, hence for any nonnegative measurable f. If $f \in L^2(\varphi)$, the Schwarz inequality gives

$$\left| \int_X f\, d\lambda \right| \leq \int_X |f|\, d\lambda \leq \int_X |f|\, d\varphi \leq \left\{ \int_X |f|^2\, d\varphi \right\}^{1/2} \{\varphi(X)\}^{1/2}.$$

Since $\varphi(X) < \infty$, we see that

$$f \to \int_X f\, d\lambda \tag{5}$$

is a bounded linear functional on $L^2(\varphi)$. We know that every bounded linear functional on a Hilbert space H is given by an inner product with an element of H. Hence there exists a $g \in L^2(\varphi)$ such that

$$\int_X f\, d\lambda = \int_X fg\, d\varphi \tag{6}$$

for every $f \in L^2(\varphi)$.

Observe how the completeness of $L^2(\varphi)$ was used to guarantee the existence of g. Observe also that although g is defined uniquely as an element of $L^2(\varphi)$, g is determined only a.e. $[\varphi]$ as a point function on X.

Put $f = \chi_E$ in (6), for any $E \in \mathfrak{M}$ with $\varphi(E) > 0$. The left side of (6) is then $\lambda(E)$, and since $0 \leq \lambda \leq \varphi$, we have

$$0 \leq \frac{1}{\varphi(E)} \int_E g\, d\varphi = \frac{\lambda(E)}{\varphi(E)} \leq 1. \tag{7}$$

Hence $g(x) \in [0, 1]$ for almost all x (with respect to φ), by Theorem 1.40. We may therefore assume that $0 \leq g(x) \leq 1$ for every $x \in X$, without affecting (6), and we rewrite (6) in the form

$$\int_X (1 - g) f \, d\lambda = \int_X fgw \, d\mu. \tag{8}$$

Put

$$A = \{x \colon 0 \leq g(x) < 1\}, \qquad B = \{x \colon g(x) = 1\}, \tag{9}$$

and define measures λ_a and λ_s by

$$\lambda_a(E) = \lambda(A \cap E), \qquad \lambda_s(E) = \lambda(B \cap E), \tag{10}$$

for all $E \in \mathfrak{M}$.

If $f = \chi_B$ in (8), the left side is 0, the right side is $\int_B w \, d\mu$. Since $w(x) > 0$ for all x, we conclude that $\mu(B) = 0$. Thus $\lambda_s \perp \mu$.

Since g is bounded, (8) holds if f is replaced by

$$(1 + g + \cdots + g^n)\chi_E$$

for $n = 1, 2, 3, \ldots, E \in \mathfrak{M}$. For such f, (8) becomes

$$\int_E (1 - g^{n+1}) \, d\lambda = \int_E g(1 + g + \cdots + g^n) w \, d\mu. \tag{11}$$

At every point of B, $g(x) = 1$, hence $1 - g^{n+1}(x) = 0$. At every point of A, $g^{n+1}(x) \to 0$ monotonically. The left side of (11) converges therefore to $\lambda(A \cap E) = \lambda_a(E)$ as $n \to \infty$.

The integrands on the right side of (11) increase monotonically to a non-negative measurable limit h, and the monotone convergence theorem shows that the right side of (11) tends to $\int_E h \, d\mu$ as $n \to \infty$.

We have thus proved that (2) holds for every $E \in \mathfrak{M}$. Taking $E = X$, we see that $h \in L^1(\mu)$, since $\lambda_a(X) < \infty$.

Finally, (2) shows that $\lambda_a \ll \mu$, and the proof is complete for positive λ.

If λ is a complex measure on $\mathfrak{M}$, then $\lambda = \lambda_1 + i\lambda_2$, with λ_1 and λ_2 real, and we can apply the preceding case to the positive and negative variations of λ_1 and λ_2. ////

If both μ and λ are positive and σ-finite, most of Theorem 6.10 is still true. We can now write $X = \bigcup X_n$, where $\mu(X_n) < \infty$ and $\lambda(X_n) < \infty$, for $n = 1, 2, 3, \ldots$. The Lebesgue decompositions of the measures $\lambda(E \cap X_n)$ still give us a Lebesgue decomposition of λ, and we still get a function h which satisfies Eq. 6.10(2); however, it is no longer true that $h \in L^1(\mu)$, although h is "locally in L^1," i.e., $\int_{X_n} h \, d\mu < \infty$ for each n.

Finally, if we go beyond σ-finiteness, we meet situations where the two theorems under consideration actually fail. For example, let μ be Lebesgue measure on $(0, 1)$, and let λ be the counting measure on the σ-algebra of all Lebesgue measurable sets in (0, 1). Then λ has no Lebesgue decomposition relative to μ, and although $\mu \ll \lambda$ and μ is bounded, there is no $h \in L^1(\lambda)$ such that $d\mu = h \, d\lambda$. We omit the easy proof.

The following theorem may explain why the word "continuity" is used in connection with the relation $\lambda \ll \mu$.

6.11 Theorem *Suppose μ and λ are measures on a σ-algebra $\mathfrak{M}$, μ is positive, and λ is complex. Then the following two conditions are equivalent:*

(a) $\lambda \ll \mu$.
(b) *To every $\epsilon > 0$ corresponds a $\delta > 0$ such that $|\lambda(E)| < \epsilon$ for all $E \in \mathfrak{M}$ with $\mu(E) < \delta$.*

Property (b) is sometimes used as the definition of absolute continuity. However, (a) does not imply (b) if λ is a positive unbounded measure. For instance, let μ be Lebesgue measure on (0, 1), and put

$$\lambda(E) = \int_E t^{-1} \, dt$$

for every Lebesgue measurable set $E \subset (0, 1)$.

PROOF Suppose (b) holds. If $\mu(E) = 0$, then $\mu(E) < \delta$ for every $\delta > 0$, hence $|\lambda(E)| < \epsilon$ for every $\epsilon > 0$, so $\lambda(E) = 0$. Thus (b) implies (a).

Suppose (b) is false. Then there exists an $\epsilon > 0$ and there exist sets $E_n \in \mathfrak{M}$ ($n = 1, 2, 3, \ldots$) such that $\mu(E_n) < 2^{-n}$ but $|\lambda(E_n)| \geq \epsilon$. Hence $|\lambda|(E_n) \geq \epsilon$. Put

$$A_n = \bigcup_{i=n}^{\infty} E_i, \qquad A = \bigcap_{n=1}^{\infty} A_n. \tag{1}$$

Then $\mu(A_n) < 2^{-n+1}$, $A_n \supset A_{n+1}$, and so Theorem 1.19(e) shows that $\mu(A) = 0$ and that

$$|\lambda|(A) = \lim_{n \to \infty} |\lambda|(A_n) \geq \epsilon > 0,$$

since $|\lambda|(A_n) \geq |\lambda|(E_n)$.

It follows that we do *not* have $|\lambda| \ll \mu$, hence (a) is false, by Proposition 6.8(e). ////

Consequences of the Radon-Nikodym Theorem

6.12 Theorem *Let μ be a complex measure on a σ-algebra $\mathfrak{M}$ in X. Then there is a measurable function h such that $|h(x)| = 1$ for all $x \in X$ and such that*

$$d\mu = h \, d|\mu|. \tag{1}$$

By analogy with the representation of a complex number as the product of its absolute value and a number of absolute value 1, Eq. (1) is sometimes referred to as the *polar representation* (or *polar decomposition*) of μ.

PROOF It is trivial that $\mu \ll |\mu|$, and therefore the Radon-Nikodym theorem guarantees the existence of some $h \in L^1(|\mu|)$ which satisfies (1).

Let $A_r = \{x: |h(x)| < r\}$, where r is some positive number, and let $\{E_j\}$ be a partition of A_r. Then

$$\sum_j |\mu(E_j)| = \sum_j \left| \int_{E_j} h \, d|\mu| \right| \leq \sum_j r|\mu|(E_j) = r|\mu|(A_r),$$

so that $|\mu|(A_r) \leq r|\mu|(A_r)$. If $r < 1$, this forces $|\mu|(A_r) = 0$. Thus $|h| \geq 1$ a.e.

On the other hand, if $|\mu|(E) > 0$, (1) shows that

$$\left| \frac{1}{|\mu|(E)} \int_E h \, d|\mu| \right| = \frac{|\mu(E)|}{|\mu|(E)} \leq 1.$$

We now apply Theorem 1.40 (with the closed unit disc in place of S) and conclude that $|h| \leq 1$ a.e.

Let $B = \{x \in X : |h(x)| \neq 1\}$. We have shown that $|\mu|(B) = 0$, and if we redefine h on B so that $h(x) = 1$ on B, we obtain a function with the desired properties. ////

6.13 Theorem *Suppose μ is a positive measure on $\mathfrak{M}$, $g \in L^1(\mu)$, and*

$$\lambda(E) = \int_E g \, d\mu \qquad (E \in \mathfrak{M}). \tag{1}$$

Then

$$|\lambda|(E) = \int_E |g| \, d\mu \qquad (E \in \mathfrak{M}). \tag{2}$$

PROOF By Theorem 6.12, there is a function h, of absolute value 1, such that $d\lambda = h \, d|\lambda|$. By hypothesis, $d\lambda = g \, d\mu$. Hence

$$h \, d|\lambda| = g \, d\mu.$$

This gives $d|\lambda| = \bar{h} g \, d\mu$. (Compare with Theorem 1.29.)

Since $|\lambda| \geq 0$ and $\mu \geq 0$, it follows that $\bar{h} g \geq 0$ a.e. $[\mu]$, so that $\bar{h} g = |g|$ a.e. $[\mu]$. ////

6.14 The Hahn Decomposition Theorem *Let μ be a real measure on a σ-algebra $\mathfrak{M}$ in a set X. Then there exist sets A and $B \in \mathfrak{M}$ such that*

$A \cup B = X$, $A \cap B = \emptyset$, and such that the positive and negative variations μ^+ and μ^- of μ satisfy

$$\mu^+(E) = \mu(A \cap E), \quad \mu^-(E) = -\mu(B \cap E) \qquad (E \in \mathfrak{M}). \tag{1}$$

In other words, X is the union of two disjoint measurable sets A and B, such that "A carries all the positive mass of μ" [since (1) implies that $\mu(E) \geq 0$ if $E \subset A$] and "B carries all the negative mass of μ" [since $\mu(E) \leq 0$ if $E \subset B$]. The pair (A, B) is called a Hahn decomposition of X, induced by μ.

Proof By Theorem 6.12, $d\mu = h \, d|\mu|$, where $|h| = 1$. Since μ is real, it follows that h is real (a.e., and therefore everywhere, by redefining on a set of measure 0), hence $h = \pm 1$. Put

$$A = \{x: h(x) = 1\}, \qquad B = \{x: h(x) = -1\}. \tag{2}$$

Since $\mu^+ = \frac{1}{2}(|\mu| + \mu)$, and since

$$\tfrac{1}{2}(1 + h) = \begin{cases} h & \text{on } A, \\ 0 & \text{on } B, \end{cases} \tag{3}$$

we have, for any $E \in \mathfrak{M}$,

$$\mu^+(E) = \frac{1}{2} \int_E (1 + h) \, d|\mu| = \int_{E \cap A} h \, d|\mu| = \mu(E \cap A). \tag{4}$$

Since $\mu(E) = \mu(E \cap A) + \mu(E \cap B)$ and since $\mu = \mu^+ - \mu^-$, the second half of (1) follows from the first. ////

Corollary *If $\mu = \lambda_1 - \lambda_2$, where λ_1 and λ_2 are positive measures, then $\lambda_1 \geq \mu^+$ and $\lambda_2 \geq \mu^-$.*

This is the minimum property of the Jordan decomposition which was mentioned in Sec. 6.6.

Proof Since $\mu \leq \lambda_1$, we have

$$\mu^+(E) = \mu(E \cap A) \leq \lambda_1(E \cap A) \leq \lambda_1(E). \qquad \text{////}$$

Bounded Linear Functionals on L^p

6.15 Let μ be a positive measure, suppose $1 \leq p \leq \infty$, and let q be the exponent conjugate to p. The Hölder inequality (Theorem 3.8) shows that if $g \in L^q(\mu)$ and if Φ_g is defined by

$$\Phi_g(f) = \int_X fg \, d\mu, \tag{1}$$

then Φ_g is a bounded linear functional on $L^p(\mu)$, of norm at most $\|g\|_q$. The question naturally arises whether all bounded linear functionals on $L^p(\mu)$ have this form, and whether the representation is unique.

For $p = \infty$, Exercise 13 shows that the answer is negative: $L^1(m)$ does not furnish all bounded linear functionals on $L^\infty(m)$. For $1 < p < \infty$, the answer is affirmative. It is also affirmative for $p = 1$, provided certain measure-theoretic pathologies are excluded. For σ-finite measure spaces, no difficulties arise, and we shall confine ourselves to this case.

6.16 Theorem *Suppose $1 \leq p < \infty$, μ is a σ-finite positive measure on X, and Φ is a bounded linear functional on $L^p(\mu)$. Then there is a unique $g \in L^q(\mu)$, where q is the exponent conjugate to p, such that*

$$\Phi(f) = \int_X fg \, d\mu \qquad (f \in L^p(\mu)). \tag{1}$$

Moreover, if Φ and g are related as in (1), we have

$$\|\Phi\| = \|g\|_q. \tag{2}$$

In other words, $L^q(\mu)$ is isometrically isomorphic to the dual space of $L^p(\mu)$, under the stated conditions.

PROOF The uniqueness of g is clear, for if g and g' satisfy (1), then the integral of $g - g'$ over any measurable set E of finite measure is 0 (as we see by taking χ_E for f), and the σ-finiteness of μ implies therefore that $g - g' = 0$ a.e.

Next, if (1) holds, Hölder's inequality implies

$$\|\Phi\| \leq \|g\|_q. \tag{3}$$

So it remains to prove that g exists and that equality holds in (3). If $\|\Phi\| = 0$, (1) and (2) hold with $g = 0$. So assume $\|\Phi\| > 0$.

We first consider the case $\mu(X) < \infty$.

For any measurable set $E \subset X$, define

$$\lambda(E) = \Phi(\chi_E).$$

Since Φ is linear, and since $\chi_{A \cup B} = \chi_A + \chi_B$ if A and B are disjoint, we see that λ is additive. To prove countable additivity, suppose E is the union of countably many disjoint measurable sets E_i, put $A_k = E_1 \cup \cdots \cup E_k$, and note that

$$\|\chi_E - \chi_{A_k}\|_p = [\mu(E - A_k)]^{1/p} \to 0 \qquad (k \to \infty); \tag{4}$$

the continuity of Φ shows now that $\lambda(A_k) \to \lambda(E)$. So λ is a complex measure. [In (4) the assumption $p < \infty$ was used.] It is clear that $\lambda(E) = 0$ if $\mu(E) = 0$, since then $\|\chi_E\|_p = 0$. Thus $\lambda \ll \mu$, and the Radon-Nikodym theorem ensures the existence of a function $g \in L^1(\mu)$ such that, for every measurable $E \subset X$,

$$\Phi(\chi_E) = \int_E g \, d\mu = \int_X \chi_E g \, d\mu. \tag{5}$$

By linearity it follows that

$$\Phi(f) = \int_X fg \, d\mu \tag{6}$$

holds for every simple measurable f, and so also for every $f \in L^\infty(\mu)$, since every $f \in L^\infty(\mu)$ is a uniform limit of simple functions f_i. Note that the uniform convergence of f_i to f implies $\|f_i - f\|_p \to 0$, hence $\Phi(f_i) \to \Phi(f)$, as $i \to \infty$.

We want to conclude that $g \in L^q(\mu)$ and that (2) holds; it is best to split the argument into two cases.

CASE 1 $p = 1$. Here (5) shows that

$$\left| \int_E g \, d\mu \right| \le \|\Phi\| \cdot \|\chi_E\|_1 = \|\Phi\| \cdot \mu(E)$$

for every $E \in \mathfrak{M}$. By Theorem 1.40, $|g(x)| \le \|\Phi\|$ a.e., so that $\|g\|_\infty \le \|\Phi\|$.

CASE 2 $1 < p < \infty$. There is a measurable function α, $|\alpha| = 1$, such that $\alpha g = |g|$ [Proposition 1.9(e)]. Let $E_n = \{x : |g(x)| \le n\}$, and define $f = \chi_{E_n} |g|^{q-1} \alpha$. Then $|f|^p = |g|^q$ on E_n, $f \in L^\infty(\mu)$, and (6) gives

$$\int_{E_n} |g|^q \, d\mu = \int_X fg \, d\mu = \Phi(f) \le \|\Phi\| \left\{ \int_{E_n} |g|^q \right\}^{1/p},$$

so that

$$\int_X \chi_{E_n} |g|^q \, d\mu \le \|\Phi\|^q \quad (n = 1, 2, 3, \ldots). \tag{7}$$

If we apply the monotone convergence theorem to (7), we obtain $\|g\|_q \le \|\Phi\|$.

Thus (2) holds and $g \in L^q(\mu)$. It follows that both sides of (6) are continuous functions on $L^p(\mu)$. They coincide on the dense subset $L^\infty(\mu)$ of $L^p(\mu)$; hence they coincide on all of $L^p(\mu)$, and this completes the proof if $\mu(X) < \infty$.

If $\mu(X) = \infty$ but μ is σ-finite, choose $w \in L^1(\mu)$ as in Lemma 6.9. Then $d\tilde{\mu} = w \, d\mu$ defines a finite measure on $\mathfrak{M}$, and

$$F \to w^{1/p} F \tag{8}$$

is a linear isometry of $L^p(\tilde{\mu})$ onto $L^p(\mu)$, because $w(x) > 0$ for every $x \in X$. Hence

$$\Psi(F) = \Phi(w^{1/p} F) \tag{9}$$

defines a bounded linear functional Ψ on $L^p(\tilde{\mu})$, with $\|\Psi\| = \|\Phi\|$.

The first part of the proof shows now that there exists $G \in L^q(\tilde{\mu})$ such that

$$\Psi(F) = \int_X FG \, d\tilde{\mu} \qquad (F \in L^p(\tilde{\mu})). \tag{10}$$

Put $g = w^{1/q}G$. (If $p = 1$, $g = G$.) Then

$$\int_X |g|^q \, d\mu = \int_X |G|^q \, d\tilde{\mu} = \|\Psi\|^q = \|\Phi\|^q \tag{11}$$

if $p > 1$, whereas $\|g\|_\infty = \|G\|_\infty = \|\Psi\| = \|\Phi\|$ if $p = 1$. Thus (2) holds, and since $G \, d\tilde{\mu} = w^{1/p} g \, d\mu$, we finally get

$$\Phi(f) = \Psi(w^{-1/p}f) = \int_X w^{-1/p} fG \, d\tilde{\mu} = \int_X fg \, d\mu \tag{12}$$

for every $f \in L^p(\mu)$. ////

6.17 Remark We have already encountered the special case $p = q = 2$ of Theorem 6.16. In fact, the proof of the general case was based on this special case, for we used the knowledge of the bounded linear functionals on $L^2(\mu)$ in the proof of the Radon-Nikodym theorem, and the latter was the key to the proof of Theorem 6.16. The special case $p = 2$, in turn, depended on the completeness of $L^2(\mu)$, on the fact that $L^2(\mu)$ is therefore a Hilbert space, and on the fact that the bounded linear functionals on a Hilbert space are given by inner products.

We now turn to the complex version of Theorem 2.14.

The Riesz Representation Theorem

6.18 Let X be a locally compact Hausdorff space. Theorem 2.14 characterizes the *positive* linear functionals on $C_c(X)$. We are now in a position to characterize the *bounded* linear functionals Φ on $C_c(X)$. Since $C_c(X)$ is a dense subspace of $C_0(X)$, relative to the supremum norm, every such Φ has a unique extension to a bounded linear functional on $C_0(X)$. Hence we may as well assume to begin with that we are dealing with the Banach space $C_0(X)$.

If μ is a complex Borel measure, Theorem 6.12 asserts that there is a complex Borel function h with $|h| = 1$ such that $d\mu = h \, d|\mu|$. It is therefore reasonable to define integration with respect to a complex measure μ by the formula

$$\int f \, d\mu = \int fh \, d|\mu|. \tag{1}$$

The relation $\int \chi_E \, d\mu = \mu(E)$ is a special case of (1). Thus

$$\int_X \chi_E \, d(\mu + \lambda) = (\mu + \lambda)(E) = \mu(E) + \lambda(E) = \int_X \chi_E \, d\mu + \int_X \chi_E \, d\lambda \tag{2}$$

whenever μ and λ are complex measures on $\mathfrak{M}$ and $E \in \mathfrak{M}$. This leads to the addition formula

$$\int_X f \, d(\mu + \lambda) = \int_X f \, d\mu + \int_X f \, d\lambda, \qquad (3)$$

which is valid (for instance) for every bounded measurable f.

We shall call a complex Borel measure μ on X *regular* if $|\mu|$ is regular in the sense of Definition 2.15. If μ is a complex Borel measure on X, it is clear that the mapping

$$f \to \int_X f \, d\mu \qquad (4)$$

is a bounded linear functional on $C_0(X)$, whose norm is no larger than $|\mu|(X)$. That *all* bounded linear functionals on $C_0(X)$ are obtained in this way is the content of the Riesz theorem:

6.19 Theorem *If X is a locally compact Hausdorff space, then every bounded linear functional Φ on $C_0(X)$ is represented by a unique regular complex Borel measure μ, in the sense that*

$$\Phi f = \int_X f \, d\mu \qquad (1)$$

for every $f \in C_0(X)$. Moreover, the norm of Φ is the total variation of μ:

$$\|\Phi\| = |\mu|(X). \qquad (2)$$

PROOF We first settle the uniqueness question. Suppose μ is a regular complex Borel measure on X and $\int f \, d\mu = 0$ for all $f \in C_0(X)$. By Theorem 6.12 there is a Borel function h, with $|h| = 1$, such that $d\mu = h \, d|\mu|$. For any sequence $\{f_n\}$ in $C_0(X)$ we then have

$$|\mu|(X) = \int_X (\bar{h} - f_n) h \, d|\mu| \le \int_X |\bar{h} - f_n| \, d|\mu|, \qquad (3)$$

and since $C_c(X)$ is dense in $L^1(|\mu|)$ (Theorem 3.14), $\{f_n\}$ can be so chosen that the last expression in (3) tends to 0 as $n \to \infty$. Thus $|\mu|(X) = 0$, and $\mu = 0$. It is easy to see that the difference of two regular complex Borel measures on X is regular. This shows that at most one μ corresponds to each Φ.

Now consider a given bounded linear functional Φ on $C_0(X)$. Assume $\|\Phi\| = 1$, without loss of generality. We shall construct a *positive* linear functional Λ on $C_c(X)$, such that

$$|\Phi(f)| \le \Lambda(|f|) \le \|f\| \qquad (f \in C_c(X)), \qquad (4)$$

where $\|f\|$ denotes the supremum norm.

Once we have this Λ, we associate with it a positive Borel measure λ, as in Theorem 2.14. The conclusion of Theorem 2.14 shows that λ is regular if $\lambda(X) < \infty$. Since

$$\lambda(X) = \sup \{\Lambda f : 0 \leq f \leq 1, f \in C_c(X)\}$$

and since $|\Lambda f| \leq 1$ if $\|f\| \leq 1$, we see that actually $\lambda(X) \leq 1$.

We also deduce from (4) that

$$|\Phi(f)| \leq \Lambda(|f|) = \int_X |f|\, d\lambda = \|f\|_1 \qquad (f \in C_c(X)). \tag{5}$$

The last norm refers to the space $L^1(\lambda)$. Thus Φ is a linear functional on $C_c(X)$ of norm at most 1, *with respect to the $L^1(\lambda)$-norm on $C_c(X)$*. There is a norm-preserving extension of Φ to a linear functional on $L^1(\lambda)$, and therefore Theorem 6.16 (the case $p = 1$) gives a Borel function g, with $|g| \leq 1$, such that

$$\Phi(f) = \int_X fg\, d\lambda \qquad (f \in C_c(X)). \tag{6}$$

Each side of (6) is a continuous functional on $C_0(X)$, and $C_c(X)$ is dense in $C_0(X)$. Hence (6) holds for all $f \in C_0(X)$, and we obtain the representation (1) with $d\mu = g\, d\lambda$.

Since $\|\Phi\| = 1$, (6) shows that

$$\int_X |g|\, d\lambda \geq \sup \{|\Phi(f)| : f \in C_0(X), \|f\| \leq 1\} = 1. \tag{7}$$

We also know that $\lambda(X) \leq 1$ and $|g| \leq 1$. These facts are compatible only if $\lambda(X) = 1$ and $|g| = 1$ a.e. $[\lambda]$. Thus $d\,|\mu| = |g|\, d\lambda = d\lambda$, by Theorem 6.13, and

$$|\mu|(X) = \lambda(X) = 1 = \|\Phi\|, \tag{8}$$

which proves (2).

So all depends on finding a positive linear functional Λ that satisfies (4). If $f \in C_c^+(X)$ [the class of all nonnegative real members of $C_c(X)$], define

$$\Lambda f = \sup \{|\Phi(h)| : h \in C_c(X), |h| \leq f\}. \tag{9}$$

Then $\Lambda f \geq 0$, Λ satisfies (4), $0 \leq f_1 \leq f_2$ implies $\Lambda f_1 \leq \Lambda f_2$, and $\Lambda(cf) = c\Lambda f$ if c is a positive constant. We have to show that

$$\Lambda(f + g) = \Lambda f + \Lambda g \qquad (f \text{ and } g \in C_c^+(X)), \tag{10}$$

and we then have to extend Λ to a linear functional on $C_c(X)$.

Fix f and $g \in C_c^+(X)$. If $\epsilon > 0$, there exist h_1 and $h_2 \in C_c(X)$ such that $|h_1| \leq f, |h_2| \leq g$, and

$$\Lambda f \leq |\Phi(h_1)| + \epsilon, \qquad \Lambda g \leq |\Phi(h_2)| + \epsilon. \tag{11}$$

There are complex numbers α_i, $|\alpha_i| = 1$, so that $\alpha_i \Phi(h_i) = |\Phi(h_i)|$, $i = 1, 2$. Then

$$\Lambda f + \Lambda g \le |\Phi(h_1)| + |\Phi(h_2)| + 2\epsilon$$
$$= \Phi(\alpha_1 h_1 + \alpha_2 h_2) + 2\epsilon$$
$$\le \Lambda(|h_1| + |h_2|) + 2\epsilon$$
$$\le \Lambda(f + g) + 2\epsilon,$$

so that the inequality $\ge$ holds in (10).

Next, choose $h \in C_c(X)$, subject only to the condition $|h| \le f + g$, let $V = \{x : f(x) + g(x) > 0\}$, and define

$$h_1(x) = \frac{f(x)h(x)}{f(x) + g(x)}, \quad h_2(x) = \frac{g(x)h(x)}{f(x) + g(x)} \quad (x \in V), \tag{12}$$
$$h_1(x) = h_2(x) = 0 \quad (x \notin V).$$

It is clear that h_1 is continuous at every point of V. If $x_0 \notin V$, then $h(x_0) = 0$; since h is continuous and since $|h_1(x)| \le |h(x)|$ for all $x \in X$, it follows that x_0 is a point of continuity of h_1. Thus $h_1 \in C_c(X)$, and the same holds for h_2. Since $h_1 + h_2 = h$ and $|h_1| \le f$, $|h_2| \le g$, we have

$$|\Phi(h)| = |\Phi(h_1) + \Phi(h_2)| \le |\Phi(h_1)| + |\Phi(h_2)| \le \Lambda f + \Lambda g.$$

Hence $\Lambda(f + g) \le \Lambda f + \Lambda g$, and we have proved (10).

If f is now a real function, $f \in C_c(X)$, then $2f^+ = |f| + f$, so that $f^+ \in C_c^+(X)$; likewise, $f^- \in C_c^+(X)$; and since $f = f^+ - f^-$, it is natural to define

$$\Lambda f = \Lambda f^+ - \Lambda f^- \quad (f \in C_c(X), f \text{ real}) \tag{13}$$

and

$$\Lambda(u + iv) = \Lambda u + i\Lambda v. \tag{14}$$

Simple algebraic manipulations, just like those which occur in the proof of Theorem 1.32, show now that our extended functional Λ is linear on $C_c(X)$.

This completes the proof. ////

Exercises

1 If μ is a complex measure on a σ-algebra $\mathfrak{M}$, and if $E \in \mathfrak{M}$, define

$$\lambda(E) = \sup \sum |\mu(E_i)|,$$

the supremum being taken over all *finite* partitions $\{E_i\}$ of E. Does it follow that $\lambda = |\mu|$?

2 Prove that the example given at the end of Sec. 6.10 has the stated properties.

3 Prove that the vector space $M(X)$ of all complex regular Borel measures on a locally compact Hausdorff space X is a Banach space if $\|\mu\| = |\mu|(X)$. *Hint:* Compare Exercise 8, Chap. 5. [That the difference of any two members of $M(X)$ is in $M(X)$ was used in the first paragraph of the proof of Theorem 6.19; supply a proof of this fact.]

4 Suppose $1 \le p \le \infty$, and q is the exponent conjugate to p. Suppose μ is a positive σ-finite measure and g is a measurable function such that $fg \in L^1(\mu)$ for every $f \in L^p(\mu)$. Prove that then $g \in L^q(\mu)$.

5 Suppose X consists of two points a and b; define $\mu(\{a\}) = 1$, $\mu(\{b\}) = \mu(X) = \infty$, and $\mu(\emptyset) = 0$. Is it true, for this μ, that $L^\infty(\mu)$ is the dual space of $L^1(\mu)$?

6 Suppose $1 < p < \infty$ and prove that $L^q(\mu)$ is the dual space of $L^p(\mu)$ even if μ is not σ-finite. (As usual, $1/p + 1/q = 1$.)

7 Suppose μ is a complex Borel measure on $[0, 2\pi)$ (or on the unit circle T), and define the Fourier coefficients of μ by

$$\hat{\mu}(n) = \int e^{-int} \, d\mu(t) \qquad (n = 0, \pm 1, \pm 2, \ldots).$$

Assume that $\hat{\mu}(n) \to 0$ as $n \to +\infty$ and prove that then $\hat{\mu}(n) \to 0$ as $n \to -\infty$.

Hint: The assumption also holds with $f \, d\mu$ in place of $d\mu$ if f is any trigonometric polynomial, hence if f is continuous, hence if f is any bounded Borel function, hence if $d\mu$ is replaced by $d|\mu|$.

8 In the terminology of Exercise 7, find all μ such that $\hat{\mu}$ is periodic, with period k. [This means that $\hat{\mu}(n + k) = \hat{\mu}(n)$ for all integers n; of course, k is also assumed to be an integer.]

9 Suppose that $\{g_n\}$ is a sequence of positive continuous functions on $I = [0, 1]$, that μ is a positive Borel measure on I, and that
   (i) $\lim_{n \to \infty} g_n(x) = 0$ a.e. $[m]$,
   (ii) $\int_I g_n \, dm = 1$ for all n,
   (iii) $\lim_{n \to \infty} \int_I f g_n \, dm = \int_I f \, d\mu$ for every $f \in C(I)$.
Does it follow that $\mu \perp m$?

10 Let $(X, \mathfrak{M}, \mu)$ be a positive measure space. Call a set $\Phi \subset L^1(\mu)$ *uniformly integrable* if to each $\epsilon > 0$ corresponds a $\delta > 0$ such that

$$\left| \int_E f \, d\mu \right| < \epsilon$$

whenever $f \in \Phi$ and $\mu(E) < \delta$.

(a) Prove that every finite subset of $L^1(\mu)$ is uniformly integrable.

(b) Prove the following convergence theorem of Vitali:
If (i) $\mu(X) < \infty$, (ii) $\{f_n\}$ is uniformly integrable, (iii) $f_n(x) \to f(x)$ a.e. as $n \to \infty$, and (iv) $|f(x)| < \infty$ a.e., then $f \in L^1(\mu)$ and

$$\lim_{n \to \infty} \int_X |f_n - f| \, d\mu = 0.$$

Suggestion: Use Egoroff's theorem.

(c) Show that (b) fails if μ is Lebesgue measure on $(-\infty, \infty)$, even if $\{\|f_n\|_1\}$ is assumed to be bounded. Hypothesis (i) can therefore not be omitted in (b).

(d) Show that hypothesis (iv) is redundant in (b) for some μ (for instance, for Lebesgue measure on a bounded interval), but that there are finite measures for which the omission of (iv) would make (b) false.

(e) Show that Vitali's theorem implies Lebesgue's dominated convergence theorem, for finite measure spaces. Construct an example in which Vitali's theorem applies although the hypotheses of Lebesgue's theorem do not hold.

(f) Construct a sequence $\{f_n\}$, say on $[0, 1]$, so that $f_n(x) \to 0$ for every x, $\int f_n \to 0$, but $\{f_n\}$ is *not* uniformly integrable (with respect to Lebesgue measure).

(g) However, the following converse of Vitali's theorem is true:
If $\mu(X) < \infty$, $f_n \in L^1(\mu)$, and

$$\lim_{n \to \infty} \int_E f_n \, d\mu$$

exists for every $E \in \mathfrak{M}$, then $\{f_n\}$ is uniformly integrable.

Prove this by completing the following outline.

Define $\rho(A, B) = \int |\chi_A - \chi_B| \, d\mu$. Then $(\mathfrak{M}, \rho)$ is a complete metric space (modulo sets of measure 0), and $E \to \int_E f_n \, d\mu$ is continuous for each n. If $\epsilon > 0$, there exist E_0, δ, N (Exercise 13, Chap. 5) so that

$$\left| \int_E (f_n - f_N) \, d\mu \right| < \epsilon \quad \text{if} \quad \rho(E, E_0) < \delta, \quad n > N. \tag{*}$$

If $\mu(A) < \delta$, (*) holds with $B = E_0 - A$ and $C = E_0 \cup A$ in place of E. Thus (*) holds with A in place of E and 2ϵ in place of ϵ. Now apply (a) to $\{f_1, \ldots, f_N\}$: There exists $\delta' > 0$ such that

$$\left| \int_A f_n \, d\mu \right| < 3\epsilon \quad \text{if} \quad \mu(A) < \delta', \quad n = 1, 2, 3, \ldots.$$

11 Suppose μ is a positive measure on X, $\mu(X) < \infty$, $f_n \in L^1(\mu)$ for $n = 1, 2, 3, \ldots$, $f_n(x) \to f(x)$ a.e., and there exists $p > 1$ and $C < \infty$ such that $\int_X |f_n|^p \, d\mu < C$ for all n. Prove that

$$\lim_{n \to \infty} \int_X |f - f_n| \, d\mu = 0.$$

Hint: $\{f_n\}$ is uniformly integrable.

12 Let $\mathfrak{M}$ be the collection of all sets E in the unit interval $[0, 1]$ such that either E or its complement is at most countable. Let μ be the counting measure on this σ-algebra $\mathfrak{M}$. If $g(x) = x$ for $0 \le x \le 1$, show that g is not $\mathfrak{M}$-measurable, although the mapping

$$f \to \sum xf(x) = \int fg \, d\mu$$

makes sense for every $f \in L^1(\mu)$ and defines a bounded linear functional on $L^1(\mu)$. Thus $(L^1)^* \ne L^\infty$ in this situation.

13 Let $L^\infty = L^\infty(m)$, where m is Lebesgue measure on $I = [0, 1]$. Show that there is a bounded linear functional $\Lambda \ne 0$ on L^∞ that is 0 on $C(I)$, and that therefore there is no $g \in L^1(m)$ that satisfies $\Lambda f = \int_I fg \, dm$ for every $f \in L^\infty$. Thus $(L^\infty)^* \ne L^1$.

CHAPTER SEVEN
DIFFERENTIATION

In elementary Calculus we learn that integration and differentiation are inverses of each other. This fundamental relation persists to a large extent in the context of the Lebesgue integral. We shall see that some of the most important facts about differentiation of integrals and integration of derivatives can be obtained with a minimum of effort by first studying derivatives of measures and the associated maximal functions. The Radon-Nikodym theorem and the Lebesgue decomposition will play a prominent role.

Derivatives of Measures

We begin with a simple theorem whose main purpose is to motivate the definitions that follow.

7.1 Theorem *Suppose μ is a complex Borel measure on R^1 and*

$$f(x) = \mu((-\infty, x)) \qquad (x \in R^1). \tag{1}$$

If $x \in R^1$ and A is a complex number, each of the following two statements implies the other:
(a) f is differentiable at x and $f'(x) = A$.
(b) To every $\epsilon > 0$ corresponds a $\delta > 0$ such that

$$\left| \frac{\mu(I)}{m(I)} - A \right| < \epsilon \tag{2}$$

for every open segment I that contains x and whose length is less than δ. Here m denotes Lebesgue measure on R^1.

7.2 Definitions Theorem 7.1 suggests that one might define the derivative of μ at x to be the limit of the quotients $\mu(I)/m(I)$, as the segments I shrink to x, and that an analogous definition might be appropriate in several variables, i.e., in R^k rather than in R^1.

Accordingly, let us fix a dimension k, denote the open ball with center $x \in R^k$ and radius $r > 0$ by

$$B(x, r) = \{y \in R^k : |y - x| < r\} \tag{1}$$

(the absolute value indicates the euclidean metric, as in Sec. 2.19), associate to any complex Borel measure μ on R^k the quotients

$$(Q_r \mu)(x) = \frac{\mu(B(x, r))}{m(B(x, r))}, \tag{2}$$

where $m = m_k$ is Lebesgue measure on R^k, and define the *symmetric derivative* of μ at x to be

$$(D\mu)(x) = \lim_{r \to 0} (Q_r \mu)(x) \tag{3}$$

at those points $x \in R^k$ at which this limit exists.

We shall study $D\mu$ by means of the *maximal function* $M\mu$. For $\mu \geq 0$, this is defined by

$$(M\mu)(x) = \sup_{0 < r < \infty} (Q_r \mu)(x), \tag{4}$$

and the maximal function of a complex Borel measure μ is, by definition, that of its total variation $|\mu|$.

The functions $M\mu: R^k \to [0, \infty]$ are *lower semicontinuous*, hence measurable.

To see this, assume $\mu \geq 0$, pick $\lambda > 0$, let $E = \{M\mu > \lambda\}$, and fix $x \in E$. Then there is an $r > 0$ such that

$$\mu(B(x, r)) = tm(B(x, r)) \tag{5}$$

for some $t > \lambda$, and there is a $\delta > 0$ that satisfies

$$(r + \delta)^k < r^k t/\lambda. \tag{6}$$

If $|y - x| < \delta$, then $B(y, r + \delta) \supset B(x, r)$, and therefore

$$\mu(B(y, r + \delta)) \geq tm(B(x, r)) = t[r/(r + \delta)]^k m(B(y, r + \delta)) > \lambda m(B(y, r + \delta)).$$

Thus $B(x, \delta) \subset E$. This proves that E is open.

Our first objective is the "maximal theorem" 7.4. The following covering lemma will be used in its proof.

7.3 Lemma *If W is the union of a finite collection of balls $B(x_i, r_i)$, $1 \leq i \leq N$, then there is a set $S \subset \{1, \ldots, N\}$ so that*

(a) *the balls $B(x_i, r_i)$ with $i \in S$ are disjoint,*
(b) $W \subset \bigcup_{i \in S} B(x_i, 3r_i)$, *and*
(c) $m(W) \leq 3^k \sum_{i \in S} m(B(x_i, r_i))$.

PROOF Order the balls $B_i = B(x_i, r_i)$ so that $r_1 \geq r_2 \geq \cdots \geq r_N$. Put $i_1 = 1$. Discard all B_j that intersect B_{i_1}. Let B_{i_2} be the first of the remaining B_j, if there are any. Discard all B_j with $j > i_2$ that intersect B_{i_2}, let B_{i_3} be the first of the remaining ones, and so on, as long as possible. This process stops after a finite number of steps and gives $S = \{i_1, i_2, \ldots\}$.

It is clear that (a) holds. Every discarded B_j is a subset of $B(x_i, 3r_i)$ for some $i \in S$, for if $r' \leq r$ and $B(x', r')$ intersects $B(x, r)$, then $B(x', r') \subset B(x, 3r)$. This proves (b), and (c) follows from (b) because

$$m(B(x, 3r)) = 3^k m(B(x, r))$$

in R^k. ////

The following theorem says, roughly speaking, that the maximal function of a measure cannot be large on a large set.

7.4 Theorem *If μ is a complex Borel measure on R^k and λ is a positive number, then*

$$m\{M\mu > \lambda\} \leq 3^k \lambda^{-1} \|\mu\|. \tag{1}$$

Here $\|\mu\| = |\mu|(R^k)$, and the left side of (1) is an abbreviation for the more cumbersome expression

$$m(\{x \in R^k : (M\mu)(x) > \lambda\}). \tag{2}$$

We shall often simplify notation in this way.

PROOF Fix μ and λ. Let K be a compact subset of the open set $\{M\mu > \lambda\}$. Each $x \in K$ is the center of an open ball B for which

$$|\mu|(B) > \lambda m(B).$$

Some finite collection of these B's covers K, and Lemma 7.3 gives us a disjoint subcollection, say $\{B_1, \ldots, B_n\}$, that satisfies

$$m(K) \leq 3^k \sum_1^n m(B_i) \leq 3^k \lambda^{-1} \sum_1^n |\mu|(B_i) \leq 3^k \lambda^{-1} \|\mu\|.$$

The disjointness of $\{B_1, \ldots, B_n\}$ was used in the last inequality.

Now (1) follows by taking the supremum over all compact $K \subset \{M\mu > \lambda\}$. ////

7.5 Weak L^1 If $f \in L^1(R^k)$ and $\lambda > 0$, then

$$m\{|f| > \lambda\} \leq \lambda^{-1}\|f\|_1 \tag{1}$$

because, putting $E = \{|f| > \lambda\}$, we have

$$\lambda m(E) \leq \int_E |f|\, dm \leq \int_{R^k} |f|\, dm = \|f\|_1. \tag{2}$$

Accordingly, any measurable function f for which

$$\lambda \cdot m\{|f| > \lambda\} \tag{3}$$

is a bounded function of λ on $(0, \infty)$ is said to belong to *weak L^1*.

Thus weak L^1 contains L^1. That it is actually larger is shown most simply by the function $1/x$ on $(0, 1)$.

We associate to each $f \in L^1(R^k)$ its *maximal function* $Mf: R^k \to [0, \infty]$, by setting

$$(Mf)(x) = \sup_{0 < r < \infty} \frac{1}{m(B_r)} \int_{B(x, r)} |f|\, dm. \tag{4}$$

[We wrote B_r in place of $B(x, r)$ because $m(B(x, r))$ depends only on the radius r.] If we identify f with the measure μ given by $d\mu = f\, dm$, we see that (4) agrees with the previously defined $M\mu$. Theorem 7.4 states therefore that the "maximal operator" M sends L^1 to weak L^1, with a bound (namely 3^k) that depends only on the space R^k:

For every $f \in L^1(R^k)$ and every $\lambda > 0$,

$$m\{Mf > \lambda\} \leq 3^k \lambda^{-1} \|f\|_1. \tag{5}$$

7.6 Lebesgue points If $f \in L^1(R^k)$, any $x \in R^k$ for which it is true that

$$\lim_{r \to 0} \frac{1}{m(B_r)} \int_{B(x, r)} |f(y) - f(x)|\, dm(y) = 0 \tag{1}$$

is called a *Lebesgue point* of f.

For example, (1) holds if f is continuous at the point x. In general, (1) means that the averages of $|f - f(x)|$ are small on small balls centered at x. The Lebesgue points of f are thus points where f does not oscillate too much, in an average sense.

It is probably far from obvious that every $f \in L^1$ has Lebesgue points. But the following remarkable theorem shows that they always exist. (See also Exercise 23.)

7.7 Theorem *If $f \in L^1(R^k)$, then almost every $x \in R^k$ is a Lebesgue point of f.*

PROOF Define

$$(T_r f)(x) = \frac{1}{m(B_r)} \int_{B(x,\,r)} |f - f(x)|\, dm \qquad (1)$$

for $x \in R^k$, $r > 0$, and put

$$(Tf)(x) = \limsup_{r \to 0} (T_r f)(x). \qquad (2)$$

We have to prove that $Tf = 0$ a.e. $[m]$.

Pick $y > 0$. Let n be a positive integer. By Theorem 3.14, there exists $g \in C(R^k)$ so that $\|f - g\|_1 < 1/n$. Put $h = f - g$.

Since g is continuous, $Tg = 0$. Since

$$(T_r h)(x) \le \frac{1}{m(B_r)} \int_{B(x,\,r)} |h|\, dm + |h(x)| \qquad (3)$$

we have

$$Th \le Mh + |h|. \qquad (4)$$

Since $T_r f \le T_r g + T_r h$, it follows that

$$Tf \le Mh + |h|. \qquad (5)$$

Therefore

$$\{Tf > 2y\} \subset \{Mh > y\} \cup \{|h| > y\}. \qquad (6)$$

Denote the union on the right of (6) by $E(y, n)$. Since $\|h\|_1 < 1/n$, Theorem 7.4 and the inequality 7.5(1) show that

$$m(E(y, n)) \le (3^k + 1)/(yn). \qquad (7)$$

The left side of (6) is independent of n. Hence

$$\{Tf > 2y\} \subset \bigcap_{n=1}^{\infty} E(y, n). \qquad (8)$$

This intersection has measure 0, by (7), so that $\{Tf > 2y\}$ is a subset of a set of measure 0. Since Lebesgue measure is complete, $\{Tf > 2y\}$ is Lebesgue measurable, and has measure 0. This holds for every positive y. Hence $Tf = 0$ a.e. $[m]$. ////

Theorem 7.7 yields interesting information, with very little effort, about topics such as

(a) differentiation of absolutely continuous measures,
(b) differentiation using sets other than balls,
(c) differentiation of indefinite integrals in R^1,
(d) metric density of measurable sets.

We shall now discuss these topics.

7.8 Theorem *Suppose μ is a complex Borel measure on R^k, and $\mu \ll m$. Let f be the Radon-Nikodym derivative of μ with respect to m. Then $D\mu = f$ a.e. $[m]$, and*

$$\mu(E) = \int_E (D\mu)\, dm \qquad (1)$$

for all Borel sets $E \subset R^k$.

In other words, the Radon-Nikodym derivative can also be obtained as a limit of the quotients $Q_r \mu$.

PROOF The Radon-Nikodym theorem asserts that (1) holds with f in place of $D\mu$. At any Lebesgue point x of f, it follows that

$$f(x) = \lim_{r \to 0} \frac{1}{m(B_r)} \int_{B(x,\, r)} f\, dm = \lim_{r \to 0} \frac{\mu(B(x, r))}{m(B(x, r))}. \qquad (2)$$

Thus $(D\mu)(x)$ exists and equals $f(x)$ at every Lebesgue point of f, hence a.e. $[m]$. ////

7.9 Nicely shrinking sets Suppose $x \in R^k$. A sequence $\{E_i\}$ of Borel sets in R^k is said to *shrink to x nicely* if there is a number $\alpha > 0$ with the following property: There is a sequence of balls $B(x, r_i)$, with $\lim r_i = 0$, such that $E_i \subset B(x, r_i)$ and

$$m(E_i) \geq \alpha \cdot m(B(x, r_i)) \qquad (1)$$

for $i = 1, 2, 3, \ldots$.

Note that it is not required that $x \in E_i$, nor even that x be in the closure of E_i. Condition (1) is a quantitative version of the requirement that each E_i must occupy a substantial portion of some spherical neighborhood of x. For example, a nested sequence of k-cells whose longest edge is at most 1,000 times as long as its shortest edge and whose diameter tends to 0 shrinks nicely. A nested sequence of rectangles (in R^2) whose edges have lengths $1/i$ and $(1/i)^2$ does not shrink nicely.

7.10 Theorem *Associate to each $x \in R^k$ a sequence $\{E_i(x)\}$ that shrinks to x nicely, and let $f \in L^1(R^k)$. Then*

$$f(x) = \lim_{i \to \infty} \frac{1}{m(E_i(x))} \int_{E_i(x)} f\, dm \qquad (1)$$

at every Lebesgue point of f, hence a.e. $[m]$.

PROOF Let x be a Lebesgue point of f and let $\alpha(x)$ and $B(x, r_i)$ be the positive number and the balls that are associated to the sequence $\{E_i(x)\}$. Then, because $E_i(x) \subset B(x, r_i)$,

$$\frac{\alpha(x)}{m(E_i(x))} \int_{E_i(x)} |f - f(x)| \, dm \le \frac{1}{m(B(x, r_i))} \int_{B(x, r_i)} |f - f(x)| \, dm.$$

The right side converges to 0 as $i \to \infty$, because $r_i \to 0$ and x is a Lebesgue point of f. Hence the left side converges to 0, and (1) follows. ////

Note that no relation of any sort was assumed to exist between $\{E_i(x)\}$ and $\{E_i(y)\}$, for different points x and y.

Note also that Theorem 7.10 leads to a correspondingly stronger form of Theorem 7.8. We omit the details.

7.11 Theorem *If $f \in L^1(R^1)$ and*

$$F(x) = \int_{-\infty}^{x} f \, dm \qquad (-\infty < x < \infty),$$

then $F'(x) = f(x)$ at every Lebesgue point of f, hence a.e. $[m]$.

(This is the easy half of the fundamental theorem of Calculus, extended to Lebesgue integrals.)

PROOF Let $\{\delta_i\}$ be a sequence of positive numbers that converges to 0. Theorem 7.10, with $E_i(x) = [x, x + \delta_i]$, shows then that the right-hand derivative of F exists at all Lebesgue points of x of f and that it is equal to $f(x)$ at these points. If we let $E_i(x)$ be $[x - \delta_i, x]$ instead, we obtain the same result for the left-hand derivative of F at x. ////

7.12 Metric density Let E be a Lebesgue measurable subset of R^k. The metric density of E at a point $x \in R^k$ is defined to be

$$\lim_{r \to 0} \frac{m(E \cap B(x, r))}{m(B(x, r))} \qquad (1)$$

provided, of course, that this limit exists.

If we let f be the characteristic function of E and apply Theorem 7.8 or Theorem 7.10, we see that *the metric density of E is 1 at almost every point of E, and that it is 0 at almost every point of the complement of E.*

Here is a rather striking consequence of this, which should be compared with Exercise 8 in Chap. 2:

If $\epsilon > 0$, there is no set $E \subset R^1$ that satisfies

$$\epsilon < \frac{m(E \cap I)}{m(I)} < 1 - \epsilon \qquad (2)$$

for every segment I.

Having dealt with differentiation of absolutely continuous measures, we now turn to those that are singular with respect to m.

7.13 Theorem *Associate to each $x \in R^k$ a sequence $\{E_i(x)\}$ that shrinks to x nicely. If μ is a complex Borel measure and $\mu \perp m$, then*

$$\lim_{i \to \infty} \frac{\mu(E_i(x))}{m(E_i(x))} = 0 \quad \text{a.e. } [m]. \tag{1}$$

PROOF The Jordan decomposition theorem shows that it suffices to prove (1) under the additional assumption that $\mu \geq 0$. In that case, arguing as in the proof of Theorem 7.10, we have

$$\frac{\alpha(x)\mu(E_i(x))}{m(E_i(x))} \leq \frac{\mu(E_i(x))}{m(B(x, r_i))} \leq \frac{\mu(B(x, r_i))}{m(B(x, r_i))}.$$

Hence (1) is a consequence of the special case

$$(D\mu)(x) = 0 \quad \text{a.e. } [m], \tag{2}$$

which will now be proved.

The upper derivative $\bar{D}\mu$, defined by

$$(\bar{D}\mu)(x) = \lim_{n \to \infty} \left[\sup_{0 < r < 1/n} (Q_r \mu)(x) \right] \quad (x \in R^k) \tag{3}$$

is a Borel function, because the quantity in brackets decreases as n increases and is, for each n, a lower semicontinuous function of x; the reasoning used in Sec. 7.2 proves this.

Choose $\lambda > 0$, $\epsilon > 0$. Since $\mu \perp m$, μ is concentrated on a set of Lebesgue measure 0. The regularity of μ (Theorem 2.18) shows therefore that there is a compact set K, with $m(K) = 0$, $\mu(K) > \|\mu\| - \epsilon$.

Define $\mu_1(E) = \mu(K \cap E)$, for any Borel set $E \subset R^k$, and put $\mu_2 = \mu - \mu_1$. Then $\|\mu_2\| < \epsilon$, and, for every x outside K,

$$(\bar{D}\mu)(x) = (\bar{D}\mu_2)(x) \leq (M\mu_2)(x). \tag{4}$$

Hence

$$\{\bar{D}\mu > \lambda\} \subset K \cup \{M\mu_2 > \lambda\}, \tag{5}$$

and Theorem 7.4 shows that

$$m\{\bar{D}\mu > \lambda\} \leq 3^k \lambda^{-1} \|\mu_2\| < 3^k \lambda^{-1} \epsilon. \tag{6}$$

Since (6) holds for every $\epsilon > 0$ and for every $\lambda > 0$, we conclude that $\bar{D}\mu = 0$ a.e. $[m]$, i.e., that (2) holds. ////

Theorems 7.10 and 7.13 can be combined in the following way:

7.14 Theorem *Suppose that to each $x \in R^k$ is associated some sequence $\{E_i(x)\}$ that shrinks to x nicely, and that μ is a complex Borel measure on R^k.*

Let $d\mu = f\, dm + d\mu_s$ be the Lebesgue decomposition of μ with respect to m. Then

$$\lim_{i \to \infty} \frac{\mu(E_i(x))}{m(E_i(x))} = f(x) \quad \text{a.e. } [m].$$

In particular, $\mu \perp m$ if and only if $(D\mu)(x) = 0$ a.e. $[m]$.

The following result contrasts strongly with Theorem 7.13:

7.15 Theorem *If μ is a positive Borel measure on R^k and $\mu \perp m$, then*

$$(D\mu)(x) = \infty \quad \text{a.e. } [\mu]. \tag{1}$$

PROOF There is a Borel set $S \subset R^k$ with $m(S) = 0$ and $\mu(R^k - S) = 0$, and there are open sets $V_j \supset S$ with $m(V_j) < 1/j$, for $j = 1, 2, 3, \ldots$.

For $N = 1, 2, 3, \ldots$, let E_N be the set of all $x \in S$ to which correspond radii $r_i = r_i(x)$, with $\lim r_i = 0$, such that

$$\mu(B(x, r_i)) < Nm(B(x, r_i)). \tag{2}$$

Then (1) holds for every $x \in S - \bigcup_N E_N$.

Fix N and j, for the moment. Every $x \in E_N$ is then the center of a ball $B_x \subset V_j$ that satisfies (2). Let β_x be the open ball with center x whose radius is $1/3$ of that of B_x. The union of these balls β_x is an open set $W_{j,N}$ that contains E_N and lies in V_j. We claim that

$$\mu(W_{j,N}) < 3^k N/j. \tag{3}$$

To prove (3), let $K \subset W_{j,N}$ be compact. Finitely many β_x cover K. Lemma 7.3 shows therefore that there is a finite set $F \subset E_N$ with the following properties:

(a) $\{\beta_x : x \in F\}$ is a disjoint collection, and
(b) $K \subset \bigcup_{x \in F} B_x$.

Thus

$$\mu(K) \leq \sum_{x \in F} \mu(B_x) < N \sum_{x \in F} m(B_x)$$

$$= 3^k N \sum_{x \in F} m(\beta_x) \leq 3^k N m(V_j) < 3^k N/j.$$

This proves (3).

Now put $\Omega_N = \bigcap_j W_{j,N}$. Then $E_N \subset \Omega_N$, Ω_N is a G_δ, $\mu(\Omega_N) = 0$, and $(D\mu)(x) = \infty$ at every point of $S - \bigcup_N \Omega_N$. ////

The Fundamental Theorem of Calculus

7.16 This theorem concerns functions defined on some compact interval $[a, b]$ in R^1. It has two parts. The first asserts, roughly speaking, that the derivative of the indefinite integral of a function is that same function. We dealt with this in Theorem 7.11. The second part goes the other way: one returns to the original function by integrating its derivative. More precisely

$$f(x) - f(a) = \int_a^x f'(t)\, dt \qquad (a \leq x \leq b). \tag{1}$$

In the elementary version of this theorem, one assumes that f is differentiable at every point of $[a, b]$ and that f' is a continuous function. The proof of (1) is then easy.

In trying to extend (1) to the setting of the Lebesgue integral, questions such as the following come up naturally:

Is it enough to assume that $f' \in L^1$, rather than that f' is continuous?

If f is continuous and differentiable at *almost all* points of $[a, b]$, must (1) then hold?

Before proving any positive results, here are two examples that show how (1) can fail.

(a) Put $f(x) = x^2 \sin(x^{-2})$ if $x \neq 0$, $f(0) = 0$. Then f is differentiable at every point, but

$$\int_0^1 |f'(t)|\, dt = \infty, \tag{2}$$

so $f' \notin L^1$.

If we interpret the integral in (1) (with $[0, 1]$ in place of $[a, b]$) as the limit, as $\epsilon \to 0$, of the integrals over $[\epsilon, 1]$, then (1) still holds for this f.

More complicated situations can arise where this kind of passage to the limit is of no use. There are integration processes, due to Denjoy and Perron (see [18], [28]), which are so designed that (1) holds whenever f is differentiable at every point. These fail to have the property that the integrability of f implies that of $|f|$, and therefore do not play such an important role in analysis.

(b) Suppose f is continuous on $[a, b]$, f is differentiable at almost every point of $[a, b]$, and $f' \in L^1$ on $[a, b]$. Do these assumptions imply that (1) holds?
Answer: No.

Choose $\{\delta_n\}$ so that $1 = \delta_0 > \delta_1 > \delta_2 > \cdots$, $\delta_n \to 0$. Put $E_0 = [0, 1]$. Suppose $n \geq 0$ and E_n is constructed so that E_n is the union of 2^n disjoint closed intervals, each of length $2^{-n}\delta_n$. Delete a segment in the center of each of these 2^n intervals, so that each of the remaining 2^{n+1} intervals has length

$2^{-n-1}\delta_{n+1}$ (this is possible, since $\delta_{n+1} < \delta_n$), and let E_{n+1} be the union of these 2^{n+1} intervals. Then $E_1 \supset E_2 \supset \cdots$, $m(E_n) = \delta_n$, and if

$$E = \bigcap_{n=1}^{\infty} E_n, \tag{3}$$

then E is compact and $m(E) = 0$. (In fact, E is perfect.) Put

$$g_n = \delta_n^{-1}\chi_{E_n} \quad \text{and} \quad f_n(x) = \int_0^x g_n(t)\,dt \quad (n = 0, 1, 2, \ldots). \tag{4}$$

Then $f_n(0) = 0$, $f_n(1) = 1$, and each f_n is a monotonic function which is constant on each segment in the complement of E_n. If I is one of the 2^n intervals whose union is E_n, then

$$\int_I g_n(t)\,dt = \int_I g_{n+1}(t)\,dt = 2^{-n}. \tag{5}$$

It follows from (5) that

$$f_{n+1}(x) = f_n(x) \quad (x \notin E_n) \tag{6}$$

and that

$$|f_n(x) - f_{n+1}(x)| \le \int_I |g_n - g_{n+1}| < 2^{-n+1} \quad (x \in E_n). \tag{7}$$

Hence $\{f_n\}$ converges uniformly to a continuous monotonic function f, with $f(0) = 0$, $f(1) = 1$, and $f'(x) = 0$ for all $x \notin E$. Since $m(E) = 0$, we have $f' = 0$ a.e.

Thus (1) fails.

If $\delta_n = (2/3)^n$, the set E is Cantor's "middle thirds" set.

Having seen what can go wrong, assume now that $f' \in L^1$ and that (1) does hold. There is then a measure μ, defined by $d\mu = f'\,dm$. Since $\mu \ll m$, Theorem 6.11 shows that there corresponds to each $\epsilon > 0$ a $\delta > 0$ so that $|\mu|(E) < \epsilon$ whenever E is a union of disjoint segments whose total length is less than δ. Since $f(y) - f(x) = \mu((x, y))$ if $a \le x < y \le b$, it follows that the absolute continuity of f, as defined below, is *necessary* for (1). Theorem 7.20 will show that this necessary condition is also sufficient.

7.17 Definition A complex function f, defined on an interval $I = [a, b]$, is said to be *absolutely continuous* on I (briefly, f is AC on I) if there corresponds to every $\epsilon > 0$ a $\delta > 0$ so that

$$\sum_{i=1}^n |f(\beta_i) - f(\alpha_i)| < \epsilon \tag{1}$$

for any n and any disjoint collection of segments $(\alpha_1, \beta_1), \ldots, (\alpha_n, \beta_n)$ in I whose lengths satisfy

$$\sum_{i=1}^{n} (\beta_i - \alpha_i) < \delta. \qquad (2)$$

Such an f is obviously continuous: simply take $n = 1$.

In the following theorem, the implication $(b) \to (c)$ is probably the most interesting. That $(a) \to (c)$ without assuming monotonicity of f is the content of Theorem 7.20.

7.18 Theorem *Let $I = [a, b]$, let $f: I \to R^1$ be continuous and nondecreasing. Each of the following three statements about f implies the other two:*

(a) *f is AC on I.*
(b) *f maps sets of measure 0 to sets of measure 0.*
(c) *f is differentiable a.e. on I, $f' \in L^1$, and*

$$f(x) - f(a) = \int_a^x f'(t)\, dt \qquad (a \le x \le b). \qquad (1)$$

Note that the functions constructed in Example 7.16(b) map certain compact sets of measure 0 onto the whole unit interval!

Exercise 12 complements this theorem.

PROOF We will show that $(a) \to (b) \to (c) \to (a)$.

Let $\mathfrak{M}$ denote the σ-algebra of all Lebesgue measurable subsets of R^1.

Assume f is AC on I, pick $E \subset I$ so that $E \in \mathfrak{M}$ and $m(E) = 0$. We have to show that $f(E) \in \mathfrak{M}$ and $m(f(E)) = 0$. Without loss of generality, assume that neither a nor b lie in E.

Choose $\epsilon > 0$. Associate $\delta > 0$ to f and ϵ, as in Definition 7.17. There is then an open set V with $m(V) < \delta$, so that $E \subset V \subset I$. Let (α_i, β_i) be the disjoint segments whose union is V. Then $\sum (\beta_i - \alpha_i) < \delta$, and our choice of δ shows that therefore

$$\sum_i (f(\beta_i) - f(\alpha_i)) \le \epsilon. \qquad (2)$$

[Definition 7.17 was stated in terms of finite sums; thus (2) holds for every partial sum of the (possibly) infinite series, hence (2) holds also for the sum of the whole series, as stated.]

Since $E \subset V$, $f(E) \subset \bigcup [f(\alpha_i), f(\beta_i)]$. The Lebesgue measure of this union is the left side of (2). This says that $f(E)$ is a subset of Borel sets of arbitrarily small measure. Since Lebesgue measure is complete, it follows that $f(E) \in \mathfrak{M}$ and $m(f(E)) = 0$.

We have now proved that (a) implies (b).

Assume next that (b) holds. Define

$$g(x) = x + f(x) \qquad (a \leq x \leq b). \tag{3}$$

If the f-image of some segment of length η has length η', then the g-image of that same segment has length $\eta + \eta'$. From this it follows easily that g satisfies (b), since f does.

Now suppose $E \subset I$, $E \in \mathfrak{M}$. Then $E = E_1 \cup E_0$ where $m(E_0) = 0$ and E_1 is an F_σ (Theorem 2.20). Thus E_1 is a countable union of compact sets, and so is $g(E_1)$, because g is continuous. Since g satisfies (b), $m(g(E_0)) = 0$. Since $g(E) = g(E_1) \cup g(E_0)$, we conclude: $g(E) \in \mathfrak{M}$.

Therefore we can define

$$\mu(E) = m(g(E)) \qquad (E \subset I, E \in \mathfrak{M}). \tag{4}$$

Since g is one-to-one (this is our reason for working with g rather than f), disjoint sets in I have disjoint g-images. The countable additivity of m shows therefore that μ is a (positive, bounded) measure on $\mathfrak{M}$. Also, $\mu \ll m$, because g satisfies (b). Thus

$$d\mu = h\, dm \tag{5}$$

for some $h \in L^1(m)$, by the Radon-Nikodym theorem.

If $E = [a, x]$, then $g(E) = [g(a), g(x)]$, and (5) gives

$$g(x) - g(a) = m(g(E)) = \mu(E) = \int_E h\, dm = \int_a^x h(t)\, dt.$$

If we now use (3), we conclude that

$$f(x) - f(a) = \int_a^x [h(t) - 1]\, dt \qquad (a \leq x \leq b). \tag{6}$$

Thus $f'(x) = h(x) - 1$ a.e. [m], by Theorem 7.11.

We have now proved that (b) implies (c).

The discussion that preceded Definition 7.17 showed that (c) implies (a).
////

7.19 Theorem *Suppose $f: I \to R^1$ is AC, $I = [a, b]$. Define*

$$F(x) = \sup \sum_{i=1}^N |f(t_i) - f(t_{i-1})| \qquad (a \leq x \leq b) \tag{1}$$

where the supremum is taken over all N and over all choices of $\{t_i\}$ such that

$$a = t_0 < t_1 < \cdots < t_N = x. \tag{2}$$

The functions F, $F + f$, $F - f$ are then nondecreasing and AC on I.

[F is called the *total variation* function of f. If f is any (complex) function on I, AC or not, and $F(b) < \infty$, then f is said to have *bounded variation* on I, and $F(b)$ is the *total variation* of f on I. Exercise 13 is relevant to this.]

PROOF If (2) holds and $x < y \leq b$, then

$$F(y) \geq |f(y) - f(x)| + \sum_{i=1}^{N} |f(t_i) - f(t_{i-1})|. \tag{3}$$

Hence $F(y) \geq |f(y) - f(x)| + F(x)$. In particular

$$F(y) \geq f(y) - f(x) + F(x) \quad \text{and} \quad F(y) \geq f(x) - f(y) + F(x). \tag{4}$$

This proves that F, $F + f$, $F - f$ are nondecreasing.

Since sums of two AC functions are obviously AC, it only remains to be proved that F is AC on I.

If $(\alpha, \beta) \subset I$ then

$$F(\beta) - F(\alpha) = \sup \sum_{1}^{n} |f(t_i) - f(t_{i-1})|, \tag{5}$$

the supremum being taken over all $\{t_i\}$ that satisfy $\alpha = t_0 < \cdots < t_n = \beta$.

Note that $\sum (t_i - t_{i-1}) = \beta - \alpha$.

Now pick $\epsilon > 0$, associate $\delta > 0$ to f and ϵ as in Definition 7.17, choose disjoint segments $(\alpha_j, \beta_j) \subset I$ with $\sum (\beta_j - \alpha_j) < \delta$, and apply (5) to each (α_j, β_j). It follows that

$$\sum_{j} (F(\beta_j) - F(\alpha_j)) \leq \epsilon, \tag{6}$$

by our choice of δ. Thus F is AC on I. ////

We have now reached our main objective:

7.20 Theorem *If f is a complex function that is AC on $I = [a, b]$, then f is differentiable at almost all points of I, $f' \in L^1(m)$, and*

$$f(x) - f(a) = \int_{a}^{x} f'(t)\, dt \qquad (a \leq x \leq b). \tag{1}$$

PROOF It is of course enough to prove this for real f. Let F be its total variation function, as in Theorem 7.19, define

$$f_1 = \tfrac{1}{2}(F + f), \qquad f_2 = \tfrac{1}{2}(F - f), \tag{2}$$

and apply the implication (a)→(c) of Theorem 7.18 to f_1 and f_2. Since

$$f = f_1 - f_2 \tag{3}$$

this yields (1). ////

The next theorem derives (1) from a different set of hypotheses, by an entirely different method of proof.

7.21 Theorem *If $f: [a, b] \to R^1$ is differentiable at every point of $[a, b]$ and $f' \in L^1$ on $[a, b]$, then*

$$f(x) - f(a) = \int_a^x f'(t)\, dt \qquad (a \le x \le b). \tag{1}$$

Note that differentiability is assumed to hold at *every* point of $[a, b]$.

PROOF It is clear that it is enough to prove this for $x = b$. Fix $\epsilon > 0$. Theorem 2.25 ensures the existence of a lower semicontinuous function g on $[a, b]$ such that $g > f'$ and

$$\int_a^b g(t)\, dt < \int_a^b f'(t)\, dt + \epsilon. \tag{2}$$

Actually, Theorem 2.25 only gives $g \ge f'$, but since $m([a, b]) < \infty$, we can add a small constant to g without affecting (2). For any $\eta > 0$, define

$$F_\eta(x) = \int_a^x g(t)\, dt - f(x) + f(a) + \eta(x - a) \qquad (a \le x \le b). \tag{3}$$

Keep η fixed for the moment. To each $x \in [a, b)$ there corresponds a $\delta_x > 0$ such that

$$g(t) > f'(x) \quad \text{and} \quad \frac{f(t) - f(x)}{t - x} < f'(x) + \eta \tag{4}$$

for all $t \in (x, x + \delta_x)$, since g is lower semicontinuous and $g(x) > f'(x)$. For any such t we therefore have

$$F_\eta(t) - F_\eta(x) = \int_x^t g(s)\, ds - [f(t) - f(x)] + \eta(t - x)$$
$$> (t - x)f'(x) - (t - x)[f'(x) + \eta] + \eta(t - x) = 0.$$

Since $F_\eta(a) = 0$ and F_η is *continuous*, there is a last point $x \in [a, b]$ at which $F_\eta(x) = 0$. If $x < b$, the preceding computation implies that $F_\eta(t) > 0$ for $t \in (x, b]$. In any case, $F_\eta(b) \ge 0$. Since this holds for every $\eta > 0$, (2) and (3) now give

$$f(b) - f(a) \le \int_a^b g(t)\, dt < \int_a^b f'(t)\, dt + \epsilon, \tag{5}$$

and since ϵ was arbitrary, we conclude that

$$f(b) - f(a) \le \int_a^b f'(t)\, dt. \tag{6}$$

If f satisfies the hypotheses of the theorem, so does $-f$; therefore (6) holds with $-f$ in place of f, and these two inequalities together give (1). ////

Differentiable Transformations

7.22 Definitions Suppose V is an open set in R^k, T maps V into R^k, and $x \in V$. If there exists a linear operator A on R^k (i.e., a linear mapping of R^k into R^k, as in Definition 2.1) such that

$$\lim_{h \to 0} \frac{|T(x+h) - T(x) - Ah|}{|h|} = 0 \tag{1}$$

(where, of course, $h \in R^k$), then we say that T is *differentiable* at x, and define

$$T'(x) = A. \tag{2}$$

The linear operator $T'(x)$ is called the *derivative* of T at x. (One shows easily that there is at most one linear A that satisfies the preceding requirements; thus it is legitimate to talk about *the* derivative of T.) The term *differential* is also often used for $T'(x)$.

The point of (1) is of course that the difference $T(x+h) - T(x)$ is approximated by $T'(x)h$, a *linear* function of h.

Since every real number α gives rise to a linear operator on R^1 (mapping h to αh), our definition of $T'(x)$ coincides with the usual one when $k = 1$.

When $A: R^k \to R^k$ is linear, Theorem 2.20(e) shows that there is a number $\Delta(A)$ such that

$$m(A(E)) = \Delta(A)m(E) \tag{3}$$

for all measurable sets $E \subset R^k$. Since

$$A'(x) = A \quad (x \in R^k) \tag{4}$$

and since every differentiable transformation T can be locally approximated by a constant plus a linear transformation, one may conjecture that

$$\frac{m(T(E))}{m(E)} \sim \Delta(T'(x)) \tag{5}$$

for suitable sets E that are close to x. This will be proved in Theorem 7.24, and furnishes the motivation for Theorem 7.26.

Recall that $\Delta(A) = |\det A|$ was proved in Sec. 2.23. When T is differentiable at x, the determinant of $T'(x)$ is called the *Jacobian* of T at x, and is denoted by $J_T(x)$. Thus

$$\Delta(T'(x)) = |J_T(x)|. \tag{6}$$

The following lemma seems geometrically obvious. Its proof depends on the Brouwer fixed point theorem. One can avoid the use of this theorem by imposing stronger hypotheses on F, for example, by assuming that F is an open mapping. But this would lead to unnecessarily strong assumptions in Theorem 7.26.

7.23 Lemma Let $S = \{x : |x| = 1\}$ be the sphere in R^k that is the boundary of the open unit ball $B = B(0, 1)$.
If $F: \bar{B} \to R^k$ is continuous, $0 < \epsilon < 1$, and

$$|F(x) - x| < \epsilon \qquad (1)$$

for all $x \in S$, then $F(B) \supset B(0, 1 - \epsilon)$.

PROOF Assume, to reach a contradiction, that some point $a \in B(0, 1 - \epsilon)$ is not in $F(B)$. By (1), $|F(x)| > 1 - \epsilon$ if $x \in S$. Thus a is not in $F(S)$, and therefore $a \neq F(x)$, for every $x \in \bar{B}$. This enables us to define a continuous map $G: \bar{B} \to \bar{B}$ by

$$G(x) = \frac{a - F(x)}{|a - F(x)|}. \qquad (2)$$

If $x \in S$, then $x \cdot x = |x|^2 = 1$, so that

$$x \cdot (a - F(x)) = x \cdot a + x \cdot (x - F(x)) - 1 < |a| + \epsilon - 1 < 0. \qquad (3)$$

This shows that $x \cdot G(x) < 0$, hence $x \neq G(x)$.
If $x \in B$, then obviously $x \neq G(x)$, simply because $G(x) \in S$.
Thus G fixes no point of $\bar{B}$, contrary to Brouwer's theorem which states that every continuous map of $\bar{B}$ into $\bar{B}$ has at least one fixed point.   ////

A proof of Brouwer's theorem that is both elementary and simple may be found on pp. 38–40 of "Dimension Theory" by Hurewicz and Wallman, Princeton University Press, 1948.

7.24 Theorem If

(a) V is open in R^k,
(b) $T: V \to R^k$ is continuous, and
(c) T is differentiable at some point $x \in V$, then

$$\lim_{r \to 0} \frac{m(T(B(x, r)))}{m(B(x, r))} = \Delta(T'(x)). \qquad (1)$$

Note that $T(B(x, r))$ is Lebesgue measurable; in fact, it is σ-compact, because $B(x, r)$ is σ-compact and T is continuous.

PROOF Assume, without loss of generality, that $x = 0$ and $T(x) = 0$. Put $A = T'(0)$.
The following elementary fact about linear operators on finite-dimensional vector spaces will be used: *A linear operator A on R^k is one-to-*

one if and only if the range of A is all of R^k. In that case, the inverse A^{-1} of A is also linear.

Accordingly, we split the proof into two cases.

CASE I *A is one-to-one.* Define

$$F(x) = A^{-1}T(x) \qquad (x \in V). \tag{2}$$

Then $F'(0) = A^{-1}T'(0) = A^{-1}A = I$, the identity operator. We shall prove that

$$\lim_{r \to 0} \frac{m(F(B(0, r)))}{m(B(0, r))} = 1. \tag{3}$$

Since $T(x) = AF(x)$, we have

$$m(T(B)) = m(A(F(B))) = \Delta(A)m(F(B)) \tag{4}$$

for every ball B, by 7.22(3). Hence (3) will give the desired result.

Choose $\epsilon > 0$. Since $F(0) = 0$ and $F'(0) = I$, there exists a $\delta > 0$ such that $0 < |x| < \delta$ implies

$$|F(x) - x| < \epsilon |x|. \tag{5}$$

We claim that the inclusions

$$B(0, (1 - \epsilon)r) \subset F(B(0, r)) \subset B(0, (1 + \epsilon)r) \tag{6}$$

hold if $0 < r < \delta$. The first of these follows from Lemma 7.23, applied to $B(0, r)$ in place of $B(0, 1)$, because $|F(x) - x| < \epsilon r$ for all x with $|x| = r$. The second follows directly from (5), since $|F(x)| < (1 + \epsilon)|x|$. It is clear that (6) implies

$$(1 - \epsilon)^k \leq \frac{m(F(B(0, r)))}{m(B(0, r))} \leq (1 + \epsilon)^k \tag{7}$$

and this proves (3).

CASE II *A is not one-to-one.* In this case, A maps R^k into a subspace of lower dimension, i.e., into a set of measure 0. Given $\epsilon > 0$, there is therefore an $\eta > 0$ such that $m(E_\eta) < \epsilon$ if E_η is the set of all points in R^k whose distance from $A(B(0, 1))$ is less than η. Since $A = T'(0)$, there is a $\delta > 0$ such that $|x| < \delta$ implies

$$|T(x) - Ax| \leq \eta |x|. \tag{8}$$

If $r < \delta$, then $T(B(0, r))$ lies therefore in the set E that consists of the points whose distance from $A(B(0, r))$ is less than ηr. Our choice of η shows that $m(E) < \epsilon r^k$. Hence

$$m(T(B(0, r))) < \epsilon r^k \qquad (0 < r < \delta). \tag{9}$$

Since $r^k = m(B(0, r))/m(B(0, 1))$, (9) implies that

$$\lim_{r \to 0} \frac{m(T(B(0, r)))}{m(B(0, r))} = 0. \tag{10}$$

This proves (1), since $\Delta(T'(0)) = \Delta(A) = 0$. ////

7.25 Lemma *Suppose $E \subset R^k$, $m(E) = 0$, T maps E into R^k, and*

$$\limsup \frac{|T(y) - T(x)|}{|y - x|} < \infty$$

for every $x \in E$, as y tends to x within E.
Then $m(T(E)) = 0$.

PROOF Fix positive integers n and p, let $F = F_{n,p}$ be the set of all $x \in E$ such that

$$|T(y) - T(x)| \le n|y - x|$$

for all $y \in B(x, 1/p) \cap E$, and choose $\epsilon > 0$. Since $m(F) = 0$, F can be covered by balls $B_i = B(x_i, r_i)$, where $x_i \in F$, $r_i < 1/p$, in such a way that $\sum m(B_i) < \epsilon$. (To do this, cover F by an open set W of small measure, decompose W into disjoint boxes of small diameter, as in Sec. 2.19, and cover each of those that intersect F by a ball whose center lies in the box and in F.)
If $x \in F \cap B_i$ then $|x_i - x| < r_i < 1/p$ and $x_i \in F$. Hence

$$|T(x_i) - T(x)| \le n|x_i - x| < nr_i$$

so that $T(F \cap B_i) \subset B(T(x_i), nr_i)$. Therefore

$$T(F) \subset \bigcup_i B(T(x_i), nr_i).$$

The measure of this union is at most

$$\sum_i m(B(T(x_i), nr_i)) = n^k \sum_i m(B_i) < n^k \epsilon.$$

Since Lebesgue measure is complete and ϵ was arbitrary, it follows that $T(F)$ is measurable and $m(T(F)) = 0$.
To complete the proof, note that E is the union of the countable collection $\{F_{n,p}\}$. ////

Here is a special case of the lemma:
If V is open in R^k and $T: V \to R^k$ is differentiable at every point of V, then T maps sets of measure 0 to sets of measure 0.
We now come to the change-of-variables theorem.

7.26 Theorem *Suppose that*

(i) $X \subset V \subset R^k$, V is open, $T: V \to R^k$ is continuous;

(ii) *X is Lebesgue measurable, T is one-to-one on X, and T is differentiable at every point of X;*
(iii) $m(T(V - X)) = 0$.

Then, setting $Y = T(X)$,

$$\int_Y f \, dm = \int_X (f \circ T)|J_T| \, dm \tag{1}$$

for every measurable $f: R^k \to [0, \infty]$.

The case $X = V$ is perhaps the most interesting one. As regards condition (iii), it holds, for instance, when $m(V - X) = 0$ and T satisfies the hypotheses of Lemma 7.25 on $V - X$.

The proof has some elements in common with that of the implication $(b) \to (c)$ in Theorem 7.18.

It will be important in this proof to distinguish between Borel sets and Lebesgue measurable sets. The σ-algebra consisting of the Lebesgue measurable subsets of R^k will be denoted by $\mathfrak{M}$.

PROOF We break the proof into the following three steps:

(I) If $E \in \mathfrak{M}$ and $E \subset V$, then $T(E) \in \mathfrak{M}$.
(II) For every $E \in \mathfrak{M}$,

$$m(T(E \cap X)) = \int_X \chi_E |J_T| \, dm.$$

(III) For every $A \in \mathfrak{M}$,

$$\int_Y \chi_A \, dm = \int_X (\chi_A \circ T)|J_T| \, dm.$$

If $E_0 \in \mathfrak{M}$, $E_0 \subset V$, and $m(E_0) = 0$, then $m(T(E_0 - X)) = 0$ by (iii), and $m(T(E_0 \cap X)) = 0$ by Lemma 7.25. Thus $m(T(E_0)) = 0$.

If $E_1 \subset V$ is an F_σ, then E_1 is σ-compact, hence $T(E_1)$ is σ-compact, because T is continuous. Thus $T(E_1) \in \mathfrak{M}$.

Since every $E \in \mathfrak{M}$ is the union of an F_σ and a set of measure 0 (Theorem 2.20), (I) is proved.

To prove (II), let n be a positive integer, and put

$$V_n = \{x \in V : |T(x)| < n\}, \qquad X_n = X \cap V_n. \tag{2}$$

Because of (I), we can define

$$\mu_n(E) = m(T(E \cap X_n)) \qquad (E \in \mathfrak{M}). \tag{3}$$

Since T is one-to-one on X_n, the countable additivity of m shows that μ_n is a measure on $\mathfrak{M}$. Also, μ_n is bounded (this was the reason for replacing X temporarily by X_n), and $\mu_n \ll m$, by another application of Lemma 7.25.

Theorem 7.8 tells us therefore that $(D\mu_n)(x)$ exists a.e. $[m]$, that $D\mu_n \in L^1(m)$, and that

$$\mu_n(E) = \int_E (D\mu_n)\, dm \qquad (E \in \mathfrak{M}). \qquad (4)$$

We claim next that

$$(D\mu_n)(x) = |J_T(x)| \qquad (x \in X_n). \qquad (5)$$

To see this, fix $x \in X_n$, and note that $B(x, r) \subset V_n$ for all sufficiently small $r > 0$, because V_n is open. Since $V_n - X_n \subset V - X$, hypothesis (iii) enables us to replace X_n by V_n in (3) without changing $\mu_n(E)$. Hence, for small $r > 0$,

$$\mu_n(B(x, r)) = m(T(B(x, r))). \qquad (6)$$

If we divide both sides of (6) by $m(B(x, r))$ and refer to Theorem 7.24 and formula 7.22(6), we obtain (5).

Since (3) implies that $\mu_n(E) = \mu_n(E \cap X_n)$, it follows from (3), (4), and (5) that

$$m(T(E \cap X_n)) = \int_{X_n} \chi_E |J_T|\, dm \qquad (E \in \mathfrak{M}). \qquad (7)$$

If we apply the monotone convergence theorem to (7), letting $n \to \infty$, we obtain (II).

We begin the proof of (III) by letting A be a Borel set in R^k. Put

$$E = T^{-1}(A) = \{x \in V : T(x) \in A\}. \qquad (8)$$

Then $\chi_E = \chi_A \circ T$. Since χ_A is a Borel function and T is continuous, χ_E is a Borel function (Theorem 1.12), hence $E \in \mathfrak{M}$. Also

$$T(E \cap X) = A \cap Y \qquad (9)$$

which implies, by (II), that

$$\int_Y \chi_A\, dm = m(T(E \cap X)) = \int_X (\chi_A \circ T)|J_T|\, dm. \qquad (10)$$

Finally, if $N \in \mathfrak{M}$ and $m(N) = 0$, there is a Borel set $A \supset N$ with $m(A) = 0$. For this A, (10) shows that $(\chi_A \circ T)|J_T| = 0$ a.e. $[m]$. Since $0 \le \chi_N \le \chi_A$, it follows that both integrals in (10) are 0 if A is replaced by N. Since every Lebesgue measurable set is the disjoint union of a Borel set and a set of measure 0, (10) holds for every $A \in \mathfrak{M}$. This proves (III).

Once we have (III), it is clear that (1) holds for every nonnegative Lebesgue measurable *simple* function f. Another application of the monotone convergence theorem completes the proof. ////

Note that we did not prove that $f \circ T$ is Lebesgue measurable for all Lebesgue measurable f. It need not be; see Exercise 8. What the proof does establish is the Lebesgue measurability of the product $(f \circ T)|J_T|$.

Here is a special case of the theorem:

Suppose $\varphi: [a, b] \to [\alpha, \beta]$ is AC, monotonic, $\varphi(a) = \alpha$, $\varphi(b) = \beta$, and $f \geq 0$ is Lebesgue measurable. Then

$$\int_\alpha^\beta f(t)\, dt = \int_a^b f(\varphi(x))\varphi'(x)\, dx. \tag{15}$$

To derive this from Theorem 7.26, put $V = (a, b)$, $T = \varphi$, let Ω be the union of the maximal segments on which φ is constant (if there are any) and let X be the set of all $x \in V - \Omega$ where $\varphi'(x)$ exists (and is finite).

Exercises

1 Show that $|f(x)| \leq (Mf)(x)$ at every Lebesgue point of f if $f \in L^1(R^k)$.

2 For $\delta > 0$, let $I(\delta)$ be the segment $(-\delta, \delta) \subset R^1$. Given α and β, $0 \leq \alpha \leq \beta \leq 1$, construct a measurable set $E \subset R^1$ so that the upper and lower limits of

$$\frac{m(E \cap I(\delta))}{2\delta}$$

are β and α, respectively, as $\delta \to 0$.

(Compare this with Section 7.12.)

3 Suppose that E is a measurable set of real numbers with arbitrarily small periods. Explicitly, this means that there are positive numbers p_i, converging to 0 as $i \to \infty$, so that

$$E + p_i = E \qquad (i = 1, 2, 3, \ldots).$$

Prove that then either E or its complement has measure 0.

Hint: Pick $\alpha \in R^1$, put $F(x) = m(E \cap [\alpha, x])$ for $x > \alpha$, show that

$$F(x + p_i) - F(x - p_i) = F(y + p_i) - F(y - p_i)$$

if $\alpha + p_i < x < y$. What does this imply about $F'(x)$ if $m(E) > 0$?

4 Call t a *period* of the function f on R^1 if $f(x + t) = f(x)$ for all $x \in R^1$. Suppose f is a real Lebesgue measurable function with periods s and t whose quotient s/t is irrational. Prove that there is a constant c such that $f(x) = c$ a.e., but that f need not be constant.

Hint: Apply Exercise 3 to the sets $\{f > \lambda\}$.

5 If $A \subset R^1$ and $B \subset R^1$, define $A + B = \{a + b: a \in A, b \in B\}$. Suppose $m(A) > 0$, $m(B) > 0$. Prove that $A + B$ contains a segment, by completing the following outline.

There are points a_0 and b_0 where A and B have metric density 1. Choose a small $\delta > 0$. Put $c_0 = a_0 + b_0$. For each ϵ, positive or negative, define B_ϵ to be the set of all $c_0 + \epsilon - b$ for which $b \in B$ and $|b - b_0| < \delta$. Then $B_\epsilon \subset (a_0 + \epsilon - \delta, a_0 + \epsilon + \delta)$. If δ was well chosen and $|\epsilon|$ is sufficiently small, it follows that A intersects B_ϵ, so that $A + B \supset (c_0 - \epsilon_0, c_0 + \epsilon_0)$ for some $\epsilon_0 > 0$.

Let C be Cantor's "middle thirds" set and show that $C + C$ is an interval, although $m(C) = 0$. (See also Exercise 19, Chap. 9.)

6 Suppose G is a subgroup of R^1 (relative to addition), $G \neq R^1$, and G is Lebesgue measurable. Prove that then $m(G) = 0$.

Hint: Use Exercise 5.

7 Construct a continuous monotonic function f on R^1 so that f is not constant on any segment although $f'(x) = 0$ a.e.

8 Let $V = (a, b)$ be a bounded segment in R^1. Choose segments $W_n \subset V$ in such a way that their union W is dense in V and the set $K = V - W$ has $m(K) > 0$. Choose continuous functions φ_n so that $\varphi_n(x) = 0$ outside W_n, $0 < \varphi_n(x) < 2^{-n}$ in W_n. Put $\varphi = \sum \varphi_n$ and define

$$T(x) = \int_a^x \varphi(t)\, dt \qquad (a < x < b).$$

Prove the following statements:

(a) T satisfies the hypotheses of Theorem 7.26, with $X = V$.

(b) T' is continuous, $T'(x) = 0$ on K, $m(T(K)) = 0$.

(c) If E is a nonmeasurable subset of K (see Theorem 2.22) and $A = T(E)$, then χ_A is Lebesgue measurable but $\chi_A \circ T$ is not.

(d) The functions φ_n can be so chosen that the resulting T is an *infinitely differentiable* homeomorphism of V onto some segment in R^1 and (c) still holds.

9 Suppose $0 < \alpha < 1$. Pick t so that $t^\alpha = 2$. Then $t > 2$, and the construction of Example (b) in Sec. 7.16 can be carried out with $\delta_n = (2/t)^n$. Show that the resulting function f belongs to Lip α on $[0, 1]$.

10 If $f \in$ Lip 1 on $[a, b]$, prove that f is absolutely continuous and that $f' \in L^\infty$.

11 Assume that $1 < p < \infty$, f is absolutely continuous on $[a, b]$, $f' \in L^p$, and $\alpha = 1/q$, where q is the exponent conjugate to p. Prove that $f \in$ Lip α.

12 Suppose $\varphi: [a, b] \to R^1$ is nondecreasing.

(a) Show that there is a left-continuous nondecreasing f on $[a, b]$ so that $\{f \neq \varphi\}$ is at most countable. [Left-continuous means: if $a < x \leq b$ and $\epsilon > 0$, then there is a $\delta > 0$ so that $|f(x) - f(x - t)| < \epsilon$ whenever $0 < t < \delta$.]

(b) Imitate the proof of Theorem 7.18 to show that there is a positive Borel measure μ on $[a, b]$ for which

$$f(x) - f(a) = \mu([a, x)) \qquad (a \leq x \leq b).$$

(c) Deduce from (b) that $f'(x)$ exists a.e. $[m]$, that $f' \in L^1(m)$, and that

$$f(x) - f(a) = \int_a^x f'(t)\, dt + s(x) \qquad (a \leq x \leq b)$$

where s is nondecreasing and $s'(x) = 0$ a.e. $[m]$.

(d) Show that $\mu \perp m$ if and only if $f'(x) = 0$ a.e. $[m]$, and that $\mu \ll m$ if and only if f is AC on $[a, b]$.

(e) Prove that $\varphi'(x) = f'(x)$ a.e. $[m]$.

13 Let BV be the class of all f on $[a, b]$ that have bounded variation on $[a, b]$, as defined after Theorem 7.19. Prove the following statements.

(a) Every monotonic bounded function on $[a, b]$ is in BV.

(b) If $f \in BV$ is real, there exist bounded monotonic functions f_1 and f_2 so that $f = f_1 - f_2$. *Hint*: Imitate the proof of Theorem 7.19.

(c) If $f \in BV$ is left-continuous then f_1 and f_2 can be chosen in (b) so as to be also left-continuous.

(d) If $f \in BV$ is left-continuous then there is a Borel measure μ on $[a, b]$ that satisfies

$$f(x) - f(a) = \mu([a, x)) \qquad (a \leq x \leq b);$$

$\mu \ll m$ if and only if f is AC on $[a, b]$.

(e) Every $f \in BV$ is differentiable a.e. $[m]$, and $f' \in L^1(m)$.

14 Show that the product of two absolutely continuous functions on $[a, b]$ is absolutely continuous. Use this to derive a theorem about integration by parts.

15 Construct a monotonic function f on R^1 so that $f'(x)$ exists (finitely) for every $x \in R^1$, but f' is not a continuous function.

16 Suppose $E \subset [a, b]$, $m(E) = 0$. Construct an absolutely continuous monotonic function f on $[a, b]$ so that $f'(x) = \infty$ at every $x \in E$.

Hint: $E \subset \bigcap V_n$, V_n open, $m(V_n) \subset 2^{-n}$. Consider the sum of the characteristic functions of these sets.

17 Suppose $\{\mu_n\}$ is a sequence of positive Borel measures on R^k and

$$\mu(E) = \sum_{n=1}^{\infty} \mu_n(E).$$

Assume $\mu(R^k) < \infty$. Show that μ is a Borel measure. What is the relation between the Lebesgue decompositions of the μ_n and that of μ?

Prove that

$$(D\mu)(x) = \sum_{n=1}^{\infty} (D\mu_n)(x) \quad \text{a.e. } [m].$$

Derive corresponding theorems for sequences $\{f_n\}$ of positive nondecreasing functions on R^1 and their sums $f = \sum f_n$.

18 Let $\varphi_0(t) = 1$ on $[0, 1)$, $\varphi_0(t) = -1$ on $[1, 2)$, extend φ_0 to R^1 so as to have period 2, and define $\varphi_n(t) = \varphi_0(2^n t)$, $n = 1, 2, 3, \ldots$.

Assume that $\sum |c_n|^2 < \infty$ and prove that the series

$$\sum_{n=1}^{\infty} c_n \varphi_n(t) \qquad (*)$$

converges then for almost every t.

Probabilistic interpretation: The series $\sum (\pm c_n)$ converges with probability 1.

Suggestion: $\{\varphi_n\}$ is orthonormal on $[0, 1]$, hence (*) is the Fourier series of some $f \in L^2$. If $a = j \cdot 2^{-N}$, $b = (j + 1) \cdot 2^{-N}$, $a < t < b$, and $s_N = c_1 \varphi_1 + \cdots + c_N \varphi_N$, then, for $n > N$,

$$s_N(t) = \frac{1}{b-a} \int_a^b s_N \, dm = \frac{1}{b-a} \int_a^b s_n \, dm,$$

and the last integral converges to $\int_a^b f \, dm$, as $n \to \infty$. Show that (*) converges to $f(t)$ at almost every Lebesgue point of f.

19 Suppose f is continuous on R^1, $f(x) > 0$ if $0 < x < 1$, $f(x) = 0$ otherwise. Define

$$h_c(x) = \sup \{n^c f(nx): n = 1, 2, 3, \ldots\}.$$

Prove that
(a) h_c is in $L^1(R^1)$ if $0 < c < 1$,
(b) h_1 is in weak L^1 but not in $L^1(R^1)$,
(c) h_c is not in weak L^1 if $c > 1$.

20 (a) For any set $E \subset R^2$, the boundary ∂E of E is, by definition, the closure of E minus the interior of E. Show that E is Lebesgue measurable whenever $m(\partial E) = 0$.

(b) Suppose that E is the union of a (*possibly uncountable*) collection of *closed* discs in R^2 whose radii are at least 1. Use (a) to show that E is Lebesgue measurable.

(c) Show that the conclusion of (b) is true even when the radii are unrestricted.

(d) Show that some unions of closed discs of radius 1 are not Borel sets. (See Sec. 2.21.)

(e) Can discs be replaced by triangles, rectangles, arbitrary polygons, etc., in all this? What is the relevant geometric property?

21 If f is a real function on $[0, 1]$ and

$$\gamma(t) = t + if(t),$$

the length of the graph of f is, by definition, the total variation of γ on $[0, 1]$. Show that this length is finite if and only if $f \in BV$. (See Exercise 13.) Show that it is equal to

$$\int_0^1 \sqrt{1 + [f'(t)]^2}\, dt$$

if f is absolutely continuous.

22 (a) Assume that both f and its maximal function Mf are in $L^1(R^k)$. Prove that then $f(x) = 0$ a.e. $[m]$. *Hint*: To every other $f \in L^1(R^k)$ corresponds a constant $c = c(f) > 0$ such that

$$(Mf)(x) \geq c|x|^{-k}$$

whenever $|x|$ is sufficiently large.

(b) Define $f(x) = x^{-1}(\log x)^{-2}$ if $0 < x < \frac{1}{2}$, $f(x) = 0$ on the rest of R^1. Then $f \in L^1(R^1)$. Show that

$$(Mf)(x) \geq |2x \log (2x)|^{-1} \qquad (0 < x < 1/4)$$

so that $\int_0^1 (Mf)(x)\, dx = \infty$.

23 The definition of Lebesgue points, as made in Sec. 7.6, applies to individual integrable functions, not to the equivalence classes discussed in Sec. 3.10. However, if $F \in L^1(R^k)$ is one of these equivalence classes, one may call a point $x \in R^k$ a *Lebesgue point of F* if there is a complex number, let us call it $(SF)(x)$, such that

$$\lim_{r \to 0} \frac{1}{m(B_r)} \int_{B(x, r)} |f - (SF)(x)|\, dm = 0$$

for one (hence for every) $f \in F$.

Define $(SF)(x)$ to be 0 at those points $x \in R^k$ that are not Lebesgue points of F.

Prove the following statement: If $f \in F$, and x is a Lebesgue point of f, then x is also a Lebesgue point of F, and $f(x) = (SF)(x)$. Hence $SF \in F$.

Thus S "selects" a member of F that has a *maximal* set of Lebesgue points.

CHAPTER EIGHT

INTEGRATION ON PRODUCT SPACES

This chapter is devoted to the proof and discussion of the theorem of Fubini concerning integration of functions of two variables. We first present the theorem in its abstract form.

Measurability on Cartesian Products

8.1 Definitions If X and Y are two sets, their *cartesian product* $X \times Y$ is the set of all ordered pairs (x, y), with $x \in X$ and $y \in Y$. If $A \subset X$ and $B \subset Y$, it follows that $A \times B \subset X \times Y$. We call any set of the form $A \times B$ a *rectangle* in $X \times Y$.

Suppose now that $(X, \mathscr{S})$ and $(Y, \mathscr{T})$ are measurable spaces. Recall that this simply means that $\mathscr{S}$ is a σ-algebra in X and $\mathscr{T}$ is a σ-algebra in Y.

A *measurable rectangle* is any set of the form $A \times B$, where $A \in \mathscr{S}$ and $B \in \mathscr{T}$.

If $Q = R_1 \cup \cdots \cup R_n$, where each R_i is a measurable rectangle and $R_i \cap R_j = \varnothing$ for $i \neq j$, we say that $Q \in \mathscr{E}$, the class of all *elementary sets*.

$\mathscr{S} \times \mathscr{T}$ is defined to be the smallest σ-algebra in $X \times Y$ which contains every measurable rectangle.

A *monotone class* $\mathfrak{M}$ is a collection of sets with the following properties: If $A_i \in \mathfrak{M}$, $B_i \in \mathfrak{M}$, $A_i \subset A_{i+1}$, $B_i \supset B_{i+1}$, for $i = 1, 2, 3, \ldots$, and if

$$A = \bigcup_{i=1}^{\infty} A_i, \qquad B = \bigcap_{i=1}^{\infty} B_i, \tag{1}$$

then $A \in \mathfrak{M}$ and $B \in \mathfrak{M}$.

If $E \subset X \times Y$, $x \in X$, $y \in Y$, we define
$$E_x = \{y: (x, y) \in E\}, \qquad E^y = \{x: (x, y) \in E\}. \tag{2}$$
We call E_x and E^y the *x-section* and *y-section*, respectively, of E. Note that $E_x \subset Y$, $E^y \subset X$.

8.2 Theorem *If $E \in \mathscr{S} \times \mathscr{T}$, then $E_x \in \mathscr{T}$ and $E^y \in \mathscr{S}$, for every $x \in X$ and $y \in Y$.*

PROOF Let Ω be the class of all $E \in \mathscr{S} \times \mathscr{T}$ such that $E_x \in \mathscr{T}$ for every $x \in X$. If $E = A \times B$, then $E_x = B$ if $x \in A$, $E_x = \emptyset$ if $x \notin A$. Therefore every measurable rectangle belongs to Ω. Since $\mathscr{T}$ is a σ-algebra, the following three statements are true. They prove that Ω is a σ-algebra and hence that $\Omega = \mathscr{S} \times \mathscr{T}$:

(a) $X \times Y \in \Omega$.
(b) If $E \in \Omega$, then $(E^c)_x = (E_x)^c$, hence $E^c \in \Omega$.
(c) If $E_i \in \Omega$ $(i = 1, 2, 3, \ldots)$ and $E = \bigcup E_i$, then $E_x = \bigcup (E_i)_x$, hence $E \in \Omega$.

The proof is the same for E^y. ////

8.3 Theorem *$\mathscr{S} \times \mathscr{T}$ is the smallest monotone class which contains all elementary sets.*

PROOF Let $\mathfrak{M}$ be the smallest monotone class which contains $\mathscr{E}$; the proof that this class exists is exactly like that of Theorem 1.10. Since $\mathscr{S} \times \mathscr{T}$ is a monotone class, we have $\mathfrak{M} \subset \mathscr{S} \times \mathscr{T}$.

The identities
$$(A_1 \times B_1) \cap (A_2 \times B_2) = (A_1 \cap A_2) \times (B_1 \cap B_2),$$
$$(A_1 \times B_1) - (A_2 \times B_2) = [(A_1 - A_2) \times B_1] \cup [(A_1 \cap A_2) \times (B_1 - B_2)]$$
show that the intersection of two measurable rectangles is a measurable rectangle and that their difference is the union of two disjoint measurable rectangles, hence is an elementary set. If $P \in \mathscr{E}$ and $Q \in \mathscr{E}$, it follows easily that $P \cap Q \in \mathscr{E}$ and $P - Q \in \mathscr{E}$. Since
$$P \cup Q = (P - Q) \cup Q$$
and $(P - Q) \cap Q = \emptyset$, we also have $P \cup Q \in \mathscr{E}$.

For any set $P \subset X \times Y$, define $\Omega(P)$ to be the class of all $Q \subset X \times Y$ such that $P - Q \in \mathfrak{M}$, $Q - P \in \mathfrak{M}$, and $P \cup Q \in \mathfrak{M}$. The following properties are obvious:

(a) $Q \in \Omega(P)$ if and only if $P \in \Omega(Q)$.
(b) Since $\mathfrak{M}$ is a monotone class, so is each $\Omega(P)$.

Fix $P \in \mathscr{E}$. Our preceding remarks about $\mathscr{E}$ show that $Q \in \Omega(P)$ for all $Q \in \mathscr{E}$, hence $\mathscr{E} \subset \Omega(P)$, and now (b) implies that $\mathfrak{M} \subset \Omega(P)$.

Next, fix $Q \in \mathfrak{M}$. We just saw that $Q \in \Omega(P)$ if $P \in \mathscr{E}$. By (a), $P \in \Omega(Q)$, hence $\mathscr{E} \subset \Omega(Q)$, and if we refer to (b) once more we obtain $\mathfrak{M} \subset \Omega(Q)$.

Summing up: If P and $Q \in \mathfrak{M}$, then $P - Q \in \mathfrak{M}$ and $P \cup Q \in \mathfrak{M}$.

It now follows that $\mathfrak{M}$ is a σ-algebra in $X \times Y$:

(i) $X \times Y \in \mathscr{E}$. Hence $X \times Y \in \mathfrak{M}$.
(ii) If $Q \in \mathfrak{M}$, then $Q^c \in \mathfrak{M}$, since the difference of any two members of $\mathfrak{M}$ is in $\mathfrak{M}$.
(iii) If $P_i \in \mathfrak{M}$ for $i = 1, 2, 3, \ldots$, and $P = \bigcup P_i$, put

$$Q_n = P_1 \cup \cdots \cup P_n.$$

Since $\mathfrak{M}$ is closed under the formation of finite unions, $Q_n \in \mathfrak{M}$.
Since $Q_n \subset Q_{n+1}$ and $P = \bigcup Q_n$, the monotonicity of $\mathfrak{M}$ shows that $P \in \mathfrak{M}$.

Thus $\mathfrak{M}$ is a σ-algebra, $\mathscr{E} \subset \mathfrak{M} \subset \mathscr{S} \times \mathscr{T}$, and (by definition) $\mathscr{S} \times \mathscr{T}$ is the smallest σ-algebra which contains $\mathscr{E}$. Hence $\mathfrak{M} = \mathscr{S} \times \mathscr{T}$. ////

8.4 Definition With each function f on $X \times Y$ and with each $x \in X$ we associate a function f_x defined on Y by $f_x(y) = f(x, y)$.

Similarly, if $y \in Y$, f^y is the function defined on X by $f^y(x) = f(x, y)$.

Since we are now dealing with three σ-algebras, $\mathscr{S}$, $\mathscr{T}$, and $\mathscr{S} \times \mathscr{T}$, we shall, for the sake of clarity, indicate in the sequel to which of these three σ-algebras the word "measurable" refers.

8.5 Theorem Let f be an $(\mathscr{S} \times \mathscr{T})$-measurable function on $X \times Y$. Then

(a) For each $x \in X$, f_x is a $\mathscr{T}$-measurable function.
(b) For each $y \in Y$, f^y is an $\mathscr{S}$-measurable function.

PROOF For any open set V, put

$$Q = \{(x, y): f(x, y) \in V\}.$$

Then $Q \in \mathscr{S} \times \mathscr{T}$, and

$$Q_x = \{y: f_x(y) \in V\}.$$

Theorem 8.2 shows that $Q_x \in \mathscr{T}$. This proves (a); the proof of (b) is similar. ////

Product Measures

8.6 Theorem Let $(X, \mathscr{S}, \mu)$ and $(Y, \mathscr{T}, \lambda)$ be σ-finite measure spaces. Suppose $Q \in \mathscr{S} \times \mathscr{T}$. If

$$\varphi(x) = \lambda(Q_x), \qquad \psi(y) = \mu(Q^y) \tag{1}$$

for every $x \in X$ and $y \in Y$, then φ is $\mathscr{S}$-measurable, ψ is $\mathscr{T}$-measurable, and

$$\int_X \varphi \, d\mu = \int_Y \psi \, d\lambda, \tag{2}$$

Notes: The assumptions on the measure spaces are, more explicitly, that μ and λ are positive measures on $\mathscr{S}$ and $\mathscr{T}$, respectively, that X is the union of countably many disjoint sets X_n with $\mu(X_n) < \infty$, and that Y is the union of countably many disjoint sets Y_m with $\lambda(Y_m) < \infty$.

Theorem 8.2 shows that the definitions (1) make sense. Since

$$\lambda(Q_x) = \int_Y \chi_Q(x, y) \, d\lambda(y) \qquad (x \in X), \tag{3}$$

with a similar statement for $\mu(Q^y)$, the conclusion (2) can be written in the form

$$\int_X d\mu(x) \int_Y \chi_Q(x, y) \, d\lambda(y) = \int_Y d\lambda(y) \int_X \chi_Q(x, y) \, d\mu(x). \tag{4}$$

PROOF Let Ω be the class of all $Q \in \mathscr{S} \times \mathscr{T}$ for which the conclusion of the theorem holds. We claim that Ω has the following four properties:

(a) Every measurable rectangle belongs to Ω.
(b) If $Q_1 \subset Q_2 \subset Q_3 \subset \cdots$, if each $Q_i \in \Omega$, and if $Q = \bigcup Q_i$, then $Q \in \Omega$.
(c) If $\{Q_i\}$ is a disjoint countable collection of members of Ω, and if $Q = \bigcup Q_i$, then $Q \in \Omega$.
(d) If $\mu(A) < \infty$ and $\lambda(B) < \infty$, if $A \times B \supset Q_1 \supset Q_2 \supset Q_3 \supset \cdots$, if $Q = \bigcap Q_i$ and $Q_i \in \Omega$ for $i = 1, 2, 3, \ldots$, then $Q \in \Omega$.

If $Q = A \times B$, where $A \in \mathscr{S}$, $B \in \mathscr{T}$, then

$$\lambda(Q_x) = \lambda(B)\chi_A(x) \quad \text{and} \quad \mu(Q^y) = \mu(A)\chi_B(y), \tag{5}$$

and therefore each of the integrals in (2) is equal to $\mu(A)\lambda(B)$. This gives (a).

To prove (b), let φ_i and ψ_i be associated with Q_i in the way in which (1) associates φ and ψ with Q. The countable additivity of μ and λ shows that

$$\varphi_i(x) \to \varphi(x), \qquad \psi_i(y) \to \psi(y) \qquad (i \to \infty), \tag{6}$$

the convergence being monotone increasing at every point. Since φ_i and ψ_i are assumed to satisfy the conclusion of the theorem, (b) follows from the monotone convergence theorem.

For finite unions of disjoint sets, (c) is clear, because the characteristic function of a union of *disjoint* sets is the sum of their characteristic functions. The general case of (c) now follows from (b).

The proof of (d) is like that of (b), except that we use the dominated convergence theorem in place of the monotone convergence theorem. This is legitimate, since $\mu(A) < \infty$ and $\lambda(B) < \infty$.

Now define

$$Q_{mn} = Q \cap (X_n \times Y_m) \qquad (m, n = 1, 2, 3, \ldots) \qquad (7)$$

and let $\mathfrak{M}$ be the class of all $Q \in \mathscr{S} \times \mathscr{T}$ such that $Q_{mn} \in \Omega$ for all choices of m and n. Then (b) and (d) show that $\mathfrak{M}$ is a monotone class; (a) and (c) show that $\mathscr{E} \subset \mathfrak{M}$; and since $\mathfrak{M} \subset \mathscr{S} \times \mathscr{T}$, Theorem 8.3 implies that $\mathfrak{M} = \mathscr{S} \times \mathscr{T}$.

Thus $Q_{mn} \in \Omega$ for every $Q \in \mathscr{S} \times \mathscr{T}$ and for all choices of m and n. Since Q is the union of the sets Q_{mn} and since these sets are disjoint, we conclude from (c) that $Q \in \Omega$. This completes the proof. ////

8.7 Definition If $(X, \mathscr{S}, \mu)$ and $(Y, \mathscr{T}, \lambda)$ are as in Theorem 8.6, and if $Q \in \mathscr{S} \times \mathscr{T}$, we define

$$(\mu \times \lambda)(Q) = \int_X \lambda(Q_x) \, d\mu(x) = \int_Y \mu(Q^y) \, d\lambda(y). \qquad (1)$$

The equality of the integrals in (1) is the content of Theorem 8.6. We call $\mu \times \lambda$ the *product* of the measures μ and λ. That $\mu \times \lambda$ is really a measure (i.e., that $\mu \times \lambda$ is countably additive on $\mathscr{S} \times \mathscr{T}$) follows immediately from Theorem 1.27.

Observe also that $\mu \times \lambda$ is σ-finite.

The Fubini Theorem

8.8 Theorem *Let $(X, \mathscr{S}, \mu)$ and $(Y, \mathscr{T}, \lambda)$ be σ-finite measure spaces, and let f be an $(\mathscr{S} \times \mathscr{T})$-measurable function on $X \times Y$.*

(a) *If $0 \leq f \leq \infty$, and if*

$$\varphi(x) = \int_Y f_x \, d\lambda, \qquad \psi(y) = \int_X f^y \, d\mu \qquad (x \in X, \, y \in Y), \qquad (1)$$

then φ is $\mathscr{S}$-measurable, ψ is $\mathscr{T}$-measurable, and

$$\int_X \varphi \, d\mu = \int_{X \times Y} f \, d(\mu \times \lambda) = \int_Y \psi \, d\lambda. \qquad (2)$$

(b) *If f is complex and if*

$$\varphi^*(x) = \int_Y |f|_x \, d\lambda \quad \text{and} \quad \int_X \varphi^* \, d\mu < \infty, \qquad (3)$$

then $f \in L^1(\mu \times \lambda)$.

(c) *If $f \in L^1(\mu \times \lambda)$, then $f_x \in L^1(\lambda)$ for almost all $x \in X$, $f^y \in L^1(\mu)$ for almost all $y \in Y$; the functions φ and ψ, defined by (1) a.e., are in $L^1(\mu)$ and $L^1(\lambda)$, respectively, and (2) holds.*

Notes: The first and last integrals in (2) can also be written in the more usual form

$$\int_X d\mu(x) \int_Y f(x, y) \, d\lambda(y) = \int_Y d\lambda(y) \int_X f(x, y) \, d\mu(x). \quad (4)$$

These are the so-called "iterated integrals" of f. The middle integral in (2) is often referred to as a *double integral*.

The combination of (b) and (c) gives the following useful result: *If f is $(\mathscr{S} \times \mathscr{T})$-measurable and if*

$$\int_X d\mu(x) \int_Y |f(x, y)| \, d\lambda(y) < \infty, \quad (5)$$

then the two iterated integrals (4) are finite and equal.

In other words, "the order of integration may be reversed" for $(\mathscr{S} \times \mathscr{T})$-measurable functions f whenever $f \geq 0$ and also whenever one of the iterated integrals of $|f|$ is finite.

PROOF We first consider (a). By Theorem 8.5, the definitions of φ and ψ make sense. Suppose $Q \in \mathscr{S} \times \mathscr{T}$ and $f = \chi_Q$. By Definition 8.7, (2) is then exactly the conclusion of Theorem 8.6. Hence (a) holds for all nonnegative simple $(\mathscr{S} \times \mathscr{T})$-measurable functions s. In the general case, there is a sequence of such functions s_n, such that $0 \leq s_1 \leq s_2 \leq \cdots$ and $s_n(x, y) \to f(x, y)$ at every point of $X \times Y$. If φ_n is associated with s_n in the same way in which φ was associated to f, we have

$$\int_X \varphi_n \, d\mu = \int_{X \times Y} s_n \, d(\mu \times \lambda) \quad (n = 1, 2, 3, \ldots). \quad (6)$$

The monotone convergence theorem, applied on $(Y, \mathscr{T}, \lambda)$, shows that $\varphi_n(x)$ increases to $\varphi(x)$, for every $x \in X$, as $n \to \infty$. Hence the monotone convergence theorem applies again, to the two integrals in (6), and the first equality (2) is obtained. The second half of (2) follows by interchanging the roles of x and y. This completes (a).

If we apply (a) to $|f|$, we see that (b) is true.

Obviously, it is enough to prove (c) for real $L^1(\mu \times \lambda)$; the complex case then follows. If f is real, (a) applies to f^+ and to f^-. Let φ_1 and φ_2 correspond to f^+ and f^- as φ corresponds to f in (1). Since $f \in L^1(\mu \times \lambda)$ and

$f^+ \leq |f|$, and since (a) holds for f^+, we see that $\varphi_1 \in L^1(\mu)$. Similarly, $\varphi_2 \in L^1(\mu)$. Since

$$f_x = (f^+)_x - (f^-)_x \qquad (7)$$

we have $f_x \in L^1(\lambda)$ for every x for which $\varphi_1(x) < \infty$ and $\varphi_2(x) < \infty$; since φ_1 and φ_2 are in $L^1(\mu)$, this happens for almost all x; and at any such x, we have $\varphi(x) = \varphi_1(x) - \varphi_2(x)$. Hence $\varphi \in L^1(\mu)$. Now (2) holds with φ_1 and f^+ and with φ_2 and f^-, in place of φ and f; if we subtract the resulting equations, we obtain one half of (c). The other half is proved in the same manner, with f^y and ψ in place of f_x and φ. ////

8.9 Counterexamples The following three examples will show that the various hypotheses in Theorems 8.6 and 8.8 cannot be dispensed with.

(a) Let $X = Y = [0, 1]$, $\mu = \lambda =$ Lebesgue measure on $[0, 1]$. Choose $\{\delta_n\}$ so that $0 = \delta_1 < \delta_2 < \delta_3 < \cdots$, $\delta_n \to 1$, and let g_n be a real continuous function with support in (δ_n, δ_{n+1}), such that $\int_0^1 g_n(t)\, dt = 1$, for $n = 1, 2, 3, \ldots$. Define

$$f(x, y) = \sum_{n=1}^{\infty} [g_n(x) - g_{n+1}(x)] g_n(y).$$

Note that at each point (x, y) at most one term in this sum is different from 0. Thus no convergence problem arises in the definition of f. An easy computation shows that

$$\int_0^1 dx \int_0^1 f(x, y)\, dy = 1 \neq 0 = \int_0^1 dy \int_0^1 f(x, y)\, dx,$$

so that the conclusion of the Fubini theorem fails, although both iterated integrals exist. Note that f is continuous in this example, except at the point $(1, 1)$, but that

$$\int_0^1 dx \int_0^1 |f(x, y)|\, dy = \infty.$$

(b) Let $X = Y = [0, 1]$, $\mu =$ Lebesgue measure on $[0, 1]$, $\lambda =$ counting measure on Y, and put $f(x, y) = 1$ if $x = y$, $f(x, y) = 0$ if $x \neq y$. Then

$$\int_X f(x, y)\, d\mu(x) = 0, \qquad \int_Y f(x, y)\, d\lambda(y) = 1$$

for all x and y in $[0, 1]$, so that

$$\int_Y d\lambda(y) \int_X f(x, y)\, d\mu(x) = 0 \neq 1 = \int_X d\mu(x) \int_Y f(x, y)\, d\lambda(y).$$

This time the failure is due to the fact that λ is not σ-finite.

Observe that our function f is $(\mathscr{S} \times \mathscr{T})$-measurable, if $\mathscr{S}$ is the class of all Lebesgue measurable sets in $[0, 1]$ and $\mathscr{T}$ consists of all subsets of $[0, 1]$.

To see this, note that $f = \chi_D$, where D is the diagonal of the unit square. Given n, put

$$I_j = \left[\frac{j-1}{n}, \frac{j}{n}\right]$$

and put

$$Q_n = (I_1 \times I_1) \cup (I_2 \times I_2) \cup \cdots \cup (I_n \times I_n).$$

Then Q_n is a finite union of measurable rectangles, and $D = \bigcap Q_n$.

(c) In examples (a) and (b), the failure of the Fubini theorem was due to the fact that either the function or the space was "too big." We now turn to the role played by the requirement that f be measurable *with respect to the σ-algebra $\mathscr{S} \times \mathscr{T}$*.

To pose the question more precisely, suppose $\mu(X) = \lambda(Y) = 1$, $0 \leq f \leq 1$ (so that "bigness" is certainly avoided); assume f_x is $\mathscr{T}$-measurable and f^y is $\mathscr{S}$-measurable, for all x and y; and assume φ is $\mathscr{S}$-measurable and ψ is $\mathscr{T}$-measurable, where φ and ψ are defined as in 8.8(1). Then $0 \leq \varphi \leq 1$, $0 \leq \psi \leq 1$, and both iterated integrals are finite. (Note that no reference to product measures is needed to *define* iterated integrals.) Does it follow that the two iterated integrals of f are equal?

The (perhaps surprising) answer is no.

In the following example (due to Sierpinski), we take

$$(X, \mathscr{S}, \mu) = (Y, \mathscr{T}, \lambda) = [0, 1]$$

with Lebesgue measure. The construction depends on the continuum hypothesis. It is a consequence of this hypothesis that there is a one-to-one mapping j of the unit interval $[0, 1]$ onto a well-ordered set W such that $j(x)$ has at most countably many predecessors in W, for each $x \in [0, 1]$. Taking this for granted, let Q be the set of all (x, y) in the unit square such that $j(x)$ precedes $j(y)$ in W. For each $x \in [0, 1]$, Q_x contains all but countably many points of $[0, 1]$; for each $y \in [0, 1]$, Q^y contains at most countably many points of $[0, 1]$. If $f = \chi_Q$, it follows that f_x and f^y are Borel measurable and that

$$\varphi(x) = \int_0^1 f(x, y)\, dy = 1, \qquad \psi(y) = \int_0^1 f(x, y)\, dx = 0$$

for all x and y. Hence

$$\int_0^1 dx \int_0^1 f(x, y)\, dy = 1 \neq 0 = \int_0^1 dy \int_0^1 f(x, y)\, dx.$$

Completion of Product Measures

8.10 If $(X, \mathscr{S}, \mu)$ and $(Y, \mathscr{T}, \lambda)$ are complete measure spaces, it need not be true that $(X \times Y, \mathscr{S} \times \mathscr{T}, \mu \times \lambda)$ is complete. There is nothing pathological about this phenomenon: Suppose that there exists an $A \in \mathscr{S}$, $A \neq \varnothing$, with $\mu(A) = 0$; and suppose that there exists a $B \subset Y$ so that $B \notin \mathscr{T}$. Then $A \times B \subset A \times Y$, $(\mu \times \lambda)(A \times Y) = 0$, but $A \times B \notin \mathscr{S} \times \mathscr{T}$. (The last assertion follows from Theorem 8.2.)

For instance, if $\mu = \lambda = m_1$ (Lebesgue measure on R^1), let A consist of any one point, and let B be any nonmeasurable set in R^1. Thus $m_1 \times m_1$ is not a complete measure; in particular, $m_1 \times m_1$ is not m_2, since the latter is complete, by its construction. However, m_2 *is the completion of* $m_1 \times m_1$. This result generalizes to arbitrary dimensions:

8.11 Theorem *Let m_k denote Lebesgue measure on R^k. If $k = r + s$, $r \geq 1$, $s \geq 1$, then m_k is the completion of the product measure $m_r \times m_s$.*

PROOF Let $\mathscr{B}_k$ and $\mathfrak{M}_k$ be the σ-algebras of all Borel sets and of all Lebesgue measurable sets in R^k, respectively. We shall first show that

$$\mathscr{B}_k \subset \mathfrak{M}_r \times \mathfrak{M}_s \subset \mathfrak{M}_k. \tag{1}$$

Every k-cell belongs to $\mathfrak{M}_r \times \mathfrak{M}_s$. The σ-algebra generated by the k-cells is $\mathscr{B}_k$. Hence $\mathscr{B}_k \subset \mathfrak{M}_r \times \mathfrak{M}_s$. Next, suppose $E \in \mathfrak{M}_r$ and $F \in \mathfrak{M}_s$. It is easy to see, by Theorem 2.20(*b*), that both $E \times R^s$ and $R^r \times F$ belong to $\mathfrak{M}_k$. The same is true of their intersection $E \times F$. It follows that $\mathfrak{M}_r \times \mathfrak{M}_s \subset \mathfrak{M}_k$.

Choose $Q \in \mathfrak{M}_r \times \mathfrak{M}_s$. Then $Q \in \mathfrak{M}_k$, so there are sets P_1 and $P_2 \in \mathscr{B}_k$ such that $P_1 \subset Q \subset P_2$ and $m_k(P_2 - P_1) = 0$. Both m_k and $m_r \times m_s$ are translation invariant Borel measures on R^k. They assign the same value to each k-cell. Hence they agree on $\mathscr{B}_k$, by Theorem 2.20(*d*). In particular,

$$(m_r \times m_s)(Q - P_1) \leq (m_r \times m_s)(P_2 - P_1) = m_k(P_2 - P_1) = 0$$

and therefore

$$(m_r \times m_s)(Q) = (m_r \times m_s)(P_1) = m_k(P_1) = m_k(Q).$$

So $m_r \times m_s$ agrees with m_k on $\mathfrak{M}_r \times \mathfrak{M}_s$.

It now follows that $\mathfrak{M}_k$ is the $(m_r \times m_s)$-completion of $\mathfrak{M}_r \times \mathfrak{M}_s$, and this is what the theorem asserts.   ////

We conclude this section with an alternative statement of Fubini's theorem which is of special interest in view of Theorem 8.11.

8.12 Theorem *Let $(X, \mathscr{S}, \mu)$ and $(Y, \mathscr{T}, \lambda)$ be complete σ-finite measure spaces. Let $(\mathscr{S} \times \mathscr{T})^*$ be the completion of $\mathscr{S} \times \mathscr{T}$, relative to the measure $\mu \times \lambda$. Let f be an $(\mathscr{S} \times \mathscr{T})^*$-measurable function on $X \times Y$. Then all conclusions of Theorem 8.8 hold, the only difference being as follows:*

The $\mathscr{T}$-measurability of f_x can be asserted only for almost all $x \in X$, so that $\varphi(x)$ is only defined a.e. $[\mu]$ by 8.8(1); a similar statement holds for f^y and ψ.

The proof depends on the following two lemmas:

Lemma 1 *Suppose v is a positive measure on a σ-algebra $\mathfrak{M}$, $\mathfrak{M}^*$ is the completion of $\mathfrak{M}$ relative to v, and f is an $\mathfrak{M}^*$-measurable function. Then there exists an $\mathfrak{M}$-measurable function g such that $f = g$ a.e. $[v]$.*

(An interesting special case of this arises when v is Lebesgue measure on R^k and $\mathfrak{M}$ is the class of all Borel sets in R^k.)

Lemma 2 *Let h be an $(\mathscr{S} \times \mathscr{T})^*$-measurable function on $X \times Y$ such that $h = 0$ a.e. $[\mu \times \lambda]$. Then for almost all $x \in X$ it is true that $h(x, y) = 0$ for almost all $y \in Y$; in particular, h_x is $\mathscr{T}$-measurable for almost all $x \in X$. A similar statement holds for h^y.*

If we assume the lemmas, the proof of the theorem is immediate: If f is as in the theorem, Lemma 1 (with $v = \mu \times \lambda$) shows that $f = g + h$, where $h = 0$ a.e. $[\mu \times \lambda]$ and g is $(\mathscr{S} \times \mathscr{T})$-measurable. Theorem 8.8 applies to g. Lemma 2 shows that $f_x = g_x$ a.e. $[\lambda]$ for almost all x and that $f^y = g^y$ a.e. $[\mu]$ for almost all y. Hence the two iterated integrals of f, as well as the double integral, are the same as those of g, and the theorem follows.

PROOF OF LEMMA 1 Suppose f is $\mathfrak{M}^*$-measurable and $f \geq 0$. There exist $\mathfrak{M}^*$-measurable simple functions $0 = s_0 \leq s_1 \leq s_2 \leq \cdots$ such that $s_n(x) \to f(x)$ for each $x \in X$, as $n \to \infty$. Hence $f = \sum (s_{n+1} - s_n)$. Since $s_{n+1} - s_n$ is a finite linear combination of characteristic functions, it follows that there are constants $c_i > 0$ and sets $E_i \in \mathfrak{M}^*$ such that

$$f(x) = \sum_{i=1}^{\infty} c_i \chi_{E_i}(x) \qquad (x \in X).$$

The definition of $\mathfrak{M}^*$ (see Theorem 1.36) shows that there are sets $A_i \in \mathfrak{M}$, $B_i \in \mathfrak{M}$, such that $A_i \subset E_i \subset B_i$ and $v(B_i - A_i) = 0$. Define

$$g(x) = \sum_{i=1}^{\infty} c_i \chi_{A_i}(x) \qquad (x \in X).$$

Then the function g is $\mathfrak{M}$-measurable, and $g(x) = f(x)$, except possibly when $x \in \bigcup (E_i - A_i) \subset \bigcup (B_i - A_i)$. Since $v(B_i - A_i) = 0$ for each i, we conclude that $g = f$ a.e. $[v]$. The general case (f real or complex) follows from this. ////

PROOF OF LEMMA 2 Let P be the set of all points in $X \times Y$ at which $h(x, y) \neq 0$. Then $P \in (\mathscr{S} \times \mathscr{T})^*$ and $(\mu \times \lambda)(P) = 0$. Hence there exists a $Q \in \mathscr{S} \times \mathscr{T}$ such that $P \subset Q$ and $(\mu \times \lambda)(Q) = 0$. By Theorem 8.6,

$$\int_X \lambda(Q_x) \, d\mu(x) = 0. \qquad (1)$$

Let N be the set of all $x \in X$ at which $\lambda(Q_x) > 0$. It follows from (1) that $\mu(N) = 0$. For every $x \notin N$, $\lambda(Q_x) = 0$. Since $P_x \subset Q_x$ and $(Y, \mathscr{T}, \lambda)$ is a complete measure space, every subset of P_x belongs to $\mathscr{T}$ if $x \notin N$. If $y \notin P_x$, then $h_x(y) = 0$. Thus we see, for every $x \notin N$, that h_x is $\mathscr{T}$-measurable and that $h_x(y) = 0$ a.e. $[\lambda]$. ////

Convolutions

8.13 It happens occasionally that one can prove that a certain set is not empty by proving that it is actually large. The word "large" may of course refer to various properties. One of these (a rather crude one) is cardinality. An example is furnished by the familiar proof that there exist transcendental numbers: There are only countably many algebraic numbers but uncountably many real numbers, hence the set of transcendental real numbers is not empty. Applications of Baire's theorem are based on a topological notion of largeness: The dense G_δ's are "large" subsets of a complete metric space. A third type of largeness is measure-theoretic: One can try to show that a certain set in a measure space is not empty by showing that it has positive measure or, better still, by showing that its complement has measure zero. Fubini's theorem often occurs in this type of argument.

For example, let f and $g \in L^1(R^1)$, assume $f \geq 0$ and $g \geq 0$ for the moment, and consider the integral

$$h(x) = \int_{-\infty}^{\infty} f(x - t)g(t)\, dt \qquad (-\infty < x < \infty). \tag{1}$$

For any fixed x, the integrand in (1) is a measurable function with range in $[0, \infty]$, so that $h(x)$ is certainly well defined by (1), and $0 \leq h(x) \leq \infty$.

But is there any x for which $h(x) < \infty$? Note that the integrand in (1) is, for each fixed x, the product of two members of L^1, and such a product is not always in L^1. [Example: $f(x) = g(x) = 1/\sqrt{x}$ if $0 < x < 1$, 0 otherwise.] The Fubini theorem will give an affirmative answer. In fact, it will show that $h \in L^1(R^1)$, hence that $h(x) < \infty$ a.e.

8.14 Theorem *Suppose* $f \in L^1(R^1)$, $g \in L^1(R^1)$. *Then*

$$\int_{-\infty}^{\infty} |f(x - y)g(y)|\, dy < \infty \tag{1}$$

for almost all x. For these x, define

$$h(x) = \int_{-\infty}^{\infty} f(x - y)g(y)\, dy. \tag{2}$$

Then $h \in L^1(R^1)$, and

$$\|h\|_1 \leq \|f\|_1 \|g\|_1, \tag{3}$$

where

$$\|f\|_1 = \int_{-\infty}^{\infty} |f(x)|\, dx. \tag{4}$$

We call h the *convolution* of f and g, and write $h = f * g$.

PROOF There exist Borel functions f_0 and g_0 such that $f_0 = f$ a.e. and $g_0 = g$ a.e. The integrals (1) and (2) are unchanged, for every x, if we replace f by f_0 and g by g_0. Hence we may assume, to begin with, that f and g are Borel functions.

To apply Fubini's theorem, we shall first prove that the function F defined by

$$F(x, y) = f(x - y)g(y) \tag{5}$$

is a Borel function on R^2.

Define $\varphi: R^2 \to R^1$ and $\psi: R^2 \to R^1$ by

$$\varphi(x, y) = x - y, \qquad \psi(x, y) = y. \tag{6}$$

Then $f(x - y) = (f \circ \varphi)(x, y)$ and $g(y) = (g \circ \psi)(x, y)$. Since φ and ψ are Borel functions, Theorem 1.12(d) shows that $f \circ \varphi$ and $g \circ \psi$ are Borel functions on R^2. Hence so is their product.

Next we observe that

$$\int_{-\infty}^{\infty} dy \int_{-\infty}^{\infty} |F(x, y)|\, dx = \int_{-\infty}^{\infty} |g(y)|\, dy \int_{-\infty}^{\infty} |f(x - y)|\, dx = \|f\|_1 \|g\|_1, \tag{7}$$

since

$$\int_{-\infty}^{\infty} |f(x - y)|\, dx = \|f\|_1 \tag{8}$$

for every $y \in R^1$, by the translation-invariance of Lebesgue measure.

Thus $F \in L^1(R^2)$, and Fubini's theorem implies that the integral (2) exists for almost all $x \in R^1$ and that $h \in L^1(R^1)$. Finally,

$$\|h\|_1 = \int_{-\infty}^{\infty} |h(x)|\, dx \le \int_{-\infty}^{\infty} dx \int_{-\infty}^{\infty} |F(x, y)|\, dy$$

$$= \int_{-\infty}^{\infty} dy \int_{-\infty}^{\infty} |F(x, y)|\, dx = \|f\|_1 \|g\|_1,$$

by (7). This gives (3), and completes the proof. ////

Convolutions will play an important role in Chap. 9.

Distribution Functions

8.15 Definition Let μ be a σ-finite positive measure on some σ-algebra in some set X. Let $f: X \to [0, \infty]$ be measurable. The function that assigns to each $t \in [0, \infty)$ the number

$$\mu\{f > t\} = \mu(\{x \in X : f(x) > t\}) \tag{1}$$

is called the *distribution function* of f. It is clearly a monotonic (nonincreasing) function of t and is therefore Borel measurable.

One reason for introducing distribution functions is that they make it possible to replace integrals over X by integrals over $[0, \infty)$; the formula

$$\int_X f \, d\mu = \int_0^\infty \mu\{f > t\} \, dt \tag{2}$$

is the special case $\varphi(t) = t$ of our next theorem. This will then be used to derive an L^p-property of the maximal functions that were introduced in Chap. 7.

8.16 Theorem *Suppose that f and μ are as above, that $\varphi: [0, \infty] \to [0, \infty]$ is monotonic, absolutely continuous on $[0, T]$ for every $T < \infty$, and that $\varphi(0) = 0$ and $\varphi(t) \to \varphi(\infty)$ as $t \to \infty$. Then*

$$\int_X (\varphi \circ f) \, d\mu = \int_0^\infty \mu\{f > t\} \varphi'(t) \, dt. \tag{1}$$

PROOF Let E be the set of all $(x, t) \in X \times [0, \infty)$ where $f(x) > t$. When f is simple, then E is a union of finitely many measurable rectangles, and is therefore measurable. In the general case, the measurability of E follows via the standard approximation of f by simple functions (Theorem 1.17). As in Sec. 8.1, put

$$E^t = \{x \in X : (x, t) \in E\} \qquad (0 \le t < \infty). \tag{2}$$

The distribution function of f is then

$$\mu(E^t) = \int_X \chi_E(x, t) \, d\mu(x). \tag{3}$$

The right side of (1) is therefore

$$\int_0^\infty \mu(E^t) \varphi'(t) \, dt = \int_X d\mu(x) \int_0^\infty \chi_E(x, t) \varphi'(t) \, dt, \tag{4}$$

by Fubini's theorem.

For each $x \in X$, $\chi_E(x, t) = 1$ if $t < f(x)$ and is 0 if $t \geq f(x)$. The inner integral in (4) is therefore

$$\int_0^{f(x)} \varphi'(t)\, dt = \varphi(f(x)) \tag{5}$$

by Theorem 7.20. Now (1) follows from (4) and (5). ////

8.17 Recall now that the maximal function Mf lies in weak L^1 when $f \in L^1(R^k)$ (Theorem 7.4). We also have the trivial estimate

$$\|Mf\|_\infty \leq \|f\|_\infty \tag{1}$$

valid for all $f \in L^\infty(R^k)$. A technique invented by Marcinkiewicz makes it possible to "interpolate" between these two extremes and to prove the following theorem of Hardy and Littlewood (which fails when $p = 1$; see Exercise 22, Chap. 7).

8.18 Theorem *If $1 < p < \infty$ and $f \in L^p(R^k)$ then $Mf \in L^p(R^k)$.*

PROOF Since $Mf = M(|f|)$ we may assume, without loss of generality, that $f \geq 0$. Theorem 7.4 shows that there is a constant A, depending only on the dimension k, such that

$$m\{Mg > t\} \leq \frac{A}{t} \|g\|_1 \tag{1}$$

for every $g \in L^1(R^k)$. Here, and in the rest of this proof, $m = m_k$, the Lebesgue measure on R^k.

Pick a constant c, $0 < c < 1$, which will be specified later so as to minimize a certain upper bound. For each $t \in (0, \infty)$, split f into a sum

$$f = g_t + h_t \tag{2}$$

where

$$g_t(x) = \begin{cases} f(x) & \text{if } f(x) > ct \\ 0 & \text{if } f(x) \leq ct. \end{cases} \tag{3}$$

Then $0 \leq h_t(x) \leq ct$ for every $x \in R^k$. Hence $h_t \in L^\infty$, $Mh_t \leq ct$, and

$$Mf \leq Mg_t + Mh_t \leq Mg_t + ct. \tag{4}$$

If $(Mf)(x) > t$ for some x, it follows that

$$(Mg_t)(x) > (1 - c)t. \tag{5}$$

Setting $E_t = \{f > ct\}$, (5), (1), and (3) imply that

$$m\{Mf > t\} \leq m\{Mg_t > (1 - c)t\} \leq \frac{A}{(1 - c)t} \|g_t\|_1 = \frac{A}{(1 - c)t} \int_{E_t} f\, dm.$$

We now use Theorem 8.16, with $X = R^k$, $\mu = m$, $\varphi(t) = t^p$, to calculate

$$\int_{R^k} (Mf)^p \, dm = p \int_0^\infty m\{Mf > t\} t^{p-1} \, dt \le \frac{Ap}{1-c} \int_0^\infty t^{p-2} \, dt \int_{E_t} f \, dm$$

$$= \frac{Ap}{1-c} \int_{R^k} f \, dm \int_0^{f/c} t^{p-2} \, dt = \frac{Apc^{1-p}}{(1-c)(p-1)} \int_{R^k} f^p \, dm.$$

This proves the theorem. However, to get a good constant, let us choose c so as to minimize that last expression. This happens when $c = (p-1)/p = 1/q$, where q is the exponent conjugate to p. For this c,

$$c^{1-p} = \left(1 + \frac{1}{p-1}\right)^{p-1} < e,$$

and the preceding computation yields

$$\|Mf\|_p \le C_p \|f\|_p \tag{6}$$

where $C_p = (Aepq)^{1/p}$. ////

Note that $C_p \to 1$ as $p \to \infty$, which agrees with formula 8.17(1), and that $C_p \to \infty$ as $p \to 1$.

Exercises

1 Find a monotone class $\mathfrak{M}$ in R^1 which is not a σ-algebra, even though $R^1 \in \mathfrak{M}$ and $R^1 - A \in \mathfrak{M}$ for every $A \in \mathfrak{M}$.

2 Suppose f is a Lebesgue measurable nonnegative real function on R^1 and $A(f)$ is the *ordinate set* of f. This is the set of all points $(x, y) \in R^2$ for which $0 < y < f(x)$.

(a) Is it true that $A(f)$ is Lebesgue measurable, in the two-dimensional sense?
(b) If the answer to (a) is affirmative, is the integral of f over R^1 equal to the measure of $A(f)$?
(c) Is the graph of f a measurable subset of R^2?
(d) If the answer to (c) is affirmative, is the measure of the graph equal to zero?

3 Find an example of a positive continuous function f in the open unit square in R^2, whose integral (relative to Lebesgue measure) is finite but such that $\varphi(x)$ (in the notation of Theorem 8.8) is infinite for some $x \in (0, 1)$.

4 Suppose $1 \le p \le \infty$, $f \in L^1(R^1)$, and $g \in L^p(R^1)$.

(a) Imitate the proof of Theorem 8.14 to show that the integral defining $(f * g)(x)$ exists for almost all x, that $f * g \in L^p(R^1)$, and that

$$\|f * g\|_p \le \|f\|_1 \|g\|_p.$$

(b) Show that equality can hold in (a) if $p = 1$ and if $p = \infty$, and find the conditions under which this happens.
(c) Assume $1 < p < \infty$, and equality holds in (a). Show that then either $f = 0$ a.e. or $g = 0$ a.e.
(d) Assume $1 \le p \le \infty$, $\epsilon > 0$, and show that there exist $f \in L^1(R^1)$ and $g \in L^p(R^1)$ such that

$$\|f * g\|_p > (1 - \epsilon) \|f\|_1 \|g\|_p.$$

5 Let M be the Banach space of all complex Borel measures on R^1. The norm in M is $\|\mu\| = |\mu|(R^1)$. Associate to each Borel set $E \subset R^1$ the set

$$E_2 = \{(x, y): x + y \in E\} \subset R^2.$$

If μ and $\lambda \in M$, define their convolution $\mu * \lambda$ to be the set function given by

$$(\mu * \lambda)(E) = (\mu \times \lambda)(E_2)$$

for every Borel set $E \subset R^1$; $\mu \times \lambda$ is as in Definition 8.7.
  (a) Prove that $\mu * \lambda \in M$ and that $\|\mu * \lambda\| \leq \|\mu\| \, \|\lambda\|$.
  (b) Prove that $\mu * \lambda$ is the unique $\nu \in M$ such that

$$\int f \, d\nu = \iint f(x + y) \, d\mu(x) \, d\lambda(y)$$

for every $f \in C_0(R^1)$. (All integrals extend over R^1.)
  (c) Prove that convolution in M is commutative, associative, and distributive with respect to addition.
  (d) Prove the formula

$$(\mu * \lambda)(E) = \int \mu(E - t) \, d\lambda(t)$$

for every μ and $\lambda \in M$ and every Borel set E. Here

$$E - t = \{x - t: x \in E\}.$$

  (e) Define μ to be *discrete* if μ is concentrated on a countable set; define μ to be *continuous* if $\mu(\{x\}) = 0$ for every point $x \in R^1$; let m be Lebesgue measure on R^1 (note that $m \notin M$). Prove that $\mu * \lambda$ is discrete if both μ and λ are discrete, that $\mu * \lambda$ is continuous if μ is continuous and $\lambda \in M$, and that $\mu * \lambda \ll m$ if $\mu \ll m$.
  (f) Assume $d\mu = f \, dm$, $d\lambda = g \, dm$, $f \in L^1(R^1)$, and $g \in L^1(R^1)$, and prove that $d(\mu * \lambda) = (f * g) \, dm$.
  (g) Properties (a) and (c) show that the Banach space M is what one calls a *commutative Banach algebra*. Show that (e) and (f) imply that the set of all discrete measures in M is a subalgebra of M, that the continuous measures form an ideal in M, and that the absolutely continuous measures (relative to m) form an ideal in M which is isomorphic (as an algebra) to $L^1(R^1)$.
  (h) Show that M has a unit, i.e., show that there exists a $\delta \in M$ such that $\delta * \mu = \mu$ for all $\mu \in M$.
  (i) Only two properties of R^1 have been used in this discussion: R^1 is a commutative group (under addition), and there exists a translation invariant Borel measure m on R^1 which is not identically 0 and which is finite on all compact subsets of R^1. Show that the same results hold if R^1 is replaced by R^k or by T (the unit circle) or by T^k (the k-dimensional torus, the cartesian product of k copies of T), as soon as the definitions are properly formulated.

6 (Polar coordinates in R^k.) Let S_{k-1} be the unit sphere in R^k, i.e., the set of all $u \in R^k$ whose distance from the origin 0 is 1. Show that every $x \in R^k$, except for $x = 0$, has a unique representation of the form $x = ru$, where r is a positive real number and $u \in S_{k-1}$. Thus $R^k - \{0\}$ may be regarded as the cartesian product $(0, \infty) \times S_{k-1}$.

Let m_k be Lebesgue measure on R^k, and define a measure σ_{k-1} on S_{k-1} as follows: If $A \subset S_{k-1}$ and A is a Borel set, let $\tilde{A}$ be the set of all points ru, where $0 < r < 1$ and $u \in A$, and define

$$\sigma_{k-1}(A) = k \cdot m_k(\tilde{A}).$$

Prove that the formula

$$\int_{R^k} f \, dm_k = \int_0^\infty r^{k-1} \, dr \int_{S_{k-1}} f(ru) \, d\sigma_{k-1}(u)$$

is valid for every nonnegative Borel function f on R^k. Check that this coincides with familiar results when $k = 2$ and when $k = 3$.

Suggestion: If $0 < r_1 < r_2$ and if A is an open subset of S_{k-1}, let E be the set of all ru with $r_1 < r < r_2$, $u \in A$, and verify that the formula holds for the characteristic function of E. Pass from these to characteristic functions of Borel sets in R^k.

7 Suppose $(X, \mathscr{S}, \mu)$ and $(Y, \mathscr{T}, \lambda)$ are σ-finite measure spaces, and suppose ψ is a measure on $\mathscr{S} \times \mathscr{T}$ such that

$$\psi(A \times B) = \mu(A)\lambda(B)$$

whenever $A \in \mathscr{S}$ and $B \in \mathscr{T}$. Prove that then $\psi(E) = (\mu \times \lambda)(E)$ for every $E \in \mathscr{S} \times \mathscr{T}$.

8 (a) Suppose f is a real function on R^2 such that each section f_x is Borel measurable and each section f^y is continuous. Prove that f is Borel measurable on R^2. Note the contrast between this and Example 8.9(c).

(b) Suppose g is a real function on R^k which is continuous in each of the k variables separately. More explicitly, for every choice of $x_2, \ldots, x_k$, the mapping $x_1 \to g(x_1, x_2, \ldots, x_k)$ is continuous, etc. Prove that g is a Borel function.

Hint: If $(i-1)/n = a_{i-1} \le x \le a_i = i/n$, put

$$f_n(x, y) = \frac{a_i - x}{a_i - a_{i-1}} f(a_{i-1}, y) + \frac{x - a_{i-1}}{a_i - a_{i-1}} f(a_i, y).$$

9 Suppose E is a dense set in R^1 and f is a real function on R^2 such that (a) f_x is Lebesgue measurable for each $x \in E$ and (b) f^y is continuous for almost all $y \in R^1$. Prove that f is Lebesgue measurable on R^2.

10 Suppose f is a real function on R^2, f_x is Lebesgue measurable for each x, and f^y is continuous for each y. Suppose $g: R^1 \to R^1$ is Lebesgue measurable, and put $h(y) = f(g(y), y)$. Prove that h is Lebesgue measurable on R^1.

Hint: Define f_n as in Exercise 8, put $h_n(y) = f_n(g(y), y)$, show that each h_n is measurable, and that $h_n(y) \to h(y)$.

11 Let $\mathscr{B}_k$ be the σ-algebra of all Borel sets in R^k. Prove that $\mathscr{B}_{m+n} = \mathscr{B}_m \times \mathscr{B}_n$. This is relevant in Theorem 8.14.

12 Use Fubini's theorem and the relation

$$\frac{1}{x} = \int_0^\infty e^{-xt} \, dt \quad (x > 0)$$

to prove that

$$\lim_{A \to \infty} \int_0^A \frac{\sin x}{x} \, dx = \frac{\pi}{2}.$$

13 If μ is a complex measure on a σ-algebra $\mathfrak{M}$, show that every set $E \in \mathfrak{M}$ has a subset A for which

$$|\mu(A)| \ge \frac{1}{\pi} |\mu|(E).$$

Suggestion: There is a measurable real function θ so that $d\mu = e^{i\theta} \, d|\mu|$. Let A_α be the subset of E where $\cos(\theta - \alpha) > 0$, show that

$$\operatorname{Re}[e^{-i\alpha}\mu(A_\alpha)] = \int_E \cos^+(\theta - \alpha) \, d|\mu|,$$

and integrate with respect to α (as in Lemma 6.3).

Show, by an example, that $1/\pi$ is the best constant in this inequality.

14 Complete the following proof of Hardy's inequality (Chap. 3, Exercise 14): Suppose $f \geq 0$ on $(0, \infty)$, $f \in L^p$, $1 < p < \infty$, and

$$F(x) = \frac{1}{x} \int_0^x f(t) \, dt.$$

Write $xF(x) = \int_0^x f(t) t^\alpha t^{-\alpha} \, dt$, where $0 < \alpha < 1/q$, use Hölder's inequality to get an upper bound for $F(x)^p$, and integrate to obtain

$$\int_0^\infty F^p(x) \, dx \leq (1 - \alpha q)^{1-p} (\alpha p)^{-1} \int_0^\infty f^p(t) \, dt.$$

Show that the best choice of α yields

$$\int_0^\infty F^p(x) \, dx \leq \left(\frac{p}{p-1}\right)^p \int_0^\infty f^p(t) \, dt.$$

15 Put $\varphi(t) = 1 - \cos t$ if $0 \leq t \leq 2\pi$, $\varphi(t) = 0$ for all other real t. For $-\infty < x < \infty$, define

$$f(x) = 1, \qquad g(x) = \varphi'(x), \qquad h(x) = \int_{-\infty}^x \varphi(t) \, dt.$$

Verify the following statements about convolutions of these functions:

(i) $(f * g)(x) = 0$ for all x.
(ii) $(g * h)(x) = (\varphi * \varphi)(x) > 0$ on $(0, 4\pi)$.
(iii) Therefore $(f * g) * h = 0$, whereas $f * (g * h)$ is a positive constant.

But convolution is supposedly associative, by Fubini's theorem (Exercise 5(c)). What went wrong?

16 Prove the following analogue of Minkowski's inequality, for $f \geq 0$:

$$\left\{\int \left[\int f(x, y) \, d\lambda(y)\right]^p d\mu(x)\right\}^{1/p} \leq \int \left[\int f^p(x, y) \, d\mu(x)\right]^{1/p} d\lambda(y).$$

Supply the required hypotheses. (Many further developments of this theme may be found in [9].)

CHAPTER
NINE

FOURIER TRANSFORMS

Formal Properties

9.1 Definitions In this chapter we shall depart from the previous notation and use the letter m not for Lebesgue measure on R^1 but for Lebesgue measure divided by $\sqrt{2\pi}$. This convention simplifies the appearance of results such as the inversion theorem and the Plancherel theorem. Accordingly, we shall use the notation

$$\int_{-\infty}^{\infty} f(x)\, dm(x) = \frac{1}{\sqrt{2\pi}} \int_{-\infty}^{\infty} f(x)\, dx, \tag{1}$$

where dx refers to ordinary Lebesgue measure, and we define

$$\|f\|_p = \left\{ \int_{-\infty}^{\infty} |f(x)|^p\, dm(x) \right\}^{1/p} \qquad (1 \le p < \infty), \tag{2}$$

$$(f * g)(x) = \int_{-\infty}^{\infty} f(x - y)g(y)\, dm(y) \qquad (x \in R^1), \tag{3}$$

and

$$\hat{f}(t) = \int_{-\infty}^{\infty} f(x)e^{-ixt}\, dm(x) \qquad (t \in R^1). \tag{4}$$

Throughout this chapter, we shall write L^p in place of $L^p(R^1)$, and C_0 will denote the space of all continuous functions on R^1 which vanish at infinity.

If $f \in L^1$, the integral (4) is well defined for every real t. The function $\hat{f}$ is called the *Fourier transform* of f. Note that the term "Fourier transform" is also applied to the *mapping* which takes f to $\hat{f}$.

The formal properties which are listed in Theorem 9.2 depend intimately on the translation-invariance of m and on the fact that for each real α the mapping $x \to e^{i\alpha x}$ is a *character* of the additive group R^1. By definition, a function φ is a character of R^1 if $|\varphi(t)| = 1$ and if

$$\varphi(s + t) = \varphi(s)\varphi(t) \tag{5}$$

for all real s and t. In other words, φ is to be a homomorphism of the additive group R^1 into the multiplicative group of the complex numbers of absolute value 1. We shall see later (in the proof of Theorem 9.23) that every continuous character of R^1 is given by an exponential.

9.2 Theorem *Suppose $f \in L^1$, and α and λ are real numbers.*

(a) *If $g(x) = f(x)e^{i\alpha x}$, then $\hat{g}(t) = \hat{f}(t - \alpha)$.*
(b) *If $g(x) = f(x - \alpha)$, then $\hat{g}(t) = \hat{f}(t)e^{-i\alpha t}$.*
(c) *If $g \in L^1$ and $h = f * g$, then $\hat{h}(t) = \hat{f}(t)\hat{g}(t)$.*

Thus the Fourier transform converts multiplication by a character into translation, and vice versa, and it converts convolutions to pointwise products.

(d) *If $g(x) = \overline{f(-x)}$, then $\hat{g}(t) = \overline{\hat{f}(t)}$.*
(e) *If $g(x) = f(x/\lambda)$ and $\lambda > 0$, then $\hat{g}(t) = \lambda \hat{f}(\lambda t)$.*
(f) *If $g(x) = -ixf(x)$ and $g \in L^1$, then $\hat{f}$ is differentiable and $\hat{f}'(t) = \hat{g}(t)$.*

PROOF (a), (b), (d), and (e) are proved by direct substitution into formula 9.1(4). The proof of (c) is an application of Fubini's theorem (see Theorem 8.14 for the required measurability proof):

$$\hat{h}(t) = \int_{-\infty}^{\infty} e^{-itx}\, dm(x) \int_{-\infty}^{\infty} f(x - y)g(y)\, dm(y)$$

$$= \int_{-\infty}^{\infty} g(y)e^{-ity}\, dm(y) \int_{-\infty}^{\infty} f(x - y)e^{-it(x-y)}\, dm(x)$$

$$= \int_{-\infty}^{\infty} g(y)e^{-ity}\, dm(y) \int_{-\infty}^{\infty} f(x)e^{-itx}\, dm(x)$$

$$= \hat{g}(t)\hat{f}(t).$$

Note how the translation-invariance of m was used.

To prove (f), note that

$$\frac{\hat{f}(s) - \hat{f}(t)}{s - t} = \int_{-\infty}^{\infty} f(x)e^{-ixt}\varphi(x, s - t)\, dm(x) \qquad (s \neq t), \tag{1}$$

where $\varphi(x, u) = (e^{-ixu} - 1)/u$. Since $|\varphi(x, u)| \leq |x|$ for all real $u \neq 0$ and since $\varphi(x, u) \to -ix$ as $u \to 0$, the dominated convergence theorem applies to (1), if s tends to t, and we conclude that

$$\hat{f}'(t) = -i \int_{-\infty}^{\infty} x f(x) e^{-ixt} \, dm(x). \tag{2}$$

////

9.3 Remarks

(a) In the preceding proof, the appeal to the dominated convergence theorem may seem to be illegitimate since the dominated convergence theorem deals only with *countable* sequences of functions. However, it does enable us to conclude that

$$\lim_{n \to \infty} \frac{\hat{f}(s_n) - \hat{f}(t)}{s_n - t} = -i \int_{-\infty}^{\infty} x f(x) e^{-ixt} \, dm(t)$$

for every sequence $\{s_n\}$ which converges to t, and this says exactly that

$$\lim_{s \to t} \frac{\hat{f}(s) - \hat{f}(t)}{s - t} = -i \int_{-\infty}^{\infty} x f(x) e^{-ixt} \, dm(t).$$

We shall encounter similar situations again, and shall apply convergence theorems to them without further comment.

(b) Theorem 9.2(b) shows that the Fourier transform of

$$[f(x + \alpha) - f(x)]/\alpha$$

is

$$\hat{f}(t) \frac{e^{i\alpha t} - 1}{\alpha}.$$

This suggests that an analogue of Theorem 9.2(f) should be true under certain conditions, namely, that the Fourier transform of f' is $it\hat{f}(t)$. If $f \in L^1, f' \in L^1$, and if f is the indefinite integral of f', the result is easily established by an integration by parts. We leave this, and some related results, as exercises. The fact that the Fourier transform converts differentiation to multiplication by ti makes the Fourier transform a useful tool in the study of differential equations.

The Inversion Theorem

9.4 We have just seen that certain operations on functions correspond nicely to operations on their Fourier transforms. The usefulness and interest of this correspondence will of course be enhanced if there is a way of returning from the transforms to the functions, that is to say, if there is an inversion formula.

Let us see what such a formula might look like, by analogy with Fourier series. If

$$c_n = \frac{1}{2\pi} \int_{-\pi}^{\pi} f(x) e^{-inx} \, dx \qquad (n \in Z), \tag{1}$$

then the inversion formula is

$$f(x) = \sum_{-\infty}^{\infty} c_n e^{inx}. \tag{2}$$

We know that (2) holds, in the sense of L^2-convergence, if $f \in L^2(T)$. We also know that (2) does not necessarily hold in the sense of pointwise convergence, even if f is continuous. Suppose now that $f \in L^1(T)$, that $\{c_n\}$ is given by (1), and that

$$\sum_{-\infty}^{\infty} |c_n| < \infty. \tag{3}$$

Put

$$g(x) = \sum_{-\infty}^{\infty} c_n e^{inx}. \tag{4}$$

By (3), the series in (4) converges uniformly (hence g is continuous), and the Fourier coefficients of g are easily computed:

$$\frac{1}{2\pi} \int_{-\pi}^{\pi} g(x) e^{-ikx} \, dx = \frac{1}{2\pi} \int_{-\pi}^{\pi} \left\{ \sum_{n=-\infty}^{\infty} c_n e^{inx} \right\} e^{-ikx} \, dx$$

$$= \sum_{n=-\infty}^{\infty} c_n \frac{1}{2\pi} \int_{-\pi}^{\pi} e^{i(n-k)x} \, dx$$

$$= c_k. \tag{5}$$

Thus f and g have the same Fourier coefficients. This implies $f = g$ a.e., so the Fourier series of f converges to $f(x)$ a.e.

The analogous assumptions in the context of Fourier transforms are that $f \in L^1$ and $\hat{f} \in L^1$, and we might then expect that a formula like

$$f(x) = \int_{-\infty}^{\infty} \hat{f}(t) e^{itx} \, dm_1(t) \tag{6}$$

is valid. Certainly, if $\hat{f} \in L^1$, the right side of (6) is well defined; call it $g(x)$; but if we want to argue as in (5), we run into the integral

$$\int_{-\infty}^{\infty} e^{i(t-s)x} \, dx, \tag{7}$$

which is meaningless as it stands. Thus even under the strong assumption that $\hat{f} \in L^1$, a proof of (6) (which *is* true) has to proceed over a more devious route.

[It should be mentioned that (6) may hold even if $\hat{f} \notin L^1$, if the integral over $(-\infty, \infty)$ is interpreted as the limit, as $A \to \infty$, of integrals over $(-A, A)$. (Analogue: a series may converge without converging absolutely.) We shall not go into this.]

9.5 Theorem *For any function f on R^1 and every $y \in R^1$, let f_y be the translate of f defined by*

$$f_y(x) = f(x - y) \qquad (x \in R^1). \tag{1}$$

If $1 \leq p < \infty$ and if $f \in L^p$, the mapping

$$y \to f_y \tag{2}$$

is a uniformly continuous mapping of R^1 into $L^p(R^1)$.

PROOF Fix $\epsilon > 0$. Since $f \in L^p$, there exists a continuous function g whose support lies in a bounded interval $[-A, A]$, such that

$$\|f - g\|_p < \epsilon$$

(Theorem 3.14). The uniform continuity of g shows that there exists a $\delta \in (0, A)$ such that $|s - t| < \delta$ implies

$$|g(s) - g(t)| < (3A)^{-1/p}\epsilon.$$

If $|s - t| < \delta$, it follows that

$$\int_{-\infty}^{\infty} |g(x - s) - g(x - t)|^p \, dx < (3A)^{-1}\epsilon^p(2A + \delta) < \epsilon^p,$$

so that $\|g_s - g_t\|_p < \epsilon$.

Note that L^p-norms (relative to Lebesgue measure) are translation-invariant: $\|f\|_p = \|f_s\|_p$. Thus

$$\|f_s - f_t\|_p \leq \|f_s - g_s\|_p + \|g_s - g_t\|_p + \|g_t - f_t\|_p$$
$$= \|(f - g)_s\|_p + \|g_s - g_t\|_p + \|(g - f)_t\|_p < 3\epsilon$$

whenever $|s - t| < \delta$. This completes the proof. ////

9.6 Theorem *If $f \in L^1$, then $\hat{f} \in C_0$ and*

$$\|\hat{f}\|_\infty \leq \|f\|_1. \tag{1}$$

PROOF The inequality (1) is obvious from 9.1(4). If $t_n \to t$, then

$$|\hat{f}(t_n) - \hat{f}(t)| \leq \int_{-\infty}^{\infty} |f(x)| \, |e^{-it_n x} - e^{-itx}| \, dm(x). \tag{2}$$

The integrand is bounded by $2|f(x)|$ and tends to 0 for every x, as $n \to \infty$. Hence $\hat{f}(t_n) \to \hat{f}(t)$, by the dominated convergence theorem. Thus $\hat{f}$ is continuous.

Since $e^{\pi i} = -1$, 9.1(4) gives

$$\hat{f}(t) = -\int_{-\infty}^{\infty} f(x) e^{-it(x+\pi/t)} \, dm(x) = -\int_{-\infty}^{\infty} f(x - \pi/t) e^{-itx} \, dm(x). \qquad (3)$$

Hence

$$2\hat{f}(t) = \int_{-\infty}^{\infty} \left\{ f(x) - f\left(x - \frac{\pi}{t}\right) \right\} e^{-itx} \, dm(x), \qquad (4)$$

so that

$$2|\hat{f}(t)| \leq \|f - f_{\pi/t}\|_1, \qquad (5)$$

which tends to 0 as $t \to \pm \infty$, by Theorem 9.5.   ////

9.7 A Pair of Auxiliary Functions In the proof of the inversion theorem it will be convenient to know a positive function H which has a positive Fourier transform whose integral is easily calculated. Among the many possibilities we choose one which is of interest in connection with harmonic functions in a half plane. (See Exercise 25, Chap. 11.)

Put

$$H(t) = e^{-|t|} \qquad (1)$$

and define

$$h_\lambda(x) = \int_{-\infty}^{\infty} H(\lambda t) e^{itx} \, dm(t) \qquad (\lambda > 0). \qquad (2)$$

A simple computation gives

$$h_\lambda(x) = \sqrt{\frac{2}{\pi}} \frac{\lambda}{\lambda^2 + x^2} \qquad (3)$$

and hence

$$\int_{-\infty}^{\infty} h_\lambda(x) \, dm(x) = 1. \qquad (4)$$

Note also that $0 < H(t) \leq 1$ and that $H(\lambda t) \to 1$ as $\lambda \to 0$.

9.8 Proposition *If $f \in L^1$, then*

$$(f * h_\lambda)(x) = \int_{-\infty}^{\infty} H(\lambda t) \hat{f}(t) e^{ixt} \, dm(t).$$

PROOF This is a simple application of Fubini's theorem.

$$\begin{aligned}(f * h_\lambda)(x) &= \int_{-\infty}^{\infty} f(x - y) \, dm(y) \int_{-\infty}^{\infty} H(\lambda t)e^{ity} \, dm(t) \\ &= \int_{-\infty}^{\infty} H(\lambda t) \, dm(t) \int_{-\infty}^{\infty} f(x - y)e^{ity} \, dm(y) \\ &= \int_{-\infty}^{\infty} H(\lambda t) \, dm(t) \int_{-\infty}^{\infty} f(y)e^{it(x - y)} \, dm(y) \\ &= \int_{-\infty}^{\infty} H(\lambda t)\hat{f}(t)e^{itx} \, dm(t).\end{aligned}$$
////

9.9 Theorem *If $g \in L^\infty$ and g is continuous at a point x, then*

$$\lim_{\lambda \to 0} (g * h_\lambda)(x) = g(x). \tag{1}$$

PROOF On account of 9.7(4), we have

$$\begin{aligned}(g * h_\lambda)(x) - g(x) &= \int_{-\infty}^{\infty} [g(x - y) - g(x)]h_\lambda(y) \, dm(y) \\ &= \int_{-\infty}^{\infty} [g(x - y) - g(x)]\lambda^{-1}h_1\left(\frac{y}{\lambda}\right) dm(y) \\ &= \int_{-\infty}^{\infty} [g(x - \lambda s) - g(x)]h_1(s) \, dm(s).\end{aligned}$$

The last integrand is dominated by $2\|g\|_\infty h_1(s)$ and converges to 0 pointwise for every s, as $\lambda \to 0$. Hence (1) follows from the dominated convergence theorem. ////

9.10 Theorem *If $1 \le p < \infty$ and $f \in L^p$, then*

$$\lim_{\lambda \to 0} \|f * h_\lambda - f\|_p = 0. \tag{1}$$

The cases $p = 1$ and $p = 2$ will be the ones of interest to us, but the general case is no harder to prove.

PROOF Since $h_\lambda \in L^q$, where q is the exponent conjugate to p, $(f * h_\lambda)(x)$ is defined for every x. (In fact, $f * h_\lambda$ is continuous; see Exercise 8.) Because of 9.7(4) we have

$$(f * h_\lambda)(x) - f(x) = \int_{-\infty}^{\infty} [f(x - y) - f(x)]h_\lambda(y) \, dm(y) \tag{2}$$

and Theorem 3.3 gives

$$|(f * h_\lambda)(x) - f(x)|^p \le \int_{-\infty}^{\infty} |f(x-y) - f(x)|^p h_\lambda(y) \, dm(y). \tag{3}$$

Integrate (3) with respect to x and apply Fubini's theorem:

$$\|f * h_\lambda - f\|_p^p \le \int_{-\infty}^{\infty} \|f_y - f\|_p^p h_\lambda(y) \, dm(y). \tag{4}$$

If $g(y) = \|f_y - f\|_p^p$, then g is bounded and continuous, by Theorem 9.5, and $g(0) = 0$. Hence the right side of (4) tends to 0 as $\lambda \to 0$, by Theorem 9.9. ////

9.11 The Inversion Theorem *If $f \in L^1$ and $\hat{f} \in L^1$, and if*

$$g(x) = \int_{-\infty}^{\infty} \hat{f}(t) e^{ixt} \, dm(t) \qquad (x \in R^1), \tag{1}$$

then $g \in C_0$ and $f(x) = g(x)$ a.e.

PROOF By Proposition 9.8,

$$(f * h_\lambda)(x) = \int_{-\infty}^{\infty} H(\lambda t) \hat{f}(t) e^{ixt} \, dm(t). \tag{2}$$

The integrands on the right side of (2) are bounded by $|\hat{f}(t)|$, and since $H(\lambda t) \to 1$ as $\lambda \to 0$, the right side of (2) converges to $g(x)$, for every $x \in R^1$, by the dominated convergence theorem.

If we combine Theorems 9.10 and 3.12, we see that there is a sequence $\{\lambda_n\}$ such that $\lambda_n \to 0$ and

$$\lim_{n \to \infty} (f * h_{\lambda_n})(x) = f(x) \text{ a.e.} \tag{3}$$

Hence $f(x) = g(x)$ a.e. That $g \in C_0$ follows from Theorem 9.6. ////

9.12 The Uniqueness Theorem *If $f \in L^1$ and $\hat{f}(t) = 0$ for all $t \in R^1$, then $f(x) = 0$ a.e.*

PROOF Since $\hat{f} = 0$ we have $\hat{f} \in L^1$, and the result follows from the inversion theorem. ////

The Plancherel Theorem

Since the Lebesgue measure of R^1 is infinite, L^2 is not a subset of L^1, and the definition of the Fourier transform by formula 9.1(4) is therefore not directly applicable to every $f \in L^2$. The definition does apply, however, if $f \in L^1 \cap L^2$, and it turns out that then $\hat{f} \in L^2$. In fact, $\|\hat{f}\|_2 = \|f\|_2$! This isometry of $L^1 \cap L^2$ into L^2 extends to an isometry of L^2 onto L^2, and this extension defines the Fourier transform (sometimes called the *Plancherel transform*) of every $f \in L^2$. The resulting L^2-theory has in fact a great deal more symmetry than is the case in L^1. In L^2, f and $\hat{f}$ play exactly the same role.

9.13 Theorem *One can associate to each $f \in L^2$ a function $\hat{f} \in L^2$ so that the following properties hold:*

(a) *If $f \in L^1 \cap L^2$, then $\hat{f}$ is the previously defined Fourier transform of f.*
(b) *For every $f \in L^2$, $\|\hat{f}\|_2 = \|f\|_2$.*
(c) *The mapping $f \to \hat{f}$ is a Hilbert space isomorphism of L^2 onto L^2.*
(d) *The following symmetric relation exists between f and $\hat{f}$: If*

$$\varphi_A(t) = \int_{-A}^{A} f(x) e^{-ixt} \, dm(x) \quad \text{and} \quad \psi_A(x) = \int_{-A}^{A} \hat{f}(t) e^{ixt} \, dm(t),$$

then $\|\varphi_A - \hat{f}\|_2 \to 0$ and $\|\psi_A - f\|_2 \to 0$ as $A \to \infty$.

Note: Since $L^1 \cap L^2$ is dense in L^2, properties (a) and (b) determine the mapping $f \to \hat{f}$ uniquely. Property (d) may be called the L^2 *inversion theorem*.

PROOF Our first objective is the relation

$$\|\hat{f}\|_2 = \|f\|_2 \qquad (f \in L^1 \cap L^2). \tag{1}$$

We fix $f \in L^1 \cap L^2$, put $\tilde{f}(x) = \overline{f(-x)}$, and define $g = f * \tilde{f}$. Then

$$g(x) = \int_{-\infty}^{\infty} f(x-y)\overline{f(-y)} \, dm(y) = \int_{-\infty}^{\infty} f(x+y)\overline{f(y)} \, dm(y), \tag{2}$$

or

$$g(x) = (f_{-x}, f), \tag{3}$$

where the inner product is taken in the Hilbert space L^2 and f_{-x} denotes a translate of f, as in Theorem 9.5. By that theorem, $x \to f_{-x}$ is a continuous mapping of R^1 into L^2, and the continuity of the inner product (Theorem 4.6) therefore implies that g is a continuous function. The Schwarz inequality shows that

$$|g(x)| \leq \|f_{-x}\|_2 \|f\|_2 = \|f\|_2^2 \tag{4}$$

so that g is bounded. Also, $g \in L^1$ since $f \in L^1$ and $\tilde{f} \in L^1$.

Since $g \in L^1$, we may apply Proposition 9.8:

$$(g * h_\lambda)(0) = \int_{-\infty}^{\infty} H(\lambda t) \hat{g}(t) \, dm(t). \tag{5}$$

Since g is continuous and bounded, Theorem 9.9 shows that

$$\lim_{\lambda \to 0} (g * h_\lambda)(0) = g(0) = \|f\|_2^2. \tag{6}$$

Theorem 9.2(d) shows that $\hat{g} = |\hat{f}|^2 \geq 0$, and since $H(\lambda t)$ increases to 1 as $\lambda \to 0$, the monotone convergence theorem gives

$$\lim_{\lambda \to 0} \int_{-\infty}^{\infty} H(\lambda t)\hat{g}(t)\, dm(t) = \int_{-\infty}^{\infty} |\hat{f}(t)|^2\, dm(t). \qquad (7)$$

Now (5), (6), and (7) shows that $\hat{f} \in L^2$ and that (1) holds.

This was the crux of the proof.

Let Y be the space of all Fourier transforms $\hat{f}$ of functions $f \in L^1 \cap L^2$. By (1), $Y \subset L^2$. We claim that Y is dense in L^2, i.e., that $Y^\perp = \{0\}$.

The functions $x \to e^{i\alpha x} H(\lambda x)$ are in $L^1 \cap L^2$, for all real α and all $\lambda > 0$. Their Fourier transforms

$$h_\lambda(\alpha - t) = \int_{-\infty}^{\infty} e^{i\alpha x} H(\lambda x) e^{-i x t}\, dm(x) \qquad (8)$$

are therefore in Y. If $w \in L^2$, $w \in Y^\perp$, it follows that

$$(h_\lambda * \bar{w})(\alpha) = \int_{-\infty}^{\infty} h_\lambda(\alpha - t)\bar{w}(t)\, dm(t) = 0 \qquad (9)$$

for all α. Hence $w = 0$, by Theorem 9.10, and therefore Y is dense in L^2.

Let us introduce the temporary notation Φf for $\hat{f}$. From what has been proved so far, we see that Φ is an L^2-isometry from one dense subspace of L^2, namely $L^1 \cap L^2$, onto another, namely Y. Elementary Cauchy sequence arguments (compare with Lemma 4.16) imply therefore that Φ extends to an isometry $\tilde{\Phi}$ of L^2 onto L^2. If we write $\hat{f}$ for $\tilde{\Phi} f$, we obtain properties (a) and (b).

Property (c) follows from (b), as in the proof of Theorem 4.18. The *Parseval formula*

$$\int_{-\infty}^{\infty} f(x)\overline{g(x)}\, dm(x) = \int_{-\infty}^{\infty} \hat{f}(t)\overline{\hat{g}(t)}\, dm(t) \qquad (10)$$

holds therefore for all $f \in L^2$ and $g \in L^2$.

To prove (d), let k_A be the characteristic function of $[-A, A]$. Then $k_A f \in L^1 \cap L^2$ if $f \in L^2$, and

$$\varphi_A = (k_A f)\hat{}. \qquad (11)$$

Since $\|f - k_A f\|_2 \to 0$ as $A \to \infty$, it follows from (b) that

$$\|\hat{f} - \varphi_A\|_2 = \|(f - k_A f)\hat{}\|_2 \to 0 \qquad (12)$$

as $A \to \infty$.

The other half of (d) is proved the same way. ////

9.14 Theorem *If $f \in L^2$ and $\hat{f} \in L^1$, then*

$$f(x) = \int_{-\infty}^{\infty} \hat{f}(t) e^{ixt}\, dm(t) \qquad a.e.$$

PROOF This is corollary of Theorem 9.13(d).   ////

9.15 Remark If $f \in L^1$, formula 9.1(4) defines $\hat{f}(t)$ unambiguously for every t. If $f \in L^2$, the Plancherel theorem defines $\hat{f}$ uniquely as an element of the Hilbert space L^2, but as a point function $\hat{f}(t)$ is only determined almost everywhere. This is an important difference between the theory of Fourier transforms in L^1 and in L^2. The indeterminacy of $\hat{f}(t)$ as a point function will cause some difficulties in the problem to which we now turn.

9.16 Translation-Invariant Subspaces of L^2 A subspace M of L^2 is said to be *translation-invariant* if $f \in M$ implies that $f_\alpha \in M$ for every real α, where $f_\alpha(x) = f(x - \alpha)$. Translations have already played an important part in our study of Fourier transforms. We now pose a problem whose solution will afford an illustration of how the Plancherel theorem can be used. (Other applications will occur in Chap. 19.) The problem is:

Describe the closed translation-invariant subspaces of L^2.

Let M be a closed translation-invariant subspace of L^2, and let $\hat{M}$ be the image of M under the Fourier transform. Then $\hat{M}$ is closed (since the Fourier transform is an L^2-isometry). If f_α is a translate of f, the Fourier transform of f_α is $\hat{f} e_\alpha$, where $e_\alpha(t) = e^{-i\alpha t}$; we proved this for $f \in L^1$ in Theorem 9.2; the result extends to L^2, as can be seen from Theorem 9.13(d). *It follows that $\hat{M}$ is invariant under multiplication by e_α, for all $\alpha \in R^1$.*

Let E be any measurable set in R^1. If $\hat{M}$ is the set of all $\varphi \in L^2$ which vanish a.e. on E, then $\hat{M}$ certainly is a subspace of L^2, which is invariant under multiplication by all e_α (note that $|e_\alpha| = 1$, so $\varphi e_\alpha \in L^2$ if $\varphi \in L^2$), and $\hat{M}$ is also closed. *Proof*: $\varphi \in \hat{M}$ if and only if φ is orthogonal to every $\psi \in L^2$ which vanishes a.e. on the complement of E.

If M is the inverse image of this $\hat{M}$, under the Fourier transform, then M is a space with the desired properties.

One may now conjecture that every one of our spaces M is obtained in this manner, from a set $E \subset R^1$. To prove this, we have to show that to every closed translation-invariant $M \subset L^2$ there corresponds a set $E \subset R^1$ such that $f \in M$ if and only if $\hat{f}(t) = 0$ a.e. on E. The obvious way of constructing E from M is to associate with each $f \in M$ the set E_f consisting of all points at which $\hat{f}(t) = 0$, and to define E as the intersection of these sets E_f. But this obvious attack runs into a serious difficulty: Each E_f is defined only up to sets of measure 0. If $\{A_i\}$ is a countable collection of sets, each determined up to sets of measure 0, then $\bigcap A_i$ is also determined up to sets of measure 0. But there are uncountably many $f \in M$, so we lose all control over $\bigcap E_f$.

This difficulty disappears entirely if we think of our functions as elements of the Hilbert space L^2, and not primarily as point functions.

We shall now prove the conjecture. Let $\hat{M}$ be the image of a closed translation-invariant subspace $M \subset L^2$, under the Fourier transform. Let P be the orthogonal projection of L^2 onto $\hat{M}$ (Theorem 4.11): To each $f \in L^2$ there corresponds a unique $Pf \in \hat{M}$ such that $f - Pf$ is orthogonal to $\hat{M}$. Hence

$$f - Pf \perp Pg \qquad (f \text{ and } g \in L^2) \qquad (1)$$

and since $\hat{M}$ is invariant under multiplication by e_α, we also have

$$f - Pf \perp (Pg)e_\alpha \qquad (f \text{ and } g \in L^2, \alpha \in R^1). \qquad (2)$$

If we recall how the inner product is defined in L^2, we see that (2) is equivalent to

$$\int_{-\infty}^{\infty} (f - Pf) \cdot \overline{Pg} \cdot e_{-\alpha} \, dm = 0 \qquad (f \text{ and } g \in L^2, \alpha \in R^1) \qquad (3)$$

and this says that the Fourier transform of

$$(f - Pf) \cdot \overline{Pg} \qquad (4)$$

is 0. The function (4) is the product of two L^2-functions, hence is in L^1, and the uniqueness theorem for Fourier transforms shows now that the function (4) is 0 a.e. This remains true if $\overline{Pg}$ is replaced by Pg. Hence

$$f \cdot Pg = (Pf) \cdot (Pg) \qquad (f \text{ and } g \in L^2). \qquad (5)$$

Interchanging the roles of f and g leads from (5) to

$$f \cdot Pg = g \cdot Pf \qquad (f \text{ and } g \in L^2). \qquad (6)$$

Now let g be a fixed positive function in L^2; for instance, put $g(t) = e^{-|t|}$. Define

$$\varphi(t) = \frac{(Pg)(t)}{g(t)}. \qquad (7)$$

$(Pg)(t)$ may only be defined a.e.; choose any one determination in (7). Now (6) becomes

$$Pf = \varphi \cdot f \qquad (f \in L^2). \qquad (8)$$

If $f \in \hat{M}$, then $Pf = f$. This says that $P^2 = P$, and it follows that $\varphi^2 = \varphi$, because

$$\varphi^2 \cdot g = \varphi \cdot Pg = P^2 g = Pg = \varphi \cdot g. \qquad (9)$$

Since $\varphi^2 = \varphi$, we have $\varphi = 0$ or 1 a.e., and if we let E be the set of all t where $\varphi(t) = 0$, then $\hat{M}$ consists precisely of those $f \in L^2$ which are 0 a.e. on E, since $f \in \hat{M}$ if and only if $f = Pf = \varphi \cdot f$.

We therefore obtain the following solution to our problem.

9.17 Theorem *Associate to each measurable set $E \subset R^1$ the space M_E of all $f \in L^2$ such that $\hat{f} = 0$ a.e. on E. Then M_E is a closed translation-invariant subspace of L^2. Every closed translation-invariant subspace of L^2 is M_E for some E, and $M_A = M_B$ if and only if*

$$m((A - B) \cup (B - A)) = 0.$$

The uniqueness statement is easily proved; we leave the details to the reader.

The above problem can of course be posed in other function spaces. It has been studied in great detail in L^1. The known results show that the situation is infinitely more complicated there than in L^2.

The Banach Algebra L^1

9.18 Definition A Banach space A is said to be a *Banach algebra* if there is a multiplication defined in A which satisfies the inequality

$$\|xy\| \leq \|x\| \, \|y\| \qquad (x \text{ and } y \in A), \tag{1}$$

the associative law $x(yz) = (xy)z$, the distributive laws

$$x(y + z) = xy + xz, \quad (y + z)x = yx + zx \quad (x, y, \text{ and } z \in A), \tag{2}$$

and the relation

$$(\alpha x)y = x(\alpha y) = \alpha(xy) \tag{3}$$

where α is any scalar.

9.19 Examples

(a) Let $A = C(X)$, where X is a compact Hausdorff space, with the supremum norm and the usual pointwise multiplication of functions: $(fg)(x) = f(x)g(x)$. This is a commutative Banach algebra ($fg = gf$) with unit (the constant function 1).
(b) $C_0(R^1)$ is a commutative Banach algebra without unit, i.e., without an element u such that $uf = f$ for all $f \in C_0(R^1)$.
(c) The set of all linear operators on R^k (or on any Banach space), with the operator norm as in Definition 5.3, and with addition and multiplication defined by

$$(A + B)(x) = Ax + Bx, \quad (AB)x = A(Bx),$$

is a Banach algebra with unit which is not commutative when $k > 1$.
(d) L^1 is a Banach algebra if we define multiplication by convolution; since

$$\|f * g\|_1 \leq \|f\|_1 \|g\|_1,$$

the norm inequality is satisfied. The associative law could be verified directly (an application of Fubini's theorem), but we can proceed as follows: We know that the Fourier transform of $f * g$ is $\hat{f} \cdot \hat{g}$, and we know that the mapping $f \to \hat{f}$ is one-to-one. For every $t \in R^1$,

$$\hat{f}(t)[\hat{g}(t)\hat{h}(t)] = [\hat{f}(t)\hat{g}(t)]\hat{h}(t),$$

by the associative law for complex numbers. It follows that

$$f * (g * h) = (f * g) * h.$$

In the same way we see immediately that $f * g = g * f$. The remaining requirements of Definition 9.18 are also easily seen to hold in L^1.

Thus L^1 is a commutative Banach algebra. The Fourier transform is an algebra isomorphism of L^1 into C_0. Hence there is no $f \in L^1$ with $\hat{f} \equiv 1$, and therefore L^1 has no unit.

9.20 Complex Homomorphisms The most important complex functions on a Banach algebra A are the homomorphisms of A into the complex field. These are precisely the linear functionals which also preserve multiplication, i.e., the functions φ such that

$$\varphi(\alpha x + \beta y) = \alpha \varphi(x) + \beta \varphi(y), \qquad \varphi(xy) = \varphi(x)\varphi(y)$$

for all x and $y \in A$ and all scalars α and β. Note that no boundedness assumption is made in this definition. It is a very interesting fact that this would be redundant:

9.21 Theorem *If φ is a complex homomorphism on a Banach algebra A, then the norm of φ, as a linear functional, is at most 1.*

PROOF Assume, to get a contradiction, that $|\varphi(x_0)| > \|x_0\|$ for some $x_0 \in A$. Put $\lambda = \varphi(x_0)$, and put $x = x_0/\lambda$. Then $\|x\| < 1$ and $\varphi(x) = 1$.

Since $\|x^n\| \le \|x\|^n$ and $\|x\| < 1$, the elements

$$s_n = -x - x^2 - \cdots - x^n \tag{1}$$

form a Cauchy sequence in A. Since A is complete, being a Banach space, there exists a $y \in A$ such that $\|y - s_n\| \to 0$, and it is easily seen that $x + s_n = xs_{n-1}$, so that

$$x + y = xy. \tag{2}$$

Hence $\varphi(x) + \varphi(y) = \varphi(x)\varphi(y)$, which is impossible since $\varphi(x) = 1$. ////

9.22 The Complex Homomorphisms of L^1 Suppose φ is a complex homomorphism of L^1, i.e., a linear functional (of norm at most 1, by Theorem 9.21) which also satisfies the relation

$$\varphi(f * g) = \varphi(f)\varphi(g) \qquad (f \text{ and } g \in L^1). \tag{1}$$

By Theorem 6.16, there exists a $\beta \in L^\infty$ such that

$$\varphi(f) = \int_{-\infty}^{\infty} f(x)\beta(x)\, dm(x) \qquad (f \in L^1). \tag{2}$$

We now exploit the relation (1) to see what else we can say about β. On the one hand,

$$\varphi(f * g) = \int_{-\infty}^{\infty} (f * g)(x)\beta(x)\, dm(x)$$

$$= \int_{-\infty}^{\infty} \beta(x)\, dm(x) \int_{-\infty}^{\infty} f(x - y)g(y)\, dm(y)$$

$$= \int_{-\infty}^{\infty} g(y)\, dm(y) \int_{-\infty}^{\infty} f_y(x)\beta(x)\, dm(x)$$

$$= \int_{-\infty}^{\infty} g(y)\varphi(f_y)\, dm(y). \tag{3}$$

On the other hand,

$$\varphi(f)\varphi(g) = \varphi(f) \int_{-\infty}^{\infty} g(y)\beta(y)\, dm(y). \tag{4}$$

Let us now assume that φ is not identically 0. Fix $f \in L^1$ so that $\varphi(f) \neq 0$. Since the last integral in (3) is equal to the right side of (4) for every $g \in L^1$, the uniqueness assertion of Theorem 6.16 shows that

$$\varphi(f)\beta(y) = \varphi(f_y) \tag{5}$$

for almost all y. But $y \to f_y$ is a continuous mapping of R^1 into L^1 (Theorem 9.5) and φ is continuous on L^1. Hence the right side of (5) is a continuous function of y, and we may assume [by changing $\beta(y)$ on a set of measure 0 if necessary, which does not affect (2)] that β is continuous. If we replace y by $x + y$ and then f by f_x in (5), we obtain

$$\varphi(f)\beta(x + y) = \varphi(f_{x+y}) = \varphi((f_x)_y) = \varphi(f_x)\beta(y) = \varphi(f)\beta(x)\beta(y),$$

so that

$$\beta(x + y) = \beta(x)\beta(y) \qquad (x \text{ and } y \in R^1). \tag{6}$$

Since β is not identically 0, (6) implies that $\beta(0) = 1$, and the continuity of β shows that there is a $\delta > 0$ such that

$$\int_0^\delta \beta(y)\, dy = c \neq 0. \tag{7}$$

Then

$$c\beta(x) = \int_0^\delta \beta(y)\beta(x)\,dy = \int_0^\delta \beta(y+x)\,dy = \int_x^{x+\delta} \beta(y)\,dy. \tag{8}$$

Since β is continuous, the last integral is a differentiable function of x; hence (8) shows that β is differentiable. Differentiate (6) with respect to y, then put $y = 0$; the result is

$$\beta'(x) = A\beta(x), \qquad A = \beta'(0). \tag{9}$$

Hence the derivative of $\beta(x)e^{-Ax}$ is 0, and since $\beta(0) = 1$, we obtain

$$\beta(x) = e^{Ax}. \tag{10}$$

But β is bounded on R^1. Therefore A must be pure imaginary, and we conclude: There exists a $t \in R^1$ such that

$$\beta(x) = e^{-itx}. \tag{11}$$

We have thus arrived at the Fourier transform.

9.23 Theorem *To every complex homomorphism φ on L^1 (except to $\varphi = 0$) there corresponds a unique $t \in R^1$ such that $\varphi(f) = \hat{f}(t)$.*

The existence of t was proved above. The uniqueness follows from the observation that if $t \neq s$ then there exists an $f \in L^1$ such that $\hat{f}(t) \neq \hat{f}(s)$; take for $f(x)$ a suitable translate of $e^{-|x|}$.

Exercises

1 Suppose $f \in L^1, f > 0$. Prove that $|\hat{f}(y)| < \hat{f}(0)$ for every $y \neq 0$.

2 Compute the Fourier transform of the characteristic function of an interval. For $n = 1, 2, 3, \ldots$, let g_n be the characteristic function of $[-n, n]$, let h be the characteristic function of $[-1, 1]$, and compute $g_n * h$ explicitly. (The graph is piecewise linear.) Show that $g_n * h$ is the Fourier transform of a function $f_n \in L^1$; except for a multiplicative constant,

$$f_n(x) = \frac{\sin x \sin nx}{x^2}$$

Show that $\|f_n\|_1 \to \infty$ and conclude that the mapping $f \to \hat{f}$ maps L^1 into a *proper* subset of C_0.
Show, however, that the range of this mapping is dense in C_0.

3 Find

$$\lim_{A \to \infty} \int_{-A}^{A} \frac{\sin \lambda t}{t} e^{itx}\,dt \qquad (-\infty < x < \infty)$$

where λ is a positive constant.

4 Give examples of $f \in L^2$ such that $f \notin L^1$ but $\hat{f} \in L^1$. Under what circumstances can this happen?

5 If $f \in L^1$ and $\int |t\hat{f}(t)|\, dm(t) < \infty$, prove that f coincides a.e. with a differentiable function whose derivative is

$$i \int_{-\infty}^{\infty} t\hat{f}(t) e^{ixt}\, dm(t).$$

6 Suppose $f \in L^1$, f is differentiable almost everywhere, and $f' \in L^1$. Does it follow that the Fourier transform of f' is $ti\hat{f}(t)$?

7 Let S be the class of all functions f on R^1 which have the following property: f is infinitely differentiable, and there are numbers $A_{mn}(f) < \infty$, for m and $n = 0, 1, 2, \ldots$, such that

$$|x^n D^m f(x)| \le A_{mn}(f) \qquad (x \in R^1).$$

Here D is the ordinary differentiation operator.

Prove that the Fourier transform maps S onto S. Find examples of members of S.

8 If p and q are conjugate exponents, $f \in L^p$, $g \in L^q$, and $h = f * g$, prove that h is uniformly continuous. If also $1 < p < \infty$, then $h \in C_0$; show that this fails for some $f \in L^1$, $g \in L^\infty$.

9 Suppose $1 \le p < \infty$, $f \in L^p$, and

$$g(x) = \int_x^{x+1} f(t)\, dt.$$

Prove that $g \in C_0$. What can you say about g if $f \in L^\infty$?

10 Let C^∞ be the class of all infinitely differentiable complex functions on R^1, and let C_c^∞ consist of all $g \in C^\infty$ whose supports are compact. Show that C_c^∞ does not consist of 0 alone.

Let L_{loc}^1 be the class of all f which belong to L^1 locally; that is, $f \in L_{\text{loc}}^1$ provided that f is measurable and $\int_I |f| < \infty$ for every bounded interval I.

If $f \in L_{\text{loc}}^1$ and $g \in C_c^\infty$, prove that $f * g \in C^\infty$.

Prove that there are sequences $\{g_n\}$ in C_c^∞ such that

$$\|f * g_n - f\|_1 \to 0$$

as $n \to \infty$, for every $f \in L^1$. (Compare Theorem 9.10.) Prove that $\{g_n\}$ can also be so chosen that $(f * g_n)(x) \to f(x)$ a.e., for every $f \in L_{\text{loc}}^1$; in fact, for suitable $\{g_n\}$ the convergence occurs at every point x at which f is the derivative of its indefinite integral.

Prove that $(f * h_\lambda)(x) \to f(x)$ a.e. if $f \in L^1$, as $\lambda \to 0$, and that $f * h_\lambda \in C^\infty$, although h_λ does not have compact support. (h_λ is defined in Sec. 9.7.)

11 Find conditions on f and/or $\hat{f}$ which ensure the correctness of the following formal argument: If

$$\varphi(t) = \frac{1}{2\pi} \int_{-\infty}^{\infty} f(x) e^{-itx}\, dx$$

and

$$F(x) = \sum_{k=-\infty}^{\infty} f(x + 2k\pi)$$

then F is periodic, with period 2π, the nth Fourier coefficient of F is $\varphi(n)$, hence $F(x) = \sum \varphi(n) e^{inx}$. In particular,

$$\sum_{k=-\infty}^{\infty} f(2k\pi) = \sum_{n=-\infty}^{\infty} \varphi(n).$$

More generally,

$$\sum_{k=-\infty}^{\infty} f(k\beta) = \alpha \sum_{n=-\infty}^{\infty} \varphi(n\alpha) \qquad \text{if } \alpha > 0,\ \beta > 0,\ \alpha\beta = 2\pi. \qquad (*)$$

What does (*) say about the limit, as $\alpha \to 0$, of the right-hand side (for "nice" functions, of course)? Is this in agreement with the inversion theorem?

[(*) is known as the Poisson summation formula.]

12 Take $f(x) = e^{-|x|}$ in Exercise 11 and derive the identity

$$\frac{e^{2\pi\alpha} + 1}{e^{2\pi\alpha} - 1} = \frac{1}{\pi} \sum_{n=-\infty}^{\infty} \frac{\alpha}{\alpha^2 + n^2}.$$

13 If $0 < c < \infty$, define $f_c(x) = \exp(-cx^2)$.
  (a) Compute $\hat{f}_c$. Hint: If $\varphi = \hat{f}_c$, an integration by parts gives $2c\varphi'(t) + t\varphi(t) = 0$.
  (b) Show that there is one (and only one) c for which $\hat{f}_c = f_c$.
  (c) Show that $f_a * f_b = \gamma f_c$; find γ and c explicitly in terms of a and b.
  (d) Take $f = f_c$ in Exercise 11. What is the resulting identity?

14 The Fourier transform can be defined for $f \in L^1(R^k)$ by

$$\hat{f}(y) = \int_{R^k} f(x) e^{-ix \cdot y} \, dm_k(x) \quad (y \in R^k),$$

where $x \cdot y = \sum \xi_i \eta_i$ if $x = (\xi_1, \ldots, \xi_k)$, $y = (\eta_1, \ldots, \eta_k)$, and m_k is Lebesgue measure on R^k, divided by $(2\pi)^{k/2}$ for convenience. Prove the inversion theorem and the Plancherel theorem in this context, as well as the analogue of Theorem 9.23.

15 If $f \in L^1(R^k)$, A is a linear operator on R^k, and $g(x) = f(Ax)$, how is $\hat{g}$ related to $\hat{f}$? If f is invariant under rotations, i.e., if $f(x)$ depends only on the euclidean distance of x from the origin, prove that the same is true of $\hat{f}$.

16 The *Laplacian* of a function f on R^k is

$$\Delta f = \sum_{j=1}^{k} \frac{\partial^2 f}{\partial x_j^2},$$

provided the partial derivatives exist. What is the relation between $\hat{f}$ and $\hat{g}$ if $g = \Delta f$ and all necessary integrability conditions are satisfied? It is clear that the Laplacian commutes with translations. Prove that it also commutes with rotations, i.e., that

$$\Delta(f \circ A) = (\Delta f) \circ A$$

whenever f has continuous second derivatives and A is a rotation of R^k. (Show that it is enough to do this under the additional assumption that f has compact support.)

17 Show that every Lebesgue measurable character of R^1 is continuous. Do the same for R^k. (Adapt part of the proof of Theorem 9.23.) Compare with Exercise 18.

18 Show (with the aid of the Hausdorff maximality theorem) that there exist real *discontinuous* functions f on R^1 such that

$$f(x + y) = f(x) + f(y) \tag{1}$$

for all x and $y \in R^1$.

  Show that if (1) holds and f is Lebesgue measurable, then f is continuous.
  Show that if (1) holds and the graph of f is not dense in the plane, then f is continuous.
  Find all continuous functions which satisfy (1).

19 Suppose A and B are measurable subsets of R^1, having finite positive measure. Show that the convolution $\chi_A * \chi_B$ is continuous and not identically 0. Use this to prove that $A + B$ contains a segment.

  (A different proof was suggested in Exercise 5, Chap. 7.)

CHAPTER
TEN

ELEMENTARY PROPERTIES OF HOLOMORPHIC FUNCTIONS

Complex Differentiation

We shall now study complex functions defined in subsets of the complex plane. It will be convenient to adopt some standard notations which will be used throughout the rest of this book.

10.1 Definitions If $r > 0$ and a is a complex number,

$$D(a; r) = \{z: |z - a| < r\} \tag{1}$$

is the open circular disc with center at a and radius r. $\bar{D}(a; r)$ is the closure of $D(a; r)$, and

$$D'(a; r) = \{z: 0 < |z - a| < r\} \tag{2}$$

is the punctured disc with center at a and radius r.

A set E in a topological space X is said to be *not connected* if E is the union of two nonempty sets A and B such that

$$\bar{A} \cap B = \emptyset = A \cap \bar{B}. \tag{3}$$

If A and B are as above, and if V and W are the complements of $\bar{A}$ and $\bar{B}$, respectively, it follows that $A \subset W$ and $B \subset V$. Hence

$$E \subset V \cup W, \quad E \cap V \neq \emptyset, \quad E \cap W \neq \emptyset, \quad E \cap V \cap W = \emptyset. \tag{4}$$

Conversely, if open sets V and W exist such that (4) holds, it is easy to see that E is not connected, by taking $A = E \cap W$, $B = E \cap V$.

If E is closed and not connected, then (3) shows that E is the union of two disjoint nonempty closed sets; for if $\bar{A} \subset A \cup B$ and $\bar{A} \cap B = \emptyset$, then $\bar{A} = A$.

ELEMENTARY PROPERTIES OF HOLOMORPHIC FUNCTIONS 197

If E is open and not connected, then (4) shows that E is the union of two disjoint nonempty open sets, namely $E \cap V$ and $E \cap W$.

Each set consisting of a single point is obviously connected. If $x \in E$, the family Φ_x of all connected subsets of E that contain x is therefore not empty. The union of all members of Φ_x is easily seen to be connected, and to be a *maximal connected subset* of E. These sets are called the *components* of E. Any two components of E are thus disjoint, and E is the union of its components.

By a *region* we shall mean a nonempty connected open subset of the complex plane. Since each open set Ω in the plane is a union of discs, and since all discs are connected, each component of Ω is open. Every plane open set is thus a union of disjoint regions. The letter Ω will from now on denote a plane open set.

10.2 Definition Suppose f is a complex function defined in Ω. If $z_0 \in \Omega$ and if

$$\lim_{z \to z_0} \frac{f(z) - f(z_0)}{z - z_0} \tag{1}$$

exists, we denote this limit by $f'(z_0)$ and call it the *derivative* of f at z_0. If $f'(z_0)$ exists for every $z_0 \in \Omega$, we say that f is *holomorphic* (or *analytic*) in Ω. The class of all holomorphic functions in Ω will be denoted by $H(\Omega)$.

To be quite explicit, $f'(z_0)$ exists if to every $\epsilon > 0$ there corresponds a $\delta > 0$ such that

$$\left| \frac{f(z) - f(z_0)}{z - z_0} - f'(z_0) \right| < \epsilon \qquad \text{for all } z \in D'(z_0; \delta). \tag{2}$$

Thus $f'(z_0)$ is a complex number, obtained as a limit of quotients of complex numbers. Note that f is a mapping of Ω into R^2 and that Definition 7.22 associates with such mappings another kind of derivative, namely, a linear operator on R^2. In our present situation, if (2) is satisfied, this linear operator turns out to be multiplication by $f'(z_0)$ (regarding R^2 as the complex field). We leave it to the reader to verify this.

10.3 Remarks If $f \in H(\Omega)$ and $g \in H(\Omega)$, then also $f + g \in H(\Omega)$ and $fg \in H(\Omega)$, so that $H(\Omega)$ is a ring; the usual differentiation rules apply.

More interesting is the fact that superpositions of holomorphic functions are holomorphic: If $f \in H(\Omega)$, if $f(\Omega) \subset \Omega_1$, if $g \in H(\Omega_1)$, and if $h = g \circ f$, then $h \in H(\Omega)$, and h' can be computed by the chain rule

$$h'(z_0) = g'(f(z_0))f'(z_0) \qquad (z_0 \in \Omega). \tag{1}$$

To prove this, fix $z_0 \in \Omega$, and put $w_0 = f(z_0)$. Then

$$f(z) - f(z_0) = [f'(z_0) + \epsilon(z)](z - z_0), \tag{2}$$

$$g(w) - g(w_0) = [g'(w_0) + \eta(w)](w - w_0), \tag{3}$$

where $\epsilon(z) \to 0$ as $z \to z_0$ and $\eta(w) \to 0$ as $w \to w_0$. Put $w = f(z)$, and substitute (2) into (3): If $z \neq z_0$,

$$\frac{h(z) - h(z_0)}{z - z_0} = [g'(f(z_0)) + \eta(f(z))][f'(z_0) + \epsilon(z)]. \tag{4}$$

The differentiability of f forces f to be continuous at z_0. Hence (1) follows from (4).

10.4 Examples For $n = 0, 1, 2, \ldots$, z^n is holomorphic in the whole plane, and the same is true of every polynomial in z. One easily verifies directly that $1/z$ is holomorphic in $\{z: z \neq 0\}$. Hence, taking $g(w) = 1/w$ in the chain rule, we see that if f_1 and f_2 are in $H(\Omega)$ and Ω_0 is an open subset of Ω in which f_2 has no zero, then $f_1/f_2 \in H(\Omega_0)$.

Another example of a function which is holomorphic in the whole plane (such functions are called *entire*) is the exponential function defined in the Prologue. In fact, we saw there that exp is differentiable everywhere, in the sense of Definition 10.2, and that $\exp'(z) = \exp(z)$ for every complex z.

10.5 Power Series From the theory of power series we shall assume only one fact as known, namely, that to each power series

$$\sum_{n=0}^{\infty} c_n(z - a)^n \tag{1}$$

there corresponds a number $R \in [0, \infty]$ such that the series converges absolutely and uniformly in $\bar{D}(a; r)$, for every $r < R$, and diverges if $z \notin \bar{D}(a; R)$. The "radius of convergence" R is given by the root test:

$$\frac{1}{R} = \limsup_{n \to \infty} |c_n|^{1/n}. \tag{2}$$

Let us say that a function f defined in Ω is *representable by power series in* Ω if to every disc $D(a; r) \subset \Omega$ there corresponds a series (1) which converges to $f(z)$ for all $z \in D(a; r)$.

10.6 Theorem *If f is representable by power series in Ω, then $f \in H(\Omega)$ and f' is also representable by power series in Ω. In fact, if*

$$f(z) = \sum_{n=0}^{\infty} c_n(z - a)^n \tag{1}$$

for $z \in D(a; r)$, then for these z we also have

$$f'(z) = \sum_{n=1}^{\infty} nc_n(z - a)^{n-1}. \tag{2}$$

PROOF If the series (1) converges in $D(a;r)$, the root test shows that the series (2) also converges there. Take $a = 0$, without loss of generality, denote the sum of the series (2) by $g(z)$, fix $w \in D(a;r)$, and choose ρ so that $|w| < \rho < r$. If $z \neq w$, we have

$$\frac{f(z)-f(w)}{z-w} - g(w) = \sum_{n=1}^{\infty} c_n \left[\frac{z^n - w^n}{z-w} - nw^{n-1} \right]. \tag{3}$$

The expression in brackets is 0 if $n = 1$, and is

$$(z-w) \sum_{k=1}^{n-1} kw^{k-1}z^{n-k-1} \tag{4}$$

if $n \geq 2$. If $|z| < \rho$, the absolute value of the sum in (4) is less than

$$\frac{n(n-1)}{2}\rho^{n-2} \tag{5}$$

so

$$\left| \frac{f(z)-f(w)}{z-w} - g(w) \right| \leq |z-w| \sum_{n=2}^{\infty} n^2 |c_n| \rho^{n-2}. \tag{6}$$

Since $\rho < r$, the last series converges. Hence the left side of (6) tends to 0 as $z \to w$. This says that $f'(w) = g(w)$, and completes the proof. ////

Corollary *Since f' is seen to satisfy the same hypothesis as f does, the theorem can be applied to f'. It follows that f has derivatives of all orders, that each derivative is representable by power series in Ω, and that*

$$f^{(k)}(z) = \sum_{n=k}^{\infty} n(n-1) \cdots (n-k+1)c_n(z-a)^{n-k} \tag{7}$$

if (1) holds. Hence (1) implies that

$$k! c_k = f^{(k)}(a) \qquad (k = 0, 1, 2, \ldots), \tag{8}$$

so that for each $a \in \Omega$ there is a unique sequence $\{c_n\}$ for which (1) holds.

We now describe a process which manufactures functions that are representable by power series. Special cases will be of importance later.

10.7 Theorem *Suppose μ is a complex (finite) measure on a measurable space X, φ is a complex measurable function on X, Ω is an open set in the plane which does not intersect $\varphi(X)$, and*

$$f(z) = \int_X \frac{d\mu(\zeta)}{\varphi(\zeta) - z} \qquad (z \in \Omega). \tag{1}$$

Then f is representable by power series in Ω.

PROOF Suppose $D(a; r) \subset \Omega$. Since

$$\left| \frac{z - a}{\varphi(\zeta) - a} \right| \leq \frac{|z - a|}{r} < 1 \tag{2}$$

for every $z \in D(a; r)$ and every $\zeta \in X$, the geometric series

$$\sum_{n=0}^{\infty} \frac{(z - a)^n}{(\varphi(\zeta) - a)^{n+1}} = \frac{1}{\varphi(\zeta) - z} \tag{3}$$

converges uniformly on X, for every fixed $z \in D(a; r)$. Hence the series (3) may be substituted into (1), and $f(z)$ may be computed by interchanging summation and integration. It follows that

$$f(z) = \sum_{0}^{\infty} c_n (z - a)^n \qquad (z \in D(a; r)) \tag{4}$$

where

$$c_n = \int_X \frac{d\mu(\zeta)}{(\varphi(\zeta) - a)^{n+1}} \qquad (n = 0, 1, 2, \ldots). \tag{5}$$

////

Note: The convergence of the series (4) in $D(a; r)$ is a consequence of the proof. We can also derive it from (5), since (5) shows that

$$|c_n| \leq \frac{|\mu|(X)}{r^{n+1}} \qquad (n = 0, 1, 2, \ldots). \tag{6}$$

Integration over Paths

Our first major objective in this chapter is the converse of Theorem 10.6: *Every $f \in H(\Omega)$ is representable by power series in Ω*. The quickest route to this is via Cauchy's theorem which leads to an important integral representation of holomorphic functions. In this section the required integration theory will be developed; we shall keep it as simple as possible, and shall regard it merely as a useful tool in the investigation of properties of holomorphic functions.

10.8 Definitions If X is a topological space, a *curve in X* is a continuous mapping γ of a compact interval $[\alpha, \beta] \subset R^1$ into X; here $\alpha < \beta$. We call $[\alpha, \beta]$ the *parameter interval* of γ and denote the range of γ by γ^*. Thus γ is a mapping, and γ^* is the set of all points $\gamma(t)$, for $\alpha \leq t \leq \beta$.

If the *initial point* $\gamma(\alpha)$ of γ coincides with its *end point* $\gamma(\beta)$, we call γ a *closed curve*.

A *path* is a piecewise continuously differentiable curve in the plane. More explicitly, a path with parameter interval $[\alpha, \beta]$ is a continuous complex function γ on $[\alpha, \beta]$, such that the following holds: There are finitely many points s_j, $\alpha = s_0 < s_1 < \cdots < s_n = \beta$, and the restriction of γ to each interval $[s_{j-1}, s_j]$ has a continuous derivative on $[s_{j-1}, s_j]$; however, at the points $s_1, \ldots, s_{n-1}$ the left- and right-hand derivatives of γ may differ.

A *closed path* is a closed curve which is also a path.

Now suppose γ is a path, and f is a continuous function on γ^*. The integral of f over γ is defined as an integral over the parameter interval $[\alpha, \beta]$ of γ:

$$\int_\gamma f(z)\, dz = \int_\alpha^\beta f(\gamma(t))\gamma'(t)\, dt. \tag{1}$$

Let φ be a continuously differentiable one-to-one mapping of an interval $[\alpha_1, \beta_1]$ onto $[\alpha, \beta]$, such that $\varphi(\alpha_1) = \alpha$, $\varphi(\beta_1) = \beta$, and put $\gamma_1 = \gamma \circ \varphi$. Then γ_1 is a path with parameter interval $[\alpha_1, \beta_1]$; the integral of f over γ_1 is

$$\int_{\alpha_1}^{\beta_1} f(\gamma_1(t))\gamma_1'(t)\, dt = \int_{\alpha_1}^{\beta_1} f(\gamma(\varphi(t)))\gamma'(\varphi(t))\varphi'(t)\, dt = \int_\alpha^\beta f(\gamma(s))\gamma'(s)\, ds,$$

so that our "reparametrization" has not changed the integral:

$$\int_{\gamma_1} f(z)\, dz = \int_\gamma f(z)\, dz. \tag{2}$$

Whenever (2) holds for a pair of paths γ and γ_1 (and for all f), we shall regard γ and γ_1 as equivalent.

It is convenient to be able to replace a path by an equivalent one, i.e., to choose parameter intervals at will. For instance, if the end point of γ_1 coincides with the initial point of γ_2, we may locate their parameter intervals so that γ_1 and γ_2 join to form one path γ, with the property that

$$\int_\gamma f = \int_{\gamma_1} f + \int_{\gamma_2} f \tag{3}$$

for every continuous f on $\gamma^* = \gamma_1^* \cup \gamma_2^*$.

However, suppose that $[0, 1]$ is the parameter interval of a path γ, and $\gamma_1(t) = \gamma(1 - t)$, $0 \le t \le 1$. We call γ_1 the path *opposite* to γ, for the following reason: For any f continuous on $\gamma_1^* = \gamma^*$, we have

$$\int_0^1 f(\gamma_1(t))\gamma_1'(t)\, dt = -\int_0^1 f(\gamma(1-t))\gamma'(1-t)\, dt = -\int_0^1 f(\gamma(s))\gamma'(s)\, ds,$$

so that

$$\int_{\gamma_1} f = -\int_\gamma f. \tag{4}$$

From (1) we obtain the inequality

$$\left| \int_\gamma f(z)\, dz \right| \le \|f\|_\infty \int_\alpha^\beta |\gamma'(t)|\, dt, \tag{5}$$

where $\|f\|_\infty$ is the maximum of $|f|$ on γ^* and the last integral in (5) is (by definition) the *length* of γ.

10.9 Special Cases

(a) If a is a complex number and $r > 0$, the path defined by

$$\gamma(t) = a + re^{it} \qquad (0 \le t \le 2\pi) \tag{1}$$

is called the *positively oriented circle* with center at a and radius r; we have

$$\int_\gamma f(z)\, dz = ir \int_0^{2\pi} f(a + re^{i\theta}) e^{i\theta}\, d\theta, \tag{2}$$

and the length of γ is $2\pi r$, as expected.

(b) If a and b are complex numbers, the path γ given by

$$\gamma(t) = a + (b - a)t \qquad (0 \le t \le 1) \tag{3}$$

is the *oriented interval* $[a, b]$; its length is $|b - a|$, and

$$\int_{[a,\,b]} f(z)\, dz = (b - a) \int_0^1 f[a + (b - a)t]\, dt. \tag{4}$$

If

$$\gamma_1(t) = \frac{a(\beta - t) + b(t - \alpha)}{\beta - \alpha} \qquad (\alpha \le t \le \beta), \tag{5}$$

we obtain an equivalent path, which we still denote by $[a, b]$. The path opposite to $[a, b]$ is $[b, a]$.

(c) Let $\{a, b, c\}$ be an ordered triple of complex numbers, let

$$\Delta = \Delta(a, b, c)$$

be the triangle with vertices at a, b, and c (Δ is the smallest convex set which contains a, b, and c), and define

$$\int_{\partial\Delta} f = \int_{[a,\,b]} f + \int_{[b,\,c]} f + \int_{[c,\,a]} f, \tag{6}$$

for any f continuous on the boundary of Δ. We may regard (6) as the definition of its left side. Or we may regard $\partial\Delta$ as a path obtained by joining $[a, b]$ to $[b, c]$ to $[c, a]$, as outlined in Definition 10.8, in which case (6) is easily proved to be true.

If $\{a, b, c\}$ is permuted cyclically, we see from (6) that the left side of (6) is unaffected. If $\{a, b, c\}$ is replaced by $\{a, c, b\}$, then the left side of (6) changes sign.

We now come to a theorem which plays a very important role in function theory.

10.10 Theorem *Let γ be a closed path, let Ω be the complement of γ^* (relative to the plane), and define*

$$\mathrm{Ind}_\gamma(z) = \frac{1}{2\pi i} \int_\gamma \frac{d\zeta}{\zeta - z} \qquad (z \in \Omega). \tag{1}$$

Then Ind_γ is an integer-valued function on Ω which is constant in each component of Ω and which is 0 in the unbounded component of Ω.

We call $\mathrm{Ind}_\gamma(z)$ the *index* of z with respect to γ. Note that γ^* is compact, hence γ^* lies in a bounded disc D whose complement D^c is connected; thus D^c lies in some component of Ω. This shows that Ω has precisely one unbounded component.

PROOF Let $[\alpha, \beta]$ be the parameter interval of γ, fix $z \in \Omega$, then

$$\mathrm{Ind}_\gamma(z) = \frac{1}{2\pi i} \int_\alpha^\beta \frac{\gamma'(s)}{\gamma(s) - z} \, ds. \tag{2}$$

Since $w/2\pi i$ is an integer if and only if $e^w = 1$, the first assertion of the theorem, namely, that $\mathrm{Ind}_\gamma(z)$ is an integer, is equivalent to the assertion that $\varphi(\beta) = 1$, where

$$\varphi(t) = \exp\left\{ \int_\alpha^t \frac{\gamma'(s)}{\gamma(s) - z} \, ds \right\} \qquad (\alpha \le t \le \beta). \tag{3}$$

Differentiation of (3) shows that

$$\frac{\varphi'(t)}{\varphi(t)} = \frac{\gamma'(t)}{\gamma(t) - z} \tag{4}$$

except possibly on a finite set S where γ is not differentiable. Therefore $\varphi/(\gamma - z)$ is a continuous function on $[\alpha, \beta]$ whose derivative is zero in $[\alpha, \beta] - S$. Since S is finite, $\varphi/(\gamma - z)$ is constant on $[\alpha, \beta]$; and since $\varphi(\alpha) = 1$, we obtain

$$\varphi(t) = \frac{\gamma(t) - z}{\gamma(\alpha) - z} \qquad (\alpha \le t \le \beta). \tag{5}$$

We now use the assumption that γ is a closed path, i.e., that $\gamma(\beta) = \gamma(\alpha)$; (5) shows that $\varphi(\beta) = 1$, and this, as we observed above, implies that $\mathrm{Ind}_\gamma(z)$ is an integer.

By Theorem 10.7, (1) shows that $\text{Ind}_\gamma \in H(\Omega)$. The image of a connected set under a continuous mapping is connected ([26], Theorem 4.22), and since Ind_γ is an integer-valued function, Ind_γ must be constant on each component of Ω.

Finally, (2) shows that $|\text{Ind}_\gamma(z)| < 1$ if $|z|$ is sufficiently large. This implies that $\text{Ind}_\gamma(z) = 0$ in the unbounded component of Ω. ////

Remark: If $\lambda(t)$ denotes the integral in (3), the preceding proof shows that $2\pi \, \text{Ind}_\gamma(z)$ is the net increase in the imaginary part of $\lambda(t)$, as t runs from α to β, and this is the same as the net increase of the argument of $\gamma(t) - z$. (We have not defined "argument" and will have no need for it.) If we divide this increase by 2π, we obtain "the number of times that γ winds around z," and this explains why the term "winding number" is frequently used for the index. One virtue of the preceding proof is that it establishes the main properties of the index without any reference to the (multiple-valued) argument of a complex number.

10.11 Theorem *If γ is the positively oriented circle with center at a and radius r, then*

$$\text{Ind}_\gamma(z) = \begin{cases} 1 & \text{if } |z - a| < r, \\ 0 & \text{if } |z - a| > r. \end{cases}$$

PROOF We take γ as in Sec. 10.9(a). By Theorem 10.10, it is enough to compute $\text{Ind}_\gamma(a)$, and 10.9(2) shows that this equals

$$\frac{1}{2\pi i} \int_\gamma \frac{dz}{z - a} = \frac{r}{2\pi} \int_0^{2\pi} (re^{it})^{-1} e^{it} \, dt = 1. \qquad ////$$

The Local Cauchy Theorem

There are several forms of Cauchy's theorem. They all assert that if γ is a closed path or cycle in Ω, and if γ and Ω satisfy certain topological conditions, then the integral of every $f \in H(\Omega)$ over γ is 0. We shall first derive a simple local version of this (Theorem 10.14) which is quite sufficient for many applications. A more general global form will be established later.

10.12 Theorem *Suppose $F \in H(\Omega)$ and F' is continuous in Ω. Then*

$$\int_\gamma F'(z) \, dz = 0$$

for every closed path γ in Ω.

PROOF If $[\alpha, \beta]$ is the parameter interval of γ, the fundamental theorem of calculus shows that

$$\int_\gamma F'(z)\, dz = \int_\alpha^\beta F'(\gamma(t))\gamma'(t)\, dt = F(\gamma(\beta)) - F(\gamma(\alpha)) = 0,$$

since $\gamma(\beta) = \gamma(\alpha)$. ////

Corollary *Since z^n is the derivative of $z^{n+1}/(n+1)$ for all integers $n \ne -1$, we have*

$$\int_\gamma z^n\, dz = 0$$

for every closed path γ if $n = 0, 1, 2, \ldots$, and for those closed paths γ for which $0 \notin \gamma^*$ if $n = -2, -3, -4, \ldots$.

The case $n = -1$ was dealt with in Theorem 10.10.

10.13 Cauchy's Theorem for a Triangle *Suppose Δ is a closed triangle in a plane open set Ω, $p \in \Omega$, f is continuous on Ω, and $f \in H(\Omega - \{p\})$. Then*

$$\int_{\partial \Delta} f(z)\, dz = 0. \tag{1}$$

For the definition of $\partial \Delta$ we refer to Sec. 10.9(c). We shall see later that our hypothesis actually implies that $f \in H(\Omega)$, i.e., that the exceptional point p is not really exceptional. However, the above formulation of the theorem will be useful in the proof of the Cauchy formula.

PROOF We assume first that $p \notin \Delta$. Let a, b, and c be the vertices of Δ, let a', b', and c' be the midpoints of $[b, c]$, $[c, a]$, and $[a, b]$, respectively, and consider the four triangles Δ^j formed by the ordered triples

$$\{a, c', b'\}, \quad \{b, a', c'\}, \quad \{c, b', a'\}, \quad \{a', b', c'\}. \tag{2}$$

If J is the value of the integral (1), it follows from 10.9(6) that

$$J = \sum_{j=1}^4 \int_{\partial \Delta^j} f(z)\, dz. \tag{3}$$

The absolute value of at least one of the integrals on the right of (3) is therefore at least $|J/4|$. Call the corresponding triangle Δ_1, repeat the argument with Δ_1 in place of Δ, and so forth. This generates a sequence of triangles Δ_n

such that $\Delta \supset \Delta_1 \supset \Delta_2 \supset \cdots$, such that the length of $\partial\Delta_n$ is $2^{-n}L$, if L is the length of $\partial\Delta$, and such that

$$|J| \leq 4^n \left| \int_{\partial\Delta_n} f(z)\,dz \right| \qquad (n = 1, 2, 3, \ldots). \tag{4}$$

There is a (unique) point z_0 which the triangles Δ_n have in common. Since Δ is compact, $z_0 \in \Delta$, so f is differentiable at z_0.

Let $\epsilon > 0$ be given. There exists an $r > 0$ such that

$$|f(z) - f(z_0) - f'(z_0)(z - z_0)| \leq \epsilon|z - z_0| \tag{5}$$

whenever $|z - z_0| < r$, and there exists an n such that $|z - z_0| < r$ for all $z \in \Delta_n$. For this n we also have $|z - z_0| \leq 2^{-n}L$ for all $z \in \Delta_n$. By the Corollary to Theorem 10.12,

$$\int_{\partial\Delta_n} f(z)\,dz = \int_{\partial\Delta_n} [f(z) - f(z_0) - f'(z_0)(z - z_0)]\,dz, \tag{6}$$

so that (5) implies

$$\left| \int_{\partial\Delta_n} f(z)\,dz \right| \leq \epsilon(2^{-n}L)^2, \tag{7}$$

and now (4) shows that $|J| \leq \epsilon L^2$. Hence $J = 0$ if $p \notin \Delta$.

Assume next that p is a vertex of Δ, say $p = a$. If a, b, and c are collinear, then (1) is trivial, for any continuous f. If not, choose points $x \in [a, b]$ and $y \in [a, c]$, both close to a, and observe that the integral of f over $\partial\Delta$ is the sum of the integrals over the boundaries of the triangles $\{a, x, y\}$, $\{x, b, y\}$, and $\{b, c, y\}$. The last two of these are 0, since these triangles do not contain p. Hence the integral over $\partial\Delta$ is the sum of the integrals over $[a, x]$, $[x, y]$, and $[y, a]$, and since these intervals can be made arbitrarily short and f is bounded on Δ, we again obtain (1).

Finally, if p is an arbitrary point of Δ, apply the preceding result to $\{a, b, p\}$, $\{b, c, p\}$, and $\{c, a, p\}$ to complete the proof. ////

10.14 Cauchy's Theorem in a Convex Set *Suppose Ω is a convex open set, $p \in \Omega$, f is continuous on Ω, and $f \in H(\Omega - \{p\})$. Then $f = F'$ for some $F \in H(\Omega)$. Hence*

$$\int_\gamma f(z)\,dz = 0 \tag{1}$$

for every closed path γ in Ω.

ELEMENTARY PROPERTIES OF HOLOMORPHIC FUNCTIONS **207**

PROOF Fix $a \in \Omega$. Since Ω is convex, Ω contains the straight line interval from a to z for every $z \in \Omega$, so we can define

$$F(z) = \int_{[a,\,z]} f(\xi)\, d\xi \qquad (z \in \Omega). \tag{2}$$

For any z and $z_0 \in \Omega$, the triangle with vertices at a, z_0, and z lies in Ω; hence $F(z) - F(z_0)$ is the integral of f over $[z_0, z]$, by Theorem 10.13. Fixing z_0, we thus obtain

$$\frac{F(z) - F(z_0)}{z - z_0} - f(z_0) = \frac{1}{z - z_0} \int_{[z_0,\,z]} [f(\xi) - f(z_0)]\, d\xi, \tag{3}$$

if $z \neq z_0$. Given $\epsilon > 0$, the continuity of f at z_0 shows that there is a $\delta > 0$ such that $|f(\xi) - f(z_0)| < \epsilon$ if $|\xi - z_0| < \delta$; hence the absolute value of the left side of (3) is less than ϵ as soon as $|z - z_0| < \delta$. This proves that $f = F'$. In particular, $F \in H(\Omega)$. Now (1) follows from Theorem 10.12. ////

10.15 Cauchy's Formula in a Convex Set *Suppose γ is a closed path in a convex open set Ω, and $f \in H(\Omega)$. If $z \in \Omega$ and $z \notin \gamma^*$, then*

$$f(z) \cdot \mathrm{Ind}_\gamma(z) = \frac{1}{2\pi i} \int_\gamma \frac{f(\xi)}{\xi - z}\, d\xi. \tag{1}$$

The case of greatest interest is, of course, $\mathrm{Ind}_\gamma(z) = 1$.

PROOF Fix z so that the above conditions hold, and define

$$g(\xi) = \begin{cases} \dfrac{f(\xi) - f(z)}{\xi - z} & \text{if } \xi \in \Omega,\ \xi \neq z, \\ f'(z) & \text{if } \xi = z. \end{cases} \tag{2}$$

Then g satisfies the hypotheses of Theorem 10.14. Hence

$$\frac{1}{2\pi i} \int_\gamma g(\xi)\, d\xi = 0. \tag{3}$$

If we substitute (2) into (3) we obtain (1). ////

The theorem concerning the representability of holomorphic functions by power series is an easy consequence of Theorem 10.15, if we take a circle for γ:

10.16 Theorem *For every open set Ω in the plane, every $f \in H(\Omega)$ is representable by power series in Ω.*

PROOF Suppose $f \in H(\Omega)$ and $D(a; R) \subset \Omega$. If γ is a positively oriented circle with center at a and radius $r < R$, the convexity of $D(a; R)$ allows us to apply Theorem 10.15; by Theorem 10.11, we obtain

$$f(z) = \frac{1}{2\pi i} \int_\gamma \frac{f(\xi)}{\xi - z} d\xi \qquad (z \in D(a; r)). \tag{1}$$

But now we can apply Theorem 10.7, with $X = [0, 2\pi]$, $\varphi = \gamma$, and $d\mu(t) = f(\gamma(t))\gamma'(t) dt$, and we conclude that there is a sequence $\{c_n\}$ such that

$$f(z) = \sum_{n=0}^\infty c_n(z - a)^n \qquad (z \in D(a; r)). \tag{2}$$

The uniqueness of $\{c_n\}$ (see the Corollary to Theorem 10.6) shows that the same power series is obtained for every $r < R$ (as long as a is fixed). Hence the representation (2) is valid for every $z \in D(a; R)$, and the proof is complete.
////

Corollary *If $f \in H(\Omega)$, then $f' \in H(\Omega)$.*

PROOF Combine Theorems 10.6 and 10.16. ////

The Cauchy theorem has a useful converse:

10.17 Morera's Theorem *Suppose f is a continuous complex function in an open set Ω such that*

$$\int_{\partial \Delta} f(z) \, dz = 0$$

for every closed triangle $\Delta \subset \Omega$. Then $f \in H(\Omega)$.

PROOF Let V be a convex open set in Ω. As in the proof of Theorem 10.14, we can construct $F \in H(V)$ such that $F' = f$. Since derivatives of holomorphic functions are holomorphic (Theorem 10.16), we have $f \in H(V)$, for every convex open $V \subset \Omega$, hence $f \in H(\Omega)$. ////

The Power Series Representation

The fact that every holomorphic function is locally the sum of a convergent power series has a large number of interesting consequences. A few of these are developed in this section.

10.18 Theorem *Suppose Ω is a region, $f \in H(\Omega)$, and*

$$Z(f) = \{a \in \Omega : f(a) = 0\}. \tag{1}$$

Then either $Z(f) = \Omega$, or $Z(f)$ has no limit point in Ω. In the latter case there corresponds to each $a \in Z(f)$ a unique positive integer $m = m(a)$ such that

$$f(z) = (z - a)^m g(z) \qquad (z \in \Omega), \qquad (2)$$

where $g \in H(\Omega)$ and $g(a) \neq 0$; furthermore, $Z(f)$ is at most countable.

(We recall that regions are *connected* open sets.)

The integer m is called the *order* of the zero which f has at the point a. Clearly, $Z(f) = \Omega$ if and only if f is identically 0 in Ω. We call $Z(f)$ the *zero set* of f. Analogous results hold of course for the set of α-points of f, i.e., the zero set of $f - \alpha$, where α is any complex number.

PROOF Let A be the set of all limit points of $Z(f)$ in Ω. Since f is continuous, $A \subset Z(f)$.

Fix $a \in Z(f)$, and choose $r > 0$ so that $D(a; r) \subset \Omega$. By Theorem 10.16,

$$f(z) = \sum_{n=0}^{\infty} c_n (z - a)^n \qquad (z \in D(a; r)). \qquad (3)$$

There are now two possibilities. Either all c_n are 0, in which case $D(a; r) \subset A$ and a is an interior point of A, or there is a smallest integer m [necessarily positive, since $f(a) = 0$] such that $c_m \neq 0$. In that case, define

$$g(z) = \begin{cases} (z - a)^{-m} f(z) & (z \in \Omega - \{a\}), \\ c_m & (z = a). \end{cases} \qquad (4)$$

Then (2) holds. It is clear that $g \in H(\Omega - \{a\})$. But (3) implies

$$g(z) = \sum_{k=0}^{\infty} c_{m+k} (z - a)^k \qquad (z \in D(a; r)). \qquad (5)$$

Hence $g \in H(D(a; r))$, so actually $g \in H(\Omega)$.

Moreover, $g(a) \neq 0$, and the continuity of g shows that there is a neighborhood of a in which g has no zero. Thus a is an isolated point of $Z(f)$, by (2).

If $a \in A$, the first case must therefore occur. So A is open. If $B = \Omega - A$, it is clear from the definition of A as a set of limit points that B is open. Thus Ω is the union of the disjoint open sets A and B. Since Ω is connected, we have either $A = \Omega$, in which case $Z(f) = \Omega$, or $A = \emptyset$. In the latter case, $Z(f)$ has at most finitely many points in each compact subset of Ω, and since Ω is σ-compact, $Z(f)$ is at most countable. ////

Corollary *If f and g are holomorphic functions in a region Ω and if $f(z) = g(z)$ for all z in some set which has a limit point in Ω, then $f(z) = g(z)$ for all $z \in \Omega$.*

In other words, a holomorphic function in a region Ω is determined by its values on any set which has a limit point in Ω. This is an important uniqueness theorem.

Note: The theorem fails if we drop the assumption that Ω is connected: If $\Omega = \Omega_0 \cup \Omega_1$, and Ω_0 and Ω_1 are disjoint open sets, put $f = 0$ in Ω_0 and $f = 1$ in Ω_1.

10.19 Definition If $a \in \Omega$ and $f \in H(\Omega - \{a\})$, then f is said to have an *isolated singularity* at the point a. If f can be so defined at a that the extended function is holomorphic in Ω, the singularity is said to be *removable*.

10.20 Theorem Suppose $f \in H(\Omega - \{a\})$ and f is bounded in $D'(a; r)$, for some $r > 0$. Then f has a removable singularity at a.

Recall that $D'(a; r) = \{z: 0 < |z - a| < r\}$.

PROOF Define $h(a) = 0$, and $h(z) = (z - a)^2 f(z)$ in $\Omega - \{a\}$. Our boundedness assumption shows that $h'(a) = 0$. Since h is evidently differentiable at every other point of Ω, we have $h \in H(\Omega)$, so

$$h(z) = \sum_{n=2}^{\infty} c_n (z - a)^n \qquad (z \in D(a; r)).$$

We obtain the desired holomorphic extension of f by setting $f(a) = c_2$, for then

$$f(z) = \sum_{n=0}^{\infty} c_{n+2} (z - a)^n \qquad (z \in D(a; r)). \qquad ////$$

10.21 Theorem If $a \in \Omega$ and $f \in H(\Omega - \{a\})$, then one of the following three cases must occur:

(a) f has a removable singularity at a.
(b) There are complex numbers $c_1, \ldots, c_m$, where m is a positive integer and $c_m \neq 0$, such that

$$f(z) - \sum_{k=1}^{m} \frac{c_k}{(z - a)^k}$$

has a removable singularity at a.
(c) If $r > 0$ and $D(a; r) \subset \Omega$, then $f(D'(a; r))$ is dense in the plane.

In case (b), f is said to have a *pole of order m* at a. The function

$$\sum_{k=1}^{m} c_k (z - a)^{-k},$$

a polynomial in $(z - a)^{-1}$, is called the *principal part* of f at a. It is clear in this situation that $|f(z)| \to \infty$ as $z \to a$.

In case (c), f is said to have an *essential singularity* at a. A statement equivalent to (c) is that to each complex number w there corresponds a sequence $\{z_n\}$ such that $z_n \to a$ and $f(z_n) \to w$ as $n \to \infty$.

PROOF Suppose (c) fails. Then there exist $r > 0$, $\delta > 0$, and a complex number w such that $|f(z) - w| > \delta$ in $D'(a; r)$. Let us write D for $D(a; r)$ and D' for $D'(a; r)$. Define

$$g(z) = \frac{1}{f(z) - w} \qquad (z \in D'). \tag{1}$$

Then $g \in H(D')$ and $|g| < 1/\delta$. By Theorem 10.20, g extends to a holomorphic function in D.

If $g(a) \neq 0$, (1) shows that f is bounded in $D'(a; \rho)$ for some $\rho > 0$. Hence (a) holds, by Theorem 10.20.

If g has a zero of order $m \geq 1$ at a, Theorem 10.18 shows that

$$g(z) = (z - a)^m g_1(z) \qquad (z \in D), \tag{2}$$

where $g_1 \in H(D)$ and $g_1(a) \neq 0$. Also, g_1 has no zero in D', by (1). Put $h = 1/g_1$ in D. Then $h \in H(D)$, h has no zero in D, and

$$f(z) - w = (z - a)^{-m} h(z) \qquad (z \in D'). \tag{3}$$

But h has an expansion of the form

$$h(z) = \sum_{n=0}^{\infty} b_n (z - a)^n \qquad (z \in D), \tag{4}$$

with $b_0 \neq 0$. Now (3) shows that (b) holds, with $c_k = b_{m-k}$, $k = 1, \ldots, m$.
This completes the proof. ////

We shall now exploit the fact that the restriction of a power series $\sum c_n(z - a)^n$ to a circle with center at a is a trigonometric series.

10.22 Theorem *If*

$$f(z) = \sum_{n=0}^{\infty} c_n (z - a)^n \qquad (z \in D(a; R)) \tag{1}$$

and if $0 < r < R$, *then*

$$\sum_{n=0}^{\infty} |c_n|^2 r^{2n} = \frac{1}{2\pi} \int_{-\pi}^{\pi} |f(a + re^{i\theta})|^2 \, d\theta. \tag{2}$$

PROOF We have

$$f(a + re^{i\theta}) = \sum_{n=0}^{\infty} c_n r^n e^{in\theta}. \qquad (3)$$

For $r < R$, the series (3) converges uniformly on $[-\pi, \pi]$. Hence

$$c_n r^n = \frac{1}{2\pi} \int_{-\pi}^{\pi} f(a + re^{i\theta}) e^{-in\theta} \, d\theta \qquad (n = 0, 1, 2, \ldots), \qquad (4)$$

and (2) is seen to be a special case of Parseval's formula. ////

Here are some consequences:

10.23 Liouville's Theorem *Every bounded entire function is constant.*

Recall that a function is *entire* if it is holomorphic in the whole plane.

PROOF Suppose f is entire, $|f(z)| < M$ for all z, and $f(z) = \sum c_n z^n$ for all z. By Theorem 10.22,

$$\sum_{n=0}^{\infty} |c_n|^2 r^{2n} < M^2$$

for all r, which is possible only if $c_n = 0$ for all $n \geq 1$. ////

10.24 The Maximum Modulus Theorem *Suppose Ω is a region, $f \in H(\Omega)$, and $\bar{D}(a; r) \subset \Omega$. Then*

$$|f(a)| \leq \max_{\theta} |f(a + re^{i\theta})|. \qquad (1)$$

Equality occurs in (1) if and only if f is constant in Ω.

Consequently, $|f|$ has no local maximum at any point of Ω, unless f is constant.

PROOF Assume that $|f(a + re^{i\theta})| \leq |f(a)|$ for all real θ. In the notation of Theorem 10.22 it follows then that

$$\sum_{n=0}^{\infty} |c_n|^2 r^{2n} \leq |f(a)|^2 = |c_0|^2.$$

Hence $c_1 = c_2 = c_3 = \cdots = 0$, which implies that $f(z) = f(a)$ in $D(a; r)$. Since Ω is connected, Theorem 10.18 shows that f is constant in Ω. ////

Corollary *Under the same hypotheses,*

$$|f(a)| \geq \min_{\theta} |f(a + re^{i\theta})| \qquad (2)$$

if f has no zero in $D(a; r)$.

PROOF If $f(a + re^{i\theta}) = 0$ for some θ then (2) is obvious. In the other case, there is a region Ω_0 that contains $\bar{D}(a; r)$ and in which f has no zero; hence (1) can be applied to $1/f$, and (2) follows. ////

10.25 Theorem *If n is a positive integer and*

$$P(z) = z^n + a_{n-1}z^{n-1} + \cdots + a_1 z + a_0,$$

where $a_0, \ldots, a_{n-1}$ are complex numbers, then P has precisely n zeros in the plane.

Of course, these zeros are counted according to their multiplicities: A zero of order m, say, is counted as m zeros. This theorem contains the fact that the complex field is algebraically closed, i.e., that every nonconstant polynomial with complex coefficients has at least one complex zero.

PROOF Choose $r > 1 + 2|a_0| + |a_1| + \cdots + |a_{n-1}|$. Then

$$|P(re^{i\theta})| > |P(0)| \qquad (0 \le \theta \le 2\pi).$$

If P had no zeros, then the function $f = 1/P$ would be entire and would satisfy $|f(0)| > |f(re^{i\theta})|$ for all θ, which contradicts the maximum modulus theorem. Thus $P(z_1) = 0$ for some z_1. Consequently, there is a polynomial Q, of degree $n-1$, such that $P(z) = (z - z_1)Q(z)$. The proof is completed by induction on n. ////

10.26 Theorem (Cauchy's Estimates) *If $f \in H(D(a; R))$ and $|f(z)| \le M$ for all $z \in D(a; R)$, then*

$$|f^{(n)}(a)| \le \frac{n!\, M}{R^n} \qquad (n = 1, 2, 3, \ldots). \tag{1}$$

PROOF For each $r < R$, each term of the series 10.22(2) is bounded above by M^2. ////

If we take $a = 0$, $R = 1$, and $f(z) = z^n$, then $M = 1$, $f^{(n)}(0) = n!$, and we see that (1) cannot be improved.

10.27 Definition A sequence $\{f_j\}$ of functions in Ω is said to *converge to f uniformly on compact subsets of* Ω if to every compact $K \subset \Omega$ and to every $\epsilon > 0$ there corresponds an $N = N(K, \epsilon)$ such that $|f_j(z) - f(z)| < \epsilon$ for all $z \in K$ if $j > N$.

For instance, the sequence $\{z^n\}$ converges to 0 uniformly on compact subsets of $D(0; 1)$, but the convergence is *not* uniform in $D(0; 1)$.

It is uniform convergence on compact subsets which arises most naturally in connection with limit operations on holomorphic functions. The term "almost uniform convergence" is sometimes used for this concept.

10.28 Theorem *Suppose $f_j \in H(\Omega)$, for $j = 1, 2, 3, \ldots$, and $f_j \to f$ uniformly on compact subsets of Ω. Then $f \in H(\Omega)$, and $f'_j \to f'$ uniformly on compact subsets of Ω.*

PROOF Since the convergence is uniform on each compact disc in Ω, f is continuous. Let Δ be a triangle in Ω. Then Δ is compact, so

$$\int_{\partial \Delta} f(z)\, dz = \lim_{j \to \infty} \int_{\partial \Delta} f_j(z)\, dz = 0,$$

by Cauchy's theorem. Hence Morera's theorem implies that $f \in H(\Omega)$.

Let K be compact, $K \subset \Omega$. There exists an $r > 0$ such that the union E of the closed discs $\bar{D}(z; r)$, for all $z \in K$, is a compact subset of Ω. Applying Theorem 10.26 to $f - f_j$, we have

$$|f'(z) - f'_j(z)| \leq r^{-1} \|f - f_j\|_E \qquad (z \in K),$$

where $\|f\|_E$ denotes the supremum of $|f|$ on E. Since $f_j \to f$ uniformly on E, it follows that $f'_j \to f'$ uniformly on K. ////

Corollary *Under the same hypothesis, $f_j^{(n)} \to f^{(n)}$ uniformly, as $j \to \infty$, on every compact set $K \subset \Omega$, and for every positive integer n.*

Compare this with the situation on the real line, where sequences of infinitely differentiable functions can converge uniformly to nowhere differentiable functions!

The Open Mapping Theorem

If Ω is a region and $f \in H(\Omega)$, then $f(\Omega)$ is either a region or a point.

This important property of holomorphic functions will be proved, in more detailed form, in Theorem 10.32.

10.29 Lemma *If $f \in H(\Omega)$ and g is defined in $\Omega \times \Omega$ by*

$$g(z, w) = \begin{cases} \dfrac{f(z) - f(w)}{z - w} & \text{if } w \neq z, \\ f'(z) & \text{if } w = z, \end{cases}$$

then g is continuous in $\Omega \times \Omega$.

PROOF The only points $(z, w) \in \Omega \times \Omega$ at which the continuity of g is possibly in doubt have $z = w$.

Fix $a \in \Omega$. Fix $\epsilon > 0$. There exists $r > 0$ such that $D(a; r) \subset \Omega$ and $|f'(\zeta) - f'(a)| < \epsilon$ for all $\zeta \in D(a; r)$. If z and w are in $D(a; r)$ and if

$$\zeta(t) = (1 - t)z + tw,$$

then $\zeta(t) \in D(a; r)$ for $0 \le t \le 1$, and

$$g(z, w) - g(a, a) = \int_0^1 [f'(\zeta(t)) - f'(a)] \, dt.$$

The absolute value of the integrand is $< \epsilon$, for every t. Thus $|g(z, w) - g(a, a)| < \epsilon$. This proves that g is continuous at (a, a).   ////

10.30 Theorem Suppose $\varphi \in H(\Omega)$, $z_0 \in \Omega$, and $\varphi'(z_0) \ne 0$. Then Ω contains a neighborhood V of z_0 such that

(a) φ is one-to-one in V,
(b) $W = \varphi(V)$ is an open set, and
(c) if $\psi: W \to V$ is defined by $\psi(\varphi(z)) = z$, then $\psi \in H(W)$.

Thus $\varphi: V \to W$ has a holomorphic inverse.

PROOF Lemma 10.29, applied to φ in place of f, shows that Ω contains a neighborhood V of z_0 such that

$$|\varphi(z_1) - \varphi(z_2)| \ge \tfrac{1}{2} |\varphi'(z_0)| \, |z_1 - z_2| \tag{1}$$

if $z_1 \in V$ and $z_2 \in V$. Thus (a) holds, and also

$$\varphi'(z) \ne 0 \quad (z \in V). \tag{2}$$

To prove (b), pick $a \in V$ and choose $r > 0$ so that $\bar{D}(a, r) \subset V$. By (1) there exists $c > 0$ such that

$$|\varphi(a + re^{i\theta}) - \varphi(a)| > 2c \quad (-\pi \le \theta \le \pi). \tag{3}$$

If $\lambda \in D(\varphi(a); c)$, then $|\lambda - \varphi(a)| < c$, hence (3) implies

$$\min_\theta |\lambda - \varphi(a + re^{i\theta})| > c. \tag{4}$$

By the corollary to Theorem 10.24, $\lambda - \varphi$ must therefore have a zero in $D(a; r)$. Thus $\lambda = \varphi(z)$ for some $z \in D(a; r) \subset V$.

This proves that $D(\varphi(a); c) \subset \varphi(V)$. Since a was an arbitrary point of V, $\varphi(V)$ is open.

To prove (c), fix $w_1 \in W$. Then $\varphi(z_1) = w_1$ for a unique $z_1 \in V$. If $w \in W$ and $\psi(w) = z \in V$, we have

$$\frac{\psi(w) - \psi(w_1)}{w - w_1} = \frac{z - z_1}{\varphi(z) - \varphi(z_1)}. \tag{5}$$

By (1), $z \to z_1$ when $w \to w_1$. Hence (2) implies that $\psi'(w_1) = 1/\varphi'(z_1)$. Thus $\psi \in H(W)$. ////

10.31 Definition For $m = 1, 2, 3, \ldots$, we denote the "m^{th} power function" $z \to z^m$ by π_m.

Each $w \neq 0$ is $\pi_m(z)$ for precisely m distinct values of z: If $w = re^{i\theta}$, $r > 0$, then $\pi_m(z) = w$ if and only if $z = r^{1/m} e^{i(\theta + 2k\pi)/m}$, $k = 1, \ldots, m$.

Note also that each π_m is an open mapping: If V is open and does not contain 0, then $\pi_m(V)$ is open by Theorem 10.30. On the other hand, $\pi_m(D(0; r)) = D(0; r^m)$.

Compositions of open mappings are clearly open. In particular, $\pi_m \circ \varphi$ is open, by Theorem 10.30, if φ' has no zero. The following theorem (which contains the more detailed version of the open mapping theorem that was mentioned prior to Lemma 10.29) states a converse: Every nonconstant holomorphic function in a region is locally of the form $\pi_m \circ \varphi$, except for an additive constant.

10.32 Theorem *Suppose Ω is a region, $f \in H(\Omega)$, f is not constant, $z_0 \in \Omega$, and $w_0 = f(z_0)$. Let m be the order of the zero which the function $f - w_0$ has at z_0.*

Then there exists a neighborhood V of z_0, $V \subset \Omega$, and there exists $\varphi \in H(V)$, such that

(a) $f(z) = w_0 + [\varphi(z)]^m$ for all $z \in V$,
(b) φ' has no zero in V and φ is an invertible mapping of V onto a disc $D(0; r)$.

Thus $f - w_0 = \pi_m \circ \varphi$ in V. It follows that f is an exactly m-to-1 mapping of $V - \{z_0\}$ onto $D'(w_0; r^m)$, and that each $w_0 \in f(\Omega)$ is an interior point of $f(\Omega)$. Hence $f(\Omega)$ is open.

PROOF Without loss of generality we may assume that Ω is a convex neighborhood of z_0 which is so small that $f(z) \neq w_0$ if $z \in \Omega - \{z_0\}$. Then

$$f(z) - w_0 = (z - z_0)^m g(z) \qquad (z \in \Omega) \qquad (1)$$

for some $g \in H(\Omega)$ which has no zero in Ω. Hence $g'/g \in H(\Omega)$. By Theorem 10.14, $g'/g = h'$ for some $h \in H(\Omega)$. The derivative of $g \cdot \exp(-h)$ is 0 in Ω. If h is modified by the addition of a suitable constant, it follows that $g = \exp(h)$. Define

$$\varphi(z) = (z - z_0) \exp \frac{h(z)}{m} \qquad (z \in \Omega). \qquad (2)$$

Then (a) holds, for all $z \in \Omega$.

Also, $\varphi(z_0) = 0$ and $\varphi'(z_0) \neq 0$. The existence of an open set V that satisfies (b) follows now from Theorem 10.30. This completes the proof. ////

The next theorem is really contained in the preceding results, but it seems advisable to state it explicitly.

10.33 Theorem *Suppose Ω is a region, $f \in H(\Omega)$, and f is one-to-one in Ω. Then $f'(z) \neq 0$ for every $z \in \Omega$, and the inverse of f is holomorphic.*

PROOF If $f'(z_0)$ were 0 for some $z_0 \in \Omega$, the hypotheses of Theorem 10.32 would hold with some $m > 1$, so that f would be m-to-1 in some deleted neighborhood of z_0. Now apply part (c) of Theorem 10.30. ////

Note that the converse of Theorem 10.33 is false: If $f(z) = e^z$, then $f'(z) \neq 0$ for every z, but f is not one-to-one in the whole complex plane.

The Global Cauchy Theorem

Before we state and prove this theorem, which will remove the restriction to convex regions that was imposed in Theorem 10.14, it will be convenient to add a little to the integration apparatus which was sufficient up to now. Essentially, it is a matter of no longer restricting ourselves to integrals over single paths, but to consider finite "sums" of paths instead. A simple instance of this occurred already in Sec. 10.9(c).

10.34 Chains and Cycles Suppose $\gamma_1, \ldots, \gamma_n$ are paths in the plane, and put $K = \gamma_1^* \cup \cdots \cup \gamma_n^*$. Each γ_i induces a linear functional $\tilde{\gamma}_i$ on the vector space $C(K)$, by the formula

$$\tilde{\gamma}_i(f) = \int_{\gamma_i} f(z)\, dz. \tag{1}$$

Define

$$\tilde{\Gamma} = \tilde{\gamma}_1 + \cdots + \tilde{\gamma}_n. \tag{2}$$

Explicitly, $\tilde{\Gamma}(f) = \tilde{\gamma}_1(f) + \cdots + \tilde{\gamma}_n(f)$ for all $f \in C(K)$. The relation (2) suggests that we introduce a "formal sum"

$$\Gamma = \gamma_1 \dotplus \cdots \dotplus \gamma_n \tag{3}$$

and define

$$\int_\Gamma f(z)\, dz = \tilde{\Gamma}(f). \tag{4}$$

Then (3) is merely an abbreviation for the statement

$$\int_\Gamma f(z)\, dz = \sum_{i=1}^n \int_{\gamma_i} f(z)\, dz \qquad (f \in C(K)). \tag{5}$$

Note that (5) serves as the definition of its left side.

The objects Γ so defined are called *chains*. If each γ_j in (3) is a *closed* path, then Γ is called a *cycle*. If each γ_j in (3) is a path in some open set Ω, we say that Γ is a *chain in* Ω.

If (3) holds, we define

$$\Gamma^* = \gamma_1^* \cup \cdots \cup \gamma_n^*. \tag{6}$$

If Γ is a cycle and $\alpha \notin \Gamma^*$, we define the *index* of α with respect to Γ by

$$\mathrm{Ind}_\Gamma(\alpha) = \frac{1}{2\pi i} \int_\Gamma \frac{dz}{z - \alpha}, \tag{7}$$

just as in Theorem 10.10. Obviously, (3) implies

$$\mathrm{Ind}_\Gamma(\alpha) = \sum_{i=1}^{n} \mathrm{Ind}_{\gamma_i}(\alpha). \tag{8}$$

If each γ_i in (3) is replaced by its opposite path (see Sec. 10.8), the resulting chain will be denoted by $-\Gamma$. Then

$$\int_{-\Gamma} f(z)\, dz = -\int_\Gamma f(z)\, dz \qquad (f \in C(\Gamma^*)). \tag{9}$$

In particular, $\mathrm{Ind}_{-\Gamma}(\alpha) = -\mathrm{Ind}_\Gamma(\alpha)$ if Γ is a cycle and $\alpha \notin \Gamma^*$.

Chains can be added and subtracted in the obvious way, by adding or subtracting the corresponding functionals: The statement $\Gamma = \Gamma_1 \dotplus \Gamma_2$ means

$$\int_\Gamma f(z)\, dz = \int_{\Gamma_1} f(z)\, dz + \int_{\Gamma_2} f(z)\, dz \tag{10}$$

for every $f \in C(\Gamma_1^* \cup \Gamma_2^*)$.

Finally, note that a chain may be represented as a sum of paths in many ways. To say that

$$\gamma_1 \dotplus \cdots \dotplus \gamma_n = \delta_1 \dotplus \cdots \dotplus \delta_k$$

means simply that

$$\sum_i \int_{\gamma_i} f(z)\, dz = \sum_j \int_{\delta_j} f(z)\, dz$$

for every f that is continuous on $\gamma_1^* \cup \cdots \cup \gamma_n^* \cup \delta_1^* \cup \cdots \cup \delta_k^*$. In particular, a cycle may very well be represented as a sum of paths that are not closed.

10.35 Cauchy's Theorem *Suppose* $f \in H(\Omega)$, *where* Ω *is an arbitrary open set in the complex plane. If* Γ *is a cycle in* Ω *that satisfies*

$$\mathrm{Ind}_\Gamma(\alpha) = 0 \qquad \text{for every } \alpha \text{ not in } \Omega, \tag{1}$$

then

$$f(z) \cdot \text{Ind}_\Gamma (z) = \frac{1}{2\pi i} \int_\Gamma \frac{f(w)}{w - z} \, dw \qquad \text{for } z \in \Omega - \Gamma^* \qquad (2)$$

and

$$\int_\Gamma f(z) \, dz = 0. \qquad (3)$$

If Γ_0 and Γ_1 are cycles in Ω such that

$$\text{Ind}_{\Gamma_0} (\alpha) = \text{Ind}_{\Gamma_1} (\alpha) \qquad \text{for every } \alpha \text{ not in } \Omega, \qquad (4)$$

then

$$\int_{\Gamma_0} f(z) \, dz = \int_{\Gamma_1} f(z) \, dz. \qquad (5)$$

PROOF The function g defined in $\Omega \times \Omega$ by

$$g(z, w) = \begin{cases} \dfrac{f(w) - f(z)}{w - z} & \text{if } w \neq z, \\ f'(z) & \text{if } w = z, \end{cases} \qquad (6)$$

is continuous in $\Omega \times \Omega$ (Lemma 10.29). Hence we can define

$$h(z) = \frac{1}{2\pi i} \int_\Gamma g(z, w) \, dw \qquad (z \in \Omega). \qquad (7)$$

For $z \in \Omega - \Gamma^*$, the Cauchy formula (2) is clearly equivalent to the assertion that

$$h(z) = 0. \qquad (8)$$

To prove (8), let us first prove that $h \in H(\Omega)$. Note that g is uniformly continuous on every compact subset of $\Omega \times \Omega$. If $z \in \Omega$, $z_n \in \Omega$, and $z_n \to z$, it follows that $g(z_n, w) \to g(z, w)$ uniformly for $w \in \Gamma^*$ (a compact subset of Ω). Hence $h(z_n) \to h(z)$. This proves that h is continuous in Ω. Let Δ be a closed triangle in Ω. Then

$$\int_{\partial \Delta} h(z) \, dz = \frac{1}{2\pi i} \int_\Gamma \left(\int_{\partial \Delta} g(z, w) \, dz \right) dw. \qquad (9)$$

For each $w \in \Omega$, $z \to g(z, w)$ is holomorphic in Ω. (The singularity at $z = w$ is removable.) The inner integral on the right side of (9) is therefore 0 for every $w \in \Gamma^*$. Morera's theorem shows now that $h \in H(\Omega)$.

Next, we let Ω_1 be the set of all complex numbers z for which $\text{Ind}_\Gamma(z) = 0$, and we define

$$h_1(z) = \frac{1}{2\pi i} \int_\Gamma \frac{f(w)}{w - z} \, dw \qquad (z \in \Omega_1). \tag{10}$$

If $z \in \Omega \cap \Omega_1$, the definition of Ω_1 makes it clear that $h_1(z) = h(z)$. Hence there is a function $\varphi \in H(\Omega \cup \Omega_1)$ whose restriction to Ω is h and whose restriction to Ω_1 is h_1.

Our hypothesis (1) shows that Ω_1 contains the complement of Ω. Thus φ is an entire function. Ω_1 also contains the unbounded component of the complement of Γ^*, since $\text{Ind}_\Gamma(z)$ is 0 there. Hence

$$\lim_{|z| \to \infty} \varphi(z) = \lim_{|z| \to \infty} h_1(z) = 0. \tag{11}$$

Liouville's theorem implies now that $\varphi(z) = 0$ for every z. This proves (8), and hence (2).

To deduce (3) from (2), pick $a \in \Omega - \Gamma^*$ and define $F(z) = (z - a)f(z)$. Then

$$\frac{1}{2\pi i} \int_\Gamma f(z) \, dz = \frac{1}{2\pi i} \int_\Gamma \frac{F(z)}{z - a} \, dz = F(a) \cdot \text{Ind}_\Gamma(a) = 0, \tag{12}$$

because $F(a) = 0$.

Finally, (5) follows from (4) if (3) is applied to the cycle $\Gamma = \Gamma_1 - \Gamma_0$. This completes the proof. ////

10.36 Remarks

(a) If γ is a closed path in a convex region Ω and if $\alpha \notin \Omega$, an application of Theorem 10.14 to $f(z) = (z - \alpha)^{-1}$ shows that $\text{Ind}_\gamma(\alpha) = 0$. Hypothesis (1) of Theorem 10.35 is therefore satisfied by every cycle in Ω if Ω is convex. This shows that Theorem 10.35 generalizes Theorems 10.14 and 10.15.

(b) The last part of Theorem 10.35 shows under what circumstances integration over one cycle can be replaced by integration over another, without changing the value of the integral. For example, let Ω be the plane with three disjoint closed discs D_i removed. If Γ, γ_1, γ_2, γ_3 are positively oriented circles in Ω such that Γ surrounds $D_1 \cup D_2 \cup D_3$ and γ_i surrounds D_i but not D_j for $j \neq i$, then

$$\int_\Gamma f(z) \, dz = \sum_{i=1}^{3} \int_{\gamma_i} f(z) \, dz$$

for every $f \in H(\Omega)$.

(c) In order to apply Theorem 10.35, it is desirable to have a reasonably efficient method of finding the index of a point with respect to a closed path. The following theorem does this for all paths that occur in practice.

It says, essentially, that the index *increases* by 1 when the path is crossed "from right to left." If we recall that $\text{Ind}_\gamma(\alpha) = 0$ if α is in the unbounded component of the complement W of γ^*, we can then successively determine $\text{Ind}_\gamma(\alpha)$ in the other components of W, provided that W has only finitely many components and that γ traverses no arc more than once.

10.37 Theorem *Suppose γ is a closed path in the plane, with parameter interval $[\alpha, \beta]$. Suppose $\alpha < u < v < \beta$, a and b are complex numbers, $|b| = r > 0$, and*

(i) $\gamma(u) = a - b$, $\gamma(v) = a + b$,
(ii) $|\gamma(s) - a| < r$ *if and only if* $u < s < v$,
(iii) $|\gamma(s) - a| = r$ *if and only if* $s = u$ *or* $s = v$.

Assume furthermore that $D(a; r) - \gamma^*$ is the union of two regions, D_+ and D_-, labeled so that $a + bi \in \bar{D}_+$ and $a - bi \in \bar{D}_-$. Then

$$\text{Ind}_\gamma(z) = 1 + \text{Ind}_\gamma(w)$$

if $x \in D_+$ and $w \in D_-$.

As $\gamma(t)$ traverses $D(a; r)$ from $a - b$ to $a + b$, D_- is "on the right" and D_+ is "on the left" of the path.

PROOF To simplify the writing, reparametrize γ so that $u = 0$ and $v = \pi$. Define

$$C(s) = a - be^{is} \quad (0 \le s \le 2\pi)$$

$$f(s) = \begin{cases} C(s) & (0 \le s \le \pi) \\ \gamma(2\pi - s) & (\pi \le s \le 2\pi) \end{cases}$$

$$g(s) = \begin{cases} \gamma(s) & (0 \le s \le \pi) \\ C(s) & (\pi \le s \le 2\pi) \end{cases}$$

$$h(s) = \begin{cases} \gamma(s) & (\alpha \le s \le 0 \text{ or } \pi \le s \le \beta) \\ C(s) & (0 \le s \le \pi). \end{cases}$$

Since $\gamma(0) = C(0)$ and $\gamma(\pi) = C(\pi)$, f, g, and h are closed paths.

If $E \subset \bar{D}(a; r)$, $|\zeta - a| = r$, and $\zeta \notin E$, then E lies in the disc $D(2a - \zeta; 2r)$ which does not contain ζ. Apply this to $E = g^*$, $\zeta = a - bi$, to see [from Remark 10.36(a)] that $\text{Ind}_g(a - bi) = 0$. Since $\bar{D}_-$ is connected and D_- does not intersect g^*, it follows that

$$\text{Ind}_g(w) = 0 \quad \text{if } w \in D_-. \tag{1}$$

The same reasoning shows that

$$\text{Ind}_f(z) = 0 \quad \text{if } z \in D_+. \tag{2}$$

We conclude that

$$\operatorname{Ind}_\gamma(z) = \operatorname{Ind}_h(z) = \operatorname{Ind}_h(w)$$
$$= \operatorname{Ind}_C(w) + \operatorname{Ind}_\gamma(w) = 1 + \operatorname{Ind}_\gamma(w).$$

The first of these equalities follows from (2), since $h = \gamma \dotplus f$. The second holds because z and w lie in $D(a; r)$, a connected set which does not intersect h^*. The third follows from (1), since $h \dotplus g = C \dotplus \gamma$, and the fourth is a consequence of Theorem 10.11. This completes the proof.   ////

We now turn to a brief discussion of another topological concept that is relevant to Cauchy's theorem.

10.38 Homotopy Suppose γ_0 and γ_1 are closed curves in a topological space X, both with parameter interval $I = [0, 1]$. We say that γ_0 and γ_1 are *X-homotopic* if there is a continuous mapping H of the unit square $I^2 = I \times I$ into X such that

$$H(s, 0) = \gamma_0(s), \qquad H(s, 1) = \gamma_1(s), \qquad H(0, t) = H(1, t) \qquad (1)$$

for all $s \in I$ and $t \in I$. Put $\gamma_t(s) = H(s, t)$. Then (1) defines a *one-parameter family of closed curves* γ_t *in* X, which *connects* γ_0 *and* γ_1. Intuitively, this means that γ_0 can be continuously deformed to γ_1, within X.

If γ_0 is X-homotopic to a constant mapping γ_1 (i.e., if γ_1^* consists of just one point), we say that γ_0 is *null-homotopic in* X. If X is connected and if every closed curve in X is null-homotopic, X is said to be *simply connected*.

For example, every convex region Ω is simply connected. To see this, let γ_0 be a closed curve in Ω, fix $z_1 \in \Omega$, and define

$$H(s, t) = (1 - t)\gamma_0(s) + tz_1 \qquad (0 \leq s \leq 1, \ 0 \leq t \leq 1). \qquad (2)$$

Theorem 10.40 will show that condition (4) of Cauchy's theorem 10.35 holds whenever Γ_0 and Γ_1 are Ω-homotopic closed paths. As a special case of this, note that *condition* (1) *of Theorem* 10.35 *holds for every closed path* Γ *in* Ω *if* Ω *is simply connected*.

10.39 Lemma *If* γ_0 *and* γ_1 *are closed paths with parameter interval* $[0, 1]$, *if* α *is a complex number, and if*

$$|\gamma_1(s) - \gamma_0(s)| < |\alpha - \gamma_0(s)| \qquad (0 \leq s \leq 1) \qquad (1)$$

then $\operatorname{Ind}_{\gamma_1}(\alpha) = \operatorname{Ind}_{\gamma_0}(\alpha)$.

PROOF Note first that (1) implies that $\alpha \notin \gamma_0^*$ and $\alpha \notin \gamma_1^*$. Hence one can define $\gamma = (\gamma_1 - \alpha)/(\gamma_0 - \alpha)$. Then

$$\frac{\gamma'}{\gamma} = \frac{\gamma_1'}{\gamma_1 - \alpha} - \frac{\gamma_0'}{\gamma_0 - \alpha} \qquad (2)$$

and $|1 - \gamma| < 1$, by (1). Hence $\gamma^* \subset D(1; 1)$, which implies that $\text{Ind}_\gamma (0) = 0$. Integration of (2) over $[0, 1]$ now gives the desired result. ////

10.40 Theorem *If Γ_0 and Γ_1 are Ω-homotopic closed paths in a region Ω, and if $\alpha \notin \Omega$, then*

$$\text{Ind}_{\Gamma_1} (\alpha) = \text{Ind}_{\Gamma_0} (\alpha). \tag{1}$$

PROOF By definition, there is a continuous $H: I^2 \to \Omega$ such that

$$H(s, 0) = \Gamma_0(s), \quad H(s, 1) = \Gamma_1(s), \quad H(0, t) = H(1, t). \tag{2}$$

Since I^2 is compact, so is $H(I^2)$. Hence there exists $\epsilon > 0$ such that

$$|\alpha - H(s, t)| > 2\epsilon \quad \text{if} \quad (s, t) \in I^2. \tag{3}$$

Since H is uniformly continuous, there is a positive integer n such that

$$|H(s, t) - H(s', t')| < \epsilon \quad \text{if} \quad |s - s'| + |t - t'| \le 1/n. \tag{4}$$

Define polygonal closed paths $\gamma_0, \ldots, \gamma_n$ by

$$\gamma_k(s) = H\left(\frac{i}{n}, \frac{k}{n}\right)(ns + 1 - i) + H\left(\frac{i-1}{n}, \frac{k}{n}\right)(i - ns) \tag{5}$$

if $i - 1 \le ns \le i$ and $i = 1, \ldots, n$. By (4) and (5),

$$|\gamma_k(s) - H(s, k/n)| < \epsilon \quad (k = 0, \ldots, n; 0 \le s \le 1). \tag{6}$$

In particular, taking $k = 0$ and $k = n$,

$$|\gamma_0(s) - \Gamma_0(s)| < \epsilon, \quad |\gamma_n(s) - \Gamma_1(s)| < \epsilon. \tag{7}$$

By (6) and (3),

$$|\alpha - \gamma_k(s)| > \epsilon \quad (k = 0, \ldots, n; 0 \le s \le 1). \tag{8}$$

On the other hand, (4) and (5) also give

$$|\gamma_{k-1}(s) - \gamma_k(s)| < \epsilon \quad (k = 1, \ldots, n; 0 \le s \le 1). \tag{9}$$

Now it follows from (7), (8), (9), and $n + 2$ applications of Lemma 10.39 that α has the same index with respect to each of the paths $\Gamma_0, \gamma_0, \gamma_1, \ldots, \gamma_n$, Γ_1. This proves the theorem. ////

Note: If $\Gamma_t(s) = H(s, t)$ in the preceding proof, then each Γ_t is a closed *curve*, but not necessarily a *path*, since H is not assumed to be differentiable. The paths γ_k were introduced for this reason. Another (and perhaps more satisfactory) way to circumvent this difficulty is to extend the definition of index to closed curves. This is sketched in Exercise 28.

The Calculus of Residues

10.41 Definition A function f is said to be *meromorphic* in an open set Ω if there is a set $A \subset \Omega$ such that

(a) A has no limit point in Ω,
(b) $f \in H(\Omega - A)$,
(c) f has a pole at each point of A.

Note that the possibility $A = \emptyset$ is not excluded. Thus every $f \in H(\Omega)$ is meromorphic in Ω.

Note also that (a) implies that no compact subset of Ω contains infinitely many points of A, and that A is therefore at most countable.

If f and A are as above, if $a \in A$, and if

$$Q(z) = \sum_{k=1}^{m} c_k (z-a)^{-k} \tag{1}$$

is the principal part of f at a, as defined in Theorem 10.21 (i.e., if $f - Q$ has a removable singularity at a), then the number c_1 is called the *residue* of f at a:

$$c_1 = \text{Res}(f; a). \tag{2}$$

If Γ is a cycle and $a \notin \Gamma^*$, (1) implies

$$\frac{1}{2\pi i} \int_\Gamma Q(z)\, dz = c_1 \, \text{Ind}_\Gamma(a) = \text{Res}(Q; a)\, \text{Ind}_\Gamma(a). \tag{3}$$

This very special case of the following theorem will be used in its proof.

10.42 The Residue Theorem *Suppose f is a meromorphic function in Ω. Let A be the set of points in Ω at which f has poles. If Γ is a cycle in $\Omega - A$ such that*

$$\text{Ind}_\Gamma(\alpha) = 0 \quad \textit{for all} \quad \alpha \notin \Omega, \tag{1}$$

then

$$\frac{1}{2\pi i} \int_\Gamma f(z)\, dz = \sum_{a \in A} \text{Res}(f; a) \, \text{Ind}_\Gamma(a). \tag{2}$$

PROOF Let $B = \{a \in A : \text{Ind}_\Gamma(a) \neq 0\}$. Let W be the complement of Γ^*. Then $\text{Ind}_\Gamma(z)$ is constant in each component V of W. If V is unbounded, or if V intersects Ω^c, (1) implies that $\text{Ind}_\Gamma(z) = 0$ for every $z \in V$. Since A has no limit point in Ω, we conclude that B is a finite set.

The sum in (2), though formally infinite, is therefore actually finite.

Let $a_1, \ldots, a_n$ be the points of B, let $Q_1, \ldots, Q_n$ be the principal parts of f at $a_1, \ldots, a_n$, and put $g = f - (Q_1 + \cdots + Q_n)$. (If $B = \emptyset$, a possibility which is not excluded, then $g = f$.) Put $\Omega_0 = \Omega - (A - B)$. Since g has removable singularities at $a_1, \ldots, a_n$, Theorem 10.35, applied to the function g and the open set Ω_0, shows that

$$\int_\Gamma g(z)\,dz = 0. \tag{3}$$

Hence

$$\frac{1}{2\pi i} \int_\Gamma f(z)\,dz = \sum_{i=1}^n \frac{1}{2\pi i} \int_\Gamma Q_k(z)\,dz = \sum_{k=1}^n \operatorname{Res}\,(Q_k;\,a_k)\,\operatorname{Ind}_\Gamma\,(a_k),$$

and since f and Q_k have the same residue at a_k, we obtain (2). ////

We conclude this chapter with two typical applications of the residue theorem. The first one concerns zeros of holomorphic functions, the second is the evaluation of a certain integral.

10.43 Theorem *Suppose γ is a closed path in a region Ω, such that $\operatorname{Ind}_\gamma(\alpha) = 0$ for every α not in Ω. Suppose also that $\operatorname{Ind}_\gamma(\alpha) = 0$ or 1 for every $\alpha \in \Omega - \gamma^*$, and let Ω_1 be the set of all α with $\operatorname{Ind}_\gamma(\alpha) = 1$.*

For any $f \in H(\Omega)$ let N_f be the number of zeros of f in Ω_1, counted according to their multiplicities.

(a) If $f \in H(\Omega)$ and f has no zeros on γ^* then

$$N_f = \frac{1}{2\pi i} \int_\gamma \frac{f'(z)}{f(z)}\,dz = \operatorname{Ind}_\Gamma\,(0) \tag{1}$$

where $\Gamma = f \circ \gamma$.
(b) If also $g \in H(\Omega)$ and

$$|f(z) - g(z)| < |f(z)| \quad \text{for all } z \in \gamma^* \tag{2}$$

then $N_g = N_f$.

Part (b) is usually called Rouché's theorem. It says that two holomorphic functions have the same number of zeros in Ω_1 if they are close together on the boundary of Ω_1, as specified by (2).

PROOF Put $\varphi = f'/f$, a meromorphic function in Ω. If $a \in \Omega$ and f has a zero of order $m = m(a)$ at a, then $f(z) = (z - a)^m h(z)$, where h and $1/h$ are holomorphic in some neighborhood V of a. In $V - \{a\}$,

$$\varphi(z) = \frac{f'(z)}{f(z)} = \frac{m}{z - a} + \frac{h'(z)}{h(z)}. \tag{3}$$

Thus

$$\operatorname{Res}\,(\varphi;\,a) = m(a). \tag{4}$$

Let $A = \{a \in \Omega_1 : f(a) = 0\}$. If our assumptions about the index of γ are combined with the residue theorem one obtains

$$\frac{1}{2\pi i} \int_\gamma \frac{f'(z)}{f(z)} \, dz = \sum_{a \in A} \text{Res}\,(\varphi; a) = \sum_{a \in A} m(a) = N_f.$$

This proves one half of (1). The other half is a matter of direct computation:

$$\text{Ind}_\Gamma(0) = \frac{1}{2\pi i} \int_\Gamma \frac{dz}{z} = \frac{1}{2\pi i} \int_0^{2\pi} \frac{\Gamma'(s)}{\Gamma(s)} \, ds$$

$$= \frac{1}{2\pi i} \int_0^{2\pi} \frac{f'(\gamma(s))}{f(\gamma(s))} \gamma'(s)\, ds = \frac{1}{2\pi i} \int_\gamma \frac{f'(z)}{f(z)} \, dz.$$

The parameter interval of γ was here taken to be $[0, 2\pi]$.

Next, (2) shows that g has no zero on γ^*. Hence (1) holds with g in place of f. Put $\Gamma_0 = g \circ \gamma$. Then it follows from (1), (2), and Lemma 10.39 that

$$N_g = \text{Ind}_{\Gamma_0}(0) = \text{Ind}_\Gamma(0) = N_f. \qquad \text{////}$$

10.44 Problem *For real t, find the limit, as $A \to \infty$, of*

$$\int_{-A}^{A} \frac{\sin x}{x} e^{ixt} \, dx. \qquad (1)$$

SOLUTION Since $z^{-1} \cdot \sin z \cdot e^{itz}$ is entire, its integral over $[-A, A]$ equals that over the path Γ_A obtained by going from $-A$ to -1 along the real axis, from -1 to 1 along the lower half of the unit circle, and from 1 to A along the real axis. This follows from Cauchy's theorem. Γ_A avoids the origin, and we may therefore use the identity

$$2i \sin z = e^{iz} - e^{-iz}$$

to see that (1) equals $\varphi_A(t+1) - \varphi_A(t-1)$, where

$$\frac{1}{\pi} \varphi_A(s) = \frac{1}{2\pi i} \int_{\Gamma_A} \frac{e^{isz}}{z} \, dz. \qquad (2)$$

Complete Γ_A to a closed path in two ways: First, by the semicircle from A to $-Ai$ to $-A$; secondly, by the semicircle from A to Ai to $-A$. The function e^{isz}/z has a single pole, at $z = 0$, where its residue is 1. It follows that

$$\frac{1}{\pi} \varphi_A(s) = \frac{1}{2\pi} \int_{-\pi}^{0} \exp\,(isAe^{i\theta}) \, d\theta \qquad (3)$$

and

$$\frac{1}{\pi} \varphi_A(s) = 1 - \frac{1}{2\pi} \int_0^{\pi} \exp\,(isAe^{i\theta}) \, d\theta. \qquad (4)$$

Note that

$$|\exp(isAe^{i\theta})| = \exp(-As\sin\theta), \qquad (5)$$

and that this is <1 and tends to 0 as $A \to \infty$ if s and $\sin\theta$ have the same sign. The dominated convergence theorem shows therefore that the integral in (3) tends to 0 if $s < 0$, and the one in (4) tends to 0 if $s > 0$. Thus

$$\lim_{A\to\infty} \varphi_A(s) = \begin{cases} \pi & \text{if } s > 0, \\ 0 & \text{if } s < 0, \end{cases} \qquad (6)$$

and if we apply (6) to $s = t+1$ and to $s = t-1$, we get

$$\lim_{A\to\infty} \int_{-A}^{A} \frac{\sin x}{x} e^{itx} \, dx = \begin{cases} \pi & \text{if } -1 < t < 1, \\ 0 & \text{if } |t| > 1. \end{cases} \qquad (7)$$

Since $\varphi_A(0) = \pi/2$, the limit in (7) is $\pi/2$ when $t = \pm 1$. ////

Note that (7) gives the Fourier transform of $(\sin x)/x$. We leave it as an exercise to check the result against the inversion theorem.

Exercises

1 The following fact was tacitly used in this chapter: If A and B are disjoint subsets of the plane, if A is compact, and if B is closed, then there exists a $\delta > 0$ such that $|\alpha - \beta| \geq \delta$ for all $\alpha \in A$ and $\beta \in B$. Prove this, with an arbitrary metric space in place of the plane.

2 Suppose that f is an entire function, and that in every power series

$$f(z) = \sum_{n=0}^{\infty} c_n(z-a)^n$$

at least one coefficient is 0. Prove that f is a polynomial.

Hint: $n! \, c_n = f^{(n)}(a)$.

3 Suppose f and g are entire functions, and $|f(z)| \leq |g(z)|$ for every z. What conclusion can you draw?

4 Suppose f is an entire function, and

$$|f(z)| \leq A + B|z|^k$$

for all z, where A, B, and k are positive numbers. Prove that f must be a polynomial.

5 Suppose $\{f_n\}$ is a uniformly bounded sequence of holomorphic functions in Ω such that $\{f_n(z)\}$ converges for every $z \in \Omega$. Prove that the convergence is uniform on every compact subset of Ω.

Hint: Apply the dominated convergence theorem to the Cauchy formula for $f_n - f_m$.

6 There is a region Ω that $\exp(\Omega) = D(1; 1)$. Show that $\exp$ is one-to-one in Ω, but that there are many such Ω. Fix one, and define $\log z$, for $|z - 1| < 1$, to be that $w \in \Omega$ for which $e^w = z$. Prove that $\log'(z) = 1/z$. Find the coefficients a_n in

$$\frac{1}{z} = \sum_{n=0}^{\infty} a_n(z-1)^n$$

and hence find the coefficients c_n in the expansion

$$\log z = \sum_{n=0}^{\infty} c_n(z-1)^n.$$

In what other discs can this be done?

7 If $f \in H(\Omega)$, the Cauchy formula for the derivatives of f,

$$f^{(n)}(z) = \frac{n!}{2\pi i} \int_\Gamma \frac{f(\zeta)}{(\zeta - z)^{n+1}} d\zeta \qquad (n = 1, 2, 3, \ldots)$$

is valid under certain conditions on z and Γ. State these, and prove the formula.

8 Suppose P and Q are polynomials, the degree of Q exceeds that of P by at least 2, and the rational function $R = P/Q$ has no pole on the real axis. Prove that the integral of R over $(-\infty, \infty)$ is $2\pi i$ times the sum of the residues of R in the upper half plane. [Replace the integral over $(-A, A)$ by one over a suitable semicircle, and apply the residue theorem.] What is the analogous statement for the lower half plane? Use this method to compute

$$\int_{-\infty}^{\infty} \frac{x^2}{1 + x^4} dx.$$

9 Compute $\int_{-\infty}^{\infty} e^{itx}/(1 + x^2) \, dx$ for real t, by the method described in Exercise 8. Check your answer against the inversion theorem for Fourier transforms.

10 Let γ be the positively oriented unit circle, and compute

$$\frac{1}{2\pi i} \int_\gamma \frac{e^z - e^{-z}}{z^4} dz.$$

11 Suppose α is a complex number, $|\alpha| \neq 1$, and compute

$$\int_0^{2\pi} \frac{d\theta}{1 - 2\alpha \cos \theta + \alpha^2}$$

by integrating $(z - \alpha)^{-1}(z - 1/\alpha)^{-1}$ over the unit circle.

12 Compute

$$\int_{-\infty}^{\infty} \left(\frac{\sin x}{x}\right)^2 e^{itx} dx \qquad \text{(for real } t\text{)}.$$

13 Compute

$$\int_0^{\infty} \frac{dx}{1 + x^n} \qquad (n = 2, 3, 4, \ldots).$$

[For even n, the method of Exercise 8 can be used. However, a different path can be chosen, which simplifies the computation and which also works for odd n: from 0 to R to $R \exp(2\pi i/n)$ to 0.]

  Answer: $(\pi/n)/\sin(\pi/n)$.

14 Suppose Ω_1 and Ω_2 are plane regions, f and g are nonconstant complex functions defined in Ω_1 and Ω_2, respectively, and $f(\Omega_1) \subset \Omega_2$. Put $h = g \circ f$. If f and g are holomorphic, we know that h is holomorphic. Suppose we know that f and h are holomorphic. Can we conclude anything about g? What if we know that g and h are holomorphic?

15 Suppose Ω is a region, $\varphi \in H(\Omega)$, φ' has no zero in Ω, $f \in H(\varphi(\Omega))$, $g = f \circ \varphi$, $z_0 \in \Omega$, and $w_0 = \varphi(z_0)$. Prove that if f has a zero of order m at w_0, then g also has a zero of order m at z_0. How is this modified if φ' has a zero of order k at z_0?

16 Suppose μ is a complex measure on a measure space X, Ω is an open set in the plane, φ is a bounded function on $\Omega \times X$ such that $\varphi(z, t)$ is a measurable function of t, for each $z \in \Omega$, and $\varphi(z, t)$ is holomorphic in Ω, for each $t \in X$. Define

$$f(z) = \int_X \varphi(z, t) \, d\mu(t)$$

for $z \in \Omega$. Prove that $f \in H(\Omega)$. *Hint*: Show that to every compact $K \subset \Omega$ there corresponds a constant $M < \infty$ such that

$$\left| \frac{\varphi(z, t) - \varphi(z_0, t)}{z - z_0} \right| < M \qquad (z \text{ and } z_0 \in K, t \in X).$$

17 Determine the regions in which the following functions are defined and holomorphic:

$$f(z) = \int_0^1 \frac{dt}{1 + tz}, \qquad g(z) = \int_0^\infty \frac{e^{tz}}{1 + t^2} \, dt, \qquad h(z) = \int_{-1}^1 \frac{e^{tz}}{1 + t^2} \, dt.$$

Hint: Either use Exercise 16, or combine Morera's theorem with Fubini's.

18 Suppose $f \in H(\Omega)$, $\bar{D}(a; r) \subset \Omega$, γ is the positively oriented circle with center at a and radius r, and f has no zero on γ^*. For $p = 0$, the integral

$$\frac{1}{2\pi i} \int_\gamma \frac{f'(z)}{f(z)} z^p \, dz$$

is equal to the number of zeros of f in $D(a; r)$. What is the value of this integral (in terms of the zeros of f) for $p = 1, 2, 3, \ldots$? What is the answer if z^p is replaced by any $\varphi \in H(\Omega)$?

19 Suppose $f \in H(U)$, $g \in H(U)$, and neither f nor g has a zero in U. If

$$\frac{f'}{f}\left(\frac{1}{n}\right) = \frac{g'}{g}\left(\frac{1}{n}\right) \qquad (n = 1, 2, 3, \ldots)$$

find another simple relation between f and g.

20 Suppose Ω is a region, $f_n \in H(\Omega)$ for $n = 1, 2, 3, \ldots$, none of the functions f_n has a zero in Ω, and $\{f_n\}$ converges to f uniformly on compact subsets of Ω. Prove that either f has no zero in Ω or $f(z) = 0$ for all $z \in \Omega$.

If Ω' is a region that contains every $f_n(\Omega)$, and if f is not constant, prove that $f(\Omega) \subset \Omega'$.

21 Suppose $f \in H(\Omega)$, Ω contains the closed unit disc, and $|f(z)| < 1$ if $|z| = 1$. How many fixed points must f have in the disc? That is, how many solutions does the equation $f(z) = z$ have there?

22 Suppose $f \in H(\Omega)$, Ω contains the closed unit disc, $|f(z)| > 2$ if $|z| = 1$, and $f(0) = 1$. Must f have a zero in the unit disc?

23 Suppose $P_n(z) = 1 + z/1! + \cdots + z^n/n!$, $Q_n(z) = P_n(z) - 1$, where $n = 1, 2, 3, \ldots$. What can you say about the location of the zeros of P_n and Q_n for large n? Be as specific as you can.

24 Prove the following general form of Rouché's theorem: Let Ω be the interior of a compact set K in the plane. Suppose f and g are continuous on K and holomorphic in Ω, and $|f(z) - g(z)| < |f(z)|$ for all $z \in K - \Omega$. Then f and g have the same number of zeros in Ω.

25 Let A be the annulus $\{z: r_1 < |z| < r_2\}$, where r_1 and r_2 are given positive numbers.
 (a) Show that the Cauchy formula

$$f(z) = \frac{1}{2\pi i} \left(\int_{\gamma_1} + \int_{\gamma_2} \right) \frac{f(\zeta)}{\zeta - z} \, d\zeta$$

is valid under the following conditions: $f \in H(A)$,

$$r_1 + \epsilon < |z| < r_2 - \epsilon,$$

and

$$\gamma_1(t) = (r_1 + \epsilon)e^{-it}, \qquad \gamma_2(t) = (r_2 - \epsilon)e^{it} \qquad (0 \le t \le 2\pi).$$

(b) Show by means of (a) that every $f \in H(A)$ can be decomposed into a sum $f = f_1 + f_2$, where f_1 is holomorphic outside $\bar{D}(0; r_1)$ and $f_2 \in H(D(0; r_2))$; the decomposition is unique if we require that $f_1(z) \to 0$ as $|z| \to \infty$.

(c) Use this decomposition to associate with each $f \in H(A)$ its so-called "Laurent series"

$$\sum_{-\infty}^{\infty} c_n z^n$$

which converges to f in A. Show that there is only one such series for each f. Show that it converges to f uniformly on compact subsets of A.

(d) If $f \in H(A)$ and f is bounded in A, show that the components f_1 and f_2 are also bounded.

(e) How much of the foregoing can you extend to the case $r_1 = 0$ (or $r_2 = \infty$, or both)?

(f) How much of the foregoing can you extend to regions bounded by finitely many (more than two) circles?

26 It is required to expand the function

$$\frac{1}{1-z^2} + \frac{1}{3-z}$$

in a series of the form $\sum_{-\infty}^{\infty} c_n z^n$.

How many such expansions are there? In which region is each of them valid? Find the coefficients c_n explicitly for each of these expansions.

27 Suppose Ω is a horizontal strip, determined by the inequalities $a < y < b$, say. Suppose $f \in H(\Omega)$, and $f(z) = f(z + 1)$ for all $z \in \Omega$. Prove that f has a Fourier expansion in Ω,

$$f(z) = \sum_{-\infty}^{\infty} c_n e^{2\pi i n z},$$

which converges uniformly in $\{z: a + \epsilon \le y \le b - \epsilon\}$, for every $\epsilon > 0$. *Hint*: The map $z \to e^{2\pi i z}$ converts f to a function in an annulus.

Find the integral formulas by means of which the coefficients c_n can be computed from f.

28 Suppose Γ is a closed curve in the plane, with parameter interval $[0, 2\pi]$. Take $\alpha \notin \Gamma^*$. Approximate Γ uniformly by trigonometric polynomials Γ_n. Show that $\mathrm{Ind}_{\Gamma_m}(\alpha) = \mathrm{Ind}_{\Gamma_n}(\alpha)$ if m and n are sufficiently large. Define this common value to be $\mathrm{Ind}_\Gamma(\alpha)$. Prove that the result does not depend on the choice of $\{\Gamma_n\}$; prove that Lemma 10.39 is now true for closed curves, and use this to give a different proof of Theorem 10.40.

29 Define

$$f(z) = \frac{1}{\pi} \int_0^1 r\, dr \int_{-\pi}^{\pi} \frac{d\theta}{re^{i\theta} + z}.$$

Show that $f(z) = \bar{z}$ if $|z| < 1$ and that $f(z) = 1/z$ if $|z| \ge 1$.

Thus f is not holomorphic in the unit disc, although the integrand is a holomorphic function of z. Note the contrast between this, on the one hand, and Theorem 10.7 and Exercise 16 on the other.

Suggestion: Compute the inner integral separately for $r < |z|$ and for $r > |z|$.

30 Let Ω be the plane minus two points, and show that some closed paths Γ in Ω satisfy assumption (1) of Theorem 10.35 without being null-homotopic in Ω.

CHAPTER ELEVEN

HARMONIC FUNCTIONS

The Cauchy-Riemann Equations

11.1 The Operators ∂ and $\bar{\partial}$ Suppose f is a complex function defined in a plane open set Ω. Regard f as a transformation which maps Ω into R^2, and assume that f has a differential at some point $z_0 \in \Omega$, in the sense of Definition 7.22. For simplicity, suppose $z_0 = f(z_0) = 0$. Our differentiability assumption is then equivalent to the existence of two complex numbers α and β (the partial derivatives of f with respect to x and y at $z_0 = 0$) such that

$$f(z) = \alpha x + \beta y + \eta(z)z \qquad (z = x + iy), \tag{1}$$

where $\eta(z) \to 0$ as $z \to 0$.

Since $2x = z + \bar{z}$ and $2iy = z - \bar{z}$, (1) can be rewritten in the form

$$f(z) = \frac{\alpha - i\beta}{2} z + \frac{\alpha + i\beta}{2} \bar{z} + \eta(z)z. \tag{2}$$

This suggests the introduction of the differential operators

$$\partial = \frac{1}{2}\left(\frac{\partial}{\partial x} - i\frac{\partial}{\partial y}\right), \qquad \bar{\partial} = \frac{1}{2}\left(\frac{\partial}{\partial x} + i\frac{\partial}{\partial y}\right). \tag{3}$$

Now (2) becomes

$$\frac{f(z)}{z} = (\partial f)(0) + (\bar{\partial}f)(0) \cdot \frac{\bar{z}}{z} + \eta(z) \qquad (z \neq 0). \tag{4}$$

For real z, $\bar{z}/z = 1$; for pure imaginary z, $\bar{z}/z = -1$. Hence $f(z)/z$ has a limit at 0 if and only if $(\bar{\partial}f)(0) = 0$, and we obtain the following characterization of holomorphic functions:

11.2 Theorem *Suppose f is a complex function in Ω that has a differential at every point of Ω. Then $f \in H(\Omega)$ if and only if the Cauchy-Riemann equation*

$$(\bar{\partial} f)(z) = 0 \tag{1}$$

holds for every $z \in \Omega$. In that case we have

$$f'(z) = (\partial f)(z) \qquad (z \in \Omega). \tag{2}$$

If $f = u + iv$, u and v real, (1) splits into the pair of equations

$$u_x = v_y, \qquad u_y = -v_x$$

where the subscripts refer to partial differentiation with respect to the indicated variable. These are the *Cauchy-Riemann equations* which must be satisfied by the real and imaginary parts of a holomorphic function.

11.3 The Laplacian Let f be a complex function in a plane open set Ω, such that f_{xx} and f_{yy} exist at every point of Ω. The *Laplacian* of f is then defined to be

$$\Delta f = f_{xx} + f_{yy}. \tag{1}$$

If f is continuous in Ω and if

$$\Delta f = 0 \tag{2}$$

at every point of Ω, then f is said to be *harmonic* in Ω.

Since the Laplacian of a real function is real (if it exists), it is clear that *a complex function is harmonic in Ω if and only if both its real part and its imaginary part are harmonic in Ω.*

Note that

$$\Delta f = 4 \partial \bar{\partial} f \tag{3}$$

provided that $f_{xy} = f_{yx}$, and that this happens for all f which have continuous second-order derivatives.

If f is holomorphic, then $\bar{\partial} f = 0$, f has continuous derivatives of all orders, and therefore (3) shows:

11.4 Theorem *Holomorphic functions are harmonic.*

We shall now turn our attention to an integral representation of harmonic functions which is closely related to the Cauchy formula for holomorphic functions. It will show, among other things, that every real harmonic function is locally the real part of a holomorphic function, and it will yield information about the boundary behavior of certain classes of holomorphic functions in open discs.

The Poisson Integral

11.5 The Poisson Kernel This is the function

$$P_r(t) = \sum_{-\infty}^{\infty} r^{|n|} e^{int} \qquad (0 \le r < 1, \; t \text{ real}). \tag{1}$$

We may regard $P_r(t)$ as a function of two variables r and t or as a family of functions of t, indexed by r.

If $z = re^{i\theta}$ ($0 \le r < 1$, θ real), a simple calculation, made in Sec. 5.24, shows that

$$P_r(\theta - t) = \operatorname{Re}\left[\frac{e^{it} + z}{e^{it} - z}\right] = \frac{1 - r^2}{1 - 2r \cos(\theta - t) + r^2}. \tag{2}$$

From (1) we see that

$$\frac{1}{2\pi} \int_{-\pi}^{\pi} P_r(t) \, dt = 1 \qquad (0 \le r < 1). \tag{3}$$

From (2) it follows that $P_r(t) > 0$, $P_r(t) = P_r(-t)$, that

$$P_r(t) < P_r(\delta) \qquad (0 < \delta < |t| \le \pi), \tag{4}$$

and that

$$\lim_{r \to 1} P_r(\delta) = 0 \qquad (0 < \delta \le \pi). \tag{5}$$

These properties are reminiscent of the trigonometric polynomials $Q_k(t)$ that were discussed in Sec. 4.24.

The open unit disc $D(0; 1)$ will from now on be denoted by U. The unit circle – the boundary of U in the complex plane – will be denoted by T. Whenever it is convenient to do so, we shall identify the spaces $L^p(T)$ and $C(T)$ with the corresponding spaces of 2π-periodic functions on R^1, as in Sec. 4.23.

One can also regard $P_r(\theta - t)$ as a function of $z = re^{i\theta}$ and e^{it}. Then (2) becomes

$$P(z, e^{it}) = \frac{1 - |z|^2}{|e^{it} - z|^2} \tag{6}$$

for $z \in U$, $e^{it} \in T$.

11.6 The Poisson Integral If $f \in L^1(T)$ and

$$F(re^{i\theta}) = \frac{1}{2\pi} \int_{-\pi}^{\pi} P_r(\theta - t) f(t) \, dt, \tag{1}$$

then the function F so defined in U is called the *Poisson integral* of f. We shall sometimes abbreviate the relation (1) to

$$F = P[f]. \tag{2}$$

If f is real, formula 11.5(2) shows that $P[f]$ is the real part of

$$\frac{1}{2\pi} \int_{-\pi}^{\pi} \frac{e^{it} + z}{e^{it} - z} f(t) \, dt, \tag{3}$$

which is a holomorphic function of $z = re^{i\theta}$ in U, by Theorem 10.7. Hence $P[f]$ is harmonic in U. Since linear combinations (with constant coefficients) of harmonic functions are harmonic, we see that the following is true:

11.7 Theorem *If $f \in L^1(T)$ then the Poisson integral $P[f]$ is a harmonic function in U.*

The following theorem shows that Poisson integrals of continuous functions behave particularly well near the boundary of U.

11.8 Theorem *If $f \in C(T)$ and if Hf is defined on the closed unit disc $\bar{U}$ by*

$$(Hf)(re^{i\theta}) = \begin{cases} f(e^{i\theta}) & \text{if } r = 1, \\ P[f](re^{i\theta}) & \text{if } 0 \le r < 1, \end{cases} \tag{1}$$

then $Hf \in C(\bar{U})$.

PROOF Since $P_r(t) > 0$, formula 11.5(3) shows, for every $g \in C(T)$, that

$$|P[g](re^{i\theta})| \le \|g\|_T \quad (0 \le r < 1), \tag{2}$$

so that

$$\|Hg\|_{\bar{U}} = \|g\|_T \quad (g \in C(T)). \tag{3}$$

(As in Sec. 5.22, we use the notation $\|g\|_E$ to denote the supremum of $|g|$ on the set E.) If

$$g(e^{i\theta}) = \sum_{n=-N}^{N} c_n e^{in\theta} \tag{4}$$

is any trigonometric polynomial, it follows from 11.5(1) that

$$(Hg)(re^{i\theta}) = \sum_{n=-N}^{N} c_n r^{|n|} e^{in\theta}, \tag{5}$$

so that $Hg \in C(\bar{U})$.

Finally, there are trigonometric polynomials g_k such that $\|g_k - f\|_T \to 0$ as $k \to \infty$. (See Sec. 4.24.) By (3), it follows that

$$\|Hg_k - Hf\|_{\bar{U}} = \|H(g_k - f)\|_{\bar{U}} \to 0 \tag{6}$$

as $k \to \infty$. This says that the functions $Hg_k \in C(\bar{U})$ converge, uniformly on $\bar{U}$, to Hf. Hence $Hf \in C(\bar{U})$. ////

Note: This theorem provides the solution of a boundary value problem (the *Dirichlet problem*): A continuous function f is given on T and it is required to find a harmonic function F in U "whose boundary values are f." The theorem exhibits a solution, by means of the Poisson integral of f, and it states the relation between f and F more precisely. The uniqueness theorem which corresponds to this existence theorem is contained in the following result.

11.9 Theorem *Suppose u is a continuous real function on the closed unit disc $\bar{U}$, and suppose u is harmonic in U. Then (in U) u is the Poisson integral of its restriction to T, and u is the real part of the holomorphic function*

$$f(z) = \frac{1}{2\pi} \int_{-\pi}^{\pi} \frac{e^{it} + z}{e^{it} - z} u(e^{it}) \, dt \qquad (z \in U). \tag{1}$$

PROOF Theorem 10.7 shows that $f \in H(U)$. If $u_1 = \operatorname{Re} f$, then (1) shows that u_1 is the Poisson integral of the boundary values of u, and the theorem will be proved as soon as we show that $u = u_1$.

Put $h = u - u_1$. Then h is continuous on $\bar{U}$ (apply Theorem 11.8 to u_1), h is harmonic in U, and $h = 0$ at all points of T. Assume (this will lead to a contradiction) that $h(z_0) > 0$ for some $z_0 \in U$. Fix ϵ so that $0 < \epsilon < h(z_0)$, and define

$$g(z) = h(z) + \epsilon |z|^2 \qquad (z \in \bar{U}). \tag{2}$$

Then $g(z_0) \geq h(z_0) > \epsilon$. Since $g \in C(\bar{U})$ and since $g = \epsilon$ at all points of T, there exists a point $z_1 \in U$ at which g has a local maximum. This implies that $g_{xx} \leq 0$ and $g_{yy} \leq 0$ at z_1. But (2) shows that the Laplacian of g is $4\epsilon > 0$, and we have a contradiction.

Thus $u - u_1 \leq 0$. The same argument shows that $u_1 - u \leq 0$. Hence $u = u_1$, and the proof is complete. ////

11.10 So far we have considered only the unit disc $U = D(0; 1)$. It is clear that the preceding work can be carried over to arbitrary circular discs, by a simple change of variables. Hence we shall merely summarize some of the results:

If u is a continuous real function on the boundary of the disc $D(a; R)$ and if u is defined in $D(a; R)$ by the Poisson integral

$$u(a + re^{i\theta}) = \frac{1}{2\pi} \int_{-\pi}^{\pi} \frac{R^2 - r^2}{R^2 - 2Rr \cos(\theta - t) + r^2} u(a + Re^{it}) \, dt \tag{1}$$

then u is continuous on $\bar{D}(a; R)$ and harmonic in $D(a; R)$.

If u is harmonic (and real) in an open set Ω and if $\bar{D}(a; R) \subset \Omega$, then u satisfies (1) in $D(a; R)$ and there is a holomorphic function f defined in $D(a; R)$ whose real part is u. This f is uniquely defined, up to a pure imaginary additive constant. For if two functions, holomorphic in the same region, have the same real part, their difference must be constant (a corollary of the open mapping theorem, or the Cauchy-Riemann equations).

We may summarize this by saying that *every real harmonic function is locally the real part of a holomorphic function.*

Consequently, every harmonic function has continuous partial derivatives of all orders.

The Poisson integral also yields information about sequences of harmonic functions:

11.11 Harnack's Theorem *Let $\{u_n\}$ be a sequence of harmonic functions in a region Ω.*

(a) *If $u_n \to u$ uniformly on compact subsets of Ω, then u is harmonic in Ω.*
(b) *If $u_1 \leq u_2 \leq u_3 \leq \cdots$, then either $\{u_n\}$ converges uniformly on compact subsets of Ω, or $u_n(z) \to \infty$ for every $z \in \Omega$.*

PROOF To prove (a), assume $\bar{D}(a; R) \subset \Omega$, and replace u by u_n in the Poisson integral 11.10(1). Since $u_n \to u$ uniformly on the boundary of $\bar{D}(a; R)$, we conclude that u itself satisfies 11.10(1) in $D(a; R)$.

In the proof of (b), we may assume that $u_1 \geq 0$. (If not, replace u_n by $u_n - u_1$.) Put $u = \sup u_n$, let $A = \{z \in \Omega : u(z) < \infty\}$, and $B = \Omega - A$. Choose $\bar{D}(a; R) \subset \Omega$. The Poisson kernel satisfies the inequalities

$$\frac{R-r}{R+r} \leq \frac{R^2 - r^2}{R^2 - 2rR \cos(\theta - t) + r^2} \leq \frac{R+r}{R-r}$$

for $0 \leq r < R$. Hence

$$\frac{R-r}{R+r} u_n(a) \leq u_n(a + re^{i\theta}) \leq \frac{R+r}{R-r} u_n(a).$$

The same inequalities hold with u in place of u_n. It follows that either $u(z) = \infty$ for all $z \in D(a; R)$ or $u(z) < \infty$ for all $z \in D(a; R)$.

Thus both A and B are open; and since Ω is connected, we have either $A = \emptyset$ (in which case there is nothing to prove) or $A = \Omega$. In the latter case, the monotone convergence theorem shows that the Poisson formula holds for u in every disc in Ω. Hence u is harmonic in Ω. Whenever a sequence of continuous functions converges monotonically to a continuous limit, the convergence is uniform on compact sets ([26], Theorem 7.13). This completes the proof. ////

The Mean Value Property

11.12 Definition We say that a continuous function u in an open set Ω has the *mean value property* if to every $z \in \Omega$ there corresponds a sequence $\{r_n\}$ such that $r_n > 0$, $r_n \to 0$ as $n \to \infty$, and

$$u(z) = \frac{1}{2\pi} \int_{-\pi}^{\pi} u(z + r_n e^{it}) \, dt \qquad (n = 1, 2, 3, \ldots). \tag{1}$$

In other words, $u(z)$ is to be equal to the mean value of u on the circles of radius r_n and with center at z.

Note that the Poisson formula shows that (1) holds for every harmonic function u, and for *every* r such that $\bar{D}(z; r) \subset \Omega$. Thus harmonic functions satisfy a much stronger mean value property than the one that we just defined. The following theorem may therefore come as a surprise:

11.13 Theorem *If a continuous function u has the mean value property in an open set Ω, then u is harmonic in Ω.*

PROOF It is enough to prove this for real u. Fix $\bar{D}(a; R) \subset \Omega$. The Poisson integral gives us a continuous function h on $\bar{D}(a; R)$ which is harmonic in $D(a; R)$ and which coincides with u on the boundary of $D(a; R)$. Put $v = u - h$, and let $m = \sup \{v(z): z \in \bar{D}(a; R)\}$. Assume $m > 0$, and let E be the set of all $z \in \bar{D}(a; R)$ at which $v(z) = m$. Since $v = 0$ on the boundary of $D(a; R)$, E is a compact subset of $D(a; R)$. Hence there exists a $z_0 \in E$ such that

$$|z_0 - a| \geq |z - a|$$

for all $z \in E$. For *all* small enough r, at least half the circle with center z_0 and radius r lies outside E, so that the corresponding mean values of v are all less than $m = v(z_0)$. But v has the mean value property, and we have a contradiction. Thus $m = 0$, so $v \leq 0$. The same reasoning applies to $-v$. Hence $v = 0$, or $u = h$ in $D(a; R)$, and since $\bar{D}(a; R)$ was an arbitrary closed disc in Ω, u is harmonic in Ω. ////

Theorem 11.13 leads to a reflection theorem for holomorphic functions. By the *upper half plane* Π^+ we mean the set of all $z = x + iy$ with $y > 0$; the *lower half plane* Π^- consists of all z whose imaginary part is negative.

11.14 Theorem (The Schwarz reflection principle) *Suppose L is a segment of the real axis, Ω^+ is a region in Π^+, and every $t \in L$ is the center of an open disc D_t such that $\Pi^+ \cap D_t$ lies in Ω^+. Let Ω^- be the reflection of Ω^+:*

$$\Omega^- = \{z: \bar{z} \in \Omega^+\}. \tag{1}$$

Suppose $f = u + iv$ is holomorphic in Ω^+, and

$$\lim_{n \to \infty} v(z_n) = 0 \qquad (2)$$

for every sequence $\{z_n\}$ in Ω^+ which converges to a point of L.

Then there is a function F, holomorphic in $\Omega^+ \cup L \cup \Omega^-$, such that $F(z) = f(z)$ in Ω^+; this F satisfies the relation

$$F(\bar{z}) = \overline{F(z)} \qquad (z \in \Omega^+ \cup L \cup \Omega^-). \qquad (3)$$

The theorem asserts that f can be extended to a function which is holomorphic in a region symmetric with respect to the real axis, and (3) states that F preserves this symmetry. Note that the continuity hypothesis (2) is merely imposed on the imaginary part of f.

PROOF Put $\Omega = \Omega^+ \cup L \cup \Omega^-$. We extend v to Ω by defining $v(z) = 0$ for $z \in L$ and $v(z) = -v(\bar{z})$ for $z \in \Omega^-$. It is then immediate that v is continuous and that v has the mean value property in Ω, so that v is harmonic in Ω, by Theorem 11.13.

Hence v is locally the imaginary part of a holomorphic function. This means that to each of the discs D_t there corresponds an $f_t \in H(D_t)$ such that Im $f_t = v$. Each f_t is determined by v up to a real additive constant. If this constant is chosen so that $f_t(z) = f(z)$ for *some* $z \in D_t \cap \Pi^+$, the same will hold for all $z \in D_t \cap \Pi^+$, since $f - f_t$ is constant in the region $D_t \cap \Pi^+$. We assume that the functions f_t are so adjusted.

The power series expansion of f_t in powers of $z - t$ has only real coefficients, since $v = 0$ on L, so that all derivatives of f_t are real at t. It follows that

$$f_t(\bar{z}) = \overline{f_t(z)} \qquad (z \in D_t). \qquad (4)$$

Next, assume that $D_s \cap D_t \neq \emptyset$. Then $f_t = f = f_s$ in $D_t \cap D_s \cap \Pi^+$; and since $D_t \cap D_s$ is connected, Theorem 10.18 shows that

$$f_t(z) = f_s(z) \qquad (z \in D_t \cap D_s). \qquad (5)$$

Thus it is consistent to define

$$F(z) = \begin{cases} f(z) & \text{for } z \in \Omega^+ \\ f_t(z) & \text{for } z \in D_t \\ \overline{f(\bar{z})} & \text{for } z \in \Omega^-. \end{cases} \qquad (6)$$

and it remains to show that F is holomorphic in Ω^-. If $D(a; r) \subset \Omega^-$, then $D(\bar{a}; r) \subset \Omega^+$, so for every $z \in D(a; r)$ we have

$$f(z) = \sum_{n=0}^{\infty} c_n(\bar{z} - \bar{a})^n. \qquad (7)$$

Hence

$$F(z) = \sum_{n=0}^{\infty} \bar{c}_n (z-a)^n \qquad (z \in D(a;r)). \tag{8}$$

This completes the proof. ////

Boundary Behavior of Poisson Integrals

11.15 Our next objective is to find analogues of Theorem 11.8 for Poisson integrals of L^p-functions and measures on T.

Let us associate to any function u in U a family of functions u_r on T, defined by

$$u_r(e^{i\theta}) = u(re^{i\theta}) \qquad (0 \le r < 1). \tag{1}$$

Thus u_r is essentially the restriction of u to the circle with radius r, center 0, but we shift the domain of u_r to T.

Using this terminology, Theorem 11.8 can be stated in the following form:

If $f \in C(T)$ and $F = P[f]$, then $F_r \to f$ uniformly on T, as $r \to 1$. In other words,

$$\lim_{r \to 1} \|F_r - f\|_\infty = 0, \tag{2}$$

which implies of course that

$$\lim_{r \to 1} F_r(e^{i\theta}) = f(e^{i\theta}) \tag{3}$$

at every point of T.

As regards (2), we shall now see (Theorem 11.16) that the corresponding norm-convergence result is just as easy in L^p. Instead of confining ourselves to investigating *radial limits*, as in (3), we shall then study *nontangential limits* of Poisson integrals of measures and L^p-functions; the differentiation theory developed in Chap. 7 will play an essential role in that study.

11.16 Theorem *If* $1 \le p \le \infty$, $f \in L^p(T)$, *and* $u = P[f]$, *then*

$$\|u_r\|_p \le \|f\|_p \qquad (0 \le r < 1). \tag{1}$$

If $1 \le p < \infty$, *then*

$$\lim_{r \to 1} \|u_r - f\|_p = 0. \tag{2}$$

PROOF If we apply Jensen's inequality (or Hölder's) to

$$u_r(e^{i\theta}) = \frac{1}{2\pi} \int_{-\pi}^{\pi} f(t) P_r(\theta - t) \, dt \tag{3}$$

we obtain

$$|u_r(e^{i\theta})|^p \le \frac{1}{2\pi}\int_{-\pi}^{\pi} |f(t)|^p P_r(\theta - t)\, dt. \tag{4}$$

If we integrate (4) with respect to θ, over $[-\pi, \pi]$ and use Fubini's theorem, we obtain (1).

Note that formula 11.5(3) was used twice in this argument.

To prove (2), choose $\epsilon > 0$, and choose $g \in C(T)$ so that $\|g - f\|_p < \epsilon$ (Theorem 3.14). Let $v = P[g]$. Then

$$u_r - f = (u_r - v_r) + (v_r - g) + (g - f). \tag{5}$$

By (1), $\|u_r - v_r\|_p = \|(u - v)_r\|_p \le \|f - g\|_p < \epsilon$. Thus

$$\|u_r - f\|_p \le 2\epsilon + \|v_r - g\|_p \tag{6}$$

for all $r < 1$. Also, $\|v_r - g\|_p \le \|v_r - g\|_\infty$, and the latter converges to 0 as $r \to 1$, by Theorem 11.8. This proves (2). ////

11.17 Poisson Integrals of Measures If μ is a complex measure on T, and if we want to replace integrals over T by integrals over intervals of length 2π in R^1, these intervals have to be taken half open, because of the possible presence of point masses in μ. To avoid this (admittedly very minor) problem, we shall keep integration on the circle in what follows, and will write the Poisson integral $u = P[d\mu]$ of μ in the form

$$u(z) = \int_T P(z, e^{it})\, d\mu(e^{it}) \qquad (z \in U) \tag{1}$$

where $P(z, e^{it}) = (1 - |z|^2)/|e^{it} - z|^2$, as in formula 11.5(6).

The reasoning that led to Theorem 11.7 applies without change to Poisson integrals of measures. *Thus u, defined by (1), is harmonic in U.*

Setting $\|\mu\| = |\mu|(T)$, the analogue of the first half of Theorem 11.16 is

$$\|u_r\|_1 = \frac{1}{2\pi}\int_{-\pi}^{\pi} |u(re^{i\theta})|\, d\theta \le \|\mu\|. \tag{2}$$

To see this, replace μ by $|\mu|$ in (1), apply Fubini's theorem, and refer to formula 11.5(3).

11.18 Approach Regions For $0 < \alpha < 1$, we define Ω_α to be the union of the disc $D(0; \alpha)$ and the line segments from $z = 1$ to points of $D(0; \alpha)$.

In other words, Ω_α is the smallest convex open set that contains $D(0; \alpha)$ and has the point 1 in its boundary. Near $z = 1$, Ω_α is an angle, bisected by the radius of U that terminates at 1, of opening 2θ, where $\alpha = \sin \theta$. Curves that approach 1 within Ω_α cannot be tangent to T. Therefore Ω_α is called a *nontangential approach region*, with vertex 1.

The regions Ω_α expand when α increases. Their union is U, their intersection is the radius $[0, 1)$.

Rotated copies of Ω_α, with vertex at e^{it}, will be denoted by $e^{it}\Omega_\alpha$.

11.19 Maximal Functions If $0 < \alpha < 1$ and u is any complex function with domain U, its *nontangential maximal function* $N_\alpha u$ is defined on T by

$$(N_\alpha u)(e^{it}) = \sup\{|u(z)|: z \in e^{it}\Omega_\alpha\}. \qquad (1)$$

Similarly, the *radial maximal function* of u is

$$(M_{\text{rad}} u)(e^{it}) = \sup\{|u(re^{it})|: 0 \le r < 1\}. \qquad (2)$$

If u is *continuous* and λ is a positive number, then the set where either of these maximal functions is $\le \lambda$ is a closed subset of T. Consequently, $N_\alpha u$ and $M_{\text{rad}} u$ are *lower semicontinuous* on T; in particular, they are measurable.

Clearly, $M_{\text{rad}} u \le N_\alpha u$, and the latter increases with α. If $u = P[d\mu]$, Theorem 11.20 will show that the size of $N_\alpha u$ is, in turn, controlled by the maximal function $M\mu$ that was defined in Sec. 7.2 (taking $k = 1$). However, it will simplify the notation if we replace ordinary Lebesgue measure m on T by $\sigma = m/2\pi$. Then σ is a rotation-invariant positive Borel measure on T, so normalized that $\sigma(T) = 1$.

Accordingly, $M\mu$ is now defined by

$$(M\mu)(e^{i\theta}) = \sup \frac{|\mu|(I)}{\sigma(I)}. \qquad (3)$$

The supremum is taken over all open arcs $I \subset T$ whose centers are at $e^{i\theta}$, including T itself (even though T is of course not an arc).

Similarly, the derivative $D\mu$ of a measure μ on T is now

$$(D\mu)(e^{i\theta}) = \lim \frac{\mu(I)}{\sigma(I)}, \qquad (4)$$

as the open arcs $I \subset T$ shrink to their center $e^{i\theta}$, and $e^{i\theta}$ is a *Lebesgue point* of $f \in L^1(T)$ if

$$\lim \frac{1}{\sigma(I)} \int_I |f - f(e^{i\theta})|\, d\sigma = 0, \qquad (5)$$

where $\{I\}$ is as in (4).

If $d\mu = f\, d\sigma + d\mu_s$ is the Lebesgue decomposition of a complex Borel measure μ on T, where $f \in L^1(T)$ and $\mu_s \perp \sigma$, Theorems 7.4, 7.7, and 7.14 assert that

$$\sigma\{M\mu > \lambda\} \le \frac{3}{\lambda} \|\mu\|, \qquad (6)$$

that almost every point of T is a Lebesgue point of f, and that $D\mu = f$, $D\mu_s = 0$ a.e. $[\sigma]$.

We will now see, for any complex Borel measure μ on T, that the nontangential and radial maximal functions of the harmonic function $P[d\mu]$ are controlled by $M\mu$. In fact, if any one of them is finite at some point of T, so are the others; this can be seen by combining Theorem 11.20 with Exercise 19.

11.20 Theorem *Assume $0 < \alpha < 1$. Then there is a constant $c_\alpha > 0$ with the following property: If μ is a positive finite Borel measure on T and $u = P[d\mu]$ is its Poisson integral, then the inequalities*

$$c_\alpha (N_\alpha u)(e^{i\theta}) \leq (M_{\mathrm{rad}} u)(e^{i\theta}) \leq (M\mu)(e^{i\theta}) \tag{1}$$

hold at every point $e^{i\theta} \in T$.

PROOF We shall prove (1) for $\theta = 0$. The general case follows then if the special case is applied to the rotated measure $\mu_\theta(E) = \mu(e^{i\theta} E)$.

Since $u(z) = \int_T P(z, e^{it}) \, d\mu(e^{it})$, the first inequality in (1) will follow if we can show that

$$c_\alpha P(z, e^{it}) \leq P(|z|, e^{it}) \tag{2}$$

holds for all $z \in \Omega_\alpha$ and all $e^{it} \in T$. By formula 11.5(6), (2) is the same as

$$c_\alpha |e^{it} - r|^2 \leq |e^{it} - z|^2 \tag{3}$$

where $r = |z|$.

The definition of Ω_α shows that $|z - r|/(1 - r)$ is bounded in Ω_α, say by γ_α. Hence

$$|e^{it} - r| \leq |e^{it} - z| + |z - r|$$
$$\leq |e^{it} - z| + \gamma_\alpha(1 - r) \leq (1 + \gamma_\alpha)|e^{it} - z|$$

so that (3) holds with $c_\alpha = (1 + \gamma_\alpha)^{-2}$. This proves the first half of (1).

For the second half, we have to prove that

$$\int_T P_r(t) \, d\mu(e^{it}) \leq (M\mu)(1) \qquad (0 \leq r \leq 1). \tag{4}$$

Fix r. Choose open arcs $I_j \subset T$, centered at 1, so that $I_1 \subset I_2 \subset \cdots \subset I_{n-1}$, put $I_n = T$. For $1 \leq j \leq n$, let χ_j be the characteristic function of I_j, and let h_j be the largest positive number for which $h_j \chi_j \leq P_r$ on T. Define

$$K = \sum_{j=1}^{n} (h_j - h_{j+1})\chi_j \tag{5}$$

where $h_{n+1} = 0$. Since $P_r(t)$ is an even function of t that decreases as t increases from 0 to π, we see that $h_j - h_{j+1} \geq 0$, that $K = h_j$ on $I_j - I_{j-1}$ (putting $I_0 = \varnothing$), and that $K \leq P_r$. The definition of $M\mu$ shows that

$$\mu(I_j) \leq (M\mu)(1)\sigma(I_j). \tag{6}$$

Hence, setting $(M\mu)(1) = M$,

$$\int_T K \, d\mu = \sum_{j=1}^n (h_j - h_{j+1})\mu(I_j) \le M \sum_{j=1}^n (h_j - h_{j+1})\sigma(I_j)$$

$$= M \int_T K \, d\sigma \le M \int_T P_r \, d\sigma = M. \qquad (7)$$

Finally, if we choose the arcs I_j so that their endpoints form a sufficiently fine partition of T, we obtain step functions K that converge to P_r, uniformly on T. Hence (4) follows from (7). ////

11.21 Nontangential Limits A function F, defined in U, is said to *have nontangential limit* λ *at* $e^{i\theta} \in T$ if, for each $\alpha < 1$,

$$\lim_{j \to \infty} F(z_j) = \lambda$$

for every sequence $\{z_j\}$ that converges to $e^{i\theta}$ and that lies in $e^{i\theta}\Omega_\alpha$.

11.22 Theorem *If μ is a positive Borel measure on T and $(D\mu)(e^{i\theta}) = 0$ for some θ, then its Poisson integral $u = P[d\mu]$ has nontangential limit 0 at $e^{i\theta}$.*

PROOF By definition, the assumption $(D\mu)(e^{i\theta}) = 0$ means that

$$\lim \mu(I)/\sigma(I) = 0 \qquad (1)$$

as the open arcs $I \subset T$ shrink to their center $e^{i\theta}$. Pick $\epsilon > 0$. One of these arcs, say I_0, is then small enough to ensure that

$$\mu(I) < \epsilon\sigma(I) \qquad (2)$$

for every $I \subset I_0$ that has $e^{i\theta}$ as center.

Let μ_0 be the restriction of μ to I_0, put $\mu_1 = \mu - \mu_0$, and let u_i be the Poisson integral of μ_i ($i = 0, 1$). Suppose z_j converges to $e^{i\theta}$ within some region $e^{i\theta}\Omega_\alpha$. Then z_j stays at a positive distance from $T - I_0$. The integrands in

$$u_1(z_j) = \int_{T-I_0} P(z_j, e^{it}) \, d\mu(e^{it}) \qquad (3)$$

converge therefore to 0 as $j \to \infty$, uniformly on $T - I_0$. Hence

$$\lim_{j \to \infty} u_1(z_j) = 0. \qquad (4)$$

Next, use (2) together with Theorem 11.20 to see that

$$c_\alpha(N_\alpha u_0)(e^{i\theta}) \le (M\mu_0)(e^{i\theta}) \le \epsilon. \qquad (5)$$

In $e^{i\theta}\Omega_\alpha$, $u_0(z) \leq (N_\alpha u_0)(e^{i\theta})$. Hence (5) implies that

$$\limsup_{j \to \infty} u_0(z_j) \leq \epsilon/c_\alpha. \tag{6}$$

Since $u = u_0 + u_1$ and ϵ was arbitrary, (4) and (6) give

$$\lim_{j \to \infty} u(z_j) = 0. \tag{7}$$

////

11.23 Theorem *If $f \in L^1(T)$, then $P[f]$ has nontangential limit $f(e^{i\theta})$ at every Lebesgue point $e^{i\theta}$ of f.*

PROOF Suppose $e^{i\theta}$ is a Lebesgue point of f. By subtracting a constant from f we may assume, without loss of generality, that $f(e^{i\theta}) = 0$. Then

$$\lim \frac{1}{\sigma(I)} \int_I |f| \, d\sigma = 0 \tag{1}$$

as the open arcs $I \subset T$ shrink to their center $e^{i\theta}$. Define a Borel measure μ on T by

$$\mu(E) = \int_E |f| \, d\sigma \tag{2}$$

Then (1) says that $(D\mu)(e^{i\theta}) = 0$; hence $P[d\mu]$ has nontangential limit 0 at $e^{i\theta}$, by Theorem 11.22. The same is true of $P[f]$, because

$$|P[f]| \leq P[|f|] = P[d\mu]. \tag{3}$$

////

The last two theorems can be combined as follows.

11.24 Theorem *If $d\mu = f \, d\sigma + d\mu_s$ is the Lebesgue decomposition of a complex Borel measure μ on T, where $f \in L^1(T)$, $\mu_s \perp \sigma$, then $P[d\mu]$ has nontangential limit $f(e^{i\theta})$ at almost all points of T.*

PROOF Apply Theorem 11.22 to the positive and negative variations of the real and imaginary parts of μ_s, and apply Theorem 11.23 to f. ////

Here is another consequence of Theorem 11.20.

11.25 Theorem *For $0 < \alpha < 1$ and $1 \leq p \leq \infty$, there are constants $A(\alpha, p) < \infty$ with the following properties:*

(a) *If μ is a complex Borel measure on T, and $u = P[d\mu]$, then*

$$\sigma\{N_\alpha u > \lambda\} \le \frac{A(\alpha, 1)}{\lambda} \|\mu\| \qquad (0 < \lambda < \infty).$$

(b) *If $1 < p \le \infty, f \in L^p(T)$, and $u = P[f]$, then*

$$\|N_\alpha u\|_p \le A(\alpha, p)\|f\|_p.$$

PROOF Combine Theorem 11.20 with Theorem 7.4 and the inequality (6) in the proof of Theorem 8.18. ////

The nontangential maximal functions $N_\alpha u$ are thus in weak L^1 if $u = P[d\mu]$, and they are in $L^p(T)$ if $u = P[f]$ for some $f \in L^p(T)$, $p > 1$. This latter result may be regarded as a strengthened form of the first part of Theorem 11.16.

Representation Theorems

11.26 How can one tell whether a harmonic function u in U is a Poisson integral or not? The preceding theorems (11.16 to 11.25) contain a number of necessary conditions. It turns out that the simplest of these, the L^p-boundedness of the family $\{u_r : 0 \le r < 1\}$ is also sufficient! Thus, in particular, the boundedness of $\|u_r\|_1$, as $r \to 1$, implies the existence of nontangential limits a.e. on T, since, as we will see in Theorem 11.30, u can then be represented as the Poisson integral of a measure.

This measure will be obtained as a so-called "weak limit" of the functions u_r. Weak convergence is an important topic in functional analysis. We will approach it through another important concept, called equicontinuity, which we will meet again later, in connection with the so-called "normal families" of holomorphic functions.

11.27 Definition Let $\mathscr{F}$ be a collection of complex functions on a metric space X with metric ρ.

We say that $\mathscr{F}$ is *equicontinuous* if to every $\epsilon > 0$ corresponds a $\delta > 0$ such that $|f(x) - f(y)| < \epsilon$ for every $f \in \mathscr{F}$ and for all pairs of points x, y with $\rho(x, y) < \delta$. (In particular, every $f \in \mathscr{F}$ is then uniformly continuous.)

We say that $\mathscr{F}$ is *pointwise bounded* if to every $x \in X$ corresponds an $M(x) < \infty$ such that $|f(x)| \le M(x)$ for every $f \in \mathscr{F}$.

11.28 Theorem (Arzela-Ascoli) *Suppose that $\mathscr{F}$ is a pointwise bounded equicontinuous collection of complex functions on a metric space X, and that X contains a countable dense subset E.*

Every sequence $\{f_n\}$ in $\mathscr{F}$ has then a subsequence that converges uniformly on every compact subset of X.

PROOF Let $x_1, x_2, x_3, \ldots$ be an enumeration of the points of E. Let S_0 be the set of all positive integers. Suppose $k \geq 1$ and an infinite set $S_{k-1} \subset S_0$ has been chosen. Since $\{f_n(x_k): n \in S_{k-1}\}$ is a bounded sequence of complex numbers, it has a convergent subsequence. In other words, there is an infinite set $S_k \subset S_{k-1}$ so that $\lim f_n(x_k)$ exists as $n \to \infty$ within S_k.

Continuing in this way, we obtain infinite sets $S_0 \supset S_1 \supset S_2 \supset \cdots$ with the property that $\lim f_n(x_j)$ exists, for $1 \leq j \leq k$, if $n \to \infty$ within S_k.

Let r_k be the kth term of S_k (with respect to the natural order of the positive integers) and put

$$S = \{r_1, r_2, r_3, \ldots\}.$$

For each k there are then at most $k - 1$ terms of S that are not in S_k.

Hence $\lim f_n(x)$ exists, for every $x \in E$, as $n \to \infty$ within S.

(The construction of S from $\{S_k\}$ is the so-called *diagonal process*.)

Now let $K \subset X$ be compact, pick $\epsilon > 0$. By equicontinuity, there is a $\delta > 0$ so that $\rho(p, q) < \delta$ implies $|f_n(p) - f_n(q)| < \epsilon$, for all n. Cover K with open balls $B_1, \ldots, B_M$ of radius $\delta/2$. Since E is dense in X there are points $p_i \in B_i \cap E$ for $1 \leq i \leq M$. Since $p_i \in E$, $\lim f_n(p_i)$ exists, as $n \to \infty$ within S. Hence there is an integer N such that

$$|f_m(p_i) - f_n(p_i)| < \epsilon$$

for $i = 1, \ldots, M$, if $m > N$, $n > N$, and m and n are in S.

To finish, pick $x \in K$. Then $x \in B_i$ for some i, and $\rho(x, p_i) < \delta$. Our choice of δ and N shows that

$$|f_m(x) - f_n(x)| \leq |f_m(x) - f_m(p_i)| + |f_m(p_i) - f_n(p_i)| + |f_n(p_i) - f_n(x)|$$
$$< \epsilon + \epsilon + \epsilon = 3\epsilon$$

if $m > N$, $n > N$, $m \in S$, $n \in S$. ////

11.29 Theorem *Suppose that*

(a) *X is a separable Banach space,*
(b) *$\{\Lambda_n\}$ is a sequence of linear functionals on X,*
(c) *$\sup_n \|\Lambda_n\| = M < \infty$.*

Then there is a subsequence $\{\Lambda_{n_i}\}$ such that the limit

$$\Lambda x = \lim_{i \to \infty} \Lambda_{n_i} x \tag{1}$$

exists for every $x \in X$. Moreover, Λ is linear, and $\|\Lambda\| \leq M$.

(In this situation, Λ is said to be the *weak limit* of $\{\Lambda_{n_i}\}$; see Exercise 18.)

PROOF To say that X is separable means, by definition, that X has a countable dense subset. The inequalities

$$|\Lambda_n x| \le M\|x\|, \qquad |\Lambda_n x' - \Lambda_n x''| \le M\|x' - x''\|$$

show that $\{\Lambda_n\}$ is pointwise bounded and equicontinuous. Since each point of X is a compact set, Theorem 11.28 implies that there is a subsequence $\{\Lambda_{n_i}\}$ such that $\{\Lambda_{n_i} x\}$ converges, for every $x \in X$, as $i \to \infty$. To finish, define Λ by (1). It is then clear that Λ is linear and that $\|\Lambda\| \le M$. ////

Let us recall, for the application that follows, that $C(T)$ and $L^p(T)$ ($p < \infty$) are *separable* Banach spaces, because the trigonometric polynomials are dense in them, and because it is enough to confine ourselves to trigonometric polynomials whose coefficients lie in some prescribed countable dense subset of the complex field.

11.30 Theorem *Suppose u is harmonic in U, $1 \le p \le \infty$, and*

$$\sup_{0 < r < 1} \|u_r\|_p = M < \infty. \tag{1}$$

(a) If $p = 1$, it follows that there is a unique complex Borel measure μ on T so that $u = P[d\mu]$.
(b) If $p > 1$, it follows that there is a unique $f \in L^p(T)$ so that $u = P[f]$.
(c) Every positive harmonic function in U is the Poisson integral of a unique positive Borel measure on T.

PROOF Assume first that $p = 1$. Define linear functionals Λ_r on $C(T)$ by

$$\Lambda_r g = \int_T g u_r \, d\sigma \qquad (0 \le r < 1). \tag{2}$$

By (1), $\|\Lambda_r\| \le M$. By Theorems 11.29 and 6.19 there is a measure μ on T, with $\|\mu\| \le M$, and a sequence $r_j \to 1$, so that

$$\lim_{j \to \infty} \int_T g u_{r_j} \, d\sigma = \int_T g \, d\mu \tag{3}$$

for every $g \in C(T)$.

Put $h_j(z) = u(r_j z)$. Then h_j is harmonic in U, continuous on $\bar{U}$, and is therefore the Poisson integral of its restriction to T (Theorem 11.9). Fix $z \in U$, and apply (3) with

$$g(e^{it}) = P(z, e^{it}). \tag{4}$$

Since $h_j(e^{it}) = u_{r_j}(e^{it})$, we obtain

$$u(z) = \lim_j u(r_j z) = \lim_j h_j(z)$$

$$= \lim_j \int_T P(z, e^{it}) h_j(e^{it}) \, d\sigma(e^{it})$$

$$= \int_T P(z, e^{it}) \, d\mu(e^{it}) = P[d\mu](z).$$

If $1 < p \leq \infty$, let q be the exponent conjugate to p. Then $L^q(T)$ is separable. Define Λ_r as in (2), but for all $g \in L^q(T)$. Again, $\|\Lambda_r\| \leq M$. Refer to Theorems 6.16 and 11.29 to deduce, as above, that there is an $f \in L^p(T)$, with $\|f\|_p \leq M$, so that (3) holds, with $f \, d\sigma$ in place of $d\mu$, for every $g \in L^q(T)$. The rest of the proof is as it was in the case $p = 1$.

This establishes the existence assertions in (a) and (b). To prove uniqueness, it suffices to show that $P[d\mu] = 0$ implies $\mu = 0$.

Pick $f \in C(T)$, put $u = P[f]$, $v = P[d\mu]$. By Fubini's theorem, and the symmetry $P(re^{i\theta}, e^{it}) = P(re^{it}, e^{i\theta})$,

$$\int_T u_r \, d\mu = \int_T v_r f \, d\sigma \qquad (0 \leq r < 1). \tag{5}$$

When $v = 0$ then $v_r = 0$, and since $u_r \to f$ uniformly, as $r \to 1$, we conclude that

$$\int_T f \, d\mu = 0 \tag{6}$$

for every $f \in C(T)$ if $P[d\mu] = 0$. By Theorem 6.19, (6) implies that $\mu = 0$.

Finally, (c) is a corollary of (a), since $u > 0$ implies (1) with $p = 1$:

$$\int_T |u_r| \, d\sigma = \int_T u_r \, d\sigma = u(0) \qquad (0 \leq r < 1) \tag{7}$$

by the mean value property of harmonic functions. The functionals Λ_r used in the proof of (a) are now positive, hence $\mu \geq 0$. ////

11.31 Since holomorphic functions are harmonic, all of the preceding results (of which Theorems 11.16, 11.24, 11.25, 11.30 are the most significant) apply to holomorphic functions in U. This leads to the study of the H^p-spaces, a topic that will be taken up in Chap. 17.

At present we shall only give one application, to functions in the space H^∞. This, by definition, is the space of all bounded holomorphic functions in U; the norm

$$\|f\|_\infty = \sup\{|f(z)| : z \in U\}$$

turns H^∞ into a Banach space.

As before, $L^\infty(T)$ is the space of all (equivalence classes of) essentially bounded functions on T, normed by the essential supremum norm, relative to Lebesgue measure. For $g \in L^\infty(T)$, $\|g\|_\infty$ stands for the essential supremum of $|g|$.

11.32 Theorem *To every $f \in H^\infty$ corresponds a function $f^* \in L^\infty(T)$, defined almost everywhere by*

$$f^*(e^{i\theta}) = \lim_{r \to 1} f(re^{i\theta}). \tag{1}$$

The equality $\|f\|_\infty = \|f^*\|_\infty$ holds.

If $f^*(e^{i\theta}) = 0$ for almost all $e^{i\theta}$ on some arc $I \subset T$, then $f(z) = 0$ for every $z \in U$.

(A considerably stronger uniqueness theorem will be obtained later, in Theorem 15.19. See also Theorem 17.18 and Sec. 17.19.)

PROOF By Theorem 11.30, there is a unique $g \in L^\infty(T)$ such that $f = P[g]$. By Theorem 11.23, (1) holds with $f^* = g$. The inequality $\|f\|_\infty \le \|f^*\|_\infty$ follows from Theorem 11.16(1); the opposite inequality is obvious.

In particular, if $f^* = 0$ a.e., then $\|f^*\|_\infty = 0$, hence $\|f\|_\infty = 0$, hence $f = 0$.

Now choose a positive integer n so that the length of I is larger than $2\pi/n$. Let $\alpha = \exp\{2\pi i/n\}$ and define

$$F(z) = \prod_{k=1}^{n} f(\alpha^k z) \qquad (z \in U). \tag{2}$$

Then $F \in H^\infty$ and $F^* = 0$ a.e. on T, hence $F(z) = 0$ for all $z \in U$. If $Z(f)$, the zero set of f in U, were at most countable, the same would be true of $Z(F)$, since $Z(F)$ is the union of n sets obtained from $Z(f)$ by rotations. But $Z(F) = U$. Hence $f = 0$, by Theorem 10.18. ////

Exercises

1 Suppose u and v are real harmonic functions in a plane region Ω. Under what conditions is uv harmonic? (Note that the answer depends strongly on the fact that the question is one about *real* functions.) Show that u^2 cannot be harmonic in Ω, unless u is constant. For which $f \in H(\Omega)$ is $|f|^2$ harmonic?

2 Suppose f is a complex function in a region Ω, and both f and f^2 are harmonic in Ω. Prove that either f or $\bar{f}$ is holomorphic in Ω.

3 If u is a harmonic function in a region Ω, what can you say about the set of points at which the gradient of u is 0? (This is the set on which $u_x = u_y = 0$.)

4 Prove that every partial derivative of every harmonic function is harmonic.

Verify, by direct computation, that $P_r(\theta - t)$ is, for each fixed t, a harmonic function of $re^{i\theta}$. Deduce (without referring to holomorphic functions) that the Poisson integral $P[d\mu]$ of every finite Borel measure μ on T is harmonic in U, by showing that every partial derivative of $P[d\mu]$ is equal to the integral of the corresponding partial derivative of the kernel.

5 Suppose $f \in H(\Omega)$ and f has no zero in Ω. Prove that $\log |f|$ is harmonic in Ω, by computing its Laplacian. Is there an easier way?

6 Suppose $f \in H(U)$, where U is the open unit disc, f is one-to-one in U, $\Omega = f(U)$, and $f(z) = \sum c_n z^n$. Prove that the area of Ω is

$$\pi \sum_{n=1}^{\infty} n |c_n|^2.$$

Hint: The Jacobian of f is $|f'|^2$.

7 (a) If $f \in H(\Omega)$, $f(z) \neq 0$ for $z \in \Omega$, and $-\infty < \alpha < \infty$, prove that

$$\Delta(|f|^\alpha) = \alpha^2 |f|^{\alpha-2} |f'|^2,$$

by proving the formula

$$\partial \bar\partial (\psi \circ (f \bar f)) = (\varphi \circ |f|^2) \cdot |f'|^2,$$

in which ψ is twice differentiable on $(0, \infty)$ and

$$\varphi(t) = t\psi''(t) + \psi'(t).$$

(b) Assume $f \in H(\Omega)$ and Φ is a complex function with domain $f(\Omega)$, which has continuous second-order partial derivatives. Prove that

$$\Delta[\Phi \circ f] = [(\Delta \Phi) \circ f] \cdot |f'|^2.$$

Show that this specializes to the result of (a) if $\Phi(w) = \Phi(|w|)$.

8 Suppose Ω is a region, $f_n \in H(\Omega)$ for $n = 1, 2, 3, \ldots$, u_n is the real part of f_n, $\{u_n\}$ converges uniformly on compact subsets of Ω, and $\{f_n(z)\}$ converges for at least one $z \in \Omega$. Prove that then $\{f_n\}$ converges uniformly on compact subsets of Ω.

9 Suppose u is a Lebesgue measurable function in a region Ω, and u is locally in L^1. This means that the integral of $|u|$ over any compact subset of Ω is finite. Prove that u is harmonic if it satisfies the following form of the mean value property:

$$u(a) = \frac{1}{\pi r^2} \iint_{D(a;r)} u(x, y) \, dx \, dy$$

whenever $\bar D(a; r) \subset \Omega$.

10 Suppose $I = [a, b]$ is an interval on the real axis, φ is a continuous function on I, and

$$f(z) = \frac{1}{2\pi i} \int_a^b \frac{\varphi(t)}{t - z} \, dt \qquad (z \notin I).$$

Show that

$$\lim_{\epsilon \to 0} [f(x + i\epsilon) - f(x - i\epsilon)] \qquad (\epsilon > 0)$$

exists for every real x, and find it in terms of φ.

How is the result affected if we assume merely that $\varphi \in L^1$? What happens then at points x at which φ has right- and left-hand limits?

11 Suppose that $I = [a, b]$, Ω is a region, $I \subset \Omega$, f is continuous in Ω, and $f \in H(\Omega - I)$. Prove that actually $f \in H(\Omega)$.

Replace I by some other sets for which the same conclusion can be drawn.

12 (Harnack's Inequalities) Suppose Ω is a region, K is a compact subset of Ω, $z_0 \in \Omega$. Prove that there exist positive numbers α and β (depending on z_0, K, and Ω) such that

$$\alpha u(z_0) \leq u(z) \leq \beta u(z_0)$$

for every positive harmonic function u in Ω and for all $z \in K$.

If $\{u_n\}$ is a sequence of positive harmonic functions in Ω and if $u_n(z_0) \to 0$, describe the behavior of $\{u_n\}$ in the rest of Ω. Do the same if $u_n(z_0) \to \infty$. Show that the assumed positivity of $\{u_n\}$ is essential for these results.

13 Suppose u is a positive harmonic function in U and $u(0) = 1$. How large can $u(\frac{1}{2})$ be? How small? Get the best possible bounds.

14 For which pairs of lines L_1, L_2 do there exist real functions, harmonic in the whole plane, that are 0 at all points of $L_1 \cup L_2$ without vanishing identically?

15 Suppose u is a positive harmonic function in U, and $u(re^{i\theta}) \to 0$ as $r \to 1$, for every $e^{i\theta} \neq 1$. Prove that there is a constant c such that

$$u(re^{i\theta}) = cP_r(\theta).$$

16 Here is an example of a harmonic function in U which is not identically 0 but *all* of whose radial limits are 0:

$$u(z) = \mathrm{Im}\left[\left(\frac{1+z}{1-z}\right)^2\right].$$

Prove that this u is not the Poisson integral of any measure on T and that it is not the difference of two positive harmonic functions in U.

17 Let Φ be the set of all positive harmonic functions u in U such that $u(0) = 1$. Show that Φ is a convex set and find the extreme points of Φ. (A point x in a convex set Φ is called an *extreme point* of Φ if x lies on no segment both of whose end points lie in Φ and are different from x.) *Hint*: If C is the convex set whose members are the positive Borel measures on T, of total variation 1, show that the extreme points of C are precisely those $\mu \in C$ whose supports consist of only one point of T.

18 Let X^* be the dual space of the Banach space X. A sequence $\{\Lambda_n\}$ in X^* is said to *converge weakly* to $\Lambda \in X^*$ if $\Lambda_n x \to \Lambda x$ as $n \to \infty$, for every $x \in X$. Note that $\Lambda_n \to \Lambda$ weakly whenever $\Lambda_n \to \Lambda$ in the norm of X^*. (See Exercise 8, Chap. 5.) The converse need not be true. For example, the functionals $f \to \hat{f}(n)$ on $L^2(T)$ tend to 0 weakly (by the Bessel inequality), but each of these functionals has norm 1.

Prove that $\{\|\Lambda_n\|\}$ must be bounded if $\{\Lambda_n\}$ converges weakly.

19 (a) Show that $\delta P_r(\delta) > 1$ if $\delta = 1 - r$.

(b) If $\mu \geq 0$, $u = P[d\mu]$, and $I_\delta \subset T$ is the arc with center 1 and length 2δ, show that

$$\mu(I_\delta) \leq \delta u(1-\delta)$$

and that therefore

$$(M\mu)(1) \leq \pi(M_{\mathrm{rad}} u)(1).$$

(c) If, furthermore, $\mu \perp m$, show that

$$u(re^{i\theta}) \to \infty \quad \text{a.e. } [\mu].$$

Hint: Use Theorem 7.15.

20 Suppose $E \subset T$, $m(E) = 0$. Prove that there is an $f \in H^\infty$, with $f(0) = 1$, that has

$$\lim_{r \to 1} f(re^{i\theta}) = 0$$

at every $e^{i\theta} \in E$.

Suggestion: Find a lower semicontinuous $\psi \in L^1(T)$, $\psi > 0$, $\psi = +\infty$ at every point of E. There is a holomorphic g whose real part is $P[\psi]$. Let $f = 1/g$.

21 Define $f \in H(U)$ and $g \in H(U)$ by $f(z) = \exp\{(1+z)/(1-z)\}$, $g(z) = (1-z)\exp\{-f(z)\}$. Prove that

$$g^*(e^{i\theta}) = \lim_{r \to 1} g(re^{i\theta})$$

exists at every $e^{i\theta} \in T$, that $g^* \in C(T)$, but that g is not in H^∞.

Suggestion: Fix s, put

$$z_t = \frac{t + is - 1}{t + is + 1} \qquad (0 < t < \infty).$$

For certain values of s, $|g(z_t)| \to \infty$ as $t \to \infty$.

22 Suppose u is harmonic in U, and $\{u_r : 0 \le r < 1\}$ is a uniformly integrable subset of $L^1(T)$. (See Exercise 10, Chap. 6.) Modify the proof of Theorem 11.30 to show that $u = P[f]$ for some $f \in L^1(T)$.

23 Put $\theta_n = 2^{-n}$ and define

$$u(z) = \sum_{n=1}^{\infty} n^{-2}\{P(z, e^{i\theta_n}) - P(z, e^{-i\theta_n})\},$$

for $z \in U$. Show that u is the Poisson integral of a measure on T, that $u(x) = 0$ if $-1 < x < 1$, but that

$$u(1 - \epsilon + i\epsilon)$$

is unbounded, as ϵ decreases to 0. (Thus u has a radial limit, but no nontangential limit, at 1.)

Hint: If $\epsilon = \sin \theta$ is small and $z = 1 - \epsilon + i\epsilon$, then

$$P(z, e^{i\theta}) - P(z, e^{-i\theta}) > 1/\epsilon.$$

24 Let $D_n(t)$ be the Dirichlet kernel, as in Sec. 5.11, define the Fejér kernel by

$$K_N = \frac{1}{N+1}(D_0 + D_1 + \cdots + D_N),$$

put $L_N(t) = \min(N, \pi^2/Nt^2)$. Prove that

$$K_{N-1}(t) = \frac{1}{N} \cdot \frac{1 - \cos Nt}{1 - \cos t} \le L_N(t)$$

and that $\int_T L_N \, d\sigma \le 2$.

Use this to prove that the arithmetic means

$$\sigma_N = \frac{S_0 + S_1 + \cdots + S_N}{N+1}$$

of the partial sums s_n of the Fourier series of a function $f \in L^1(T)$ converge to $f(e^{i\theta})$ at every Lebesgue point of f. (Show that $\sup |\sigma_N|$ is dominated by Mf, then proceed as in the proof of Theorem 11.23.)

25 If $1 \le p \le \infty$ and $f \in L^p(R^1)$, prove that $(f * h_\lambda)(x)$ is a harmonic function of $x + i\lambda$ in the upper half plane. (h_λ is defined in Sec. 9.7; it is the Poisson kernel for the half plane.)

CHAPTER
TWELVE

THE MAXIMUM MODULUS PRINCIPLE

Introduction

12.1 The maximum modulus theorem (10.24) asserts that the constants are the only homomorphic functions in a region Ω whose absolute values have a local maximum at any point of Ω.

Here is a restatement: *If K is the closure of a bounded region Ω, if f is continuous on K and holomorphic in Ω, then*

$$|f(z)| \leq \|f\|_{\partial\Omega} \tag{1}$$

for every $z \in \Omega$. If equality holds at one point $z \in \Omega$, then f is constant.

[The right side of (1) is the supremum of $|f|$ on the boundary $\partial\Omega$ of Ω.]

For if $|f(z)| \geq \|f\|_{\partial\Omega}$ at some $z \in \Omega$, then the maximum of $|f|$ on K (which is attained at some point of K, since K is compact) is actually attained at some point of Ω, so f is constant, by Theorem 10.24.

The equality $\|f\|_\infty = \|f^*\|_\infty$, which is part of Theorem 11.32, implies that

$$|f(z)| \leq \|f^*\|_\infty \qquad (z \in U, f \in H^\infty(U)). \tag{2}$$

This says (roughly speaking) that $|f(z)|$ is no larger than the supremum of the boundary values of f, a statement similar to (1). But this time boundedness on U is enough; we do not need continuity on $\bar{U}$.

This chapter contains further generalizations of the maximum modulus theorem, as well as some rather striking applications of it, and it concludes with a theorem which shows that the maximum property "almost" characterizes the class of holomorphic functions.

The Schwarz Lemma

This is the name usually given to the following theorem. We use the notation established in Sec. 11.31.

12.2 Theorem *Suppose $f \in H^\infty$, $\|f\|_\infty \leq 1$, and $f(0) = 0$. Then*

$$|f(z)| \leq |z| \qquad (z \in U), \tag{1}$$

$$|f'(0)| \leq 1; \tag{2}$$

if equality holds in (1) for one $z \in U - \{0\}$, or if equality holds in (2), then $f(z) = \lambda z$, where λ is a constant, $|\lambda| = 1$.

In geometric language, the hypothesis is that f is a holomorphic mapping of U into U which keeps the origin fixed; part of the conclusion is that either f is a rotation or f moves each $z \in U - \{0\}$ closer to the origin than it was.

PROOF Since $f(0) = 0$, $f(z)/z$ has a removable singularity at $z = 0$. Hence there exists $g \in H(U)$ such that $f(z) = zg(z)$. If $z \in U$ and $|z| < r < 1$, then

$$|g(z)| \leq \max_\theta \frac{|f(re^{i\theta})|}{r} \leq \frac{1}{r}.$$

Letting $r \to 1$, we see that $|g(z)| \leq 1$ at every $z \in U$. This gives (1). Since $f'(0) = g(0)$, (2) follows. If $|g(z)| = 1$ for some $z \in U$, then g is constant, by another application of the maximum modulus theorem.   ////

Many variants of the Schwarz lemma can be obtained with the aid of the following mappings of U onto U:

12.3 Definition *For any $\alpha \in U$, define*

$$\varphi_\alpha(z) = \frac{z - \alpha}{1 - \bar\alpha z}.$$

12.4 Theorem *Fix $\alpha \in U$. Then φ_α is a one-to-one mapping which carries T onto T, U onto U, and α to 0. The inverse of φ_α is $\varphi_{-\alpha}$. We have*

$$\varphi'_\alpha(0) = 1 - |\alpha|^2, \qquad \varphi'_\alpha(\alpha) = \frac{1}{1 - |\alpha|^2}. \tag{1}$$

PROOF φ_α is holomorphic in the whole plane, except for a pole at $1/\bar\alpha$ which lies outside $\bar U$. Straightforward substitution shows that

$$\varphi_{-\alpha}(\varphi_\alpha(z)) = z. \tag{2}$$

Thus φ_α is one-to-one, and $\varphi_{-\alpha}$ is its inverse. Since, for real t,

$$\left|\frac{e^{it} - \alpha}{1 - \bar{\alpha}e^{it}}\right| = \frac{|e^{it} - \alpha|}{|e^{-it} - \bar{\alpha}|} = 1 \tag{3}$$

(z and $\bar{z}$ have the same absolute value), φ_α maps T into T; the same is true of $\varphi_{-\alpha}$; hence $\varphi_\alpha(T) = T$. It now follows from the maximum modulus theorem that $\varphi_\alpha(U) \subset U$, and consideration of $\varphi_{-\alpha}$ shows that actually $\varphi_\alpha(U) = U$. ////

12.5 An Extremal Problem Suppose α and β are complex numbers, $|\alpha| < 1$, and $|\beta| < 1$. *How large can $|f'(\alpha)|$ be if f is subject to the conditions $f \in H^\infty$, $\|f\|_\infty \leq 1$, and $f(\alpha) = \beta$?*

To solve this, put

$$g = \varphi_\beta \circ f \circ \varphi_{-\alpha}. \tag{1}$$

Since $\varphi_{-\alpha}$ and φ_β map U onto U, we see that $g \in H^\infty$ and $\|g\|_\infty \leq 1$; also, $g(0) = 0$. The passage from f to g has reduced our problem to the Schwarz lemma, which gives $|g'(0)| \leq 1$. By (1), the chain rule gives

$$g'(0) = \varphi'_\beta(\beta) f'(\alpha) \varphi'_{-\alpha}(0). \tag{2}$$

If we use Eqs. 12.4(1), we obtain the inequality

$$|f'(\alpha)| \leq \frac{1 - |\beta|^2}{1 - |\alpha|^2}. \tag{3}$$

This solves our problem, since equality can occur in (3). This happens if and only if $|g'(0)| = 1$, in which case g is a rotation (Theorem 12.2), so that

$$f(z) = \varphi_{-\beta}(\lambda \varphi_\alpha(z)) \qquad (z \in U) \tag{4}$$

for some constant λ with $|\lambda| = 1$.

A remarkable feature of the solution should be stressed. We imposed no smoothness conditions (such as continuity on $\bar{U}$, for instance) on the behavior of f near the boundary of U. Nevertheless, it turns out that the functions f which maximize $|f'(\alpha)|$ under the stated restrictions are actually rational functions. Note also that these extremal functions map U onto U (not just into) and that they are one-to-one. This observation may serve as the motivation for the proof of the Riemann mapping theorem in Chap. 14.

At present, we shall merely show how this extremal problem can be used to characterize the one-to-one holomorphic mappings of U onto U.

12.6 Theorem *Suppose $f \in H(U)$, f is one-to-one, $f(U) = U$, $\alpha \in U$, and $f(\alpha) = 0$. Then there is a constant λ, $|\lambda| = 1$, such that*

$$f(z) = \lambda \varphi_\alpha(z) \qquad (z \in U). \tag{1}$$

In other words, we obtain f by composing the mapping φ_α with a rotation.

PROOF Let g be the inverse of f, defined by $g(f(z)) = z$, $z \in U$. Since f is one-to-one, f' has no zero in U, so $g \in H(U)$, by Theorem 10.33. By the chain rule,

$$g'(0)f'(\alpha) = 1. \tag{2}$$

The solution of 12.5, applied to f and to g, yields the inequalities

$$|f'(\alpha)| \leq \frac{1}{1-|\alpha|^2}, \qquad |g'(0)| \leq 1 - |\alpha|^2. \tag{3}$$

By (2), equality must hold in (3). As we observed in the preceding problem (with $\beta = 0$), this forces f to satisfy (1). ////

The Phragmen-Lindelöf Method

12.7 For a bounded region Ω, we saw in Sec. 12.1 that if f is continuous on the closure of Ω and if $f \in H(\Omega)$, the maximum modulus theorem implies

$$\|f\|_\Omega = \|f\|_{\partial\Omega}. \tag{1}$$

For unbounded regions, this is no longer true.

To see an example, let

$$\Omega = \left\{ z = x + iy : -\frac{\pi}{2} < y < \frac{\pi}{2} \right\}; \tag{2}$$

Ω is the open strip bounded by the parallel lines $y = \pm\pi/2$; its boundary $\partial\Omega$ is the union of these two lines. Put

$$f(z) = \exp(\exp(z)). \tag{3}$$

For real x,

$$f\left(x \pm \frac{\pi i}{2}\right) = \exp(\pm ie^x) \tag{4}$$

since $\exp(\pi i/2) = i$, so $|f(z)| = 1$ for $z \in \partial\Omega$. But $f(z) \to \infty$ very rapidly as $x \to \infty$ along the positive real axis, which lies in Ω.

"Very" is the key word in the preceding sentence. A method developed by Phragmen and Lindelöf makes it possible to prove theorems of the following kind: If $f \in H(\Omega)$ and if $|f| < g$, where $g(z) \to \infty$ "slowly" as $z \to \infty$ in Ω (just what "slowly" means depends on Ω), then f is actually bounded in Ω, and this usually implies further conclusions about f, by the maximum modulus theorem.

Rather than describe the method by a theorem which would cover a large number of situations, we shall show how it works in two cases. In both, Ω will be a strip. In the first, f will be assumed to be bounded, and the theorem will improve the bound; in the second, a growth condition will be imposed on f which just excludes the function (3). In view of later applications, Ω will be a vertical strip in Theorem 12.8.

First, however, let us mention another example which also has this general flavor: *Suppose f is an entire function and*

$$|f(z)| < 1 + |z|^{1/2} \tag{5}$$

for all z. Then f is constant.

This follows immediately from the Cauchy estimates 10.26, since they show that $f^{(n)}(0) = 0$ for $n = 1, 2, 3, \ldots$.

12.8 Theorem *Suppose*

$$\Omega = \{x + iy : a < x < b\}, \qquad \bar{\Omega} = \{x + iy : a \le x \le b\}, \tag{1}$$

f is continuous on $\bar{\Omega}$, $f \in H(\Omega)$, and suppose that $|f(z)| < B$ for all $z \in \bar{\Omega}$ and some fixed $B < \infty$. If

$$M(x) = \sup\{|f(x + iy)| : -\infty < y < \infty\} \qquad (a \le x \le b) \tag{2}$$

then we actually have

$$M(x)^{b-a} \le M(a)^{b-x} M(b)^{x-a} \qquad (a < x < b). \tag{3}$$

Note: The conclusion (3) implies that the inequality $|f| < B$ can be replaced by $|f| \le \max(M(a), M(b))$, so that $|f|$ is no larger in Ω than the supremum of $|f|$ on the *boundary* of Ω.

If we apply the theorem to strips bounded by lines $x = \alpha$ and $x = \beta$, where $a \le \alpha < \beta \le b$, the conclusion can be stated in the following way:

Corollary *Under the hypotheses of the theorem, $\log M$ is a convex function on (a, b).*

PROOF We assume first that $M(a) = M(b) = 1$. In this case we have to prove that $|f(z)| \le 1$ for all $z \in \bar{\Omega}$.

For each $\epsilon > 0$, we define an auxiliary function

$$h_\epsilon(z) = \frac{1}{1 + \epsilon(z - a)} \qquad (z \in \bar{\Omega}). \tag{4}$$

Since $\operatorname{Re}\{1 + \epsilon(z - a)\} = 1 + \epsilon(x - a) \ge 1$ in $\bar{\Omega}$, we have $|h_\epsilon| \le 1$ in $\bar{\Omega}$, so that

$$|f(z)h_\epsilon(z)| \le 1 \qquad (z \in \partial\Omega). \tag{5}$$

Also, $|1 + \epsilon(z - a)| \ge \epsilon|y|$, so that

$$|f(z)h_\epsilon(z)| \le \frac{B}{\epsilon|y|} \qquad (z = x + iy \in \bar{\Omega}). \tag{6}$$

Let R be the rectangle cut off from $\bar{\Omega}$ by the lines $y = \pm B/\epsilon$. By (5) and (6), $|fh_\epsilon| \le 1$ on ∂R, hence $|fh_\epsilon| \le 1$ on R, by the maximum modulus theorem. But (6) shows that $|fh_\epsilon| \le 1$ on the rest of $\bar{\Omega}$. Thus $|f(z)h_\epsilon(z)| \le 1$

for all $z \in \Omega$ and all $\epsilon > 0$. If we fix $z \in \Omega$ and then let $\epsilon \to 0$, we obtain the desired result $|f(z)| \leq 1$.

We now turn to the general case. Put

$$g(z) = M(a)^{(b-z)/(b-a)} M(b)^{(z-a)/(b-a)}, \tag{7}$$

where, for $M > 0$ and w complex, M^w is defined by

$$M^w = \exp(w \log M), \tag{8}$$

and $\log M$ is real. Then g is entire, g has no zero, $1/g$ is bounded in $\bar{\Omega}$,

$$|g(a + iy)| = M(a), \qquad |g(b + iy)| = M(b), \tag{9}$$

and hence f/g satisfies our previous assumptions. Thus $|f/g| \leq 1$ in Ω, and this gives (3). (See Exercise 7.) ////

12.9 Theorem *Suppose*

$$\Omega = \left\{x + iy : |y| < \frac{\pi}{2}\right\}, \qquad \bar{\Omega} = \left\{x + iy : |y| \leq \frac{\pi}{2}\right\}. \tag{1}$$

Suppose f is continuous on $\bar{\Omega}$, $f \in H(\Omega)$, there are constants $\alpha < 1$, $A < \infty$, such that

$$|f(z)| < \exp\{A \exp(\alpha |x|)\} \qquad (z = x + iy \in \Omega), \tag{2}$$

and

$$\left|f\left(x \pm \frac{\pi i}{2}\right)\right| \leq 1 \qquad (-\infty < x < \infty). \tag{3}$$

Then $|f(z)| \leq 1$ for all $z \in \Omega$.

Note that the conclusion does not follow if $\alpha = 1$, as is shown by the function $\exp(\exp z)$.

PROOF Choose $\beta > 0$ so that $\alpha < \beta < 1$. For $\epsilon > 0$, define

$$h_\epsilon(z) = \exp\{-\epsilon(e^{\beta z} + e^{-\beta z})\}. \tag{4}$$

For $z \in \bar{\Omega}$,

$$\operatorname{Re}[e^{\beta z} + e^{-\beta z}] = (e^{\beta x} + e^{-\beta x}) \cos \beta y \geq \delta(e^{\beta x} + e^{-\beta x}) \tag{5}$$

where $\delta = \cos(\beta \pi/2) > 0$, since $|\beta| < 1$. Hence

$$|h_\epsilon(z)| \leq \exp\{-\epsilon\delta(e^{\beta x} + e^{-\beta x})\} < 1 \qquad (z \in \bar{\Omega}). \tag{6}$$

It follows that $|fh_\epsilon| \leq 1$ on $\partial \Omega$ and that

$$|f(z)h_\epsilon(z)| \leq \exp\{Ae^{\alpha|x|} - \epsilon\delta(e^{\beta x} + e^{-\beta x})\} \qquad (z \in \bar{\Omega}). \tag{7}$$

Fix $\epsilon > 0$. Since $\epsilon\delta > 0$ and $\beta > \alpha$, the exponent in (7) tends to $-\infty$ as $x \to \pm\infty$. Hence there exists an x_0 so that the right side of (7) is less than 1 if $|x| \geq x_0$. Since $|fh_\epsilon| \leq 1$ on the boundary of the rectangle whose vertices are $\pm x_0 \pm (\pi i/2)$, the maximum modulus theorem shows that actually $|fh_\epsilon| \leq 1$ on this rectangle. Thus $|fh_\epsilon| \leq 1$ at every point of Ω, for every $\epsilon > 0$. As $\epsilon \to 0$, $h_\epsilon(z) \to 1$ for every z, so we conclude that $|f(z)| \leq 1$ for all $z \in \Omega$.

////

Here is a slightly different application of the same method. It will be used in the proof of Theorem 14.18.

12.10 Lindelöf's Theorem *Suppose Γ is a curve, with parameter interval $[0, 1]$, such that $|\Gamma(t)| < 1$ if $t < 1$ and $\Gamma(1) = 1$. If $g \in H^\infty$ and*

$$\lim_{t \to 1} g(\Gamma(t)) = L, \tag{1}$$

then g has radial limit L at 1.

(It follows from Exercise 14, Chap. 14, that g actually has nontangential limit L at 1.)

PROOF Assume $|g| < 1$, $L = 0$, without loss of generality. Let $\epsilon > 0$ be given. There exists $t_0 < 1$ so that, setting $r_0 = \operatorname{Re} \Gamma(t_0)$, we have

$$|g(\Gamma(t))| < \epsilon \quad \text{and} \quad \operatorname{Re} \Gamma(t) > r_0 > \frac{1}{2} \tag{2}$$

as soon as $t_0 < t < 1$.

Pick r, $r_0 < r < 1$.

Define h in $\Omega = D(0; 1) \cap D(2r; 1)$ by

$$h(z) = g(z)\overline{g(\bar{z})}g(2r - z)\overline{g(2r - \bar{z})}. \tag{3}$$

Then $h \in H(\Omega)$ and $|h| < 1$. We claim that

$$|h(r)| < \epsilon. \tag{4}$$

Since $h(r) = |g(r)|^4$, the theorem follows from (4).

To prove (4), let $E_1 = \Gamma([t_1, 1])$, where t_1 is the largest t for which $\operatorname{Re} \Gamma(t) = r$, let E_2 be the reflection of E_1 in the real axis, and let E be the union of $E_1 \cup E_2$ and its reflection in the line $x = r$. Then (2) and (3) imply that

$$|h(z)| < \epsilon \quad \text{if} \quad z \in \Omega \cap E. \tag{5}$$

Pick $c > 0$, define

$$h_c(z) = h(z)(1 - z)^c(2r - 1 - z)^c \tag{6}$$

for $z \in \Omega$, and put $h_c(1) = h_c(2r - 1) = 0$. If K is the union of E and the bounded components of the complement of E, then K is compact, h_c is continuous on K, holomorphic in the interior of K, and (5) implies that $|h_c| < \epsilon$ on the boundary of K. Since the construction of E shows that $r \in K$, the maximum modulus theorem implies that $|h_c(r)| < \epsilon$.

Letting $c \to 0$, we obtain (4). ////

An Interpolation Theorem

12.11 The convexity theorem 12.8 can sometimes be used to prove that certain linear transformations are bounded with respect to certain L^p-norms. Rather than discuss this in full generality, let us look at a particular situation of this kind.

Let X be a measure space, with a positive measure μ, and suppose $\{\psi_n\}$ ($n = 1, 2, 3, \ldots$) is an orthonormal set of functions in $L^2(\mu)$; we recall what this means:

$$\int_X \psi_n \bar{\psi}_m \, d\mu = \begin{cases} 1 & \text{if } m = n, \\ 0 & \text{if } m \neq n. \end{cases} \tag{1}$$

Let us also assume that $\{\psi_n\}$ is a bounded sequence in $L^\infty(\mu)$: There exists an $M < \infty$ such that

$$|\psi_n(x)| \leq M \quad (n = 1, 2, 3, \ldots; x \in X). \tag{2}$$

Then for any $f \in L^p(\mu)$, where $1 \leq p \leq 2$, the integrals

$$\hat{f}(n) = \int_X f \bar{\psi}_n \, d\mu \quad (n = 1, 2, 3, \ldots) \tag{3}$$

exist and define a function $\hat{f}$ on the set of all positive integers.

There are now two very easy theorems: For $f \in L^1(\mu)$, (2) gives

$$\|\hat{f}\|_\infty \leq M \|f\|_1, \tag{4}$$

and for $f \in L^2(\mu)$, the Bessel inequality gives

$$\|\hat{f}\|_2 \leq \|f\|_2, \tag{5}$$

where the norms are defined as usual:

$$\|f\|_p = \left[\int |f|^p \, d\mu\right]^{1/p}, \quad \|\hat{f}\|_q = \left[\sum |\hat{f}(n)|^q\right]^{1/q}, \tag{6}$$

and $\|\hat{f}\|_\infty = \sup_n |\hat{f}(n)|$.

Since $(1, \infty)$ and $(2, 2)$ are pairs of conjugate exponents, one may conjecture that $\|\hat{f}\|_q$ is finite whenever $f \in L^p(\mu)$ and $1 < p < 2$, $q = p/(p - 1)$. This is indeed true and can be proved by "interpolation" between the preceding trivial cases $p = 1$ and $p = 2$.

12.12 The Hausdorff-Young Theorem *Under the above assumptions, the inequality*

$$\|\hat{f}\|_q \le M^{(2-p)/p}\|f\|_p \tag{1}$$

holds if $1 \le p \le 2$ and if $f \in L^p(\mu)$.

PROOF We first prove a reduced form of the theorem.

Fix p, $1 < p < 2$. Let f be a *simple* complex function such that $\|f\|_p = 1$, and let $b_1, \ldots, b_N$ be complex numbers such that $\sum |b_n|^{p'} = 1$. Our objective is the inequality

$$\left| \sum_{n=1}^{N} b_n \hat{f}(n) \right| \le M^{(2-p)/p}. \tag{2}$$

Put $F = |f|^p$, and put $B_n = |b_n|^{p'}$. Then there is a function φ and there are complex numbers $\beta_1, \ldots, \beta_N$ such that

$$f = F^{1/p}\varphi, \quad |\varphi| = 1, \quad \int_X F \, d\mu = 1, \tag{3}$$

and

$$b_n = B_n^{1/p'}\beta_n, \quad |\beta_n| = 1, \quad \sum_{n=1}^N B_n = 1. \tag{4}$$

If we use these relations and the definition of $\hat{f}(n)$ given in Sec. 12.11, we obtain

$$\sum_{n=1}^N b_n \hat{f}(n) = \sum_{n=1}^N B_n^{1/p'}\beta_n \int_X F^{1/p}\varphi\bar{\psi}_n \, d\mu. \tag{5}$$

Now replace $1/p$ by z in (5), and define

$$\Phi(z) = \sum_{n=1}^N B_n^z \beta_n \int_X F^z \varphi\bar{\psi}_n \, d\mu \tag{6}$$

for any complex number z. Recall that $A^z = \exp(z \log A)$ if $A > 0$; if $A = 0$, we agree that $A^z = 0$. Since F is simple, since $F \ge 0$, and since $B_n \ge 0$, we see that Φ is a finite linear combination of such exponentials, so Φ is an entire function which is bounded on

$$\{z: a \le \text{Re}(z) \le b\}$$

for any finite a and b. We shall take $a = \frac{1}{2}$ and $b = 1$, shall estimate Φ on the edges of this strip, and shall then apply Theorem 12.8 to estimate $\Phi(1/p)$.

For $-\infty < y < \infty$, define

$$c_n(y) = \int_X F^{1/2} F^{iy} \varphi\bar{\psi}_n \, d\mu. \tag{7}$$

The Bessel inequality gives

$$\sum_{n=1}^{N} |c_n(y)|^2 \le \int_X |F^{1/2} F^{iy} \varphi|^2 \, d\mu = \int_X |F| \, d\mu = 1, \tag{8}$$

and then the Schwarz inequality shows that

$$|\Phi(\tfrac{1}{2} + iy)| = \left| \sum_{n=1}^{N} B_n^{1/2} B_n^{iy} \beta_n c_n \right| \le \left\{ \sum_{n=1}^{N} B_n \cdot \sum_{n=1}^{N} |c_n|^2 \right\}^{1/2} \le 1. \tag{9}$$

The estimate

$$|\Phi(1 + iy)| \le M \quad (-\infty < y < \infty) \tag{10}$$

follows trivially from (3), (4), and (6), since $\|\psi_n\|_\infty \le M$.

We now conclude from (9), (10), and Theorem 12.8 that

$$|\Phi(x + iy)| \le M^{2x-1} \quad (\tfrac{1}{2} \le x \le 1, -\infty < y < \infty). \tag{11}$$

With $x = 1/p$ and $y = 0$, this gives the desired inequality (2).

The proof is now easily completed. Note first that

$$\left\{ \sum_{n=1}^{N} |\hat{f}(n)|^q \right\}^{1/q} = \sup \left| \sum_{n=1}^{N} b_n \hat{f}(n) \right|, \tag{12}$$

the supremum being taken over all $\{b_1, \ldots, b_N\}$ with $\sum |b_n|^p = 1$, since the L^q norm of any function on any measure space is equal to its norm as a linear functional on L^p. Hence (2) shows that

$$\left\{ \sum_{n=1}^{N} |\hat{f}(n)|^q \right\}^{1/q} \le M^{(2-p)/p} \|f\|_p \tag{13}$$

for every simple complex $f \in L^p(\mu)$.

If now $f \in L^p(\mu)$, there are simple functions f_j such that $\|f_j - f\|_p \to 0$ as $j \to \infty$. Then $\hat{f}_j(n) \to \hat{f}(n)$ for every n, because $\psi_n \in L^q(\mu)$. Thus since (13) holds for each f_j, it also holds for f. Since N was arbitrary, we finally obtain (1). ////

A Converse of the Maximum Modulus Theorem

We now come to the theorem which was alluded to in the introduction of the present chapter.

The letter j will denote the identity function: $j(z) = z$.

The function which assigns the number 1 to each $z \in \bar{U}$ will be denoted by **1**.

12.13 Theorem *Suppose M is a vector space of continuous complex functions on the closed unit disc $\bar{U}$, with the following properties:*

(a) $\mathbf{1} \in M$.
(b) *If $f \in M$, then also $jf \in M$.*
(c) *If $f \in M$, then $\|f\|_U = \|f\|_T$.*

Then every $f \in M$ is holomorphic in U.

Note that (c) is a rather weak form of the maximum modulus principle; (c) asserts only that the overall maximum of $|f|$ on $\bar{U}$ is attained at some point of the boundary T, but (c) does not a priori exclude the existence of *local* maxima of $|f|$ in U.

PROOF By (a) and (b), M contains all polynomials. In conjunction with (c), this shows that M satisfies the hypotheses of Theorem 5.25. Thus every $f \in M$ is harmonic in U. We shall use (b) to show that every $f \in M$ actually satisfies the Cauchy-Riemann equation.

Let ∂ and $\bar{\partial}$ be the differential operators introduced in Sec. 11.1. The product rule for differentiation gives

$$(\partial\bar{\partial})(fg) = f \cdot (\partial\bar{\partial}g) + (\partial f) \cdot (\bar{\partial}g) + (\bar{\partial}f) \cdot (\partial g) + (\partial\bar{\partial}f) \cdot g.$$

Fix $f \in M$, and take $g = j$. Then $fj \in M$. Hence f and fj are harmonic, so $\partial\bar{\partial}f = 0$ and $(\partial\bar{\partial})(fj) = 0$. Also, $\bar{\partial}j = 0$ and $\partial j = 1$. The above identity therefore reduces to $\bar{\partial}f = 0$. Thus $f \in H(U)$. ////

This result will be used in the following proof.

12.14 Radó's Theorem *Assume $f \in C(\bar{U})$, Ω is the set of all $z \in U$ at which $f(z) \neq 0$, and f is holomorphic in Ω. Then f is holomorphic in U.*

In particular, the theorem asserts that $U - \Omega$ is at most countable, unless $\Omega = \varnothing$.

PROOF Assume $\Omega \neq \varnothing$. We shall first prove that Ω is dense in U. If not, there exist $\alpha \in \Omega$ and $\beta \in U - \bar{\Omega}$ such that $2|\beta - \alpha| < 1 - |\beta|$. Choose n so that $2^n |f(\alpha)| > \|f\|_T$. Define $h(z) = (z - \beta)^{-n} f(z)$, for $z \in \bar{\Omega}$. If $z \in U \cap \partial\Omega$, then $f(z) = 0$, hence $h(z) = 0$. If $z \in T \cap \partial\Omega$, then

$$|h(z)| \leq (1 - |\beta|)^{-n} \|f\|_T < |\alpha - \beta|^{-n} |f(\alpha)| = |h(\alpha)|.$$

This contradicts the maximum modulus theorem.

Thus Ω is dense in U.

Next, let M be the vector space of all $g \in C(\bar{U})$ that are holomorphic in Ω. Fix $g \in M$. For $n = 1, 2, 3, \ldots$, $fg^n = 0$ on $U \cap \partial\Omega$. The maximum modulus theorem implies therefore, for every $\alpha \in \Omega$, that

$$|f(\alpha)| \, |g(\alpha)|^n \leq \|fg^n\|_{\partial\Omega} = \|fg^n\|_T \leq \|f\|_T \|g\|_T^n.$$

If we take nth roots and then let $n \to \infty$, we see that $|g(\alpha)| \leq \|g\|_T$, for every $\alpha \in \Omega$. Since Ω is dense in U, $\|g\|_U = \|g\|_T$.

It follows that M satisfies the hypotheses of Theorem 12.13. Since $f \in M$, f is holomorphic in U. ////

Exercises

1 Suppose Δ is a closed equilateral triangle in the plane, with vertices a, b, c. Find max $(|z - a| |z - b| |z - c|)$ as z ranges over Δ.

2 Suppose $f \in H(\Pi^+)$, where Π^+ is the upper half plane, and $|f| \leq 1$. How large can $|f'(i)|$ be? Find the extremal functions. (Compare the discussion in Sec. 12.5.)

3 Suppose $f \in H(\Omega)$. Under what conditions can $|f|$ have a local *minimum* in Ω?

4 (a) Suppose Ω is a region, D is a disc, $\bar{D} \subset \Omega$, $f \in H(\Omega)$, f is not constant, and $|f|$ is constant on the boundary of D. Prove that f has at least one zero in D.
 (b) Find all entire functions f such that $|f(z)| = 1$ whenever $|z| = 1$.

5 Suppose Ω is a bounded region, $\{f_n\}$ is a sequence of continuous functions on $\bar{\Omega}$ which are holomorphic in Ω, and $\{f_n\}$ converges uniformly on the boundary of Ω. Prove that $\{f_n\}$ converges uniformly on $\bar{\Omega}$.

6 Suppose $f \in H(\Omega)$, Γ is a cycle in Ω such that $\text{Ind}_\Gamma(\alpha) = 0$ for all $\alpha \notin \Omega$, $|f(\zeta)| \leq 1$ for every $\zeta \in \Gamma^*$, and $\text{Ind}_\Gamma(z) \neq 0$. Prove that $|f(z)| \leq 1$.

7 In the proof of Theorem 12.8 it was tacitly assumed that $M(a) > 0$ and $M(b) > 0$. Show that the theorem is true if $M(a) = 0$, and that then $f(z) = 0$ for all $z \in \Omega$.

8 If $0 < R_1 < R_2 < \infty$, let $A(R_1, R_2)$ denote the annulus

$$\{z : R_1 < |z| < R_2\}.$$

There is a vertical strip which the exponential function maps onto $A(R_1, R_2)$. Use this to prove Hadamard's *three-circle theorem*: If $f \in H(A(R_1, R_2))$, if

$$M(r) = \max_\theta |f(re^{i\theta})| \qquad (R_1 < r < R_2),$$

and if $R_1 < a < r < b < R_2$, then

$$\log M(r) \leq \frac{\log (b/r)}{\log (b/a)} \log M(a) + \frac{\log (r/a)}{\log (b/a)} \log M(b).$$

[In other words, $\log M(r)$ is a convex function of $\log r$.] For which f does equality hold in this inequality?

9 Let Π be the open right half plane ($z \in \Pi$ if and only if Re $z > 0$). Suppose f is continuous on the closure of Π (Re $z \geq 0$), $f \in H(\Pi)$, and there are constants $A < \infty$ and $\alpha < 1$ such that

$$|f(z)| < A \exp(|z|^\alpha)$$

for all $z \in \Pi$. Furthermore, $|f(iy)| \leq 1$ for all real y. Prove that $|f(z)| \leq 1$ in Π.

Show that the conclusion is false for $\alpha = 1$.

How does the result have to be modified if Π is replaced by a region bounded by two rays through the origin, at an angle not equal to π?

10 Let Π be the open right half plane. Suppose that $f \in H(\Pi)$, that $|f(z)| < 1$ for all $z \in \Pi$, and that there exists α, $-\pi/2 < \alpha < \pi/2$, such that

$$\frac{\log |f(re^{i\alpha})|}{r} \to -\infty \quad \text{as} \quad r \to \infty.$$

Prove that $f = 0$.

Hint: Put $g_n(z) = f(z)e^{nz}$, $n = 1, 2, 3, \ldots$. Apply Exercise 9 to the two angular regions defined by $-\pi/2 < \theta < \alpha$, $\alpha < \theta < \pi/2$. Conclude that each g_n is bounded in Π, and hence that $|g_n| < 1$ in Π, for all n.

11 Suppose Γ is the boundary of an unbounded region Ω, $f \in H(\Omega)$, f is continuous on $\Omega \cup \Gamma$, and there are constants $B < \infty$ and $M < \infty$ such that $|f| \leq M$ on Γ and $|f| \leq B$ in Ω. Prove that we then actually have $|f| \leq M$ in Ω.

Suggestion: Show that it involves no loss of generality to assume that $U \cap \Omega = \emptyset$. Fix $z_0 \in \Omega$, let n be a large integer, let V be a large disc with center at 0, and apply the maximum modulus theorem to the function $f^n(z)/z$ in the component of $V \cap \Omega$ which contains z_0.

12 Let f be an entire function. If there is a continuous mapping γ of $[0, 1)$ into the complex plane such that $\gamma(t) \to \infty$ and $f(\gamma(t)) \to \alpha$ as $t \to 1$, we say that α is an *asymptotic value* of f. [In the complex plane, "$\gamma(t) \to \infty$ as $t \to 1$" means that to each $R < \infty$ there corresponds a $t_R < 1$ such that $|\gamma(t)| > R$ if $t_R < t < 1$.] Prove that every nonconstant entire function has ∞ as an asymptotic value.

Suggestion: Let $E_n = \{z : |f(z)| > n\}$. Each component of E_n is unbounded (proof?) and contains a component of E_{n+1}, by Exercise 11.

13 Show that exp has exactly two asymptotic values: 0 and ∞. How about sin and cos? *Note*: sin z and cos z are defined, for all complex z, by

$$\sin z = \frac{e^{iz} - e^{-iz}}{2i}, \qquad \cos z = \frac{e^{iz} + e^{-iz}}{2}.$$

14 If f is entire and if α is not in the range of f, prove that α is an asymptotic value of f unless f is constant.

15 Suppose $f \in H(U)$. Prove that there is a sequence $\{z_n\}$ in U such that $|z_n| \to 1$ and $\{f(z_n)\}$ is bounded.

16 Suppose Ω is a bounded region, $f \in H(\Omega)$, and

$$\limsup_{n \to \infty} |f(z_n)| \leq M$$

for every sequence $\{z_n\}$ in Ω which converges to a boundary point of Ω. Prove that $|f(z)| \leq M$ for all $z \in \Omega$.

17 Let Φ be the set of all $f \in H(U)$ such that $0 < |f(z)| < 1$ for $z \in U$, and let Φ_c be the set of all $f \in \Phi$ that have $f(0) = c$. Define

$$M(c) = \sup\{|f'(0)| : f \in \Phi_c\}, \qquad M = \sup\{|f'(0)| : f \in \Phi\}.$$

Find M, and $M(c)$ for $0 < c < 1$. Find an $f \in \Phi$ with $f'(0) = M$, or prove that there is no such f.

Suggestion: log f maps U into the left half plane. Compose log f with a properly chosen map that takes this half plane to U. Apply the Schwarz lemma.

CHAPTER
THIRTEEN

APPROXIMATIONS BY RATIONAL FUNCTIONS

Preparation

13.1 The Riemann Sphere It is often convenient in the study of holomorphic functions to compactify the complex plane by the adjunction of a new point called ∞. The resulting set S^2 (the *Riemann sphere*, the union of R^2 and $\{\infty\}$) is topologized in the following manner. For any $r > 0$, let $D'(\infty; r)$ be the set of all complex numbers z such that $|z| > r$, put $D(\infty; r) = D'(\infty; r) \cup \{\infty\}$, and declare a subset of S^2 to be open if and only if it is the union of discs $D(a; r)$, where the a's are arbitrary points of S^2 and the r's are arbitrary positive numbers. On $S^2 - \{\infty\}$, this gives of course the ordinary topology of the plane. It is easy to see that S^2 is homeomorphic to a sphere (hence the notation). In fact, a homeomorphism φ of S^2 onto the unit sphere in R^3 can be explicitly exhibited: Put $\varphi(\infty) = (0, 0, 1)$, and put

$$\varphi(re^{i\theta}) = \left(\frac{2r \cos \theta}{r^2 + 1}, \frac{2r \sin \theta}{r^2 + 1}, \frac{r^2 - 1}{r^2 + 1}\right) \tag{1}$$

for all complex numbers $re^{i\theta}$. We leave it to the reader to construct the geometric picture that goes with (1).

If f is holomorphic in $D'(\infty; r)$, we say that f has an isolated singularity at ∞. The nature of this singularity is the same as that which the function $\tilde{f}$, defined in $D'(0; 1/r)$ by $\tilde{f}(z) = f(1/z)$, has at 0.

Thus if f is bounded in $D'(\infty; r)$, then $\lim_{z \to \infty} f(z)$ exists and is a complex number (as we see if we apply Theorem 10.20 to $\tilde{f}$), we define $f(\infty)$ to be this limit, and we thus obtain a function in $D(\infty; r)$ which we call holomorphic: note that this is defined in terms of the behavior of $\tilde{f}$ near 0, and not in terms of differentiability of f at ∞.

If $\tilde{f}$ has a pole of order m at 0, then f is said to have a pole of order m at ∞; the *principal part* of f at ∞ is then an ordinary polynomial of degree m (compare Theorem 10.21), and if we subtract this polynomial from f, we obtain a function with a removable singularity at ∞.

Finally, if $\tilde{f}$ has an essential singularity at 0, then f is said to have an essential singularity at ∞. For instance, every entire function which is not a polynomial has an essential singularity at ∞.

Later in this chapter we shall encounter the condition "$S^2 - \Omega$ is connected," where Ω is an open set in the plane. Note that this is not equivalent to the condition "the complement of Ω relative to the plane is connected." For example, if Ω consists of all complex $z = x + iy$ with $0 < y < 1$, the complement of Ω relative to the plane has two components, but $S^2 - \Omega$ is connected.

13.2 Rational Functions A rational function f is, by definition, a quotient of two polynomials P and Q: $f = P/Q$. It follows from Theorem 10.25 that every nonconstant polynomial is a product of factors of degree 1. We may assume that P and Q have no such factors in common. Then f has a pole at each zero of Q (the pole of f has the same order as the zero of Q). If we subtract the corresponding principal parts, we obtain a rational function whose only singularity is at ∞ and which is therefore a polynomial.

Every rational function $f = P/Q$ has thus a representation of the form

$$f(z) = A_0(z) + \sum_{j=1}^{k} A_j((z - a_j)^{-1}) \tag{1}$$

where $A_0, A_1, \ldots, A_k$ are polynomials, $A_1, \ldots, A_k$ have no constant term, and $a_1, \ldots, a_k$ are the distinct zeros of Q; (1) is called the *partial fractions decomposition* of f.

We turn to some topological considerations. We know that every open set in the plane is a countable union of compact sets (closed discs, for instance). However, it will be convenient to have some additional properties satisfied by these compact sets:

13.3 Theorem *Every open set Ω in the plane is the union of a sequence $\{K_n\}$, $n = 1, 2, 3, \ldots$, of compact sets such that*

(a) K_n *lies in the interior of* K_{n+1}, *for* $n = 1, 2, 3, \ldots$.
(b) *Every compact subset of Ω lies in some K_n.*
(c) *Every component of $S^2 - K_n$ contains a component of $S^2 - \Omega$, for $n = 1, 2, 3, \ldots$.*

Property (c) is, roughly speaking, that K_n has no holes except those which are forced upon it by the holes in Ω. Note that Ω is not assumed to be connected. The *interior* of a set E is, by definition, the largest open subset of E.

PROOF For $n = 1, 2, 3, \ldots$, put

$$V_n = D(\infty; n) \cup \bigcup_{a \notin \Omega} D\left(a; \frac{1}{n}\right) \qquad (1)$$

and put $K_n = S^2 - V_n$. [Of course, $a \neq \infty$ in (1).] Then K_n is a closed and bounded (hence compact) subset of Ω, and $\Omega = \bigcup K_n$. If $z \in K_n$ and $r = n^{-1} - (n+1)^{-1}$, one verifies easily that $D(z; r) \subset K_{n+1}$. This gives (a). Hence Ω is the union of the interiors W_n of K_n. If K is a compact subset of Ω, then $K \subset W_1 \cup \cdots \cup W_N$ for some N, hence $K \subset K_N$.

Finally, each of the discs in (1) intersects $S^2 - \Omega$; each disc is connected; hence each component of V_n intersects $S^2 - \Omega$; since $V_n \supset S^2 - \Omega$, no component of $S^2 - \Omega$ can intersect two components of V_n. This gives (c).   ////

13.4 Sets of Oriented Intervals Let Φ be a finite collection of oriented intervals in the plane. For each point p, let $m_I(p)[m_E(p)]$ be the number of members of Φ that have initial point [end point] p. If $m_I(p) = m_E(p)$ for every p, we shall say that Φ is *balanced*.

If Φ is balanced (and nonempty), the following construction can be carried out.

Pick $\gamma_1 = [a_0, a_1] \in \Phi$. Assume $k \geq 1$, and assume that distinct members $\gamma_1, \ldots, \gamma_k$ of Φ have been chosen in such a way that $\gamma_i = [a_{i-1}, a_i]$ for $1 \leq i \leq k$. If $a_k = a_0$, stop. If $a_k \neq a_0$, and if precisely r of the intervals $\gamma_1, \ldots, \gamma_k$ have a_k as end point, then only $r - 1$ of them have a_k as initial point; since Φ is balanced, Φ contains at least one other interval, say γ_{k+1}, whose initial point is a_k. Since Φ is finite, we must return to a_0 eventually, say at the nth step.

Then $\gamma_1, \ldots, \gamma_n$ join (in this order) to form a closed path.

The remaining members of Φ still form a balanced collection to which the above construction can be applied. It follows that the members of Φ can be so numbered that they form finitely many closed paths. The sum of these paths is a cycle. The following conclusion is thus reached.

If $\Phi = \{\gamma_1, \ldots, \gamma_N\}$ is a balanced collection of oriented intervals, and if

$$\Gamma = \gamma_1 \dot{+} \cdots \dot{+} \gamma_N$$

then Γ is a cycle.

13.5 Theorem *If K is a compact subset of a plane open set Ω ($\neq \emptyset$), then there is a cycle Γ in $\Omega - K$ such that the Cauchy formula*

$$f(z) = \frac{1}{2\pi i} \int_\Gamma \frac{f(\zeta)}{\zeta - z} d\zeta \qquad (1)$$

holds for every $f \in H(\Omega)$ and for every $z \in K$.

PROOF Since K is compact and Ω is open, there exists an $\eta > 0$ such that the distance from any point of K to any point outside Ω is at least 2η. Construct a grid of horizontal and vertical lines in the plane, such that the distance between any two adjacent horizontal lines is η, and likewise for the vertical lines. Let $Q_1, \ldots, Q_m$ be those squares (closed 2-cells) of edge η which are formed by this grid and which intersect K. Then $Q_r \subset \Omega$ for $r = 1, \ldots, m$.

If a_r is the center of Q_r and $a_r + b$ is one of its vertices, let γ_{rk} be the oriented interval

$$\gamma_{rk} = [a_r + i^k b, a_r + i^{k+1} b] \tag{2}$$

and define

$$\partial Q_r = \gamma_{r1} \dotplus \gamma_{r2} \dotplus \gamma_{r3} \dotplus \gamma_{r4} \qquad (r = 1, \ldots, m). \tag{3}$$

It is then easy to check (for example, as a special case of Theorem 10.37, or by means of Theorems 10.11 and 10.40) that

$$\mathrm{Ind}_{\partial Q_r}(\alpha) = \begin{cases} 1 & \text{if } \alpha \text{ is in the interior of } Q_r, \\ 0 & \text{if } \alpha \text{ is not in } Q_r. \end{cases} \tag{4}$$

Let Σ be the collection of all γ_{rk} ($1 \leq r \leq m$, $1 \leq k \leq 4$). It is clear that Σ is balanced. Remove those members of Σ whose opposites (see Sec. 10.8) also belong to Σ. Let Φ be the collection of the remaining members of Σ. Then Φ is balanced. Let Γ be the cycle constructed from Φ, as in Sec. 13.4.

If an edge E of some Q_r intersects K, then the two squares in whose boundaries E lies intersect K. Hence Σ contains two oriented intervals which are each other's opposites and whose range is E. These intervals do not occur in Φ. Thus Γ is a cycle in $\Omega - K$.

The construction of Φ from Σ shows also that

$$\mathrm{Ind}_\Gamma(\alpha) = \sum_{r=1}^{m} \mathrm{Ind}_{\partial Q_r}(\alpha) \tag{5}$$

if α is not in the boundary of any Q_r. Hence (4) implies

$$\mathrm{Ind}_\Gamma(\alpha) = \begin{cases} 1 & \text{if } \alpha \text{ is in the interior of some } Q_r, \\ 0 & \text{if } \alpha \text{ lies in no } Q_r. \end{cases} \tag{6}$$

If $z \in K$, then $z \notin \Gamma^*$, and z is a limit point of the interior of some Q_r. Since the left side of (6) is constant in each component of the complement of Γ^*, (6) gives

$$\mathrm{Ind}_\Gamma(z) = \begin{cases} 1 & \text{if } z \in K, \\ 0 & \text{if } z \notin \Omega. \end{cases} \tag{7}$$

Now (1) follows from Cauchy's theorem 10.35. ////

Runge's Theorem

The main objective of this section is Theorem 13.9. We begin with a slightly different version in which the emphasis is on uniform approximation on one compact set.

13.6 Theorem *Suppose K is a compact set in the plane and $\{\alpha_j\}$ is a set which contains one point in each component of $S^2 - K$. If Ω is open, $\Omega \supset K, f \in H(\Omega)$, and $\epsilon > 0$, there exists a rational function R, all of whose poles lie in the prescribed set $\{\alpha_j\}$, such that*

$$|f(z) - R(z)| < \epsilon \tag{1}$$

for every $z \in K$.

Note that $S^2 - K$ has at most countably many components. Note also that the preassigned point in the unbounded component of $S^2 - K$ may very well be ∞; in fact, this happens to be the most interesting choice.

PROOF We consider the Banach space $C(K)$ whose members are the continuous complex functions on K, with the supremum norm. Let M be the subspace of $C(K)$ which consists of the restrictions to K of those rational functions which have all their poles in $\{\alpha_j\}$. The theorem asserts that f is in the closure of M. By Theorem 5.19 (a consequence of the Hahn-Banach theorem), this is equivalent to saying that every bounded linear functional on $C(K)$ which vanishes on M also vanishes at f, and hence the Riesz representation theorem (Theorem 6.19) shows that we must prove the following assertion:

If μ is a complex Borel measure on K such that

$$\int_K R \, d\mu = 0 \tag{2}$$

for every rational function R with poles only in the set $\{\alpha_j\}$, and if $f \in H(\Omega)$, then we also have

$$\int_K f \, d\mu = 0. \tag{3}$$

So let us assume that μ satisfies (2). Define

$$h(z) = \int_K \frac{d\mu(\zeta)}{\zeta - z} \qquad (z \in S^2 - K). \tag{4}$$

By Theorem 10.7 (with $X = K$, $\varphi(\zeta) = \zeta$), $h \in H(S^2 - K)$.

Let V_j be the component of $S^2 - K$ which contains α_j, and suppose $D(\alpha_j; r) \subset V_j$. If $\alpha_j \neq \infty$ and if z is fixed in $D(\alpha_j; r)$, then

$$\frac{1}{\zeta - z} = \lim_{N \to \infty} \sum_{n=0}^{N} \frac{(z - \alpha_j)^n}{(\zeta - \alpha_j)^{n+1}} \qquad (5)$$

uniformly for $\zeta \in K$. Each of the functions on the right of (5) is one to which (2) applies. Hence $h(z) = 0$ for all $z \in D(\alpha_j; r)$. This implies that $h(z) = 0$ for all $z \in V_j$, by the uniqueness theorem 10.18.

If $\alpha_j = \infty$, (5) is replaced by

$$\frac{1}{\zeta - z} = -\lim_{N \to \infty} \sum_{n=0}^{N} z^{-n-1} \zeta^n \qquad (\zeta \in K, |z| > r), \qquad (6)$$

which implies again that $h(z) = 0$ in $D(\infty; r)$, hence in V_j. We have thus proved from (2) that

$$h(z) = 0 \qquad (z \in S^2 - K). \qquad (7)$$

Now choose a cycle Γ in $\Omega - K$, as in Theorem 13.5, and integrate this Cauchy integral representation of f with respect to μ. An application of Fubini's theorem (legitimate, since we are dealing with Borel measures and continuous functions on compact spaces), combined with (7), gives

$$\int_K f \, d\mu = \int_K d\mu(\zeta) \left[\frac{1}{2\pi i} \int_\Gamma \frac{f(w)}{w - \zeta} \, dw \right]$$

$$= \frac{1}{2\pi i} \int_\Gamma f(w) \, dw \int_K \frac{d\mu(\zeta)}{w - \zeta}$$

$$= -\frac{1}{2\pi i} \int_\Gamma f(w) h(w) \, dw = 0.$$

The last equality depends on the fact that $\Gamma^* \subset \Omega - K$, where $h(w) = 0$.

Thus (3) holds, and the proof is complete. ////

The following special case is of particular interest.

13.7 Theorem *Suppose K is a compact set in the plane, $S^2 - K$ is connected, and $f \in H(\Omega)$, where Ω is some open set containing K. Then there is a sequence $\{P_n\}$ of polynomials such that $P_n(z) \to f(z)$ uniformly on K.*

PROOF Since $S^2 - K$ has now only one component, we need only one point α_j to apply Theorem 13.6, and we may take $\alpha_j = \infty$. ////

13.8 Remark The preceding result is false for *every* compact K in the plane such that $S^2 - K$ is not connected. For in that case $S^2 - K$ has a bounded component V. Choose $\alpha \in V$, put $f(z) = (z - \alpha)^{-1}$, and put

$m = \max \{|z - \alpha| : z \in K\}$. Suppose P is a polynomial, such that $|P(z) - f(z)| < 1/m$ for all $z \in K$. Then

$$|(z - \alpha)P(z) - 1| < 1 \qquad (z \in K). \tag{1}$$

In particular, (1) holds if z is in the boundary of V; since the closure of V is compact, the maximum modulus theorem shows that (1) holds for every $z \in V$; taking $z = \alpha$, we obtain $1 < 1$. Hence the uniform approximation is not possible.

The same argument shows that none of the α_j can be dispensed with in Theorem 13.6.

We now apply the preceding approximation theorems to approximation in open sets. Let us emphasize that K was *not* assumed to be connected in Theorems 13.6 and 13.7 and that Ω will *not* be assumed to be connected in the theorem which follows.

13.9 Theorem *Let Ω be an open set in the plane, let A be a set which has one point in each component of $S^2 - \Omega$, and assume $f \in H(\Omega)$. Then there is a sequence $\{R_n\}$ of rational functions, with poles only in A, such that $R_n \to f$ uniformly on compact subsets of Ω.*

In the special case in which $S^2 - \Omega$ is connected, we may take $A = \{\infty\}$ and thus obtain polynomials P_n such that $P_n \to f$ uniformly on compact subsets of Ω.

Observe that $S^2 - \Omega$ may have uncountably many components; for instance, we may have $S^2 - \Omega = \{\infty\} \cup C$, where C is a Cantor set.

PROOF Choose a sequence of compact sets K_n in Ω, with the properties specified in Theorem 13.3. Fix n, for the moment. Since each component of $S^2 - K_n$ contains a component of $S^2 - \Omega$, each component of $S^2 - K_n$ contains a point of A, so Theorem 13.6 gives us a rational function R_n with poles in A such that

$$|R_n(z) - f(z)| < \frac{1}{n} \qquad (z \in K_n). \tag{1}$$

If now K is any compact set in Ω, there exists an N such that $K \subset K_n$ for all $n \geq N$. It follows from (1) that

$$|R_n(z) - f(z)| < \frac{1}{n} \qquad (z \in K, n \geq N), \tag{2}$$

which completes the proof. ////

The Mittag-Leffler Theorem

Runge's theorem will now be used to prove that meromorphic functions can be constructed with arbitrarily preassigned poles.

13.10 Theorem *Suppose Ω is an open set in the plane, $A \subset \Omega$, A has no limit point in Ω, and to each $\alpha \in A$ there are associated a positive integer $m(\alpha)$ and a rational function*

$$P_\alpha(z) = \sum_{j=1}^{m(\alpha)} c_{j,\alpha}(z - \alpha)^{-j}.$$

Then there exists a meromorphic function f in Ω, whose principal part at each $\alpha \in A$ is P_α and which has no other poles in Ω.

PROOF We choose a sequence $\{K_n\}$ of compact sets in Ω, as in Theorem 13.3: For $n = 1, 2, 3, \ldots$, K_n lies in the interior of K_{n+1}, every compact subset of Ω lies in some K_n, and every component of $S^2 - K_n$ contains a component of $S^2 - \Omega$. Put $A_1 = A \cap K_1$, and $A_n = A \cap (K_n - K_{n-1})$ for $n = 2, 3, 4, \ldots$. Since $A_n \subset K_n$ and A has no limit point in Ω (hence none in K_n), each A_n is a finite set. Put

$$Q_n(z) = \sum_{\alpha \in A_n} P_\alpha(z) \qquad (n = 1, 2, 3, \ldots). \tag{1}$$

Since each A_n is finite, each Q_n is a rational function. The poles of Q_n lie in $K_n - K_{n-1}$, for $n \geq 2$. In particular, Q_n is holomorphic in an open set containing K_{n-1}. It now follows from Theorem 13.6 that there exist rational functions R_n, all of whose poles are in $S^2 - \Omega$, such that

$$|R_n(z) - Q_n(z)| < 2^{-n} \qquad (z \in K_{n-1}). \tag{2}$$

We claim that

$$f(z) = Q_1(z) + \sum_{n=2}^{\infty} (Q_n(z) - R_n(z)) \qquad (z \in \Omega) \tag{3}$$

has the desired properties.

Fix N. On K_N, we have

$$f = Q_1 + \sum_{n=2}^{N} (Q_n - R_n) + \sum_{N+1}^{\infty} (Q_n - R_n). \tag{4}$$

By (2), each term in the last sum in (4) is less than 2^{-n} on K_N; hence this last series converges uniformly on K_N, to a function which is holomorphic in the interior of K_N. Since the poles of each R_n are outside Ω,

$$f - (Q_1 + \cdots + Q_N)$$

is holomorphic in the interior of K_N. Thus f has precisely the prescribed principal parts in the interior of K_N, and hence in Ω, since N was arbitrary.

////

Simply Connected Regions

We shall now summarize some properties of simply connected regions (see Sec. 10.38) which illustrate the important role that they play in the theory of holomorphic functions. Of these properties, (a) and (b) are what one might call internal topological properties of Ω; (c) and (d) refer to the way in which Ω is embedded in S^2; properties (e) to (h) are analytic in character; (j) is an algebraic statement about the ring $H(\Omega)$. The Riemann mapping theorem 14.8 is another very important property of simply connected regions. In fact, we shall use it to prove the implication (j) → (a).

13.11 Theorem *For a plane region Ω, each of the following nine conditions implies all the others.*

(a) Ω *is homeomorphic to the open unit disc U.*
(b) Ω *is simply connected.*
(c) $\mathrm{Ind}_\gamma (\alpha) = 0$ *for every closed path γ in Ω and for every $\alpha \in S^2 - \Omega$.*
(d) $S^2 - \Omega$ *is connected.*
(e) *Every $f \in H(\Omega)$ can be approximated by polynomials, uniformly on compact subsets of Ω.*
(f) *For every $f \in H(\Omega)$ and every closed path γ in Ω,*

$$\int_\gamma f(z)\, dz = 0.$$

(g) *To every $f \in H(\Omega)$ corresponds an $F \in H(\Omega)$ such that $F' = f$.*
(h) *If $f \in H(\Omega)$ and $1/f \in H(\Omega)$, there exists a $g \in H(\Omega)$ such that $f = \exp(g)$.*
(j) *If $f \in H(\Omega)$ and $1/f \in H(\Omega)$, there exists a $\varphi \in H(\Omega)$ such that $f = \varphi^2$.*

The assertion of (h) is that f has a "holomorphic logarithm" g in Ω; (j) asserts that f has a "holomorphic square root" φ in Ω; and (f) says that the Cauchy theorem holds for every closed path in a simply connected region.

In Chapter 16 we shall see that the monodromy theorem describes yet another characteristic property of simply connected regions.

PROOF (a) implies (b). To say that Ω is homeomorphic to U means that there is a continuous one-to-one mapping ψ of Ω onto U whose inverse ψ^{-1} is also continuous. If γ is a closed curve in Ω, with parameter interval $[0, 1]$, put

$$H(s, t) = \psi^{-1}(t\psi(\gamma(s))).$$

Then $H: I^2 \to \Omega$ is continuous; $H(s, 0) = \psi^{-1}(0)$, a constant; $H(s, 1) = \gamma(s)$; and $H(0, t) = H(1, t)$ because $\gamma(0) = \gamma(1)$. Thus Ω is simply connected.

(b) implies (c). If (b) holds and γ is a closed path in Ω, then γ is (by definition of "simply connected") Ω-homotopic to a constant path. Hence (c) holds, by Theorem 10.40.

(c) implies (d). Assume (d) is false. Then $S^2 - \Omega$ is a closed subset of S^2 which is not connected. As noted in Sec. 10.1, it follows that $S^2 - \Omega$ is the union of two nonempty disjoint closed sets H and K. Let H be the one that contains ∞. Let W be the complement of H, relative to the plane. Then $W = \Omega \cup K$. Since K is compact, Theorem 13.5 (with $f = 1$) shows that there is a cycle Γ in $W - K = \Omega$ such that $\text{Ind}_\Gamma(z) = 1$ for $z \in K$. Since $K \neq \emptyset$, (c) fails.

(d) implies (e). This is part of Theorem 13.9.

(e) implies (f). Choose $f \in H(\Omega)$, let γ be a closed path in Ω, and choose polynomials P_n which converge to f, uniformly on γ^*. Since $\int_\gamma P_n(z)\, dz = 0$ for all n, we conclude that (f) holds.

(f) implies (g). Assume (f) holds, fix $z_0 \in \Omega$, and put

$$F(z) = \int_{\Gamma(z)} f(\zeta)\, d\zeta \qquad (z \in \Omega) \tag{1}$$

where $\Gamma(z)$ is any path in Ω from z_0 to z. This defines a function F in Ω. For if $\Gamma_1(z)$ is another path from z_0 to z (in Ω), then Γ followed by the opposite of Γ_1 is a closed path in Ω, the integral of f over this closed path is 0, so (1) is not affected if $\Gamma(z)$ is replaced by $\Gamma_1(z)$. We now verify that $F' = f$. Fix $a \in \Omega$. There exists an $r > 0$ such that $D(a; r) \subset \Omega$. For $z \in D(a; r)$ we can compute $F(z)$ by integrating f over a path $\Gamma(a)$, followed by the interval $[a, z]$. Hence, for $z \in D'(a; r)$,

$$\frac{F(z) - F(a)}{z - a} = \frac{1}{z - a} \int_{[a,\, z]} f(\zeta)\, d\zeta, \tag{2}$$

and the continuity of f at a implies now that $F'(a) = f(a)$, as in the proof of Theorem 10.14.

(g) implies (h). If $f \in H(\Omega)$ and f has no zero in Ω, then $f'/f \in H(\Omega)$, and (g) implies that there exists a $g \in H(\Omega)$ so that $g' = f'/f$. We can add a constant to g, so that $\exp\{g(z_0)\} = f(z_0)$ for some $z_0 \in \Omega$. Our choice of g shows that the derivative of fe^{-g} is 0 in Ω, hence fe^{-g} is constant (since Ω is connected), and it follows that $f = e^g$.

(h) implies (j). By (h), $f = e^g$. Put $\varphi = \exp(\frac{1}{2}g)$.

(j) implies (a). If Ω is the whole plane, then Ω is homeomorphic to U: map z to $z/(1 + |z|)$.

If Ω is a proper subregion of the plane which satisfies (j), then there actually exists a *holomorphic* homeomorphism of Ω onto U (a conformal mapping). This assertion is the Riemann mapping theorem, which is the main objective of the next chapter. Hence the proof of Theorem 13.11 will be complete as soon as the Riemann mapping theorem is proved. (See the note following the statement of Theorem 14.8.) ////

The fact that (h) holds in every simply connected region has the following consequence (which can also be proved by quite elementary means):

13.12 Theorem *If $f \in H(\Omega)$, where Ω is any open set in the plane, and if f has no zero in Ω, then $\log |f|$ is harmonic in Ω.*

PROOF To every disc $D \subset \Omega$ there corresponds a function $g \in H(D)$ such that $f = e^g$ in D. If $u = \operatorname{Re} g$, then u is harmonic in D, and $|f| = e^u$. Thus $\log |f|$ is harmonic in every disc in Ω, and this gives the desired conclusion. ////

Exercises

1 Prove that every meromorphic function on S^2 is rational.

2 Let $\Omega = \{z : |z| < 1 \text{ and } |2z - 1| > 1\}$, and suppose $f \in H(\Omega)$.
 (a) Must there exist a sequence of polynomials P_n such that $P_n \to f$ uniformly on compact subsets of Ω?
 (b) Must there exist such a sequence which converges to f uniformly in Ω?
 (c) Is the answer to (b) changed if we require more of f, namely, that f be holomorphic in some open set which contains the closure of Ω?

3 Is there a sequence of polynomials P_n such that $P_n(0) = 1$ for $n = 1, 2, 3, \ldots$, but $P_n(z) \to 0$ for every $z \neq 0$, as $n \to \infty$?

4 Is there a sequence of polynomials P_n such that

$$\lim_{n \to \infty} P_n(z) = \begin{cases} 1 & \text{if } \operatorname{Im} z > 0, \\ 0 & \text{if } z \text{ is real}, \\ -1 & \text{if } \operatorname{Im} z < 0? \end{cases}$$

5 For $n = 1, 2, 3, \ldots$, let Δ_n be a closed disc in U, and let L_n be an arc (a homeomorphic image of $[0, 1]$) in $U - \Delta_n$ which intersects every radius of U. There are polynomials P_n which are very small on Δ_n and more or less arbitrary on L_n. Show that $\{\Delta_n\}$, $\{L_n\}$, and $\{P_n\}$ can be so chosen that the series $f = \Sigma P_n$ defines a function $f \in H(U)$ which has no radial limit at any point of T. In other words, for no real θ does $\lim_{r \to 1} f(re^{i\theta})$ exist.

6 Here is another construction of such a function. Let $\{n_k\}$ be a sequence of integers such that $n_1 > 1$ and $n_{k+1} > 2kn_k$. Define

$$h(z) = \sum_{k=1}^{\infty} 5^k z^{n_k}.$$

Prove that the series converges if $|z| < 1$ and prove that there is a constant $c > 0$ such that $|h(z)| > c \cdot 5^m$ for all z with $|z| = 1 - (1/n_m)$. [*Hint*: For such z the mth term in the series defining $h(z)$ is much larger than the sum of all the others.]
 Hence h has no finite radial limits.
 Prove also that h must have infinitely many zeros in U. (Compare with Exercise 15, Chap. 12.) In fact, prove that to every complex number α there correspond infinitely many $z \in U$ at which $h(z) = \alpha$.

7 Show that in Theorem 13.9 we need not assume that A intersects each component of $S^2 - \Omega$. It is enough to assume that the closure of A intersects each component of $S^2 - \Omega$.

8 Prove the Mittag-Leffler theorem for the case in which Ω is the whole plane, by a direct argument which makes no appeal to Runge's theorem.

9 Suppose Ω is a simply connected region, $f \in H(\Omega)$, f has no zero in Ω, and n is a positive integer. Prove that there exists a $g \in H(\Omega)$ such that $g^n = f$.

10 Suppose Ω is a region, $f \in H(\Omega)$, and $f \neq 0$. Prove that f has a holomorphic logarithm in Ω if and only if f has holomorphic nth roots in Ω for every positive integer n.

11 Suppose that $f_n \in H(\Omega)$ ($n = 1, 2, 3, \ldots$), f is a complex function in Ω, and $f(z) = \lim_{n \to \infty} f_n(z)$ for every $z \in \Omega$. Prove that Ω has a dense open subset V on which f is holomorphic. *Hint*: Put $\varphi = \sup |f_n|$. Use Baire's theorem to prove that every disc in Ω contains a disc on which φ is bounded. Apply Exercise 5, Chap. 10. (In general, $V \neq \Omega$. Compare Exercises 3 and 4.)

12 Suppose, however, that f is *any* complex-valued measurable function defined in the complex plane, and prove that there is a sequence of holomorphic polynomials P_n such that $\lim_{n \to \infty} P_n(z) = f(z)$ for *almost* every z (with respect to two-dimensional Lebesgue measure).

CHAPTER FOURTEEN

CONFORMAL MAPPING

Preservation of Angles

14.1 Definition Each complex number $z \neq 0$ determines a *direction* from the origin, defined by the point

$$A[z] = \frac{z}{|z|} \qquad (1)$$

on the unit circle.

Suppose f is a mapping of a region Ω into the plane, $z_0 \in \Omega$, and z_0 has a deleted neighborhood $D'(z_0; r) \subset \Omega$ in which $f(z) \neq f(z_0)$. We say that f preserves angles at z_0 if

$$\lim_{r \to 0} e^{-i\theta} A[f(z_0 + re^{i\theta}) - f(z_0)] \qquad (r > 0) \qquad (2)$$

exists and *is independent of* θ.

In less precise language, the requirement is that for any two rays L and L', starting at z_0, the angle which their images $f(L)$ and $f(L')$ make at $f(z_0)$ is the same as that made by L and L', in size as well as in orientation.

The property of preserving angles at each point of a region is characteristic of holomorphic functions whose derivative has no zero in that region. This is a corollary of Theorem 14.2 and is the reason for calling holomorphic functions with nonvanishing derivative *conformal mappings*.

14.2 Theorem Let f map a region Ω into the plane. If $f'(z_0)$ exists at some $z_0 \in \Omega$ and $f'(z_0) \neq 0$, then f preserves angles at z_0. Conversely, if the differential of f exists and is different from 0 at z_0, and if f preserves angles at z_0, then $f'(z_0)$ exists and is different from 0.

Here $f'(z_0) = \lim [f(z) - f(z_0)]/(z - z_0)$, as usual. The differential of f at z_0 is a linear transformation L of R^2 into R^2 such that, writing $z_0 = (x_0, y_0)$,

$$f(x_0 + x, y_0 + y) = f(x_0, y_0) + L(x, y) + (x^2 + y^2)^{1/2}\eta(x, y), \tag{1}$$

where $\eta(x, y) \to 0$ as $x \to 0$ and $y \to 0$, as in Definition 7.22.

PROOF Take $z_0 = f(z_0) = 0$, for simplicity. If $f'(0) = a \neq 0$, then it is immediate that

$$e^{-i\theta}A[f(re^{i\theta})] = \frac{e^{-i\theta}f(re^{i\theta})}{|f(re^{i\theta})|} \to \frac{a}{|a|} \quad (r \to 0), \tag{2}$$

so f preserves angles at 0. Conversely, if the differential of f exists at 0 and is different from 0, then (1) can be rewritten in the form

$$f(z) = \alpha z + \beta \bar{z} + |z|\eta(z), \tag{3}$$

where $\eta(z) \to 0$ as $z \to 0$, and α and β are complex numbers, not both 0. If f also preserves angles at 0, then

$$\lim_{r \to 0} e^{-i\theta}A[f(re^{i\theta})] = \frac{\alpha + \beta e^{-2i\theta}}{|\alpha + \beta e^{-2i\theta}|} \tag{4}$$

exists and is independent of θ. We may exclude those θ for which the denominator in (4) is 0; there are at most two such θ in $[0, 2\pi)$. For all other θ, we conclude that $\alpha + \beta e^{-2i\theta}$ lies on a fixed ray through 0, and this is possible only when $\beta = 0$. Hence $\alpha \neq 0$, and (3) implies that $f'(0) = \alpha$. ////

Note: No holomorphic function preserves angles at any point where its derivative is 0. We omit the easy proof of this. However, the differential of a transformation may be 0 at a point where angles are preserved. Example: $f(z) = |z|z$, $z_0 = 0$.

Linear Fractional Transformations

14.3 If a, b, c, and d are complex numbers such that $ad - bc \neq 0$, the mapping

$$z \to \frac{az + b}{cz + d} \tag{1}$$

is called a *linear fractional transformation*. It is convenient to regard (1) as a mapping of the sphere S^2 into S^2, with the obvious conventions concerning the point ∞. For instance, $-d/c$ maps to ∞ and ∞ maps to a/c, if $c \neq 0$. It is then easy to see that each linear fractional transformation is a one-to-one mapping of S^2 onto S^2. Furthermore, each is obtained by a superposition of transformations of the following types:

(a) Translations: $z \to z + b$.
(b) Rotations: $z \to az$, $|a| = 1$.

(c) Homotheties: $z \to rz$, $r > 0$.
(d) Inversion: $z \to 1/z$.

If $c = 0$ in (1), this is obvious. If $c \neq 0$, it follows from the identity

$$\frac{az+b}{cz+d} = \frac{a}{c} + \frac{\lambda}{cz+d}, \qquad \lambda = \frac{bc-ad}{c}. \tag{2}$$

The first three types evidently carry lines to lines and circles to circles. This is not true of (d). But if we let $\mathscr{F}$ be the family consisting of all straight lines and all circles, then $\mathscr{F}$ is preserved by (d), and hence we have the important result that *$\mathscr{F}$ is preserved by every linear fractional transformation.* [It may be noted that when $\mathscr{F}$ is regarded as a family of subsets of S^2, then $\mathscr{F}$ consists of all circles on S^2, via the stereographic projection 13.1(1); we shall not use this property of $\mathscr{F}$ and omit its proof.]

The proof that $\mathscr{F}$ is preserved by inversion is quite easy. Elementary analytic geometry shows that every member of $\mathscr{F}$ is the locus of an equation

$$\alpha z \bar{z} + \beta z + \bar{\beta} \bar{z} + \gamma = 0, \tag{3}$$

where α and γ are real constants and β is a complex constant, provided that $\beta \bar{\beta} > \alpha \gamma$. If $\alpha \neq 0$, (3) defines a circle; $\alpha = 0$ gives the straight lines. Replacement of z by $1/z$ transforms (3) into

$$\alpha + \beta \bar{z} + \bar{\beta} z + \gamma z \bar{z} = 0, \tag{4}$$

which is an equation of the same type.

Suppose a, b, and c are distinct complex numbers. We construct a linear fractional transformation φ which maps the ordered triple $\{a, b, c\}$ into $\{0, 1, \infty\}$, namely,

$$\varphi(z) = \frac{(b-c)(z-a)}{(b-a)(z-c)} \tag{5}$$

There is only one such φ. For if $\varphi(a) = 0$, we must have $z - a$ in the numerator; if $\varphi(c) = \infty$, we must have $z - c$ in the denominator; and if $\varphi(b) = 1$, we are led to (5). If a or b or c is ∞, formulas analogous to (5) can easily be written down. If we follow (5) by the inverse of a transformation of the same type, we obtain the following result:

For any two ordered triples $\{a, b, c\}$ and $\{a', b', c'\}$ in S^2 there is one and only one linear fractional transformation which maps a to a', b to b', and c to c'.

(It is of course assumed that $a \neq b$, $a \neq c$, and $b \neq c$, and likewise for a', b', and c'.)

We conclude from this that every circle can be mapped onto every circle by a linear fractional transformation. Of more interest is the fact that every circle can be mapped onto every straight line (if ∞ is regarded as part of the line), and hence that *every open disc can be conformally mapped onto every open half plane.*

Let us discuss one such mapping more explicitly, namely,

$$\varphi(z) = \frac{1+z}{1-z}. \tag{6}$$

This φ maps $\{-1, 0, 1\}$ to $\{0, 1, \infty\}$; the segment $(-1, 1)$ maps onto the positive real axis. The unit circle T passes through -1 and 1; hence $\varphi(T)$ is a straight line through $\varphi(-1) = 0$. Since T makes a right angle with the real axis at -1, $\varphi(T)$ makes a right angle with the real axis at 0. Thus $\varphi(T)$ is the imaginary axis. Since $\varphi(0) = 1$, it follows that φ is *a conformal one-to-one mapping of the open unit disc onto the open right half plane*.

The role of linear fractional transformations in the theory of conformal mapping is also well illustrated by Theorem 12.6.

14.4 Linear fractional transformations make it possible to transfer theorems concerning the behavior of holomorphic functions near straight lines to situations where circular arcs occur instead. It will be enough to illustrate the method with an informal discussion of the reflection principle.

Suppose Ω is a region in U, bounded in part by an arc L on the unit circle, and f is continuous on $\bar{\Omega}$, holomorphic in Ω, and real on L. The function

$$\psi(z) = \frac{z-i}{z+i} \tag{1}$$

maps the upper half plane onto U. If $g = f \circ \psi$, Theorem 11.14 gives us a holomorphic extension G of g, and then $F = G \circ \psi^{-1}$ gives a holomorphic extension F of f which satisfies the relation

$$f(z^*) = \overline{F(z)}, \tag{2}$$

where $z^* = 1/\bar{z}$.

The last assertion follows from a property of ψ: If $w = \psi(z)$ and $w_1 = \psi(\bar{z})$, then $w_1 = w^*$, as is easily verified by computation.

Exercises 2 to 5 furnish other applications of this technique.

Normal Families

The Riemann mapping theorem will be proved by exhibiting the mapping function as the solution of a certain extremum problem. The existence of this solution depends on a very useful compactness property of certain families of holomorphic functions which we now formulate.

14.5 Definition Suppose $\mathscr{F} \subset H(\Omega)$, for some region Ω. We call $\mathscr{F}$ a *normal family* if every sequence of members of $\mathscr{F}$ contains a subsequence which converges uniformly on compact subsets of Ω. The limit function is not required to belong to $\mathscr{F}$.

(Sometimes a wider definition is adopted, by merely requiring that every sequence in $\mathscr{F}$ either converges or tends to ∞, uniformly on compact subsets of Ω. This is well adapted for dealing with meromorphic functions.)

14.6 Theorem *Suppose $\mathscr{F} \subset H(\Omega)$ and $\mathscr{F}$ is uniformly bounded on each compact subset of the region Ω. Then $\mathscr{F}$ is a normal family.*

PROOF The hypothesis means that to each compact $K \subset \Omega$ there corresponds a number $M(K) < \infty$ such that $|f(z)| \leq M(K)$ for all $f \in \mathscr{F}$ and all $z \in K$.

Let $\{K_n\}$ be a sequence of compact sets whose union is Ω, such that K_n lies in the interior of K_{n+1}; such a sequence was constructed in Theorem 13.3. Then there exist positive numbers δ_n such that

$$D(z; 2\delta_n) \subset K_{n+1} \qquad (z \in K_n). \tag{1}$$

Consider two points z' and z'' in K_n, such that $|z' - z''| < \delta_n$, let γ be the positively oriented circle with center at z' and radius $2\delta_n$, and estimate $|f(z') - f(z'')|$ by the Cauchy formula. Since

$$\frac{1}{\zeta - z'} - \frac{1}{\zeta - z''} = \frac{z' - z''}{(\zeta - z')(\zeta - z'')},$$

we have

$$f(z') - f(z'') = \frac{z' - z''}{2\pi i} \int_{\gamma} \frac{f(\zeta)}{(\zeta - z')(\zeta - z'')}\, d\zeta, \tag{2}$$

and since $|\zeta - z'| = 2\delta_n$ and $|\zeta - z''| > \delta_n$ for all $\zeta \in \gamma^*$, (2) gives the inequality

$$|f(z') - f(z'')| < \frac{M(K_{n+1})}{\delta_n} |z' - z''|, \tag{3}$$

valid for all $f \in \mathscr{F}$ and all z' and $z'' \in K_n$, provided that $|z' - z''| < \delta_n$.

This was the crucial step in the proof: We have proved, for each K_n, that the restrictions of the members of $\mathscr{F}$ to K_n form an *equicontinuous family*.

If $f_j \in \mathscr{F}$, for $j = 1, 2, 3, \ldots$, Theorem 11.28 implies therefore that there are infinite sets S_n of positive integers, $S_1 \supset S_2 \supset S_3 \supset \cdots$, so that $\{f_j\}$ converges uniformly on K_n as $j \to \infty$ within S_n. The diagonal process yields then an infinite set S such that $\{f_j\}$ converges uniformly on every K_n (and hence on every compact $K \subset \Omega$) as $j \to \infty$ within S. ////

The Riemann Mapping Theorem

14.7 Conformal Equivalence We call two regions Ω_1 and Ω_2 *conformally equivalent* if there exists a $\varphi \in H(\Omega_1)$ such that φ is one-to-one in Ω_1 and such that $\varphi(\Omega_1) = \Omega_2$, i.e., if there exists a conformal one-to-one mapping of Ω_1 onto Ω_2.

Under these conditions, the inverse of φ is holomorphic in Ω_2 (Theorem 10.33) and hence is a conformal mapping of Ω_2 onto Ω_1.

It follows that conformally equivalent regions are homeomorphic. But there is a much more important relation between conformally equivalent regions: If φ is as above, $f \to f \circ \varphi$ is a one-to-one mapping of $H(\Omega_2)$ onto $H(\Omega_1)$ which preserves sums and products, i.e., which is a ring isomorphism of $H(\Omega_2)$ onto $H(\Omega_1)$. If Ω_1 has a simple structure, problems about $H(\Omega_2)$ can be transferred to problems in $H(\Omega_1)$ and the solutions can be carried back to $H(\Omega_2)$ with the aid of the mapping function φ. The most important case of this is based on the Riemann mapping theorem (where Ω_1 is the unit disc U), which reduces the study of $H(\Omega)$ to the study of $H(U)$, for any simply connected proper subregion of the plane. Of course, for explicit solutions of problems, it may be necessary to have rather precise information about the mapping function.

14.8 Theorem *Every simply connected region Ω in the plane (other than the plane itself) is conformally equivalent to the open unit disc U.*

Note: The case of the plane clearly has to be excluded, by Liouville's theorem. Thus the plane is not conformally equivalent to U, although the two regions are homeomorphic.

The only property of simply connected regions which will be used in the proof is that every holomorphic function which has no zero in such a region has a holomorphic square root there. This will furnish the conclusion "(j) implies (a)" in Theorem 13.11 and will thus complete the proof of that theorem.

PROOF Suppose Ω is a simply connected region in the plane and let w_0 be a complex number, $w_0 \notin \Omega$. Let Σ be the class of all $\psi \in H(\Omega)$ which are one-to-one in Ω and which map Ω into U. We have to prove that some $\psi \in \Sigma$ maps Ω *onto* U.

We first prove that Σ *is not empty*. Since Ω is simply connected, there exists a $\varphi \in H(\Omega)$ so that $\varphi^2(z) = z - w_0$ in Ω. If $\varphi(z_1) = \varphi(z_2)$, then also $\varphi^2(z_1) = \varphi^2(z_2)$, hence $z_1 = z_2$; thus φ is one-to-one. The same argument shows that there are no two distinct points z_1 and z_2 in Ω such that $\varphi(z_1) = -\varphi(z_2)$. Since φ is an open mapping, $\varphi(\Omega)$ contains a disc $D(a; r)$, with $0 < r < |a|$. The disc $D(-a; r)$ therefore fails to intersect $\varphi(\Omega)$, and if we define $\psi = r/(\varphi + a)$, we see that $\psi \in \Sigma$.

The next step consists in showing that *if $\psi \in \Sigma$, if $\psi(\Omega)$ does not cover all of U, and if $z_0 \in \Omega$, then there exists a $\psi_1 \in \Sigma$ with*

$$|\psi_1'(z_0)| > |\psi'(z_0)|.$$

It will be convenient to use the functions φ_α defined by

$$\varphi_\alpha(z) = \frac{z - \alpha}{1 - \bar{\alpha}z}.$$

For $\alpha \in U$, φ_α is a one-to-one mapping of U onto U; its inverse is $\varphi_{-\alpha}$ (Theorem 12.4).

Suppose $\psi \in \Sigma$, $\alpha \in U$, and $\alpha \notin \psi(\Omega)$. Then $\varphi_\alpha \circ \psi \in \Sigma$, and $\varphi_\alpha \circ \psi$ has no zero in Ω; hence there exists a $g \in H(\Omega)$ such that $g^2 = \varphi_\alpha \circ \psi$. We see that g is one-to-one (as in the proof that $\Sigma \neq \emptyset$), hence $g \in \Sigma$; and if $\psi_1 = \varphi_\beta \circ g$, where $\beta = g(z_0)$, it follows that $\psi_1 \in \Sigma$. With the notation $w^2 = s(w)$, we now have

$$\psi = \varphi_{-\alpha} \circ s \circ g = \varphi_{-\alpha} \circ s \circ \varphi_{-\beta} \circ \psi_1.$$

Since $\psi_1(z_0) = 0$, the chain rule gives

$$\psi'(z_0) = F'(0)\psi_1'(z_0),$$

where $F = \varphi_{-\alpha} \circ s \circ \varphi_{-\beta}$. We see that $F(U) \subset U$ and that F is not one-to-one in U. Therefore $|F'(0)| < 1$, by the Schwarz lemma (see Sec. 12.5), so that $|\psi'(z_0)| < |\psi_1'(z_0)|$. [Note that $\psi'(z_0) \neq 0$, since ψ is one-to-one in Ω.]

Fix $z_0 \in \Omega$, and put

$$\eta = \sup \{|\psi'(z_0)| : \psi \in \Sigma\}.$$

The foregoing makes it clear that *any $h \in \Sigma$ for which $|h'(z_0)| = \eta$ will map Ω onto U.* Hence the proof will be completed as soon as we prove the existence of such an h.

Since $|\psi(z)| < 1$ for all $\psi \in \Sigma$ and $z \in \Omega$, Theorem 14.6 shows that Σ is a normal family. The definition of η shows that there is a sequence $\{\psi_n\}$ in Σ such that $|\psi_n'(z_0)| \to \eta$, and by normality of Σ we can extract a subsequence (again denoted by $\{\psi_n\}$, for simplicity) which converges, uniformly on compact subsets of Ω, to a limit $h \in H(\Omega)$. By Theorem 10.28, $|h'(z_0)| = \eta$. Since $\Sigma \neq \emptyset$, $\eta > 0$, so h is not constant. Since $\psi_n(\Omega) \subset U$, for $n = 1, 2, 3, \ldots$, we have $h(\Omega) \subset \bar{U}$, but the open mapping theorem shows that actually $h(\Omega) \subset U$.

So all that remains to be shown is that h is one-to-one. Fix distinct points z_1 and $z_2 \in \Omega$; put $\alpha = h(z_1)$ and $\alpha_n = \psi_n(z_1)$ for $n = 1, 2, 3, \ldots$; and let $\bar{D}$ be a closed circular disc in Ω with center at z_2, such that $z_1 \notin \bar{D}$ and such that $h - \alpha$ has no zero on the boundary of $\bar{D}$. This is possible, since the zeros of $h - \alpha$ have no limit point in Ω. The functions $\psi_n - \alpha_n$ converge to $h - \alpha$, uniformly on $\bar{D}$; they have no zero in D since they are one-to-one and have a zero at z_1; it now follows from Rouché's theorem that $h - \alpha$ has no zero in D; in particular, $h(z_2) \neq h(z_1)$.

Thus $h \in \Sigma$, and the proof is complete. ////

A more constructive proof is outlined in Exercise 26.

CONFORMAL MAPPING 285

14.9 Remarks The preceding proof also shows that $h(z_0) = 0$. For if $h(z_0) = \beta$ and $\beta \neq 0$, then $\varphi_\beta \circ h \in \Sigma$, and

$$|(\varphi_\beta \circ h)'(z_0)| = |\varphi'_\beta(\beta) h'(z_0)| = \frac{|h'(z_0)|}{1 - |\beta|^2} > |h'(z_0)|.$$

It is interesting to observe that although h was obtained by maximizing $|\psi'(z_0)|$ for $\psi \in \Sigma$, h also maximizes $|f'(z_0)|$ if f is allowed to range over the class consisting of all holomorphic mappings of Ω into U (not necessarily one-to-one). For if f is such a function, then $g = f \circ h^{-1}$ maps U into U, hence $|g'(0)| \leq 1$, with equality holding (by the Schwarz lemma) if and only if g is a rotation, so the chain rule gives the following result:

If $f \in H(\Omega)$, $f(\Omega) \subset U$, and $z_0 \in \Omega$, then $|f'(z_0)| \leq |h'(z_0)|$. Equality holds if and only if $f(z) = \lambda h(z)$, for some constant λ with $|\lambda| = 1$.

The Class $\mathscr{S}$

14.10 Definition $\mathscr{S}$ is the class of all $f \in H(U)$ which are *one-to-one* in U and which satisfy

$$f(0) = 0, \qquad f'(0) = 1. \tag{1}$$

Thus every $f \in \mathscr{S}$ has a power series expansion

$$f(z) = z + \sum_{n=2}^\infty a_n z^n \qquad (z \in U). \tag{2}$$

The class $\mathscr{S}$ is not closed under addition or multiplication, but has many other interesting properties. We shall develop only a few of these in this section. Theorem 14.15 will be used in the proof of Mergelyan's theorem, in Chap. 20.

14.11 Example If $|\alpha| \leq 1$ and

$$f_\alpha(z) = \frac{z}{(1 - \alpha z)^2} = \sum_{n=1}^\infty n\alpha^{n-1} z^n$$

then $f_\alpha \in \mathscr{S}$.

For if $f_\alpha(z) = f_\alpha(w)$, then $(z - w)(1 - \alpha^2 zw) = 0$, and the second factor is not 0 if $|z| < 1$ and $|w| < 1$.

When $|\alpha| = 1$, f_α is called a Koebe function. We leave it as an exercise to find the regions $f_\alpha(U)$.

14.12 Theorem (a) If $f \in \mathscr{S}$, $|\alpha| = 1$, and $g(z) = \bar\alpha f(\alpha z)$, then $g \in \mathscr{S}$. (b) If $f \in \mathscr{S}$ there exists a $g \in \mathscr{S}$ such that

$$g^2(z) = f(z^2) \qquad (z \in U). \tag{1}$$

PROOF (a) is clear. To prove (b), write $f(z) = z\varphi(z)$. Then $\varphi \in H(U)$, $\varphi(0) = 1$, and φ has no zero in U, since f has no zero in $U - \{0\}$. Hence there exists an $h \in H(U)$ with $h(0) = 1$, $h^2(z) = \varphi(z)$. Put

$$g(z) = zh(z^2) \qquad (z \in U). \tag{2}$$

Then $g^2(z) = z^2 h^2(z^2) = z^2 \varphi(z^2) = f(z^2)$, so that (1) holds. It is clear that $g(0) = 0$ and $g'(0) = 1$. We have to show that g is one-to-one.

Suppose z and $w \in U$ and $g(z) = g(w)$. Since f is one-to-one, (1) implies that $z^2 = w^2$. So either $z = w$ (which is what we want to prove) or $z = -w$. In the latter case, (2) shows that $g(z) = -g(w)$; it follows that $g(z) = g(w) = 0$, and since g has no zero in $U - \{0\}$, we have $z = w = 0$. ////

14.13 Theorem *If $F \in H(U - \{0\})$, F is one-to-one in U, and*

$$F(z) = \frac{1}{z} + \sum_{n=0}^{\infty} \alpha_n z^n \qquad (z \in U), \tag{1}$$

then

$$\sum_{n=1}^{\infty} n |\alpha_n|^2 \leq 1. \tag{2}$$

This is usually called the *area theorem*, for reasons which will become apparent in the proof.

PROOF The choice of α_0 is clearly irrelevant. So assume $\alpha_0 = 0$. Neither the hypothesis nor the conclusion is affected if we replace $F(z)$ by $\lambda F(\lambda z)$ ($|\lambda| = 1$). So we may assume that α_1 is real.
Put $U_r = \{z : |z| < r\}$, $C_r = \{z : |z| = r\}$, and $V_r = \{z : r < |z| < 1\}$, for $0 < r < 1$. Then $F(U_r)$ is a neighborhood of ∞ (by the open mapping theorem, applied to $1/F$); the sets $F(U_r)$, $F(C_r)$, and $F(V_r)$ are disjoint, since F is one-to-one. Write

$$F(z) = \frac{1}{z} + \alpha_1 z + \varphi(z) \qquad (z \in U), \tag{3}$$

$F = u + iv$, and

$$A = \frac{1}{r} + \alpha_1 r, \qquad B = \frac{1}{r} - \alpha_1 r. \tag{4}$$

For $z = re^{i\theta}$, we then obtain

$$u = A \cos \theta + \operatorname{Re} \varphi \quad \text{and} \quad v = -B \sin \theta + \operatorname{Im} \varphi. \tag{5}$$

Divide Eqs. (5) by A and B, respectively, square, and add:

$$\frac{u^2}{A^2} + \frac{v^2}{B^2} = 1 + \frac{2\cos\theta}{A}\operatorname{Re}\varphi + \left(\frac{\operatorname{Re}\varphi}{A}\right)^2 - \frac{2\sin\theta}{B}\operatorname{Im}\varphi + \left(\frac{\operatorname{Im}\varphi}{B}\right)^2.$$

By (3), φ has a zero of order at least 2 at the origin. If we keep account of (4), it follows that there exists an $\eta > 0$ such that, for all sufficiently small r,

$$\frac{u^2}{A^2} + \frac{v^2}{B^2} < 1 + \eta r^3 \qquad (z = re^{i\theta}). \tag{6}$$

This says that $F(C_r)$ is in the interior of the ellipse E_r whose semiaxes are $A\sqrt{1+\eta r^3}$ and $B\sqrt{1+\eta r^3}$, and which therefore bounds an area

$$\pi AB(1+\eta r^3) = \pi\left(\frac{1}{r} + \alpha_1 r\right)\left(\frac{1}{r} - \alpha_1 r\right)(1+\eta r^3) \leq \frac{\pi}{r^2}(1+\eta r^3). \tag{7}$$

Since $F(C_r)$ is in the interior of E_r, we have $E_r \subset F(U_r)$; hence $F(V_r)$ is in the interior of E_r, so the area of $F(V_r)$ is no larger than (7). The Cauchy-Riemann equations show that the Jacobian of the mapping $(x, y) \to (u, v)$ is $|F'|^2$. Theorem 7.26 therefore gives the following result:

$$\frac{\pi}{r^2}(1+\eta r^3) \geq \iint_{V_r} |F'|^2$$

$$= \int_r^1 t\, dt \int_0^{2\pi} \left| -t^{-2}e^{-2i\theta} + \sum_1^\infty n\alpha_n t^{n-1} e^{i(n-1)\theta} \right|^2 d\theta$$

$$= 2\pi \int_r^1 \left(t^{-3} + \sum_1^\infty n^2|\alpha_n|^2 t^{2n-1}\right) dt$$

$$= \pi\left\{r^{-2} - 1 + \sum_1^\infty n|\alpha_n|^2(1 - r^{2n})\right\}. \tag{8}$$

If we divide (8) by π and then subtract r^{-2} from each side, we obtain

$$\sum_{n=1}^N n|\alpha_n|^2(1 - r^{2n}) \leq 1 + \eta r \tag{9}$$

for all sufficiently small r and for all positive integers N. Let $r \to 0$ in (9), then let $N \to \infty$. This gives (2). ////

Corollary *Under the same hypothesis, $|\alpha_1| \leq 1$.*

That this is in fact best possible is shown by $F(z) = (1/z) + \alpha z$, $|\alpha| = 1$, which is one-to-one in U.

14.14 Theorem *If $f \in \mathscr{S}$, and*

$$f(z) = z + \sum_{n=2}^{\infty} a_n z^n,$$

then (a) $|a_2| \leq 2$, and (b) $f(U) \supset D(0; \tfrac{1}{4})$.

The second assertion is that $f(U)$ contains all w with $|w| < \tfrac{1}{4}$.

PROOF By Theorem 14.12, there exists a $g \in \mathscr{S}$ so that $g^2(z) = f(z^2)$. If $G = 1/g$, then Theorem 14.13 applies to G, and this will give (a). Since

$$f(z^2) = z^2(1 + a_2 z^2 + \cdots),$$

we have

$$g(z) = z(1 + \tfrac{1}{2}a_2 z^2 + \cdots),$$

and hence

$$G(z) = \frac{1}{z}(1 - \tfrac{1}{2}a_2 z^2 + \cdots) = \frac{1}{z} - \frac{a_2}{2} z + \cdots.$$

The Corollary to Theorem 14.13 shows now that $|a_2| \leq 2$.

To prove (b), suppose $w \notin f(U)$. Define

$$h(z) = \frac{f(z)}{1 - f(z)/w}.$$

Then $h \in H(U)$, h is one-to-one in U, and

$$h(z) = (z + a_2 z^2 + \cdots)\left(1 + \frac{z}{w} + \cdots\right) = z + \left(a_2 + \frac{1}{w}\right) z^2 + \cdots,$$

so that $h \in \mathscr{S}$. Apply (a) to h: We have $|a_2 + (1/w)| \leq 2$, and since $|a_2| \leq 2$, we finally obtain $|1/w| \leq 4$. So $|w| \geq \tfrac{1}{4}$ for every $w \notin f(U)$. This completes the proof. ////

Example 14.11 shows that both (a) and (b) are best possible.

Moreover, given any $\alpha \neq 0$, one can find entire functions f, with $f(0) = 0$, $f'(0) = 1$, that omit the value α. For example,

$$f(z) = \alpha(1 - e^{-z/\alpha}).$$

Of course, no such f can be one-to-one in U if $|\alpha| < \tfrac{1}{4}$.

14.15 Theorem *Suppose $F \in H(U - \{0\})$, F is one-to-one in U, F has a pole of order 1 at $z = 0$, with residue 1, and neither w_1 nor w_2 are in $F(U)$.*

Then $|w_1 - w_2| \leq 4$.

PROOF If $f = 1/(F - w_1)$, then $f \in \mathscr{S}$, hence $f(U) \supset D(0, \frac{1}{4})$, so the image of U under $F - w_1$ contains all w with $|w| > 4$. Since $w_2 - w_1$ is not in this image, we have $|w_2 - w_1| \le 4$. ////

Note that this too is best possible: If $F(z) = z^{-1} + z$, then $F(U)$ does not contain the points 2, -2. In fact, the complement of $F(U)$ is precisely the interval $[-2, 2]$ on the real axis.

Continuity at the Boundary

Under certain conditions, every conformal mapping of a simply connected region Ω onto U can be extended to a homeomorphism of its closure $\bar{\Omega}$ onto $\bar{U}$. The nature of the boundary of Ω plays a decisive role here.

14.16 Definition A boundary point β of a simply connected plane region Ω will be called a *simple boundary point of* Ω if β has the following property: To every sequence $\{\alpha_n\}$ in Ω such that $\alpha_n \to \beta$ as $n \to \infty$ there corresponds a curve γ, with parameter interval $[0, 1]$, and a sequence $\{t_n\}$, $0 < t_1 < t_2 < \cdots$, $t_n \to 1$, such that $\gamma(t_n) = \alpha_n$ ($n = 1, 2, 3, \ldots$) and $\gamma(t) \in \Omega$ for $0 \le t < 1$.

In other words, there is a curve in Ω which passes through the points α_n and which ends at β.

14.17 Examples Since examples of simple boundary points are obvious, let us look at some that are not simple.

If Ω is $U - \{x : 0 \le x < 1\}$, then Ω is simply connected; and if $0 < \beta \le 1$, β is a boundary point of Ω which is not simple.

To get a more complicated example, let Ω_0 be the interior of the square with vertices at the points 0, 1, $1 + i$, and i. Remove the intervals

$$\left[\frac{1}{2n}, \frac{1}{2n} + \frac{n-1}{n} i\right] \quad \text{and} \quad \left[\frac{1}{2n+1} + \frac{i}{n}, \frac{1}{2n+1} + i\right]$$

from Ω_0. The resulting region Ω is simply connected. If $0 \le y \le 1$, then iy is a boundary point which is not simple.

14.18 Theorem *Let Ω be a bounded simply connected region in the plane, and let f be a conformal mapping of Ω onto U.*

(a) *If β is a simple boundary point of Ω, then f has a continuous extension to $\Omega \cup \{\beta\}$. If f is so extended, then $|f(\beta)| = 1$.*

(b) *If β_1 and β_2 are distinct simple boundary points of Ω and if f is extended to $\Omega \cup \{\beta_1\} \cup \{\beta_2\}$ as in (a), then $f(\beta_1) \ne f(\beta_2)$.*

PROOF Let g be the inverse of f. Then $g \in H(U)$, by Theorem 10.33, $g(U) = \Omega$, g is one-to-one, and $g \in H^\infty$, since Ω is bounded.

Suppose (a) is false. Then there is a sequence $\{\alpha_n\}$ in Ω such that $\alpha_n \to \beta$, $f(\alpha_{2n}) \to w_1$, $f(\alpha_{2n+1}) \to w_2$, and $w_1 \neq w_2$. Choose γ as in Definition 14.16, and put $\Gamma(t) = f(\gamma(t))$, for $0 \leq t < 1$. Put $K_r = g(\bar{D}(0; r))$, for $0 < r < 1$. Then K_r is a compact subset of Ω. Since $\gamma(t) \to \beta$ as $t \to 1$, there exists a $t^* < 1$ (depending on r) such that $\gamma(t) \notin K_r$ if $t^* < t < 1$. Thus $|\Gamma(t)| > r$ if $t^* < t < 1$. This says that $|\Gamma(t)| \to 1$ as $t \to 1$. Since $\Gamma(t_{2n}) \to w_1$ and $\Gamma(t_{2n+1}) \to w_2$, we also have $|w_1| = |w_2| = 1$.

It now follows that one of the two open arcs J whose union is $T - (\{w_1\} \cup \{w_2\})$ has the property that every radius of U which ends at a point of J intersects the range of Γ in a set which has a limit point on T. Note that $g(\Gamma(t)) = \gamma(t)$ for $0 \leq t < 1$ and that g has radial limits a.e. on T, since $g \in H^\infty$. Hence

$$\lim_{r \to 1} g(re^{it}) = \beta \qquad \text{(a.e. on } J\text{)}, \tag{1}$$

since $g(\Gamma(t)) \to \beta$ as $t \to 1$. By Theorem 11.32, applied to $g - \beta$, (1) shows that g is constant. But g is one-to-one in U, and we have a contradiction. Thus $w_1 = w_2$, and (a) is proved.

Suppose (b) is false. If we multiply f by a suitable constant of absolute value 1, we then have $\beta_1 \neq \beta_2$ but $f(\beta_1) = f(\beta_2) = 1$.

Since β_1 and β_2 are simple boundary points of Ω, there are curves γ_i with parameter interval $[0, 1]$ such that $\gamma_i([0, 1)) \subset \Omega$ for $i = 1$ and 2 and $\gamma_i(1) = \beta_i$. Put $\Gamma_i(t) = f(\gamma_i(t))$. Then $\Gamma_i([0, 1)) \subset U$, and $\Gamma_1(1) = \Gamma_2(1) = 1$. Since $g(\Gamma_i(t)) = \gamma_i(t)$ on $[0, 1)$, we have

$$\lim_{t \to 1} g(\Gamma_i(t)) = \beta_i \qquad (i = 1, 2). \tag{2}$$

Theorem 12.10 implies therefore that the radial limit of g at 1 is β_1 as well as β_2. This is impossible if $\beta_1 \neq \beta_2$. ////

14.19 Theorem *If Ω is a bounded simply connected region in the plane and if every boundary point of Ω is simple, then every conformal mapping of Ω onto U extends to a homeomorphism of $\bar{\Omega}$ onto $\bar{U}$.*

PROOF Suppose $f \in H(\Omega)$, $f(\Omega) = U$, and f is one-to-one. By Theorem 14.18 we can extend f to a mapping of $\bar{\Omega}$ into $\bar{U}$ such that $f(\alpha_n) \to f(z)$ whenever $\{\alpha_n\}$ is a sequence in Ω which converges to z. If $\{z_n\}$ is a sequence in $\bar{\Omega}$ which converges to z, there exist points $\alpha_n \in \Omega$ such that $|\alpha_n - z_n| < 1/n$ and $|f(\alpha_n) - f(z_n)| < 1/n$. Thus $\alpha_n \to z$, hence $f(\alpha_n) \to f(z)$, and this shows that $f(z_n) \to f(z)$.

We have now proved that our extension of f is continuous on $\bar{\Omega}$. Also $U \subset f(\bar{\Omega}) \subset \bar{U}$. The compactness of $\bar{U}$ implies that $f(\bar{\Omega})$ is compact. Hence $f(\bar{\Omega}) = \bar{U}$.

Theorem 14.18(b) shows that f is one-to-one on $\bar{\Omega}$. Since every continuous one-to-one mapping of a compact set has a continuous inverse ([26], Theorem 4.17), the proof is complete. ////

14.20 Remarks
(a) *The preceding theorem has a purely topological corollary: If every boundary point of a bounded simply connected plane region Ω is simple, then the boundary of Ω is a Jordan curve, and $\bar{\Omega}$ is homeomorphic to $\bar{U}$.*

(A Jordan curve is, by definition, a homeomorphic image of the unit circle.)

The converse is true, but we shall not prove it: If the boundary of Ω is a Jordan curve, then every boundary point of Ω is simple.

(b) Suppose f is as in Theorem 14.19, a, b, and c are distinct boundary points of Ω, and A, B, and C are distinct points of T. There is a linear fractional transformation φ which maps the triple $\{f(a), f(b), f(c)\}$ to $\{A, B, C\}$; suppose the orientation of $\{A, B, C\}$ agrees with that of $\{f(a), f(b), f(c)\}$; then $\varphi(U) = U$, and the function $g = \varphi \circ f$ is a homeomorphism of $\bar{\Omega}$ onto $\bar{U}$ which is holomorphic in Ω and which maps $\{a, b, c\}$ to prescribed values $\{A, B, C\}$. It follows from Sec. 14.3 that g is uniquely determined by these requirements.

(c) Theorem 14.19, as well as the above remark (b), extends without difficulty to simply connected regions Ω in the Riemann sphere S^2, all of whose boundary points are simple, provided that $S^2 - \Omega$ has a nonempty interior, for then a linear fractional transformation brings us back to the case in which Ω is a bounded region in the plane. Likewise, U can be replaced, for instance, by a half plane.

(d) More generally, if f_1 and f_2 map Ω_1 and Ω_2 onto U, as in Theorem 14.19, then $f = f_2^{-1} \circ f_1$ is a homeomorphism of $\bar{\Omega}_1$ onto $\bar{\Omega}_2$ which is holomorphic in Ω_1.

Conformal Mapping of an Annulus

14.21 It is a consequence of the Riemann mapping theorem that any two simply connected proper subregions of the plane are conformally equivalent, since each of them is conformally equivalent to the unit disc. This is a very special property of simply connected regions. One may ask whether it extends to the next simplest situation, i.e., whether any two annuli are conformally equivalent. The answer is negative.

For $0 < r < R$, let

$$A(r, R) = \{z : r < |z| < R\} \quad (1)$$

be the annulus with inner radius r and outer radius R. If $\lambda > 0$, the mapping $z \to \lambda z$ maps $A(r, R)$ onto $A(\lambda r, \lambda R)$. Hence $A(r, R)$ and $A(r_1, R_1)$ are *conformally equivalent if $R/r = R_1/r_1$.* The surprising fact is that this sufficient condition is also necessary; thus among the annuli there is a different conformal type associated with each real number greater than 1.

14.22 Theorem $A(r_1, R_1)$ and $A(r_2, R_2)$ are conformally equivalent if and only if $R_1/r_1 = R_2/r_2$.

PROOF Assume $r_1 = r_2 = 1$, without loss of generality. Put

$$A_1 = A(1, R_1), \qquad A_2 = A(1, R_2), \tag{1}$$

and assume there exists $f \in H(A_1)$ such that f is one-to-one and $f(A_1) = A_2$. Let K be the circle with center at 0 and radius $r = \sqrt{R_2}$. Since $f^{-1}: A_2 \to A_1$ is also holomorphic, $f^{-1}(K)$ is compact. Hence

$$A(1, 1 + \epsilon) \cap f^{-1}(K) = \emptyset \tag{2}$$

for some $\epsilon > 0$. Then $V = f(A(1, 1 + \epsilon))$ is a connected subset of A_2 which does not intersect K, so that $V \subset A(1, r)$ or $V \subset A(r, R_2)$. In the latter case, replace f by R_2/f. So we can assume that $V \subset A(1, r)$. If $1 < |z_n| < 1 + \epsilon$ and $|z_n| \to 1$, then $f(z_n) \in V$ and $\{f(z_n)\}$ has no limit point in A_2 (since f^{-1} is continuous); thus $|f(z_n)| \to 1$. In the same manner we see that $|f(z_n)| \to R_2$ if $|z_n| \to R_1$.

Now define

$$\alpha = \frac{\log R_2}{\log R_1} \tag{3}$$

and

$$u(z) = 2 \log |f(z)| - 2\alpha \log |z| \qquad (z \in A_1). \tag{4}$$

Let ∂ be one of the Cauchy-Riemann operators. Since $\partial \bar{f} = 0$ and $\partial f = f'$, the chain rule gives

$$\partial(2 \log |f|) = \partial(\log(\bar{f}f)) = f'/f, \tag{5}$$

so that

$$(\partial u)(z) = \frac{f'(z)}{f(z)} - \frac{\alpha}{z} \qquad (z \in A_1). \tag{6}$$

Thus u is a harmonic function in A_1 which, by the first paragraph of this proof, extends to a continuous function on $\bar{A}_1$ which is 0 on the boundary of A_1. Since nonconstant harmonic functions have no local maxima or minima, we conclude that $u = 0$. Thus

$$\frac{f'(z)}{f(z)} = \frac{\alpha}{z} \qquad (z \in A_1). \tag{7}$$

Put $\gamma(t) = \sqrt{R_1}\, e^{it}$ $(-\pi \le t \le \pi)$; put $\Gamma = f \circ \gamma$. As in the proof of Theorem 10.43, (7) gives

$$\alpha = \frac{1}{2\pi i} \int_\gamma \frac{f'(z)}{f(z)}\, dz = \text{Ind}_\Gamma\, (0). \tag{8}$$

Thus α is an integer. By (3), $\alpha > 0$. By (7), the derivative of $z^{-\alpha} f(z)$ is 0 in A_1. Thus $f(z) = cz^\alpha$. Since f is one-to-one in A_1, $\alpha = 1$. Hence $R_2 = R_1$. ////

Exercises

1 Find necessary and sufficient conditions which the complex numbers a, b, c, and d have to satisfy so that the linear fractional transformation $z \to (az + b)/(cz + d)$ maps the upper half plane onto itself.

2 In Theorem 11.14 the hypotheses were, in simplified form, that $\Omega \subset \Pi^+$, L is on the real axis, and $\text{Im}\, f(z) \to 0$ as $z \to L$. Use this theorem to establish analogous reflection theorems under the following hypotheses:

(a) $\Omega \subset \Pi^+$, L on real axis, $|f(z)| \to 1$ as $z \to L$.
(b) $\Omega \subset U$, $L \subset T$, $|f(z)| \to 1$ as $z \to L$.
(c) $\Omega \subset U$, $L \subset T$, $\text{Im}\, f(z) \to 0$ as $z \to L$.

In case (b), if f has a zero at $\alpha \in \Omega$, show that its extension has a pole at $1/\bar{\alpha}$. What are the analogues of this in cases (a) and (c)?

3 Suppose R is a rational function such that $|R(z)| = 1$ if $|z| = 1$. Prove that

$$R(z) = cz^m \prod_{n=1}^{k} \frac{z - \alpha_n}{1 - \bar{\alpha}_n z}$$

where c is a constant, m is an integer, and $\alpha_1, \ldots, \alpha_k$ are complex numbers such that $\alpha_n \ne 0$ and $|\alpha_n| \ne 1$. Note that each of the above factors has absolute value 1 if $|z| = 1$.

4 Obtain an analogous description of those rational functions which are positive on T.

Hint: Such a function must have the same number of zeros as poles in U. Consider products of factors of the form

$$\frac{(z - \alpha)(1 - \bar{\alpha} z)}{(z - \beta)(1 - \bar{\beta} z)}$$

where $|\alpha| < 1$ and $|\beta| < 1$.

5 Suppose f is a trigonometric polynomial,

$$f(\theta) = \sum_{k=-n}^{n} a_k e^{ik\theta},$$

and $f(\theta) > 0$ for all real θ. Prove that there is a polynomial $P(z) = c_0 + c_1 z + \cdots + c_n z^n$ such that

$$f(\theta) = |P(e^{i\theta})|^2 \quad (\theta \text{ real}).$$

Hint: Apply Exercise 4 to the rational function $\Sigma a_k z^k$. Is the result still valid if we assume $f(\theta) \ge 0$ instead of $f(\theta) > 0$?

6 Find the fixed points of the mappings φ_α (Definition 12.3). Is there a straight line which φ_α maps to itself?

7 Find all complex numbers α for which f_α is one-to-one in U, where

$$f_\alpha(z) = \frac{z}{1 + \alpha z^2}.$$

Describe $f_\alpha(U)$ for all these cases.

8 Suppose $f(z) = z + (1/z)$. Describe the families of ellipses and hyperbolas onto which f maps circles with center at 0 and rays through 0.

9 (a) Suppose $\Omega = \{z: -1 < \text{Re } z < 1\}$. Find an explicit formula for the one-to-one conformal mapping f of Ω onto U for which $f(0) = 0$ and $f'(0) > 0$. Compute $f'(0)$.

(b) Note that the inverse of the function constructed in (a) has its real part bounded in U, whereas its imaginary part is unbounded. Show that this implies the existence of a continuous real function u on $\bar{U}$ which is harmonic in U and whose harmonic conjugate v is unbounded in U. [v is the function which makes $u + iv$ holomorphic in U; we can determine v uniquely by the requirement $v(0) = 0$.]

(c) Suppose $g \in H(U)$, $|\text{Re } g| < 1$ in U, and $g(0) = 0$. Prove that

$$|g(re^{i\theta})| \le \frac{2}{\pi} \log \frac{1+r}{1-r}.$$

Hint: See Exercise 10.

(d) Let Ω be the strip that occurs in Theorem 12.9. Fix a point $\alpha + i\beta$ in Ω. Let h be a conformal one-to-one mapping of Ω onto Ω that carries $\alpha + i\beta$ to 0. Prove that

$$|h'(\alpha + i\beta)| = 1/\cos \beta.$$

10 Suppose f and g are holomorphic mappings of U into Ω, f is one-to-one, $f(U) = \Omega$, and $f(0) = g(0)$. Prove that

$$g(D(0; r)) \subset f(D(0; r)) \qquad (0 < r < 1).$$

11 Let Ω be the upper half of the unit disc U. Find the conformal mapping f of Ω onto U that carries $\{-1, 0, 1\}$ to $\{-1, -i, 1\}$. Find $z \in \Omega$ such that $f(z) = 0$. Find $f(i/2)$. *Hint*: $f = \varphi \circ s \circ \psi$, where φ and ψ are linear fractional transformations and $s(\lambda) = \lambda^2$.

12 Suppose Ω is a convex region, $f \in H(\Omega)$, and $\text{Re } f'(z) > 0$ for all $z \in \Omega$. Prove that f is one-to-one in Ω. Is the result changed if the hypothesis is weakened to $\text{Re } f'(z) \ge 0$? (Exclude the trivial case $f = $ constant.) Show by an example that "convex" cannot be replaced by "simply connected."

13 Suppose Ω is a region, $f_n \in H(\Omega)$ for $n = 1, 2, 3, \ldots$, each f_n is one-to-one in Ω, and $f_n \to f$ uniformly on compact subsets of Ω. Prove that f is either constant or one-to-one in Ω. Show that both cases can occur.

14 Suppose $\Omega = \{x + iy: -1 < y < 1\}, f \in H(\Omega), |f| < 1$, and $f(x) \to 0$ as $x \to \infty$. Prove that

$$\lim_{x \to \infty} f(x + iy) = 0 \qquad (-1 < y < 1)$$

and that the passage to the limit is uniform if y is confined to an interval $[-\alpha, \alpha]$, where $\alpha < 1$. *Hint*: Consider the sequence $\{f_n\}$, where $f_n(z) = z + n$, in the square $|x| < 1, |y| < 1$.

What does this theorem tell about the behavior of a function $g \in H^\infty$ near a boundary point of U at which the radial limit of g exists?

15 Let $\mathscr{F}$ be the class of all $f \in H(U)$ such that $\text{Re } f > 0$ and $f(0) = 1$. Show that $\mathscr{F}$ is a normal family. Can the condition "$f(0) = 1$" be omitted? Can it be replaced by "$|f(0)| \le 1$"?

16 Let $\mathscr{F}$ be the class of all $f \in H(U)$ for which

$$\iint_U |f(z)|^2 \, dx \, dy \le 1.$$

Is this a normal family?

17 Suppose Ω is a region, $f_n \in H(\Omega)$ for $n = 1, 2, 3, \ldots, f_n \to f$ uniformly on compact subsets of Ω, and f is one-to-one in Ω. Does it follow that to each compact $K \subset \Omega$ there corresponds an integer $N(K)$ such that f_n is one-to-one on K for all $n > N(K)$? Give proof or counterexample.

18 Suppose Ω is a simply connected region, $z_0 \in \Omega$, and f and g are one-to-one conformal mappings of Ω onto U which carry z_0 to 0. What relation exists between f and g? Answer the same question if $f(z_0) = g(z_0) = a$, for some $a \in U$.

19 Find a homeomorphism of U onto U which cannot be extended to a continuous function on $\bar{U}$.

20 If $f \in \mathscr{S}$ (Definition 14.10) and n is a positive integer, prove that there exists a $g \in \mathscr{S}$ such that $g^n(z) = f(z^n)$ for all $z \in U$.

21 Find all $f \in \mathscr{S}$ such that $(a) f(U) \supset U$, $(b) f(U) \supset \bar{U}$, $(c) |a_2| = 2$.

22 Suppose f is a one-to-one conformal mapping of U onto a square with center at 0, and $f(0) = 0$. Prove that $f(iz) = if(z)$. If $f(z) = \Sigma c_n z^n$, prove that $c_n = 0$ unless $n - 1$ is a multiple of 4. Generalize this: Replace the square by other simply connected regions with rotational symmetry.

23 Let Ω be a bounded region whose boundary consists of two nonintersecting circles. Prove that there is a one-to-one conformal mapping of Ω onto an annulus. (This is true for every region Ω such that $S^2 - \Omega$ has exactly two components, each of which contains more than one point, but this general situation is harder to handle.)

24 Complete the details in the following proof of Theorem 14.22. Suppose $1 < R_2 < R_1$ and f is a one-to-one conformal mapping of $A(1, R_1)$ onto $A(1, R_2)$. Define $f_1 = f$ and $f_n = f \circ f_{n-1}$. Then a subsequence of $\{f_n\}$ converges uniformly on compact subsets of $A(1, R_1)$ to a function g. Show that the range of g cannot contain any nonempty open set (by the three-circle theorem, for instance). On the other hand, show that g cannot be constant on the circle $\{z : |z|^2 = R_1\}$. Hence f cannot exist.

25 Here is yet another proof of Theorem 14.22. If f is as in 14.22, repeated use of the reflection principle extends f to an entire function such that $|f(z)| = 1$ whenever $|z| = 1$. This implies $f(z) = \alpha z^n$, where $|\alpha| = 1$ and n is an integer. Complete the details.

26 Iteration of Step 2 in the proof of Theorem 14.8 leads to a proof (due to Koebe) of the Riemann mapping theorem which is constructive in the sense that it makes no appeal to the theory of normal families and so does not depend on the existence of some unspecified subsequence. For the final step of the proof it is convenient to assume that Ω has property (h) of Theorem 13.11. Then any region conformally equivalent to Ω will satisfy (h). Recall also that (h) implies (j), trivially.

By Step 1 in Theorem 14.8 we may assume, without loss of generality, that $0 \in \Omega$, $\Omega \subset U$, and $\Omega \neq U$. Put $\Omega = \Omega_0$. The proof consists in the construction of regions $\Omega_1, \Omega_2, \Omega_3, \ldots$ and of functions $f_1, f_2, f_3, \ldots$, so that $f_n(\Omega_{n-1}) = \Omega_n$ and so that the functions $f_n \circ f_{n-1} \circ \cdots \circ f_2 \circ f_1$ converge to a conformal mapping of Ω onto U.

Complete the details in the following outline.

(a) Suppose Ω_{n-1} is constructed, let r_n be the largest number such that $D(0; r_n) \subset \Omega_{n-1}$, let α_n be a boundary point of Ω_{n-1} with $|\alpha_n| = r_n$, choose β_n so that $\beta_n^2 = -\alpha_n$, and put

$$F_n = \varphi_{-\alpha_n} \circ s \circ \varphi_{-\beta_n}.$$

(The notation is as in the proof of Theorem 14.8.) Show that F_n has a holomorphic inverse G_n in Ω_{n-1}, and put $f_n = \lambda_n G_n$, where $\lambda_n = |c|/c$ and $c = G'_n(0)$. (This f_n is the *Koebe mapping* associated with Ω_{n-1}. Note that f_n is an elementary function. It involves only two linear fractional transformations and a square root.)

(b) Compute that $f'_n(0) = (1 + r_n)/2\sqrt{r_n} > 1$.

(c) Put $\psi_0(z) = z$ and $\psi_n(z) = f_n(\psi_{n-1}(z))$. Show that ψ_n is a one-to-one mapping of Ω onto a region $\Omega_n \subset U$, that $\{\psi'_n(0)\}$ is bounded, that

$$\psi'_n(0) = \prod_{k=1}^{n} \frac{1 + r_k}{2\sqrt{r_k}},$$

and that therefore $r_n \to 1$ as $n \to \infty$.

(d) Write $\psi_n(z) = z h_n(z)$, for $z \in \Omega$, show that $|h_n| \leq |h_{n+1}|$, apply Harnack's theorem and Exercise 8 of Chap. 11 to $\{\log |h_n|\}$ to prove that $\{\psi_n\}$ converges uniformly on compact subsets of Ω, and show that $\lim \psi_n$ is a one-to-one mapping of Ω onto U.

27 Prove that $\sum_{n=1}^{\infty} (1 - r_n)^2 < \infty$, where $\{r_n\}$ is the sequence which occurs in Exercise 26. *Hint*:

$$\frac{1+r}{2\sqrt{r}} = 1 + \frac{(1-\sqrt{r})^2}{2\sqrt{r}}.$$

28 Suppose that in Exercise 26 we choose $\alpha_n \in U - \Omega_{n-1}$ without insisting that $|\alpha_n| = r_n$. For example, insist only that

$$|\alpha_n| \le \frac{1 + r_n}{2}.$$

Will the resulting sequence $\{\psi_n\}$ still converge to the desired mapping function?

29 Suppose Ω is a bounded region, $a \in \Omega, f \in H(\Omega), f(\Omega) \subset \Omega$, and $f(a) = a$.
 (a) Put $f_1 = f$ and $f_n = f \circ f_{n-1}$, compute $f'_n(a)$, and conclude that $|f'(a)| \le 1$.
 (b) If $f'(a) = 1$, prove that $f(z) = z$ for all $z \in \Omega$. *Hint*: If

$$f(z) = z + c_m(z - a)^m + \cdots,$$

compute the coefficient of $(z - a)^m$ in the expansion of $f_n(z)$.
 (c) If $|f'(a)| = 1$, prove that f is one-to-one and that $f(\Omega) = \Omega$.
 Hint: If $\gamma = f'(a)$, there are integers $n_k \to \infty$ such that $\gamma^{n_k} \to 1$ and $f_{n_k} \to g$. Then $g'(a) = 1$, $g(\Omega) \subset \Omega$ (by Exercise 20, Chap. 10), hence $g(z) = z$, by part (b). Use g to draw the desired conclusions about f.

30 Let Λ be the set of all linear fractional transformations.
 If $\{\alpha, \beta, \gamma, \delta\}$ is an ordered quadruple of distinct complex numbers, its *cross ratio* is defined to be

$$[\alpha, \beta, \gamma, \delta] = \frac{(\alpha - \beta)(\gamma - \delta)}{(\alpha - \delta)(\gamma - \beta)}.$$

If one of these numbers is ∞, the definition is modified in the obvious way, by continuity. The same applies if α coincides with β or γ or δ.
 (a) If $\varphi(z) = [z, \alpha, \beta, \gamma]$, show that $\varphi \in \Lambda$ and φ maps $\{\alpha, \beta, \gamma\}$ to $\{0, 1, \infty\}$.
 (b) Show that the equation $[w, a, b, c] = [z, \alpha, \beta, \gamma]$ can be solved in the form $w = \varphi(z)$; then $\varphi \in \Lambda$ maps $\{\alpha, \beta, \gamma\}$ to $\{a, b, c\}$.
 (c) If $\varphi \in \Lambda$, show that

$$[\varphi(\alpha), \varphi(\beta), \varphi(\gamma), \varphi(\delta)] = [\alpha, \beta, \gamma, \delta].$$

 (d) Show that $[\alpha, \beta, \gamma, \delta]$ is real if and only if the four points lie on the same circle or straight line.
 (e) Two points z and z^* are said to be *symmetric* with respect to the circle (or straight line) C through α, β, and γ if $[z^*, \alpha, \beta, \gamma]$ is the complex conjugate of $[z, \alpha, \beta, \gamma]$. If C is the unit circle, find a simple geometric relation between z and z^*. Do the same if C is a straight line.
 (f) Suppose z and z^* are symmetric with respect to C. Show that $\varphi(z)$ and $\varphi(z^*)$ are symmetric with respect to $\varphi(C)$, for every $\varphi \in \Lambda$.

31 (a) Show that Λ (see Exercise 30) is a group, with composition as group operation. That is, if $\varphi \in \Lambda$ and $\psi \in \Lambda$, show that $\varphi \circ \psi \in \Lambda$ and that the inverse φ^{-1} of φ is in Λ. Show that Λ is not commutative.
 (b) Show that each member of Λ (other than the identity mapping) has either one or two fixed points on S^2. [A fixed point of φ is a point α such that $\varphi(\alpha) = \alpha$.]
 (c) Call two mappings φ and $\varphi_1 \in \Lambda$ *conjugate* if there exists a $\psi \in \Lambda$ such that $\varphi_1 = \psi^{-1} \circ \varphi \circ \psi$. Prove that every $\varphi \in \Lambda$ with a unique fixed point is conjugate to the mapping $z \to z + 1$. Prove that every $\varphi \in \Lambda$ with two distinct fixed points is conjugate to the mapping $z \to \alpha z$, where α is a complex number; to what extent is α determined by φ?

(d) Let α be a complex number. Show that to every $\varphi \in \Lambda$ which has α for its unique fixed point there corresponds a β such that

$$\frac{1}{\varphi(z) - \alpha} = \frac{1}{z - \alpha} + \beta.$$

Let G_α be the set of all these φ, plus the identity transformation. Prove that G_α is a subgroup of Λ and that G_α is isomorphic to the additive group of all complex numbers.

(e) Let α and β be distinct complex numbers, and let $G_{\alpha,\beta}$ be the set of all $\varphi \in \Lambda$ which have α and β as fixed points. Show that every $\varphi \in G_{\alpha,\beta}$ is given by

$$\frac{\varphi(z) - \alpha}{\varphi(z) - \beta} = \gamma \cdot \frac{z - \alpha}{z - \beta},$$

where γ is a complex number. Show that $G_{\alpha,\beta}$ is a subgroup of Λ which is isomorphic to the multiplicative group of all nonzero complex numbers.

(f) If φ is as in (d) or (e), for which circles C is it true that $\varphi(C) = C$? The answer should be in terms of the parameters α, β, and γ.

32 For $z \in \bar{U}$, $z^2 \neq 1$, define

$$f(z) = \exp\left\{i \log \frac{1+z}{1-z}\right\},$$

choosing the branch of log that has $\log 1 = 0$. Describe $f(E)$ if E is
  (a) U,
  (b) the upper half of T,
  (c) the lower half of T,
  (d) any circular arc (in U) from -1 to 1,
  (e) the radius $[0, 1)$,
  (f) any disc $\{z: |z - r| < 1 - r\}$, $0 < r < 1$.
  (g) any curve in U tending to 1.

33 If φ_α is as in Definition 12.3, show that

(a) $\dfrac{1}{\pi} \displaystyle\int_U |\varphi'_\alpha|^2 \, dm = 1,$

(b) $\dfrac{1}{\pi} \displaystyle\int_U |\varphi'_\alpha| \, dm = \dfrac{1 - |\alpha|^2}{|\alpha|^2} \log \dfrac{1}{1 - |\alpha|^2}.$

Here m denotes Lebesgue measure in R^2.

CHAPTER
FIFTEEN

ZEROS OF HOLOMORPHIC FUNCTIONS

Infinite Products

15.1 So far we have met only one result concerning the zero set $Z(f)$ of a nonconstant holomorphic function f in a region Ω, namely, $Z(f)$ has no limit point in Ω. We shall see presently that this is all that can be said about $Z(f)$, if no other conditions are imposed on f, because of the theorem of Weierstrass (Theorem 15.11) which asserts that every $A \subset \Omega$ without limit point in Ω is $Z(f)$ for some $f \in H(\Omega)$. If $A = \{\alpha_n\}$, a natural way to construct such an f is to choose functions $f_n \in H(\Omega)$ so that f_n has only one zero, at α_n, and to consider the limit of the products

$$p_n = f_1 f_2 \cdots f_n,$$

as $n \to \infty$. One has to arrange it so that the sequence $\{p_n\}$ converges to some $f \in H(\Omega)$ *and* so that the limit function f is not 0 except at the prescribed points α_n. It is therefore advisable to begin by studying some general properties of infinite products.

15.2 Definition Suppose $\{u_n\}$ is a sequence of complex numbers,

$$p_n = (1 + u_1)(1 + u_2) \cdots (1 + u_n), \tag{1}$$

and $p = \lim_{n \to \infty} p_n$ exists. Then we write

$$p = \prod_{n=1}^{\infty} (1 + u_n). \tag{2}$$

The p_n are the *partial products* of the *infinite product* (2). We shall say that the infinite product (2) converges if the sequence $\{p_n\}$ converges.

In the study of infinite series Σa_n it is of significance whether the a_n approach 0 rapidly. Analogously, in the study of infinite products it is of interest whether the factors are or are not close to 1. This accounts for the above notation: $1 + u_n$ is close to 1 if u_n is close to 0.

15.3 Lemma *If $u_1, \ldots, u_N$ are complex numbers, and if*

$$p_N = \prod_{n=1}^{N} (1 + u_n), \qquad p_N^* = \prod_{n=1}^{N} (1 + |u_n|), \tag{1}$$

then

$$p_N^* \leq \exp(|u_1| + \cdots + |u_N|) \tag{2}$$

and

$$|p_N - 1| \leq p_N^* - 1. \tag{3}$$

PROOF For $x \geq 0$, the inequality $1 + x \leq e^x$ is an immediate consequence of the expansion of e^x in powers of x. Replace x by $|u_1|, \ldots, |u_N|$ and multiply the resulting inequalities. This gives (2). For $N = 1$, (3) is trivial. The general case follows by induction: For $k = 1, \ldots, N - 1$,

$$p_{k+1} - 1 = p_k(1 + u_{k+1}) - 1 = (p_k - 1)(1 + u_{k+1}) + u_{k+1},$$

so that if (3) holds with k in place of N, then also

$$|p_{k+1} - 1| \leq (p_k^* - 1)(1 + |u_{k+1}|) + |u_{k+1}| = p_{k+1}^* - 1. \qquad ////$$

15.4 Theorem *Suppose $\{u_n\}$ is a sequence of bounded complex functions on a set S, such that $\Sigma |u_n(s)|$ converges uniformly on S. Then the product*

$$f(s) = \prod_{n=1}^{\infty} (1 + u_n(s)) \tag{1}$$

converges uniformly on S, and $f(s_0) = 0$ at some $s_0 \in S$ if and only if $u_n(s_0) = -1$ for some n.

Furthermore, if $\{n_1, n_2, n_3, \ldots\}$ is any permutation of $\{1, 2, 3, \ldots\}$, then we also have

$$f(s) = \prod_{k=1}^{\infty} (1 + u_{n_k}(s)) \qquad (s \in S). \tag{2}$$

PROOF The hypothesis implies that $\Sigma |u_n(s)|$ is bounded on S, and if p_N denotes the Nth partial product of (1), we conclude from Lemma 15.3 that there is a constant $C < \infty$ such that $|p_N(s)| \leq C$ for all N and all s.

Choose ϵ, $0 < \epsilon < \frac{1}{2}$. There exists an N_0 such that

$$\sum_{n=N_0}^{\infty} |u_n(s)| < \epsilon \qquad (s \in S). \tag{3}$$

Let $\{n_1, n_2, n_3, \ldots\}$ be a permutation of $\{1, 2, 3, \ldots\}$. If $N \geq N_0$, if M is so large that

$$\{1, 2, \ldots, N\} \subset \{n_1, n_2, \ldots, n_M\}, \tag{4}$$

and if $q_M(s)$ denotes the Mth partial product of (2), then

$$q_M - p_N = p_N\{\prod (1 + u_{n_k}) - 1\}. \tag{5}$$

The n_k which occur in (5) are all distinct and are larger than N_0. Therefore (3) and Lemma 15.3 show that

$$|q_M - p_N| \leq |p_N|(e^\epsilon - 1) \leq 2|p_N|\epsilon \leq 2C\epsilon. \tag{6}$$

If $n_k = k$ ($k = 1, 2, 3, \ldots$), then $q_M = p_M$, and (6) shows that $\{p_N\}$ converges uniformly to a limit function f. Also, (6) shows that

$$|p_M - p_{N_0}| \leq 2|p_{N_0}|\epsilon \qquad (M > N_0), \tag{7}$$

so that $|p_M| \geq (1 - 2\epsilon)|p_{N_0}|$. Hence

$$|f(s)| \geq (1 - 2\epsilon)|p_{N_0}(s)| \qquad (s \in S), \tag{8}$$

which shows that $f(s) = 0$ if and only if $p_{N_0}(s) = 0$.

Finally, (6) also shows that $\{q_M\}$ converges to the same limit as $\{p_N\}$. ////

15.5 Theorem *Suppose $0 \leq u_n < 1$. Then*

$$\prod_{n=1}^{\infty} (1 - u_n) > 0 \quad \text{if and only if} \quad \sum_{n=1}^{\infty} u_n < \infty.$$

PROOF If $p_N = (1 - u_1) \cdots (1 - u_N)$, then $p_1 \geq p_2 \geq \cdots$, $p_N > 0$, hence $p = \lim p_N$ exists. If $\Sigma u_n < \infty$, Theorem 15.4 implies $p > 0$. On the other hand,

$$p \leq p_N = \prod_1^N (1 - u_n) \leq \exp\{-u_1 - u_2 - \cdots - u_N\},$$

and the last expression tends to 0 as $N \to \infty$, if $\Sigma u_n = \infty$. ////

We shall frequently use the following consequence of Theorem 15.4:

15.6 Theorem *Suppose $f_n \in H(\Omega)$ for $n = 1, 2, 3, \ldots$, no f_n is identically 0 in any component of Ω, and*

$$\sum_{n=1}^{\infty} |1 - f_n(z)| \tag{1}$$

converges uniformly on compact subsets of Ω. Then the product

$$f(z) = \prod_{n=1}^{\infty} f_n(z) \qquad (2)$$

converges uniformly on compact subsets of Ω. Hence $f \in H(\Omega)$. Furthermore, we have

$$m(f; z) = \sum_{n=1}^{\infty} m(f_n; z) \qquad (z \in \Omega), \qquad (3)$$

where $m(f; z)$ is defined to be the multiplicity of the zero of f at z. [If $f(z) \neq 0$, then $m(f; z) = 0$.]

PROOF The first part follows immediately from Theorem 15.4. For the second part, observe that each $z \in \Omega$ has a neighborhood V in which at most finitely many of the f_n have a zero, by (1). Take these factors first. The product of the remaining ones has no zero in V, by Theorem 15.4, and this gives (3). Incidentally, we see also that at most finitely many terms in the series (3) can be positive for any given $z \in \Omega$. ////

The Weierstrass Factorization Theorem

15.7 Definition Put $E_0(z) = 1 - z$, and for $p = 1, 2, 3, \ldots$,

$$E_p(z) = (1 - z) \exp\left\{z + \frac{z^2}{2} + \cdots + \frac{z^p}{p}\right\}.$$

These functions, introduced by Weierstrass, are sometimes called *elementary factors*. Their only zero is at $z = 1$. Their utility depends on the fact that they are close to 1 if $|z| < 1$ and p is large, although $E_p(1) = 0$.

15.8 Lemma *For $|z| \leq 1$ and $p = 0, 1, 2, \ldots$,*

$$|1 - E_p(z)| \leq |z|^{p+1}.$$

PROOF For $p = 0$, this is obvious. For $p \geq 1$, direct computation shows that

$$-E'_p(z) = z^p \exp\left\{z + \frac{z^2}{2} + \cdots + \frac{z^p}{p}\right\}.$$

So $-E'_p$ has a zero of order p at $z = 0$, and the expansion of $-E'_p$ in powers of z has nonnegative real coefficients. Since

$$1 - E_p(z) = -\int_{[0, z]} E'_p(w)\, dw,$$

$1 - E_p$ has a zero of order $p + 1$ at $z = 0$, and if

$$\varphi(z) = \frac{1 - E_p(z)}{z^{p+1}},$$

then $\varphi(z) = \Sigma a_n z^n$, with all $a_n \geq 0$. Hence $|\varphi(z)| \leq \varphi(1) = 1$ if $|z| \leq 1$, and this gives the assertion of the lemma. ////

15.9 Theorem *Let $\{z_n\}$ be a sequence of complex numbers such that $z_n \neq 0$ and $|z_n| \to \infty$ as $n \to \infty$. If $\{p_n\}$ is a sequence of nonnegative integers such that*

$$\sum_{n=1}^{\infty} \left(\frac{r}{r_n}\right)^{1+p_n} < \infty \quad (1)$$

for every positive r (where $r_n = |z_n|$), then the infinite product

$$P(z) = \prod_{n=1}^{\infty} E_{p_n}\left(\frac{z}{z_n}\right) \quad (2)$$

defines an entire function P which has a zero at each point z_n and which has no other zeros in the plane.

More precisely, if α occurs m times in the sequence $\{z_n\}$, then P has a zero of order m at α.

Condition (1) is always satisfied if $p_n = n - 1$, for instance.

PROOF For every r, $r_n > 2r$ for all but finitely many n, hence $r/r_n < \frac{1}{2}$ for these n, so (1) holds with $1 + p_n = n$.

Now fix r. If $|z| \leq r$, Lemma 15.8 shows that

$$\left|1 - E_{p_n}\left(\frac{z}{z_n}\right)\right| \leq \left|\frac{z}{z_n}\right|^{1+p_n} \leq \left(\frac{r}{r_n}\right)^{1+p_n}$$

if $r_n \geq r$, which holds for all but finitely many n. It now follows from (1) that the series

$$\sum_{n=1}^{\infty} \left|1 - E_{p_n}\left(\frac{z}{z_n}\right)\right|$$

converges uniformly on compact sets in the plane, and Theorem 15.6 gives the desired conclusion. ////

Note: For certain sequences $\{r_n\}$, (1) holds for a constant sequence $\{p_n\}$. It is of interest to take this constant as small as possible; the resulting function (2) is then called the *canonical product* corresponding to $\{z_n\}$. For instance, if $\Sigma 1/r_n < \infty$, we can take $p_n = 0$, and the canonical product is simply

$$\prod_{n=1}^{\infty} \left(1 - \frac{z}{z_n}\right).$$

If $\Sigma 1/r_n = \infty$ but $\Sigma 1/r_n^2 < \infty$, the canonical product is

$$\prod_{n=1}^{\infty} \left(1 - \frac{z}{z_n}\right) e^{z/z_n}.$$

Canonical products are of great interest in the study of entire functions of finite order. (See Exercise 2 for the definition.)

We now state the Weierstrass factorization theorem.

15.10 Theorem *Let f be an entire function, suppose $f(0) \neq 0$, and let $z_1, z_2, z_3, \ldots$ be the zeros of f, listed according to their multiplicities. Then there exist an entire function g and a sequence $\{p_n\}$ of nonnegative integers, such that*

$$f(z) = e^{g(z)} \prod_{n=1}^{\infty} E_{p_n}\left(\frac{z}{z_n}\right). \tag{1}$$

Note: (a) If f has a zero of order k at $z = 0$, the preceding applies to $f(z)/z^k$. (b) The factorization (1) is not unique; a unique factorization can be associated with those f whose zeros satisfy the condition required for the convergence of a canonical product.

PROOF Let P be the product in Theorem 15.9, formed with the zeros of f. Then f/P has only removable singularities in the plane, hence is (or can be extended to) an entire function. Also, f/P has no zero, and since the plane is simply connected, $f/P = e^g$ for some entire function g. ////

The proof of Theorem 15.9 is easily adapted to any open set:

15.11 Theorem *Let Ω be an open set in S^2, $\Omega \neq S^2$. Suppose $A \subset \Omega$ and A has no limit point in Ω. With each $\alpha \in A$ associate a positive integer $m(\alpha)$. Then there exists an $f \in H(\Omega)$ all of whose zeros are in A, and such that f has a zero of order $m(\alpha)$ at each $\alpha \in A$.*

PROOF It simplifies the argument, and causes no loss of generality, to assume that $\infty \in \Omega$ but $\infty \notin A$. (If this is not so, a linear fractional transformation will make it so.) Then $S^2 - \Omega$ is a nonempty compact subset of the plane, and ∞ is not a limit point of A.

If A is finite, we can take a rational function for f.

If A is infinite, then A is countable (otherwise there would be a limit point in Ω). Let $\{\alpha_n\}$ be a sequence whose terms are in A and in which each $\alpha \in A$ is listed precisely $m(\alpha)$ times. Associate with each α_n a point $\beta_n \in S^2 - \Omega$ such that $|\beta_n - \alpha_n| \leq |\beta - \alpha_n|$ for all $\beta \in S^2 - \Omega$; this is possible since $S^2 - \Omega$ is compact. Then

$$|\beta_n - \alpha_n| \to 0$$

as $n \to \infty$; otherwise A would have a limit point in Ω. We claim that

$$f(z) = \prod_{n=1}^{\infty} E_n\left(\frac{\alpha_n - \beta_n}{z - \beta_n}\right)$$

has the desired properties.

Put $r_n = 2|\alpha_n - \beta_n|$. Let K be a compact subset of Ω. Since $r_n \to 0$, there exists an N such that $|z - \beta_n| > r_n$ for all $z \in K$ and all $n \geq N$. Hence

$$\left|\frac{\alpha_n - \beta_n}{z - \beta_n}\right| \leq \frac{1}{2},$$

which implies, by Lemma 15.8, that

$$\left|1 - E_n\left(\frac{\alpha_n - \beta_n}{z - \beta_n}\right)\right| \leq \left(\frac{1}{2}\right)^{n+1} \qquad (z \in K, n \geq N),$$

and this again completes the proof, by Theorem 15.6. ////

As a consequence, we can now obtain a characterization of meromorphic functions (see Definition 10.41):

15.12 Theorem *Every meromorphic function in an open set Ω is a quotient of two functions which are holomorphic in Ω.*

The converse is obvious: If $g \in H(\Omega)$, $h \in H(\Omega)$, and h is not identically 0 in any component of Ω, then g/h is meromorphic in Ω.

PROOF Suppose f is meromorphic in Ω; let A be the set of all poles of f in Ω; and for each $\alpha \in A$, let $m(\alpha)$ be the order of the pole of f at α. By Theorem 15.11 there exists an $h \in H(\Omega)$ such that h has a zero of multiplicity $m(\alpha)$ at each $\alpha \in A$, and h has no other zeros. Put $g = fh$. The singularities of g at the points of A are removable, hence we can extend g so that $g \in H(\Omega)$. Clearly, $f = g/h$ in $\Omega - A$. ////

An Interpolation Problem

The Mittag-Leffler theorem may be combined with the Weierstrass theorem 15.11 to give a solution of the following problem: Can we take an arbitrary set $A \subset \Omega$, without limit point in Ω, and find a function $f \in H(\Omega)$ which has prescribed values at every point of A? The answer is affirmative. In fact, we can do even better, and also prescribe finitely many derivatives at each point of A:

15.13 Theorem *Suppose Ω is an open set in the plane, $A \subset \Omega$, A has no limit point in Ω, and to each $\alpha \in A$ there are associated a nonnegative integer $m(\alpha)$*

and complex numbers $w_{n,\alpha}$, $0 \le n \le m(\alpha)$. Then there exists an $f \in H(\Omega)$ such that

$$f^{(n)}(\alpha) = n! \, w_{n,\alpha} \qquad (\alpha \in A,\ 0 \le n \le m(\alpha)). \tag{1}$$

PROOF By Theorem 15.11, there exists a $g \in H(\Omega)$ whose only zeros are in A and such that g has a zero of order $m(\alpha) + 1$ at each $\alpha \in A$. We claim we can associate to each $\alpha \in A$ a function P_α of the form

$$P_\alpha(z) = \sum_{j=1}^{1+m(\alpha)} c_{j,\alpha}(z - \alpha)^{-j} \tag{2}$$

such that gP_α has the power series expansion

$$g(z)P_\alpha(z) = w_{0,\alpha} + w_{1,\alpha}(z-\alpha) + \cdots + w_{m(\alpha),\alpha}(z-\alpha)^{m(\alpha)} + \cdots \tag{3}$$

in some disc with center at α.

To simplify the writing, take $\alpha = 0$ and $m(\alpha) = m$, and omit the subscripts α. For z near 0, we have

$$g(z) = b_1 z^{m+1} + b_2 z^{m+2} + \cdots, \tag{4}$$

where $b_1 \ne 0$. If

$$P(z) = c_1 z^{-1} + \cdots + c_{m+1} z^{-m-1}, \tag{5}$$

then

$$g(z)P(z) = (c_{m+1} + c_m z + \cdots + c_1 z^m)(b_1 + b_2 z + b_3 z^2 + \cdots). \tag{6}$$

The b's are given, and we want to choose the c's so that

$$g(z)P(z) = w_0 + w_1 z + \cdots + w_m z^m + \cdots. \tag{7}$$

If we compare the coefficients of $1, z, \ldots, z^m$ in (6) and (7), we can solve the resulting equations successively for $c_{m+1}, c_m, \ldots, c_1$, since $b_1 \ne 0$.

In this way we obtain the desired P_α's. The Mittag-Leffler theorem now gives us a meromorphic h in Ω whose principal parts are these P_α's, and if we put $f = gh$ we obtain a function with the desired properties. ////

The solution of this interpolation problem can be used to determine the structure of all finitely generated ideals in the rings $H(\Omega)$.

15.14 Definition The *ideal* $[g_1, \ldots, g_n]$ generated by the functions $g_1, \ldots, g_n \in H(\Omega)$ is the set of all functions of the form $\Sigma f_i g_i$, where $f_i \in H(\Omega)$ for $i = 1, \ldots, n$. A *principal ideal* is one that is generated by a single function. Note that $[1] = H(\Omega)$.

If $f \in H(\Omega)$, $\alpha \in \Omega$, and f is not identically 0 in a neighborhood of α, the multiplicity of the zero of f at α will be denoted by $m(f; \alpha)$. If $f(\alpha) \ne 0$, then $m(f; \alpha) = 0$, as in Theorem 15.6.

15.15 Theorem *Every finitely generated ideal in $H(\Omega)$ is principal.*

More explicitly: If $g_1, \ldots, g_n \in H(\Omega)$, then there exist functions $g, f_i, h_i \in H(\Omega)$ such that

$$g = \sum_{i=1}^{n} f_i g_i \quad \text{and} \quad g_i = h_i g \qquad (1 \leq i \leq n).$$

PROOF We shall assume that Ω is a region. This is done to avoid problems posed by functions that are identically 0 in some components of Ω but not in all. Once the theorem is proved for regions, that case can be applied to each component of an arbitrary open set Ω, and the full theorem can be deduced. We leave the details of this as an exercise.

Let $P(n)$ be the following proposition:

If $g_1, \ldots, g_n \in H(\Omega)$, if no g_i is identically 0, and if no point of Ω is a zero of every g_i, then $[g_1, \ldots, g_n] = [1]$.

$P(1)$ is trivial. Assume that $n > 1$ and that $P(n-1)$ is true. Take $g_1, \ldots, g_n \in H(\Omega)$, without common zero. By the Weierstrass theorem 15.11 there exists $\varphi \in H(\Omega)$ such that

$$m(\varphi; \alpha) = \min \{m(g_i; \alpha): 1 \leq i \leq n-1\} \qquad (\alpha \in \Omega). \tag{1}$$

The functions $f_i = g_i/\varphi$ ($1 \leq i \leq n-1$) are in $H(\Omega)$ and have no common zero in Ω. Since $P(n-1)$ is true, $[f_1, \ldots, f_{n-1}] = [1]$. Hence

$$[g_1, \ldots, g_{n-1}, g_n] = [\varphi, g_n]. \tag{2}$$

Moreover, our choice of φ shows that $g_n(\alpha) \neq 0$ at every point of the set $A = \{\alpha \in \Omega : \varphi(\alpha) = 0\}$. Hence it follows from Theorem 15.13 that there exists $h \in H(\Omega)$ such that

$$m(1 - hg_n; \alpha) \geq m(\varphi; \alpha) \qquad (\alpha \in \Omega). \tag{3}$$

Such an h is obtained by a suitable choice of the prescribed values of $h^{(k)}(\alpha)$ for $\alpha \in A$ and for $0 \leq k \leq m(\varphi; \alpha)$.

By (3), $(1 - hg_n)/\varphi$ has removable singularities. Thus

$$1 = hg_n + f\varphi \tag{4}$$

for some $f \in H(\Omega)$. By (2) and (4), $1 \in [g_1, \ldots, g_n]$.

We have shown that $P(n-1)$ implies $P(n)$. Hence $P(n)$ is true for all n.

Finally, suppose $G_1, \ldots, G_n \in H(\Omega)$, and no G_i is identically 0. (This involves no loss of generality.) Another application of Theorem 15.11 yields $\varphi \in H(\Omega)$ with $m(\varphi; \alpha) = \min m(G_i; \alpha)$ for all $\alpha \in \Omega$. Put $g_i = G_i/\varphi$. Then $g_i \in H(\Omega)$, and the functions $g_1, \ldots, g_n$ have no common zeros in Ω. By $P(n)$, $[g_1, \ldots, g_n] = [1]$. Hence $[G_1, \ldots, G_n] = [\varphi]$. This completes the proof. ////

Jensen's Formula

15.16 As we see from Theorem 15.11, the location of the zeros of a holomorphic function in a region Ω is subject to no restriction except the obvious one concerning the absence of limit points in Ω. The situation is quite different if we replace $H(\Omega)$ by certain subclasses which are defined by certain growth conditions. In those situations the distribution of the zeros has to satisfy certain quantitative conditions. The basis of most of these theorems is Jensen's formula (Theorem 15.18). We shall apply it to certain classes of entire functions and to certain subclasses of $H(U)$.

The following lemma affords an opportunity to apply Cauchy's theorem to the evaluation of a definite integral.

15.17 Lemma $\quad \dfrac{1}{2\pi} \displaystyle\int_0^{2\pi} \log |1 - e^{i\theta}|\, d\theta = 0.$

PROOF Let $\Omega = \{z \colon \operatorname{Re} z < 1\}$. Since $1 - z \neq 0$ in Ω and Ω is simply connected, there exists an $h \in H(\Omega)$ such that

$$\exp\{h(z)\} = 1 - z$$

in Ω, and this h is uniquely determined if we require that $h(0) = 0$. Since $\operatorname{Re}(1 - z) > 0$ in Ω, we then have

$$\operatorname{Re} h(z) = \log|1 - z|, \qquad |\operatorname{Im} h(z)| < \frac{\pi}{2} \qquad (z \in \Omega). \tag{1}$$

For small $\delta > 0$, let Γ be the path

$$\Gamma(t) = e^{it} \qquad (\delta \le t \le 2\pi - \delta), \tag{2}$$

and let γ be the circular arc whose center is at 1 and which passes from $e^{i\delta}$ to $e^{-i\delta}$ within U. Then

$$\frac{1}{2\pi}\int_\delta^{2\pi-\delta} \log|1 - e^{i\theta}|\, d\theta = \operatorname{Re}\left[\frac{1}{2\pi i}\int_\Gamma h(z)\,\frac{dz}{z}\right] = \operatorname{Re}\left[\frac{1}{2\pi i}\int_\gamma h(z)\,\frac{dz}{z}\right]. \tag{3}$$

The last equality depended on Cauchy's theorem; note that $h(0) = 0$.

The length of γ is less than $\pi\delta$, so (1) shows that the absolute value of the last integral in (3) is less than $C\delta \log(1/\delta)$, where C is a constant. This gives the result if $\delta \to 0$ in (3). ////

15.18 Theorem *Suppose $\Omega = D(0; R)$, $f \in H(\Omega)$, $f(0) \neq 0$, $0 < r < R$, and $\alpha_1, \ldots, \alpha_N$ are the zeros of f in $\bar{D}(0; r)$, listed according to their multiplicities. Then*

$$|f(0)| \prod_{n=1}^N \frac{r}{|\alpha_n|} = \exp\left\{\frac{1}{2\pi}\int_{-\pi}^\pi \log|f(re^{i\theta})|\, d\theta\right\}. \tag{1}$$

This is known as *Jensen's formula*. The hypothesis $f(0) \neq 0$ causes no harm in applications, for if f has a zero of order k at 0, the formula can be applied to $f(z)/z^k$.

PROOF Order the points α_j so that $\alpha_1, \ldots, \alpha_m$ are in $D(0; r)$ and $|\alpha_{m+1}| = \cdots = |\alpha_N| = r$. (Of course, we may have $m = N$ or $m = 0$.) Put

$$g(z) = f(z) \prod_{n=1}^{m} \frac{r^2 - \bar{\alpha}_n z}{r(\alpha_n - z)} \prod_{n=m+1}^{N} \frac{\alpha_n}{\alpha_n - z}. \tag{2}$$

Then $g \in H(D)$, where $D = D(0; r + \epsilon)$ for some $\epsilon > 0$, g has no zero in D, hence $\log |g|$ is harmonic in D (Theorem 13.12), and so

$$\log |g(0)| = \frac{1}{2\pi} \int_{-\pi}^{\pi} \log |g(re^{i\theta})| \, d\theta. \tag{3}$$

By (2),

$$|g(0)| = |f(0)| \prod_{n=1}^{m} \frac{r}{|\alpha_n|}. \tag{4}$$

For $1 \leq n \leq m$, the factors in (2) have absolute value 1 if $|z| = r$. If $\alpha_n = re^{i\theta_n}$ for $m < n \leq N$, it follows that

$$\log |g(re^{i\theta})| = \log |f(re^{i\theta})| - \sum_{n=m+1}^{N} \log |1 - e^{i(\theta - \theta_n)}|. \tag{5}$$

Lemma 15.17 shows therefore that the integral in (3) is unchanged if g is replaced by f. Comparison with (4) now gives (1). ////

Jensen's formula gives rise to an inequality which involves the boundary values of bounded holomorphic functions in U (we recall that the class of these functions has been denoted by H^∞):

15.19 Theorem *If $f \in H^\infty$, f not identically 0, define*

$$\mu_r(f) = \frac{1}{2\pi} \int_{-\pi}^{\pi} \log |f(re^{i\theta})| \, d\theta \qquad (0 < r < 1) \tag{1}$$

and

$$\mu^*(f) = \frac{1}{2\pi} \int_{-\pi}^{\pi} \log |f^*(e^{i\theta})| \, d\theta \tag{2}$$

where f^* is the radial limit function of f, as in Theorem 11.32. Then

$$\mu_r(f) \leq \mu_s(f) \quad \text{if} \quad 0 < r < s < 1, \tag{3}$$

$$\mu_r(f) \to \log |f(0)| \quad \text{as} \quad r \to 0, \tag{4}$$

and

$$\mu_r(f) \le \mu^*(f) \quad \text{if} \quad 0 < r < 1. \tag{5}$$

Note the following consequence: One can choose r so that $f(z) \ne 0$ if $|z| = r$; then $\mu_r(f)$ is finite, and so is $\mu^*(f)$, by (5). Thus $\log |f^*| \in L^1(T)$, and $f^*(e^{i\theta}) \ne 0$ at almost every point of T.

PROOF There is an integer $m \ge 0$ such that $f(z) = z^m g(z)$, $g \in H^\infty$, and $g(0) \ne 0$. Apply Jensen's formula 15.18(1) to g in place of f. Its left side obviously cannot decrease if r increases. Thus $\mu_r(g) \le \mu_s(g)$ if $r < s$. Since

$$\mu_r(f) = \mu_r(g) + m \log r,$$

we have proved (3).

Let us now assume, without loss of generality, that $|f| \le 1$. Write $f_r(e^{i\theta})$ in place of $f(re^{i\theta})$. Then $f_r \to f(0)$ as $r \to 0$, and $f_r \to f^*$ a.e. as $r \to 1$. Since $\log(1/|f_r|) \ge 0$, two applications of Fatou's lemma, combined with (3), give (4) and (5). ////

15.20 Zeros of Entire Functions Suppose f is an entire function,

$$M(r) = \sup_\theta |f(re^{i\theta})| \quad (0 < r < \infty), \tag{1}$$

and $n(r)$ is the number of zeros of f in $\bar{D}(0; r)$. Assume $f(0) = 1$, for simplicity. Jensen's formula gives

$$M(2r) \ge \exp\left\{\frac{1}{2\pi}\int_{-\pi}^\pi \log |f(2re^{i\theta})|\, d\theta\right\} = \prod_{n=1}^{n(2r)} \frac{2r}{|\alpha_n|} \ge \prod_{n=1}^{n(r)} \frac{2r}{|\alpha_n|} \ge 2^{n(r)},$$

if $\{\alpha_n\}$ is the sequence of zeros of f, arranged so that $|\alpha_1| \le |\alpha_2| \le \cdots$. Hence

$$n(r) \log 2 \le \log M(2r). \tag{2}$$

Thus the rapidity with which $n(r)$ can increase (i.e., the density of the zeros of f) is controlled by the rate of growth of $M(r)$. Suppose, to look at a more specific situation, that for large r

$$M(r) < \exp\{Ar^k\} \tag{3}$$

where A and k are given positive numbers. Then (2) leads to

$$\limsup_{r \to \infty} \frac{\log n(r)}{\log r} \le k. \tag{4}$$

For example, if k is a positive integer and

$$f(z) = 1 - e^{z^k}, \tag{5}$$

then $n(r)$ is about $\pi^{-1}kr^k$, so that

$$\lim_{r \to \infty} \frac{\log n(r)}{\log r} = k. \tag{6}$$

This shows that the estimate (4) cannot be improved.

Blaschke Products

Jensen's formula makes it possible to determine the precise conditions which the zeros of a nonconstant $f \in H^\infty$ must satisfy.

15.21 Theorem *If $\{\alpha_n\}$ is a sequence in U such that $\alpha_n \neq 0$ and*

$$\sum_{n=1}^{\infty} (1 - |\alpha_n|) < \infty, \tag{1}$$

if k is a nonnegative integer, and if

$$B(z) = z^k \prod_{n=1}^{\infty} \frac{\alpha_n - z}{1 - \bar{\alpha}_n z} \cdot \frac{|\alpha_n|}{\alpha_n} \qquad (z \in U), \tag{2}$$

then $B \in H^\infty$, and B has no zeros except at the points α_n (and at the origin, if $k > 0$).

We call this function B a *Blaschke product*. Note that some of the α_n may be repeated, in which case B has multiple zeros at those points. Note also that each factor in (2) has absolute value 1 on T.

The term "Blaschke product" will also be used if there are only finitely many factors, and even if there are none, in which case $B(z) = 1$.

PROOF The nth term in the series

$$\sum_{n=1}^{\infty} \left| 1 - \frac{\alpha_n - z}{1 - \bar{\alpha}_n z} \cdot \frac{|\alpha_n|}{\alpha_n} \right|$$

is

$$\left| \frac{\alpha_n + |\alpha_n| z}{(1 - \bar{\alpha}_n z)\alpha_n} \right| (1 - |\alpha_n|) \le \frac{1 + r}{1 - r} (1 - |\alpha_n|)$$

if $|z| \le r$. Hence Theorem 15.6 shows that $B \in H(U)$ and that B has only the prescribed zeros. Since each factor in (2) has absolute value less than 1 in U, it follows that $|B(z)| < 1$, and the proof is complete. ////

15.22 The preceding theorem shows that

$$\sum_{n=1}^{\infty} (1 - |\alpha_n|) < \infty \tag{1}$$

is a *sufficient* condition for the existence of an $f \in H^\infty$ which has only the prescribed zeros $\{\alpha_n\}$. This condition also turns out to be necessary: *If $f \in H^\infty$ and f is not identically zero, the zeros of f must satisfy* (1). This is a special case of Theorem 15.23. It is interesting that (1) is a necessary condition in a much larger class of functions, which we now describe.

For any real number t, define $\log^+ t = \log t$ if $t \geq 1$ and $\log^+ t = 0$ if $t < 1$. We let N (for Nevanlinna) be the class of all $f \in H(U)$ for which

$$\sup_{0 < r < 1} \frac{1}{2\pi} \int_{-\pi}^{\pi} \log^+ |f(re^{i\theta})| \, d\theta < \infty. \tag{2}$$

It is clear that $H^\infty \subset N$. Note that (2) imposes a restriction on the rate of growth of $|f(z)|$ as $|z| \to 1$, whereas the boundedness of the integrals

$$\frac{1}{2\pi} \int_{-\pi}^{\pi} \log |f(re^{i\theta})| \, d\theta \tag{3}$$

imposes no such restriction. For instance, (3) is independent of r if $f = e^g$ for any $g \in H(U)$. The point is that (3) can stay small because $\log |f|$ assumes large negative values as well as large positive ones, whereas $\log^+ |f| \geq 0$. The class N will be discussed further in Chap. 17.

15.23 Theorem *Suppose $f \in N$, f is not identically 0 in U, and $\alpha_1, \alpha_2, \alpha_3, \ldots$ are the zeros of f, listed according to their multiplicities. Then*

$$\sum_{n=1}^{\infty} (1 - |\alpha_n|) < \infty. \tag{1}$$

(We tacitly assume that f has infinitely many zeros in U. If there are only finitely many, the above sum has only finitely many terms, and there is nothing to prove. Also, $|\alpha_n| \leq |\alpha_{n+1}|$.)

PROOF If f has a zero of order m at the origin, and $g(z) = z^{-m} f(z)$, then $g \in N$, and g has the same zeros as f, except at the origin. Hence we may assume, without loss of generality, that $f(0) \neq 0$. Let $n(r)$ be the number of zeros of f in $\bar{D}(0; r)$, fix k, and take $r < 1$ so that $n(r) > k$. Then Jensen's formula

$$|f(0)| \prod_{n=1}^{n(r)} \frac{r}{|\alpha_n|} = \exp\left\{\frac{1}{2\pi} \int_{-\pi}^{\pi} \log |f(re^{i\theta})| \, d\theta\right\} \tag{2}$$

implies that

$$|f(0)| \prod_{n=1}^{k} \frac{r}{|\alpha_n|} \leq \exp\left\{\frac{1}{2\pi} \int_{-\pi}^{\pi} \log^+ |f(re^{i\theta})| \, d\theta\right\}. \tag{3}$$

Our assumption that $f \in N$ is equivalent to the existence of a constant $C < \infty$ which exceeds the right side of (3) for all r, $0 < r < 1$. It follows that

$$\prod_{n=1}^{k} |\alpha_n| \geq C^{-1} |f(0)| r^k. \tag{4}$$

The inequality persists, for every k, as $r \to 1$. Hence

$$\prod_{n=1}^{\infty} |\alpha_n| \geq C^{-1} |f(0)| > 0. \tag{5}$$

By Theorem 15.5, (5) implies (1). ////

Corollary If $f \in H^\infty$ (or even if $f \in N$), if $\alpha_1, \alpha_2, \alpha_3, \ldots$ are the zeros of f in U, and if $\Sigma(1 - |\alpha_n|) = \infty$, then $f(z) = 0$ for all $z \in U$.

For instance, no nonconstant bounded holomorphic function in U can have a zero at each of the points $(n-1)/n$ $(n = 1, 2, 3, \ldots)$.

We conclude this section with a theorem which describes the behavior of a Blaschke product near the boundary of U. Recall that as a member of H^∞, B has radial limits $B^*(e^{i\theta})$ at almost all points of T.

15.24 Theorem If B is a Blaschke product, then $|B^*(e^{i\theta})| = 1$ a.e. and

$$\lim_{r \to 1} \frac{1}{2\pi} \int_{-\pi}^{\pi} \log |B(re^{i\theta})| \, d\theta = 0. \tag{1}$$

PROOF The existence of the limit is a consequence of the fact that the integral is a monotonic function of r. Suppose $B(z)$ is as in Theorem 15.21, and put

$$B_N(z) = \prod_{n=N}^{\infty} \frac{\alpha_n - z}{1 - \bar{\alpha}_n z} \cdot \frac{|\alpha_n|}{\alpha_n} \tag{2}$$

Since $\log(|B/B_N|)$ is continuous in an open set containing T, the limit (1) is unchanged if B is replaced by B_N. If we apply Theorem 15.19 to B_N we therefore obtain

$$\log |B_N(0)| \leq \lim_{r \to 1} \frac{1}{2\pi} \int_{-\pi}^{\pi} \log |B(re^{i\theta})| \, d\theta \leq \frac{1}{2\pi} \int_{-\pi}^{\pi} \log |B^*(e^{i\theta})| \, d\theta \leq 0. \tag{3}$$

As $N \to \infty$, the first term in (3) tends to 0. This gives (1), and shows that $\int \log |B^*| = 0$. Since $\log |B^*| \leq 0$ a.e., Theorem 1.39(a) now implies that $\log |B^*| = 0$ a.e. ////

The Müntz-Szasz Theorem

15.25 A classical theorem of Weierstrass ([26], Theorem 7.26) states that the polynomials are dense in $C(I)$, the space of all continuous complex functions on the closed interval $I = [0, 1]$, with the supremum norm. In other words, the set of all finite linear combinations of the functions

$$1, t, t^2, t^3, \ldots \tag{1}$$

is dense in $C(I)$. This is sometimes expressed by saying that the functions (1) *span* $C(I)$.

This suggests a question: If $0 < \lambda_1 < \lambda_2 < \lambda_3 < \cdots$, under what conditions is it true that the functions

$$1, t^{\lambda_1}, t^{\lambda_2}, t^{\lambda_3}, \ldots \tag{2}$$

span $C(I)$?

It turns out that this problem has a very natural connection with the problem of the distribution of the zeros of a bounded holomorphic function in a half plane (or in a disc; the two are conformally equivalent). The surprisingly neat answer is that *the functions* (2) *span* $C(I)$ *if and only if* $\Sigma 1/\lambda_n = \infty$.

Actually, the proof gives an even more precise conclusion:

15.26 Theorem *Suppose* $0 < \lambda_1 < \lambda_2 < \lambda_3 < \cdots$ *and let* X *be the closure in* $C(I)$ *of the set of all finite linear combinations of the functions*

$$1, t^{\lambda_1}, t^{\lambda_2}, t^{\lambda_3}, \ldots.$$

(a) *If* $\Sigma 1/\lambda_n = \infty$, *then* $X = C(I)$.
(b) *If* $\Sigma 1/\lambda_n < \infty$, *and if* $\lambda \notin \{\lambda_n\}$, $\lambda \neq 0$, *then* X *does not contain the function* t^λ.

PROOF It is a consequence of the Hahn-Banach theorem (Theorem 5.19) that $\varphi \in C(I)$ but $\varphi \notin X$ if and only if there is a bounded linear functional on $C(I)$ which does not vanish at φ but which vanishes on all of X. Since every bounded linear functional on $C(I)$ is given by integration with respect to a complex Borel measure on I, (a) will be a consequence of the following proposition:

If $\Sigma 1/\lambda_n = \infty$ *and if* μ *is a complex Borel measure on* I *such that*

$$\int_I t^{\lambda_n} \, d\mu(t) = 0 \quad (n = 1, 2, 3, \ldots), \tag{1}$$

then also

$$\int_I t^k \, d\mu(t) = 0 \quad (k = 1, 2, 3, \ldots). \tag{2}$$

For if this is proved, the preceding remark shows that X contains all functions t^k; since $1 \in X$, all polynomials are then in X, and the Weierstrass theorem therefore implies that $X = C(I)$.

So assume that (1) holds. Since the integrands in (1) and (2) vanish at 0, we may as well assume that μ is concentrated on $(0, 1]$. We associate with μ the function

$$f(z) = \int_I t^z \, d\mu(t). \tag{3}$$

For $t > 0$, $t^z = \exp(z \log t)$, by definition. We claim that f is holomorphic in the right half plane. The continuity of f is easily checked, and we can then apply Morera's theorem. Furthermore, if $z = x + iy$, if $x > 0$, and if $0 < t \le 1$, then $|t^z| = t^x \le 1$. Thus f is bounded in the right half plane, and (1) says that $f(\lambda_n) = 0$, for $n = 1, 2, 3, \ldots$. Define

$$g(z) = f\left(\frac{1+z}{1-z}\right) \qquad (z \in U). \tag{4}$$

Then $g \in H^\infty$ and $g(\alpha_n) = 0$, where $\alpha_n = (\lambda_n - 1)/(\lambda_n + 1)$. A simple computation shows that $\Sigma(1 - |\alpha_n|) = \infty$ if $\Sigma 1/\lambda_n = \infty$. The Corollary to Theorem 15.23 therefore tells us that $g(z) = 0$ for all $z \in U$. Hence $f = 0$. In particular, $f(k) = 0$ for $k = 1, 2, 3, \ldots$, and this is (2). We have thus proved part (a) of the theorem.

To prove (b) it will be enough to construct a measure μ on I such that (3) defines a function f which is holomorphic in the half plane Re $z > -1$ (anything negative would do here), which is 0 at 0, $\lambda_1, \lambda_2, \lambda_3, \ldots$ and which has no other zeros in this half plane. For the functional induced by this measure μ will then vanish on X but will not vanish at any function t^λ if $\lambda \ne 0$ and $\lambda \notin \{\lambda_n\}$.

We begin by constructing a function f which has these prescribed zeros, and we shall then show that this f can be represented in the form (3). Define

$$f(z) = \frac{z}{(2+z)^3} \prod_{n=1}^\infty \frac{\lambda_n - z}{2 + \lambda_n + z}. \tag{5}$$

Since

$$1 - \frac{\lambda_n - z}{2 + \lambda_n + z} = \frac{2z + 2}{2 + \lambda_n + z},$$

the infinite product in (5) converges uniformly on every compact set which contains none of the points $-\lambda_n - 2$. It follows that f is a meromorphic function in the whole plane, with poles at -2 and $-\lambda_n - 2$, and with zeros at $0, \lambda_1, \lambda_2, \lambda_3, \ldots$. Also, each factor in the infinite product (5) is less than 1 in absolute value if Re $z > -1$. Thus $|f(z)| \le 1$ if Re $z \ge -1$. The factor $(2 + z)^3$ ensures that the restriction of f to the line Re $z = -1$ is in L^1.

Fix z so that Re $z > -1$, and consider the Cauchy formula for $f(z)$, where the path of integration consists of the semicircle with center at -1, radius $R > 1 + |z|$, from $-1 - iR$ to $-1 + R$ to $-1 + iR$, followed by the interval from $-1 + iR$ to $-1 - iR$. The integral over the semicircle tends to 0 as $R \to \infty$, so we are left with

$$f(z) = -\frac{1}{2\pi} \int_{-\infty}^{\infty} \frac{f(-1 + is)}{-1 + is - z} ds \qquad (\text{Re } z > -1). \qquad (6)$$

But

$$\frac{1}{1 + z - is} = \int_0^1 t^{z - is} dt \qquad (\text{Re } z > -1). \qquad (7)$$

Hence (6) can be rewritten in the form

$$f(z) = \int_0^1 t^z \left\{ \frac{1}{2\pi} \int_{-\infty}^{\infty} f(-1 + is) e^{-is \log t} ds \right\} dt. \qquad (8)$$

The interchange in the order of integration was legitimate: If the integrand in (8) is replaced by its absolute value, a finite integral results.

Put $g(s) = f(-1 + is)$. Then the inner integral in (8) is $\hat{g}(\log t)$, where $\hat{g}$ is the Fourier transform of g. This is a bounded continuous function on $(0, 1]$, and if we set $d\mu(t) = \hat{g}(\log t) dt$ we obtain a measure which represents f in the desired form (3).

This completes the proof. ////

15.27 Remark The theorem implies that whenever $\{1, t^{\lambda_1}, t^{\lambda_2}, \ldots\}$ spans $C(I)$, then some infinite subcollection of the t^{λ_i} can be removed without altering the span. In particular, $C(I)$ contains no minimal spanning sets of this type. This is in marked contrast to the behavior of orthonormal sets in a Hilbert space: if any element is removed from an orthonormal set, its span is diminished. Likewise, if $\{1, t^{\lambda_1}, t^{\lambda_2} \ldots\}$ does not span $C(I)$, removal of any of its elements will diminish the span; this follows from Theorem 15.26(b).

Exercises

1 Suppose $\{a_n\}$ and $\{b_n\}$ are sequences of complex numbers such that $\Sigma |a_n - b_n| < \infty$. On what sets will the product

$$\prod_{n=1}^{\infty} \frac{z - a_n}{z - b_n}$$

converge uniformly? Where will it define a holomorphic function?

2 Suppose f is entire, λ is a positive number, and the inequality

$$|f(z)| < \exp(|z|^\lambda)$$

holds for all large enough $|z|$. (Such functions f are said to be of *finite order*. The greatest lower bound of all λ for which the above condition holds is the *order* of f.) If $f(z) = \Sigma a_n z^n$, prove that the inequality

$$|a_n| \leq \left(\frac{e\lambda}{n}\right)^{n/\lambda}$$

holds for all large enough n. Consider the functions $\exp(z^k)$, $k = 1, 2, 3, \ldots$, to determine whether the above bound on $|a_n|$ is close to best possible.

3 Find all complex z for which $\exp(\exp(z)) = 1$. Sketch them as points in the plane. Show that there is no entire function of finite order which has a zero at each of these points (except, of course, $f \equiv 0$).

4 Show that the function

$$\pi \cot \pi z = \pi i \frac{e^{\pi i z} + e^{-\pi i z}}{e^{\pi i z} - e^{-\pi i z}}$$

has a simple pole with residue 1 at each integer. The same is true of the function

$$f(z) = \frac{1}{z} + \sum_{n=1}^{\infty} \frac{2z}{z^2 - n^2} = \lim_{N \to \infty} \sum_{n=-N}^{N} \frac{1}{z - n}.$$

Show that both functions are periodic $[f(z + 1) = f(z)]$, that their difference is a bounded entire function, hence a constant, and that this constant is actually 0, since

$$\lim_{y \to \infty} f(iy) = -2i \int_0^{\infty} \frac{dt}{1 + t^2} = -\pi i.$$

This gives the partial fractions decomposition

$$\pi \cot \pi z = \frac{1}{z} + \sum_{1}^{\infty} \frac{2z}{z^2 - n^2}.$$

(Compare with Exercise 12, Chap. 9.) Note that $\pi \cot \pi z$ is $(g'/g)(z)$ if $g(z) = \sin \pi z$. Deduce the product representation

$$\frac{\sin \pi z}{\pi z} = \prod_{n=1}^{\infty} \left(1 - \frac{z^2}{n^2}\right).$$

5 Suppose k is a positive integer, $\{z_n\}$ is a sequence of complex numbers such that $\Sigma |z_n|^{-k-1} < \infty$, and

$$f(z) = \prod_{n=1}^{\infty} E_k\left(\frac{z}{z_n}\right).$$

(See Definition 15.7.) What can you say about the rate of growth of

$$M(r) = \max_{\theta} |f(re^{i\theta})| ?$$

6 Suppose f is entire, $f(0) \neq 0$, $|f(z)| < \exp(|z|^p)$ for large $|z|$, and $\{z_n\}$ is the sequence of zeros of f, counted according to their multiplicities. Prove that $\Sigma |z_n|^{-p-\epsilon} < \infty$ for every $\epsilon > 0$. (Compare with Sec. 15.20.)

7 Suppose f is an entire function, $f(\sqrt{n}) = 0$ for $n = 1, 2, 3, \ldots$, and there is a positive constant α such that $|f(z)| < \exp(|z|^{\alpha})$ for all large enough $|z|$. For which α does it follow that $f(z) = 0$ for all z? [Consider $\sin(\pi z^2)$.]

8 Let $\{z_n\}$ be a sequence of distinct complex numbers, $z_n \neq 0$, such that $z_n \to \infty$ as $n \to \infty$, and let $\{m_n\}$ be a sequence of positive integers. Let g be a meromorphic function in the plane, which has a simple pole with residue m_n at each z_n and which has no other poles. If $z \notin \{z_n\}$, let $\gamma(z)$ be any path from 0 to z which passes through none of the points z_n, and define

$$f(z) = \exp\left\{\int_{\gamma(z)} g(\zeta) \, d\zeta\right\}.$$

Prove that $f(z)$ is independent of the choice of $\gamma(z)$ (although the integral itself is not), that f is holomorphic in the complement of $\{z_n\}$, that f has a removable singularity at each of the points z_n, and that the extension of f has a zero of order m_n at z_n.

The existence theorem contained in Theorem 15.9 can thus be deduced from the Mittag-Leffler theorem.

9 Suppose $0 < \alpha < 1$, $0 < \beta < 1$, $f \in H(U)$, $f(U) \subset U$, and $f(0) = \alpha$. How many zeros can f have in the disc $\bar{D}(0; \beta)$? What is the answer if (a) $\alpha = \frac{1}{2}$, $\beta = \frac{1}{2}$; (b) $\alpha = \frac{1}{4}$, $\beta = \frac{1}{2}$; (c) $\alpha = \frac{2}{3}$, $\beta = \frac{1}{3}$; (d) $\alpha = 1/1{,}000$, $\beta = 1/10$?

10 For $N = 1, 2, 3, \ldots$, define

$$g_N(z) = \prod_{n=N}^{\infty} \left(1 - \frac{z^2}{n^2}\right).$$

Prove that the ideal generated by $\{g_N\}$ in the ring of entire functions is not a principal ideal.

11 Under what conditions on a sequence of real numbers y_n does there exist a bounded holomorphic function in the open right half plane which is not identically zero but which has a zero at each point $1 + iy_n$? In particular, can this happen if (a) $y_n = \log n$, (b) $y_n = \sqrt{n}$, (c) $y_n = n$, (d) $y_n = n^2$?

12 Suppose $0 < |\alpha_n| < 1$, $\Sigma(1 - |\alpha_n|) < \infty$, and B is the Blaschke product with zeros at the points α_n. Let E be the set of all points $1/\bar{\alpha}_n$ and let Ω be the complement of the closure of E. Prove that the product actually converges uniformly on every compact subset of Ω, so that $B \in H(\Omega)$, and that B has a pole at each point of E. (This is of particular interest in those cases in which Ω is connected.)

13 Put $\alpha_n = 1 - n^{-2}$, for $n = 1, 2, 3, \ldots$, let B be the Blaschke product with zeros at these points α, and prove that $\lim_{r \to 1} B(r) = 0$. (It is understood that $0 < r < 1$.)

More precisely, show that the estimate

$$|B(r)| < \prod_{1}^{N-1} \frac{r - \alpha_n}{1 - \alpha_n r} < \prod_{1}^{N-1} \frac{\alpha_N - \alpha_n}{1 - \alpha_n} < 2e^{-N/3}$$

is valid if $\alpha_{N-1} < r < \alpha_N$.

14 Prove that there is a sequence $\{\alpha_n\}$ with $0 < \alpha_n < 1$, which tends to 1 so rapidly that the Blaschke product with zeros at the points α_n satisfies the condition

$$\limsup_{r \to 1} |B(r)| = 1.$$

Hence this B has no radial limit at $z = 1$.

15 Let φ be a linear fractional transformation which maps U onto U. For any $z \in U$ define the φ-orbit of z to be the set $\{\varphi_n(z)\}$, where $\varphi_0(z) = z$, $\varphi_n(z) = \varphi(\varphi_{n-1}(z))$, $n = 1, 2, 3, \ldots$. Ignore the case $\varphi(z) = z$.

(a) For which φ is it true that the φ-orbits satisfy the Blaschke condition $\Sigma(1 - |\varphi_n(z)|) < \infty$? [The answer depends in part on the location of the fixed points of φ. There may be one fixed point in U, or one fixed point on T, or two fixed points on T. In the last two cases it is advantageous to transfer the problem to (say) the upper half plane, and to consider transformations on it which either leave only ∞ fixed or leave 0 and ∞ fixed.]

(b) For which φ do there exist nonconstant functions $f \in H^\infty$ which are invariant under φ, i.e., which satisfy the relation $f(\varphi(z)) = f(z)$ for all $z \in U$?

16 Suppose $|\alpha_1| \leq |\alpha_2| \leq |\alpha_3| \leq \cdots < 1$, and let $n(r)$ be the number of terms in the sequence $\{\alpha_j\}$ such that $|\alpha_j| \leq r$. Prove that

$$\int_0^1 n(r)\, dr = \sum_{j=1}^{\infty} (1 - |\alpha_j|).$$

17 If $B(z) = \Sigma c_k z^k$ is a Blaschke product with at least one zero off the origin, is it possible to have $c_k \geq 0$ for $k = 0, 1, 2, \ldots$?

18 Suppose B is a Blaschke product all of whose zeros lie on the segment $(0, 1)$ and

$$f(z) = (z - 1)^2 B(z).$$

Prove that the derivative of f is bounded in U.

19 Put $f(z) = \exp[-(1 + z)/(1 - z)]$. Using the notation of Theorem 15.19, show that

$$\lim_{r \to 1} \mu_r(f) < \mu^*(f),$$

although $f \in H^\infty$. Note the contrast with Theorem 15.24.

20 Suppose $\lambda_1 > \lambda_2 > \cdots$, and $\lambda_n \to 0$ in the Müntz-Szasz theorem. What is the conclusion of the theorem, under these conditions?

21 Prove an analogue of the Müntz-Szasz theorem, with $L^2(I)$ in place of $C(I)$.

22 Put $f_n(t) = t^n e^{-t}$ ($0 \le t < \infty$, $n = 0, 1, 2, \ldots$) and prove that the set of all finite linear combinations of the functions f_n is dense in $L^2(0, \infty)$. *Hint*: If $g \in L^2(0, \infty)$ is orthogonal to each f_n and if

$$F(z) = \int_0^\infty e^{-tz} \overline{g(t)}\, dt \qquad (\text{Re } z > 0),$$

then all derivatives of F are 0 at $z = 1$. Consider $F(1 + iy)$.

23 Suppose $\Omega \supset \bar{U}$, $f \in H(\Omega)$, $|f(e^{i\theta})| \ge 3$ for all real θ, $f(0) = 0$, and $\lambda_1, \lambda_2, \ldots, \lambda_N$ are the zeros of $1 - f$ in U, counted according to their multiplicities. Prove that

$$|\lambda_1 \lambda_2 \cdots \lambda_N| < \tfrac{1}{2}.$$

Suggestion: Look at $B/(1 - f)$, where B is a certain Blaschke product.

CHAPTER
SIXTEEN

ANALYTIC CONTINUATION

In this chapter we shall be concerned with questions which arise because functions which are defined and holomorphic in some region can frequently be extended to holomorphic functions in some larger region. Theorem 10.18 shows that these extensions are uniquely determined by the given functions. The extension process is called *analytic continuation*. It leads in a very natural way to the consideration of functions which are defined on Riemann surfaces rather than in plane regions. This device makes it possible to replace "multiple-valued functions" (such as the square-root function or the logarithm) by functions. A systematic treatment of Riemann surfaces would take us too far afield, however, and we shall restrict the discussion to plane regions.

Regular Points and Singular Points

16.1 Definition Let D be an open circular disc, suppose $f \in H(D)$, and let β be a boundary point of D. We call β a *regular point* of f if there exists a disc D_1 with center at β and a function $g \in H(D_1)$ such that $g(z) = f(z)$ for all $z \in D \cap D_1$. Any boundary point of D which is not a regular point of f is called a *singular point* of f.

It is clear from the definition that the set of all regular points of f is an open (possibly empty) subset of the boundary of D.

In the following theorems we shall take the unit disc U for D, without any loss of generality.

16.2 Theorem *Suppose $f \in H(U)$, and the power series*

$$f(z) = \sum_{n=0}^{\infty} a_n z^n \qquad (z \in U) \qquad (1)$$

has radius of convergence 1. Then f has at least one singular point on the unit circle T.

PROOF Suppose, on the contrary, that every point of T is a regular point of f. The compactness of T implies then that there are open discs $D_1, \ldots, D_n$ and functions $g_j \in H(D_j)$ such that the center of each D_j is on T, such that $T \subset D_1 \cup \cdots \cup D_n$, and such that $g_j(z) = f(z)$ in $D_j \cap U$, for $j = 1, \ldots, n$.

If $D_i \cap D_j \neq \emptyset$ and $V_{ij} = D_i \cap D_j \cap U$, then $V_{ij} \neq \emptyset$ (since the centers of the D_j are on T), and $g_i = f = g_j$ in V_{ij}. Since $D_i \cap D_j$ is connected, it follows from Theorem 10.18 that $g_i = g_j$ in $D_i \cap D_j$. Hence we may define a function h in $\Omega = U \cup D_1 \cup \cdots \cup D_n$ by

$$h(z) = \begin{cases} f(z) & (z \in U), \\ g_i(z) & (z \in D_i), \end{cases} \qquad (2)$$

Since $\Omega \supset \bar{U}$ and Ω is open, there exists an $\epsilon > 0$ such that the disc $D(0; 1 + \epsilon) \subset \Omega$. But $h \in H(\Omega)$, $h(z)$ is given by (1) in U, and now Theorem 10.16 implies that the radius of convergence of (1) is at least $1 + \epsilon$, contrary to our assumption. ////

16.3 Definition If $f \in H(U)$ and if every point of T is a singular point of f, then T is said to be the *natural boundary* of f. In this case, f has no holomorphic extension to any region which properly contains U.

16.4 Remark It is very easy to see that there exist $f \in H(U)$ for which T is a natural boundary. In fact, if Ω is *any* region, it is easy to find an $f \in H(\Omega)$ which has no holomorphic extension to any larger region. To see this, let A be any countable set in Ω which has no limit point in Ω but such that every boundary point of Ω is a limit point of A. Apply Theorem 15.11 to get a function $f \in H(\Omega)$ which is 0 at every point of A but is not identically 0. If $g \in H(\Omega_1)$, where Ω_1 is a region which properly contains Ω, and if $g = f$ in Ω, the zeros of g would have a limit point in Ω_1, and we have a contradiction.

A simple explicit example is furnished by

$$f(z) = \sum_{n=0}^{\infty} z^{2^n} = z + z^2 + z^4 + z^8 + \cdots \qquad (z \in U). \qquad (1)$$

This f satisfies the functional equation

$$f(z^2) = f(z) - z, \qquad (2)$$

from which it follows (we leave the details to the reader) that f is unbounded on every radius of U which ends at $\exp\{2\pi i k/2^n\}$, where k and n are positive integers. These points form a dense subset of T; and since the set of all singular points of f is closed, f has T as its natural boundary.

That this example is a power series with large gaps (i.e., with many zero coefficients) is no accident. The example is merely a special case of Theorem 16.6, due to Hadamard, which we shall derive from the following theorem of Ostrowski:

16.5 Theorem *Suppose λ, p_k, and q_k are positive integers,*

$$p_1 < p_2 < p_3 < \cdots,$$

and

$$\lambda q_k > (\lambda + 1)p_k \qquad (k = 1, 2, 3, \ldots). \tag{1}$$

Suppose

$$f(z) = \sum_{n=0}^{\infty} a_n z^n \tag{2}$$

has radius of convergence 1, and $a_n = 0$ whenever $p_k < n < q_k$ for some k. If $s_p(z)$ is the pth partial sum of (2), and if β is a regular point of f on T, then the sequence $\{s_{p_k}(z)\}$ converges in some neighborhood of β.

Note that the full sequence $\{s_p(z)\}$ cannot converge at any point outside $\bar{U}$. The gap condition (1) ensures the existence of a subsequence which converges in a neighborhood of β, hence at some points outside $\bar{U}$. This phenomenon is called *overconvergence*.

PROOF If $g(z) = f(\beta z)$, then g also satisfies the gap condition. Hence we may assume, without loss of generality, that $\beta = 1$. Then f has a holomorphic extension to a region Ω which contains $U \cup \{1\}$. Put

$$\varphi(w) = \tfrac{1}{2}(w^\lambda + w^{\lambda+1}) \tag{3}$$

and define $F(w) = f(\varphi(w))$ for all w such that $\varphi(w) \in \Omega$. If $|w| \leq 1$ but $w \neq 1$, then $|\varphi(w)| < 1$, since $|1 + w| < 2$. Also, $\varphi(1) = 1$. It follows that there exists an $\epsilon > 0$ such that $\varphi(D(0; 1 + \epsilon)) \subset \Omega$. Note that the region $\varphi(D(0; 1 + \epsilon))$ contains the point 1. The series

$$F(w) = \sum_{m=0}^{\infty} b_m w^m \tag{4}$$

converges if $|w| < 1 + \epsilon$.

The highest and lowest powers of w in $[\varphi(w)]^n$ have exponents $(\lambda + 1)n$ and λn. Hence the highest exponent in $[\varphi(w)]^{p_k}$ is less than the lowest exponent in $[\varphi(w)]^{q_k}$, by (1). Since

$$F(w) = \sum_{n=0}^{\infty} a_n [\varphi(w)]^n \qquad (|w| < 1), \tag{5}$$

the gap condition satisfied by $\{a_n\}$ now implies that

$$\sum_{n=0}^{p_k} a_n [\varphi(w)]^n = \sum_{m=0}^{(\lambda+1)p_k} b_m w^m \qquad (k = 1, 2, 3, \ldots). \tag{6}$$

The right side of (6) converges, as $k \to \infty$, whenever $|w| < 1 + \epsilon$. Hence $\{s_{p_k}(z)\}$ converges for all $z \in \varphi(D(0; 1 + \epsilon))$. This is the desired conclusion. ////

Note: Actually, $\{s_{p_k}(z)\}$ converges uniformly in some neighborhood of β. We leave it to the reader to verify this by a more careful examination of the preceding proof.

16.6 Theorem *Suppose λ is a positive integer, $\{p_k\}$ is a sequence of positive integers such that*

$$p_{k+1} > \left(1 + \frac{1}{\lambda}\right) p_k \qquad (k = 1, 2, 3, \ldots), \tag{1}$$

and the power series

$$f(z) = \sum_{k=1}^{\infty} c_k z^{p_k} \tag{2}$$

has radius of convergence 1. Then f has T as its natural boundary.

PROOF The subsequence $\{s_{p_k}\}$ of Theorem 16.5 is now the same (except for repetitions) as the full sequence of partial sums of (2). The latter cannot converge at any point outside $\bar{U}$; hence Theorem 16.5 implies that no point of T can be a regular point of f. ////

16.7 Example Put $a_n = 1$ if n is a power of 2, put $a_n = 0$ otherwise, put $\eta_n = \exp(-\sqrt{n})$, and define

$$f(z) = \sum_{n=0}^{\infty} a_n \eta_n z^n. \tag{1}$$

Since

$$\limsup_{n \to \infty} |a_n \eta_n|^{1/n} = 1, \tag{2}$$

the radius of convergence of (1) is 1. By Hadamard's theorem, f has T as its natural boundary. Nevertheless, the power series of each derivative of f,

$$f^{(k)}(z) = \sum_{n=k}^{\infty} n(n-1) \cdots (n-k+1) a_n \eta_n z^{n-k}, \qquad (3)$$

converges uniformly on the closed unit disc. Each $f^{(k)}$ is therefore uniformly continuous on $\bar{U}$, and the restriction of f to T is infinitely differentiable, as a function of θ, in spite of the fact that T is the natural boundary of f.

The example demonstrates rather strikingly that the presence of singularities, in the sense of Definition 16.1, does not imply the presence of discontinuities or (stated less precisely) of any lack of smoothness.

This seems to be the natural place to insert a theorem in which continuity does preclude the existence of singularities:

16.8 Theorem *Suppose Ω is a region, L is a straight line or a circular arc, $\Omega - L$ is the union of two regions Ω_1 and Ω_2, f is continuous in Ω, and f is holomorphic in Ω_1 and in Ω_2. Then f is holomorphic in Ω.*

PROOF The use of linear fractional transformations shows that the general case follows if we prove the theorem for straight lines L. By Morera's theorem, it is enough to show that the integral of f over the boundary $\partial \Delta$ is 0 for every triangle Δ in Ω. The Cauchy theorem implies that the integral of f vanishes over every closed path γ in $\Delta \cap \Omega_1$ or in $\Delta \cap \Omega_2$. The continuity of f shows that this is still true if part of γ is in L, and the integral over $\partial \Delta$ is the sum of at most two terms of this sort. ////

Continuation along Curves

16.9 Definitions A *function element* is an ordered pair (f, D), where D is an open circular disc and $f \in H(D)$. Two function elements (f_0, D_0) and (f_1, D_1) are *direct continuations* of each other if two conditions hold: $D_0 \cap D_1 \neq \emptyset$, and $f_0(z) = f_1(z)$ for all $z \in D_0 \cap D_1$. In this case we write

$$(f_0, D_0) \sim (f_1, D_1). \qquad (1)$$

A *chain* is a finite sequence $\mathscr{C}$ of discs, say $\mathscr{C} = \{D_0, D_1, \ldots, D_n\}$, such that $D_{i-1} \cap D_i \neq \emptyset$ for $i = 1, \ldots, n$. If (f_0, D_0) is given and if there exist elements (f_i, D_i) such that $(f_{i-1}, D_{i-1}) \sim (f_i, D_i)$ for $i = 1, \ldots, n$, then (f_n, D_n) is said to be the *analytic continuation of (f_0, D_0) along $\mathscr{C}$*. Note that f_n is uniquely determined by f_0 and $\mathscr{C}$ (if it exists at all). To see this, suppose (1) holds, and suppose (1) also holds with g_1 in place of f_1. Then $g_1 = f_0 = f_1$ in $D_0 \cap D_1$; and since D_1 is connected, we have $g_1 = f_1$ in D_1. The uniqueness of f_n now follows by induction on the number of terms in $\mathscr{C}$.

If (f_n, D_n) is the continuation of (f_0, D_0) along $\mathscr{C}$, and if $D_n \cap D_0 \neq \varnothing$, it need not be true that $(f_0, D_0) \sim (f_n, D_n)$; in other words, the relation $\sim$ is not transitive. The simplest example of this is furnished by the square-root function: Let D_0, D_1, and D_2 be discs of radius 1, with centers 1, ω, and ω^2, where $\omega^3 = 1$, choose $f_j \in H(D_j)$ so that $f_j^2(z) = z$ and so that $(f_0, D_0) \sim (f_1, D_1), (f_1, D_1) \sim (f_2, D_2)$. In $D_0 \cap D_2$ we have $f_2 = -f_0 \neq f_0$.

A chain $\mathscr{C} = \{D_0, \ldots, D_n\}$ is said to *cover* a curve γ with parameter interval $[0, 1]$ if there are numbers $0 = s_0 < s_1 < \cdots < s_n = 1$ such that $\gamma(0)$ is the center of D_0, $\gamma(1)$ is the center of D_n, and

$$\gamma([s_i, s_{i+1}]) \subset D_i \qquad (i = 0, \ldots, n-1). \tag{2}$$

If (f_0, D_0) can be continued along this $\mathscr{C}$ to (f_n, D_n), we call (f_n, D_n) an *analytic continuation* of (f_0, D_0) along γ (uniqueness will be proved in Theorem 16.11); (f_0, D_0) is then said to *admit* an analytic continuation along γ.

Although the relation (1) is not transitive, a restricted form of transitivity does hold. It supplies the key to the proof of Theorem 16.11.

16.10 Proposition *Suppose that $D_0 \cap D_1 \cap D_2 \neq \varnothing$, $(D_0, f_0) \sim (D_1, f_1)$, and $(D_1, f_1) \sim (D_2, f_2)$. Then $(D_0, f_0) \sim (D_2, f_2)$.*

PROOF By assumption, $f_0 = f_1$ in $D_0 \cap D_1$ and $f_1 = f_2$ in $D_1 \cap D_2$. Hence $f_0 = f_2$ in the nonempty open set $D_0 \cap D_1 \cap D_2$. Since f_0 and f_2 are holomorphic in $D_0 \cap D_2$ and $D_0 \cap D_2$ is connected, it follows that $f_0 = f_2$ in $D_0 \cap D_2$. ////

16.11 Theorem *If (f, D) is a function element and if γ is a curve which starts at the center of D, then (f, D) admits at most one analytic continuation along γ.*

Here is a more explicit statement of what the theorem asserts: If γ is covered by chains $\mathscr{C}_1 = \{A_0, A_1, \ldots, A_m\}$ and $\mathscr{C}_2 = \{B_0, B_1, \ldots B_n\}$, where $A_0 = B_0 = D$, if (f, D) can be analytically continued along $\mathscr{C}_1$ to a function element (g_m, A_m), and if (f, D) can be analytically continued along $\mathscr{C}_2$ to (h_n, B_n), then $g_m = h_n$ in $A_m \cap B_n$.

Since A_m and B_n are, by assumption, discs with the same center $\gamma(1)$, it follows that g_m and h_n have the same expansion in powers of $z - \gamma(1)$, and we may as well replace A_m and B_n by whichever is the larger one of the two. With this agreement, the conclusion is that $g_m = h_n$.

PROOF Let $\mathscr{C}_1$ and $\mathscr{C}_2$ be as above. There are numbers

$$0 = s_0 < s_1 < \cdots < s_m = 1 = s_{m+1}$$

and $0 = \sigma_0 < \sigma_1 < \cdots < \sigma_n = 1 = \sigma_{n+1}$ such that

$$\gamma([s_i, s_{i+1}]) \subset A_i, \qquad \gamma([\sigma_j, \sigma_{j+1}]) \subset B_j \qquad (0 \leq i \leq m, 0 \leq j \leq n). \tag{1}$$

There are function elements $(g_i, A_i) \sim (g_{i+1}, A_{i+1})$ and $(h_j, B_j) \sim (h_{j+1}, B_{j+1})$, for $0 \le i \le m - 1$ and $0 \le j \le n - 1$. Here $g_0 = h_0 = f$.

We claim that if $0 \le i \le m$ and $0 \le j \le n$, and if $[s_i, s_{i+1}]$ intersects $[\sigma_j, \sigma_{j+1}]$, then $(g_i, A_i) \sim (h_j, B_j)$.

Assume there are pairs (i, j) for which this is wrong. Among them there is one for which $i + j$ is minimal. It is clear that then $i + j > 0$. Suppose $s_i \ge \sigma_j$. Then $i \ge 1$, and since $[s_i, s_{i+1}]$ intersects $[\sigma_j, \sigma_{j+1}]$, we see that

$$\gamma(s_i) \in A_{i-1} \cap A_i \cap B_j. \tag{2}$$

The minimality of $i + j$ shows that $(g_{i-1}, A_{i-1}) \sim (h_j, B_j)$; and since $(g_{i-1}, A_{i-1}) \sim (g_i, A_i)$, Proposition 16.10 implies that $(g_i, A_i) \sim (h_j, B_j)$. This contradicts our assumption. The possibility $s_i \le \sigma_j$ is ruled out in the same way.

So our claim is established. In particular, it holds for the pair (m, n), and this is what we had to prove. ////

16.12 Definition Suppose α and β are points in a topological space X and φ is a continuous mapping of the unit square $I^2 = I \times I$ (where $I = [0, 1]$) into X such that $\varphi(0, t) = \alpha$ and $\varphi(1, t) = \beta$ for all $t \in I$. The curves γ_t defined by

$$\gamma_t(s) = \varphi(s, t) \qquad (s \in I, t \in I) \tag{1}$$

are then said to form a *one-parameter family* $\{\gamma_t\}$ *of curves from α to β in X.*

We now come to a very important property of analytic continuation:

16.13 Theorem *Suppose $\{\gamma_t\}$ $(0 \le t \le 1)$ is a one-parameter family of curves from α to β in the plane, D is a disc with center at α, and the function element (f, D) admits analytic continuation along each γ_t, to an element (g_t, D_t). Then $g_1 = g_0$.*

The last equality is to be interpreted as in Theorem 16.11:

$$(g_1, D_1) \sim (g_0, D_0),$$

and D_0 and D_1 are discs with the same center, namely, β.

PROOF Fix $t \in I$. There is a chain $\mathscr{C} = \{A_0, \ldots, A_n\}$ which covers γ_t, with $A_0 = D$, such that (g_t, D_t) is obtained by continuation of (f, D) along $\mathscr{C}$. There are numbers $0 = s_0 < \cdots < s_n = 1$ such that

$$E_i = \gamma_t([s_i, s_{i+1}]) \subset A_i \qquad (i = 0, 1, \ldots, n - 1). \tag{1}$$

There exists an $\epsilon > 0$ which is less than the distance from any of the compact sets E_i to the complement of the corresponding open disc A_i. The uniform continuity of φ on I^2 (see Definition 16.12) shows that there exists a $\delta > 0$ such that

$$|\gamma_t(s) - \gamma_u(s)| < \epsilon \qquad \text{if } s \in I, u \in I, |u - t| < \delta. \tag{2}$$

Suppose u satisfies these conditions. Then (2) shows that $\mathscr{C}$ covers γ_u, and therefore Theorem 16.11 shows that both g_t and g_u are obtained by continuation of (f, D) along this same chain $\mathscr{C}$. Hence $g_t = g_u$.

Thus each $t \in I$ is covered by a segment J_t such that $g_u = g_t$ for all $u \in I \cap J_t$. Since I is compact, I is covered by finitely many J_t; and since I is connected, we see in a finite number of steps that $g_1 = g_0$. ////

Our next item is an intuitively obvious topological fact.

16.14 Theorem *Suppose Γ_0 and Γ_1 are curves in a topological space X, with common initial point α and common end point β. If X is simply connected, then there exists a one-parameter family $\{\gamma_t\}$ ($0 \le t \le 1$) of curves from α to β in X, such that $\gamma_0 = \Gamma_0$ and $\gamma_1 = \Gamma_1$.*

PROOF Let $[0, \pi]$ be the parameter interval of Γ_0 and Γ_1. Then

$$\Gamma(s) = \begin{cases} \Gamma_0(s) & (0 \le s \le \pi) \\ \Gamma_1(2\pi - s) & (\pi \le s \le 2\pi) \end{cases} \tag{1}$$

defines a *closed* curve in X. Since X is simply connected, Γ is null-homotopic in X. Hence there is a continuous $H: [0, 2\pi] \times [0, 1] \to X$ such that

$$H(s, 0) = \Gamma(s), \quad H(s, 1) = c \in X, \quad H(0, t) = H(2\pi, t). \tag{2}$$

If $\Phi: \bar{U} \to X$ is defined by

$$\Phi(re^{i\theta}) = H(\theta, 1 - r) \quad (0 \le r \le 1, 0 \le \theta \le 2\pi),$$

(2) implies that Φ is continuous. Put

$$\gamma_t(\theta) = \Phi[(1 - t)e^{i\theta} + te^{-i\theta}] \quad (0 \le \theta \le \pi, 0 \le t \le 1).$$

Since $\Phi(e^{i\theta}) = H(\theta, 0) = \Gamma(\theta)$, it follows that

$$\gamma_t(0) = \Phi(1) = \Gamma(0) = \alpha \quad (0 \le t \le 1),$$
$$\gamma_t(\pi) = \Phi(-1) = \Gamma(\pi) = \beta \quad (0 \le t \le 1),$$
$$\gamma_0(\theta) = \Phi(e^{i\theta}) = \Gamma(\theta) = \Gamma_0(\theta) \quad (0 \le \theta \le \pi)$$

and

$$\gamma_1(\theta) = \Phi(e^{-i\theta}) = \Phi(e^{i(2\pi - \theta)}) = \Gamma(2\pi - \theta) = \Gamma_1(\theta) \quad (0 \le \theta \le \pi).$$

This completes the proof. ////

The Monodromy Theorem

The preceding considerations have essentially proved the following important theorem.

16.15 Theorem *Suppose Ω is a simply connected region, (f, D) is a function element, $D \subset \Omega$, and (f, D) can be analytically continued along every curve in Ω that starts at the center of D. Then there exists $g \in H(\Omega)$ such that $g(z) = f(z)$ for all $z \in D$.*

PROOF Let Γ_0 and Γ_1 be two curves in Ω from the center α of D to some point $\beta \in \Omega$. It follows from Theorems 16.13 and 16.14 that the analytic continuations of (f, D) along Γ_0 and Γ_1 lead to the same element (g_β, D_β), where D_β is a disc with center at β. If D_{β_1} intersects D_β, then $(g_{\beta_1}, D_{\beta_1})$ can be obtained by first continuing (f, D) to β, then along the straight line from β to β_1. This shows that $g_{\beta_1} = g_\beta$ in $D_{\beta_1} \cap D_\beta$.

The definition

$$g(z) = g_\beta(z) \qquad (z \in D_\beta)$$

is therefore consistent and gives the desired holomorphic extension of f.   ////

16.16 Remark Let Ω be a plane region, fix $w \notin \Omega$, let D be a disc in Ω. Since D is simply connected, there exists $f \in H(D)$ such that $\exp[f(z)] = z - w$. Note that $f'(z) = (z - w)^{-1}$ in D, and that the latter function is holomorphic in all of Ω. This implies that (f, D) can be analytically continued along every path γ in Ω that starts at the center α of D: If γ goes from α to β, if $D_\beta = D(\beta; r) \subset \Omega$, if

$$\Gamma_z = \gamma \dotplus [\beta, z] \qquad (z \in D_\beta) \tag{1}$$

and if

$$g_\beta(z) = \int_{\Gamma_z} (\zeta - w)^{-1} \, d\zeta + f(\alpha) \qquad (z \in D_\beta), \tag{2}$$

then (g_β, D_β) is the continuation of (f, D) along γ.

Note that $g'_\beta(z) = (z - w)^{-1}$ in D_β.

Assume now that there exists $g \in H(\Omega)$ such that $g(z) = f(z)$ in D. Then $g'(z) = (z - w)^{-1}$ for all $z \in \Omega$. If Γ is a *closed* path in Ω, it follows that

$$\operatorname{Ind}_\Gamma(w) = \frac{1}{2\pi i} \int_\Gamma g'(z) \, dz = 0. \tag{3}$$

We conclude (with the aid of Theorem 13.11) *that the monodromy theorem fails in every plane region that is not simply connected.*

Construction of a Modular Function

16.17 The Modular Group This is the set G of all linear fractional transformations φ of the form

$$\varphi(z) = \frac{az + b}{cz + d} \tag{1}$$

where a, b, c, and d are integers and $ad - bc = 1$.

Since a, b, c, and d are real, each $\varphi \in G$ maps the real axis onto itself (except for ∞). The imaginary part of $\varphi(i)$ is $(c^2 + d^2)^{-1} > 0$. Hence

$$\varphi(\Pi^+) = \Pi^+ \qquad (\varphi \in G), \tag{2}$$

where Π^+ is the open upper half plane. If φ is given by (1), then

$$\varphi^{-1}(w) = \frac{dw - b}{-cw + a} \tag{3}$$

so that $\varphi^{-1} \in G$. Also $\varphi \circ \psi \in G$ if $\varphi \in G$ and $\psi \in G$.

Thus G is a group, with composition as group operation. In view of (2) it is customary to regard G as a group of transformations on Π^+.

The transformations $z \to z + 1$ $(a = b = d = 1,\ c = 0)$ and $z \to -1/z$ $(a = d = 0,\ b = -1,\ c = 1)$ belong to G. In fact, they generate G (i.e., there is no proper subgroup of G which contains these two transformations). This can be proved by the same method which will be used in Theorem 16.19(c).

A *modular function* is a holomorphic (or meromorphic) function f on Π^+ which is invariant under G or at least under some nontrivial subgroup Γ of G. This means that $f \circ \varphi = f$ for every $\varphi \in \Gamma$.

16.18 A Subgroup We shall take for Γ the group generated by σ and τ, where

$$\sigma(z) = \frac{z}{2z + 1}, \qquad \tau(z) = z + 2. \tag{1}$$

One of our objectives is the construction of a certain function λ which is invariant under Γ and which leads to a quick proof of the Picard theorem. Actually, it is the mapping properties of λ which are important in this proof, not its invariance, and a quicker construction (using just the Riemann mapping theorem and the reflection principle) can be given. But it is instructive to study the action of Γ on Π^+, in geometric terms, and we shall proceed along this route.

Let Q be the set of all z which satisfy the following four conditions, where $z = x + iy$:

$$y > 0, \quad -1 \leq x < 1, \quad |2z + 1| \geq 1, \quad |2z - 1| > 1. \tag{2}$$

Q is bounded by the vertical lines $x = -1$ and $x = 1$ and is bounded below by two semicircles of radius $\frac{1}{2}$, with centers at $-\frac{1}{2}$ and at $\frac{1}{2}$. Q contains those of its boundary points which lie in the left half of Π^+. Q contains no point of the real axis.

We claim that Q is a *fundamental domain of* Γ. This means that statements (a) and (b) of the following theorem are true.

16.19 Theorem *Let* Γ *and* Q *be as above.*

(a) *If* φ_1 *and* $\varphi_2 \in \Gamma$ *and* $\varphi_1 \neq \varphi_2$, *then* $\varphi_1(Q) \cap \varphi_2(Q) = \emptyset$.
(b) $\bigcup_{\varphi \in \Gamma} \varphi(Q) = \Pi^+$.
(c) Γ *contains all transformations* $\varphi \in G$ *of the form*

$$\varphi(z) = \frac{az+b}{cz+d} \tag{1}$$

for which a and d are odd integers, b and c are even.

PROOF Let Γ_1 be the set of all $\varphi \in G$ described in (c). It is easily verified that Γ_1 is a subgroup of G. Since $\sigma \in \Gamma_1$ and $\tau \in \Gamma_1$, it follows that $\Gamma \subset \Gamma_1$. To show that $\Gamma = \Gamma_1$, i.e., to prove (c), it is enough to prove that (a') and (b) hold, where (a') is the statement obtained from (a) by replacing Γ by Γ_1. For if (a') and (b) hold, it is clear that Γ cannot be a proper subset of Γ_1.

We shall need the relation

$$\operatorname{Im} \varphi(z) = \frac{\operatorname{Im} z}{|cz+d|^2} \tag{2}$$

which is valid for every $\varphi \in G$ given by (1). The proof of (2) is a matter of straightforward computation, and depends on the relation $ad - bc = 1$.

We now prove (a'). Suppose φ_1 and $\varphi_2 \in \Gamma_1$, $\varphi_1 \neq \varphi_2$, and define $\varphi = \varphi_1^{-1} \circ \varphi_2$. If $z \in \varphi_1(Q) \cap \varphi_2(Q)$, then $\varphi_1^{-1}(z) \in Q \cap \varphi(Q)$. It is therefore enough to show that

$$Q \cap \varphi(Q) = \emptyset \tag{3}$$

if $\varphi \in \Gamma_1$ and φ is not the identity transformation.

The proof of (3) splits into three cases.

If $c = 0$ in (1), then $ad = 1$, and since a and d are integers, we have $a = d = \pm 1$. Hence $\varphi(z) = z + 2n$ for some integer $n \neq 0$, and the description of Q makes it evident that (3) holds.

If $c = 2d$, then $c = \pm 2$ and $d = \pm 1$ (since $ad - bc = 1$). Therefore $\varphi(z) = \sigma(z) + 2m$, where m is an integer. Since $\sigma(Q) \subset \bar{D}(\frac{1}{2}; \frac{1}{2})$, (3) holds.

If $c \neq 0$ and $c \neq 2d$, we claim that $|cz+d| > 1$ for all $z \in Q$. Otherwise, the disc $\bar{D}(-d/c; 1/|c|)$ would intersect Q. The description of Q shows that if $\alpha \neq -\frac{1}{2}$ is a real number and if $\bar{D}(\alpha; r)$ intersects Q, then at least one of the points $-1, 0, 1$ lies in $D(\alpha; r)$. Hence $|cw + d| < 1$, for $w = -1$ or 0 or 1. But for these w, $cw + d$ is an odd integer whose absolute value cannot be less than 1. So $|cz+d| > 1$, and it now follows from (2) that $\operatorname{Im} \varphi(z) < \operatorname{Im} z$ for every $z \in Q$. If it were true for some $z \in Q$ that $\varphi(z) \in Q$, the same argument would apply to φ^{-1} and would show that

$$\text{Im } z = \text{Im } \varphi^{-1}(\varphi(z)) \leq \text{Im } \varphi(z). \tag{4}$$

This contradiction shows that (3) holds.

Hence (*a'*) is proved.

To prove (*b*), let Σ be the union of the sets $\varphi(Q)$, for $\varphi \in \Gamma$. It is clear that $\Sigma \subset \Pi^+$. Also, Σ contains the sets $\tau^n(Q)$, for $n = 0, \pm 1, \pm 2, \ldots$, where $\tau^n(z) = z + 2n$. Since σ maps the circle $|2z + 1| = 1$ onto the circle $|2z - 1| = 1$, we see that Σ contains every $z \in \Pi^+$ which satisfies all inequalities

$$|2z - (2m + 1)| \geq 1 \qquad (m = 0, \pm 1, \pm 2, \ldots). \tag{5}$$

Fix $w \in \Pi^+$. Since $\text{Im } w > 0$, there are only finitely many pairs of integers c and d such that $|cw + d|$ lies below any given bound, and we can choose $\varphi_0 \in \Gamma$ so that $|cw + d|$ is minimized. By (2), this means that

$$\text{Im } \varphi(w) \leq \text{Im } \varphi_0(w) \qquad (\varphi \in \Gamma). \tag{6}$$

Put $z = \varphi_0(w)$. Then (6) becomes

$$\text{Im } \varphi(z) \leq \text{Im } z \qquad (\varphi \in \Gamma). \tag{7}$$

Apply (7) to $\varphi = \sigma\tau^{-n}$ and to $\varphi = \sigma^{-1}\tau^{-n}$. Since

$$(\sigma\tau^{-n})(z) = \frac{z - 2n}{2z - 4n + 1}, \qquad (\sigma^{-1}\tau^{-n})(z) = \frac{z - 2n}{-2z + 4n + 1}, \tag{8}$$

it follows from (2) and (7) that

$$|2z - 4n + 1| \geq 1, \qquad |2z - 4n - 1| \geq 1 \qquad (n = 0, \pm 1, \pm 2, \ldots). \tag{9}$$

Thus z satisfies (5), hence $z \in \Sigma$; and since $w = \varphi_0^{-1}(z)$ and $\varphi_0^{-1} \in \Gamma$, we have $w \in \Sigma$.

This completes the proof. ////

The following theorem summarizes some of the properties of the modular function λ which was mentioned in Sec. 16.18 and which will be used in Theorem 16.22.

16.20 Theorem *If Γ and Q are as described in Sec. 16.18, there exists a function $\lambda \in H(\Pi^+)$ such that*

(a) $\lambda \circ \varphi = \lambda$ *for every* $\varphi \in \Gamma$.
(b) λ *is one-to-one on* Q.
(c) *The range Ω of λ [which is the same as $\lambda(Q)$, by (a)], is the region consisting of all complex numbers different from 0 and 1.*
(d) λ *has the real axis as its natural boundary.*

PROOF Let Q_0 be the right half of Q. More precisely, Q_0 consists of all $z \in \Pi^+$ such that

$$0 < \operatorname{Re} z < 1, \qquad |2z - 1| > 1. \tag{1}$$

By Theorem 14.19 (and Remarks 14.20) there is a continuous function h on $\bar{Q}_0$ which is one-to-one on $\bar{Q}_0$ and holomorphic in Q_0, such that $h(Q_0) = \Pi^+$, $h(0) = 0$, $h(1) = 1$, and $h(\infty) = \infty$. The reflection principle (Theorem 11.14) shows that the formula

$$h(-x + iy) = \overline{h(x + iy)} \tag{2}$$

extends h to a continuous function on the closure $\bar{Q}$ of Q which is a conformal mapping of the interior of Q onto the complex plane minus the nonnegative real axis. We also see that h is one-to-one on Q, that $h(Q)$ is the region Ω described in (c), that

$$h(-1 + iy) = h(1 + iy) = h(\tau(-1 + iy)) \qquad (0 < y < \infty), \tag{3}$$

and that

$$h(-\tfrac{1}{2} + \tfrac{1}{2}e^{i\theta}) = h(\tfrac{1}{2} + \tfrac{1}{2}e^{i(\pi - \theta)}) = h(\sigma(-\tfrac{1}{2} + \tfrac{1}{2}e^{i\theta})) \qquad (0 < \theta < \pi). \tag{4}$$

Since h is real on the boundary of Q, (3) and (4) follow from (2) and the definitions of σ and τ.

We now define the function λ:

$$\lambda(z) = h(\varphi^{-1}(z)) \qquad (z \in \varphi(Q),\ \varphi \in \Gamma). \tag{5}$$

By Theorem 16.19, each $z \in \Pi^+$ lies in $\varphi(Q)$ for one and only one $\varphi \in \Gamma$. Thus (5) defines $\lambda(z)$ for $z \in \Pi^+$, and we see immediately that λ has properties (a) to (c) and that λ is holomorphic in the interior of each of the sets $\varphi(Q)$.

It follows from (3) and (4) that λ is continuous on

$$Q \cup \tau^{-1}(Q) \cup \sigma^{-1}(Q),$$

hence on an open set V which contains Q. Theorem 16.8 now shows that λ is holomorphic in V. Since Π^+ is covered by the union of the sets $\varphi(V)$, $\varphi \in \Gamma$, and since $\lambda \circ \varphi = \lambda$, we conclude that $\lambda \in H(\Pi^+)$.

Finally, the set of all numbers $\varphi(0) = b/d$ is dense on the real axis. If λ could be analytically continued to a region which properly contains Π^+, the zeros of λ would have a limit point in this region, which is impossible since λ is not constant. ////

The Picard Theorem

16.21 The so-called "little Picard theorem" asserts that every nonconstant entire function attains each value, with one possible exception. This is the theorem which is proved below. There is a stronger version: *Every entire function which is not a polynomial attains each value infinitely many times, again with one possible exception.* That one exception can occur is shown by $f(z) = e^z$, which omits the value 0. The latter theorem is actually true in a local situation: *If f has an isolated singularity at a point z_0 and if f omits two values in some neighborhood of z_0, then z_0 is a removable singularity or a pole of f.* This so-called "big Picard theorem" is a remarkable strengthening of the theorem of Weierstrass (Theorem 10.21) which merely asserts that the image of every neighborhood of z_0 is dense in the plane if f has an essential singularity at z_0. We shall not prove it here.

16.22 Theorem *If f is an entire function and if there are two distinct complex numbers α and β which are not in the range of f, then f is constant.*

PROOF Without loss of generality we assume that $\alpha = 0$ and $\beta = 1$; if not, replace f by $(f - \alpha)/(\beta - \alpha)$. Then f maps the plane into the region Ω described in Theorem 16.20.

With each disc $D_1 \subset \Omega$ there is associated a region $V_1 \subset \Pi^+$ (in fact, there are infinitely many such V_1, one for each $\varphi \in \Gamma$) such that λ is one-to-one on V_1 and $\lambda(V_1) = D_1$; each such V_1 intersects at most two of the domains $\varphi(Q)$. Corresponding to each choice of V_1 there is a function $\psi_1 \in H(D_1)$ such that $\psi_1(\lambda(z)) = z$ for all $z \in V_1$.

If D_2 is another disc in Ω and if $D_1 \cap D_2 \ne \varnothing$, we can choose a corresponding V_2 so that $V_1 \cap V_2 \ne \varnothing$. The function elements (ψ_1, D_1) and (ψ_2, D_2) will then be direct analytic continuations of each other. Note that $\psi_i(D_i) \subset \Pi^+$.

Since the range of f is in Ω, there is a disc A_0 with center at 0 so that $f(A_0)$ lies in a disc D_0 in Ω. Choose $\psi_0 \in H(D_0)$, as above, put $g(z) = \psi_0(f(z))$ for $z \in A_0$, and let γ be any curve in the plane which starts at 0. The range of $f \circ \gamma$ is a compact subset of Ω. Hence γ can be covered by a chain of discs, $A_0, \ldots, A_n$, so that each $f(A_i)$ lies in a disc D_i in Ω, and we can choose $\psi_i \in H(D_i)$ so that (ψ_i, D_i) is a direct analytic continuation of (ψ_{i-1}, D_{i-1}), for $i = 1, \ldots n$. This gives an analytic continuation of the function element (g, A_0) along the chain $\{A_0, \ldots, A_n\}$; note that $\psi_n \circ f$ has positive imaginary part.

Since (g, A_0) can be analytically continued along every curve in the plane and since the plane is simply connected, the monodromy theorem implies that g extends to an entire function. Also, the range of g is in Π^+, hence $(g - i)/(g + i)$ is bounded, hence constant, by Liouville's theorem. This shows that g is constant, and since ψ_0 was one-to-one on $f(A_0)$ and A_0 was a nonempty open set, we conclude that f is constant.   ////

Exercises

1 Suppose $f(z) = \Sigma a_n z^n$, $a_n \ge 0$, and the radius of convergence of the series is 1. Prove that f has a singularity at $z = 1$. *Hint*: Expand f in powers of $z - \frac{1}{2}$. If 1 were a regular point of f, the new series would converge at some $x > 1$. What would this imply about the original series?

2 Suppose (f, D) and (g, D) are function elements, P is a polynomial in two variables, and $P(f, g) = 0$ in D. Suppose f and g can be analytically continued along a curve γ, to f_1 and g_1. Prove that

$P(f_1, g_1) = 0$. Extend this to more than two functions. Is there such a theorem for some class of functions P which is larger than the polynomials?

3 Suppose Ω is a simply connected region, and u is a real harmonic function in Ω. Prove that there exists an $f \in H(\Omega)$ such that $u = \operatorname{Re} f$. Show that this fails in every region which is not simply connected.

4 Suppose X is the closed unit square in the plane, f is a continuous complex function on X, and f has no zero in X. Prove that there is a continuous function g on X such that $f = e^g$. For what class of spaces X (other than the above square) is this also true?

5 Prove that the transformations $z \to z + 1$ and $z \to -1/z$ generate the full modular group G. Let R consist of all $z = x + iy$ such that $|x| < \frac{1}{2}$, $y > 0$, and $|z| > 1$, plus those limit points which have $x \leq 0$. Prove that R is a fundamental domain of G.

6 Prove that G is also generated by the transformations φ and ψ, where

$$\varphi(z) = -\frac{1}{z}, \qquad \psi(z) = \frac{z-1}{z}.$$

Show that φ has period 2, ψ has period 3.

7 Find the relation between composition of linear fractional transformations and matrix multiplication. Try to use this to construct an algebraic proof of Theorem 16.19(c) or of the first part of Exercise 5.

8 Let E be a compact set on the real axis, of positive Lebesgue measure, let Ω be the complement of E, relative to the plane, and define

$$f(z) = \int_E \frac{dt}{t-z} \qquad (z \in \Omega).$$

Answer the following questions:
    (a) Is f constant?
    (b) Can f be extended to an entire function?
    (c) Does $\lim zf(z)$ exist as $z \to \infty$? If so, what is it?
    (d) Does f have a holomorphic square root in Ω?
    (e) Is the real part of f bounded in Ω?
    (f) Is the imaginary part of f bounded in Ω?
[If "yes" in (e) or (f), give a bound.]
    (g) What is $\int_\gamma f(z)\, dz$ if γ is a positively oriented circle which has E in its interior?
    (h) Does there exist a bounded holomorphic function φ in Ω which is not constant?

9 Check your answers in Exercise 8 against the special case

$$E = [-1, 1].$$

10 Call a compact set E in the plane *removable* if there are no nonconstant bounded holomorphic functions in the complement of E.
    (a) Prove that every countable compact set is removable.
    (b) If E is a compact subset of the real axis, and $m(E) = 0$, prove that E is removable. *Hint*: E can be surrounded by curves of arbitrarily small total length. Apply Cauchy's formula, as in Exercise 25, Chap 10.
    (c) Suppose E is removable, Ω is a region, $E \subset \Omega$, $f \in H(\Omega - E)$, and f is bounded. Prove that f can be extended to a holomorphic function in Ω.
    (d) Formulate and prove an analogue of part (b) for sets E which are not necessarily on the real axis.
    (e) Prove that no compact connected subset of the plane (with more than one point) is removable.

11 For each positive number α, let Γ_α be the path with parameter interval $(-\infty, \infty)$ defined by

$$\Gamma_\alpha(t) = \begin{cases} -t - \pi i & (-\infty < t \leq -\alpha), \\ \alpha + \dfrac{\pi i t}{\alpha} & (-\alpha \leq t \leq \alpha), \\ t + \pi i & (\alpha \leq t < \infty). \end{cases}$$

Let Ω_α be the component of the complement of Γ_α^* which contains the origin, and define

$$f_\alpha(z) = \frac{1}{2\pi i} \int_{\Gamma_\alpha} \frac{\exp(e^w)}{w - z} \, dw \qquad (z \in \Omega_\alpha).$$

Prove that f_β is an analytic continuation of f_α if $\alpha < \beta$. Prove that therefore there is an entire function f whose restriction to Ω_α is f_α. Prove that

$$\lim_{r \to \infty} f(re^{i\theta}) = 0$$

for every $e^{i\theta} \neq 1$. (Here r is positive and θ is real, as usual.) Prove that f is not constant. [*Hint*: Look at $f(r)$.] If

$$g = f \exp(-f),$$

prove that

$$\lim_{r \to \infty} g(re^{i\theta}) = 0$$

for every $e^{i\theta}$.

Show that there exists an entire function h such that

$$\lim_{n \to \infty} h(nz) = \begin{cases} 1 & \text{if } z = 0, \\ 0 & \text{if } z \neq 0. \end{cases}$$

12 Suppose

$$f(z) = \sum_{k=1}^{\infty} \left(\frac{z - z^2}{2}\right)^{3^k} = \sum_{n=1}^{\infty} a_n z^n.$$

Find the regions in which the two series converge. Show that this illustrates Theorem 16.5. Find the singular point of f which is nearest to the origin.

13 Let $\Omega = \{z : \tfrac{1}{2} < |z| < 2\}$. For $n = 1, 2, 3, \ldots$ let X_n be the set of all $f \in H(\Omega)$ that are nth derivatives of some $g \in H(\Omega)$. [In other words, X_n is the range of the differential operator D^n with domain $H(\Omega)$.]

(a) Show that $f \in X_1$ if and only if $\int_\gamma f(z) \, dz = 0$, where γ is the positively oriented unit circle.

(b) Show that $f \in X_n$ for every n if and only if f extends to a holomorphic function in $D(0; 2)$.

14 Suppose Ω is a region, $p \in \Omega$, $R < \infty$. Let $\mathscr{F}$ be the class of all $f \in H(\Omega)$ such that $|f(p)| \leq R$ and $f(\Omega)$ contains neither 0 nor 1. Prove that $\mathscr{F}$ is a normal family.

15 Show that Theorem 16.2 leads to a very simple proof of the special case of the monodromy theorem (16.15) in which Ω and D are concentric discs. Combine this special case with the Riemann mapping theorem to prove Theorem 16.15 in the generality in which it is stated.

CHAPTER SEVENTEEN

H^p-SPACES

This chapter is devoted to the study of certain subspaces of $H(U)$ which are defined by certain growth conditions; in fact, they are all contained in the class N defined in Chap. 15. These so-called H^p-spaces (named for G. H. Hardy) have a large number of interesting properties concerning factorizations, boundary values, and Cauchy-type representations in terms of measures on the boundary of U. We shall merely give some of the highlights, such as the theorem of F. and M. Riesz on measures μ whose Fourier coefficients $\hat{\mu}(n)$ are 0 for all $n < 0$, Beurling's classification of the invariant subspaces of H^2, and M. Riesz's theorem on conjugate functions.

A convenient approach to the subject is via subharmonic functions, and we begin with a brief outline of their properties.

Subharmonic Functions

17.1 Definition A function u defined in an open set Ω in the plane is said to be *subharmonic* if it has the following four properties.

(a) $-\infty \leq u(z) < \infty$ for all $z \in \Omega$.
(b) u is upper semicontinuous in Ω.
(c) Whenever $\bar{D}(a; r) \subset \Omega$, then

$$u(a) \leq \frac{1}{2\pi} \int_{-\pi}^{\pi} u(a + re^{i\theta})\, d\theta.$$

(d) None of the integrals in (c) is $-\infty$.

Note that the integrals in (c) always exist and are not $+\infty$, since (a) and (b) imply that u is bounded above on every compact $K \subset \Omega$. [*Proof*: If K_n is the set of all $z \in K$ at which $u(z) \geq n$, then $K \supset K_1 \supset K_2 \cdots$, so either $K_n = \emptyset$ for some n, or $\bigcap K_n \neq \emptyset$, in which case $u(z) = \infty$ for some $z \in K$.] Hence (d) says that the integrands in (c) belong to $L^1(T)$.

Every real harmonic function is obviously subharmonic.

17.2 Theorem *If u is subharmonic in Ω, and if φ is a monotonically increasing convex function on R^1, then $\varphi \circ u$ is subharmonic.*

[To have $\varphi \circ u$ defined at all points of Ω, we put $\varphi(-\infty) = \lim \varphi(x)$ as $x \to -\infty$.]

PROOF First, $\varphi \circ u$ is upper semicontinuous, since φ is increasing and continuous. Next, if $\bar{D}(a; r) \subset \Omega$, we have

$$\varphi(u(a)) \leq \varphi\left(\frac{1}{2\pi} \int_{-\pi}^{\pi} u(a + re^{i\theta})\, d\theta\right) \leq \frac{1}{2\pi} \int_{-\pi}^{\pi} \varphi(u(a + re^{i\theta}))\, d\theta.$$

The first of these inequalities holds since φ is increasing and u is subharmonic; the second follows from the convexity of φ, by Theorem 3.3. ////

17.3 Theorem *If Ω is a region, $f \in H(\Omega)$, and f is not identically 0, then $\log |f|$ is subharmonic in Ω, and so are $\log^+ |f|$ and $|f|^p$ $(0 < p < \infty)$.*

PROOF It is understood that $\log |f(z)| = -\infty$ if $f(z) = 0$. Then $\log |f|$ is upper semicontinuous in Ω, and Theorem 15.19 implies that $\log |f|$ is subharmonic.

The other assertions follow if we apply Theorem 17.2 to $\log |f|$ in place of u, with

$$\varphi(t) = \max(0, t) \quad \text{and} \quad \varphi(t) = e^{pt}. \qquad ////$$

17.4 Theorem *Suppose u is a continuous subharmonic function in Ω, K is a compact subset of Ω, h is a continuous real function on K which is harmonic in the interior V of K, and $u(z) \leq h(z)$ at all boundary points of K. Then $u(z) \leq h(z)$ for all $z \in K$.*

This theorem accounts for the term "subharmonic." Continuity of u is not necessary here, but we shall not need the general case and leave it as an exercise.

PROOF Put $u_1 = u - h$, and assume, to get a contradiction, that $u_1(z) > 0$ for some $z \in V$. Since u_1 is continuous on K, u_1 attains its maximum m on K; and since $u_1 \leq 0$ on the boundary of K, the set $E = \{z \in K : u_1(z) = m\}$ is a nonempty compact subset of V. Let z_0 be a boundary point of E. Then for some $r > 0$ we have $\bar{D}(z_0; r) \subset V$, but some subarc of the boundary of $\bar{D}(z_0; r)$ lies in the complement of E. Hence

$$u_1(z_0) = m > \frac{1}{2\pi} \int_{-\pi}^{\pi} u_1(z_0 + re^{i\theta})\, d\theta,$$

and this means that u_1 is not subharmonic in V. But if u is subharmonic, so is $u - h$, by the mean value property of harmonic functions, and we have our contradiction. ////

17.5 Theorem *Suppose u is a continuous subharmonic function in U, and*

$$m(r) = \frac{1}{2\pi} \int_{-\pi}^{\pi} u(re^{i\theta})\, d\theta \qquad (0 \le r < 1). \tag{1}$$

If $r_1 < r_2$, then $m(r_1) \le m(r_2)$.

PROOF Let h be the continuous function on $\bar{D}(0; r_2)$ which coincides with u on the boundary of $\bar{D}(0; r_2)$ and which is harmonic in $D(0; r_2)$. By Theorem 17.4, $u \le h$ in $D(0; r_2)$. Hence

$$m(r_1) \le \frac{1}{2\pi} \int_{-\pi}^{\pi} h(r_1 e^{i\theta})\, d\theta = h(0) = \frac{1}{2\pi} \int_{-\pi}^{\pi} h(r_2 e^{i\theta})\, d\theta = m(r_2). \quad ////$$

The Spaces H^p and N

17.6 Notation As in Secs 11.15 and 11.19, we define f_r on T by

$$f_r(e^{i\theta}) = f(re^{i\theta}) \qquad (0 \le r < 1) \tag{1}$$

if f is any continuous function with domain U, and we let σ denote Lebesgue measure on T, so normalized that $\sigma(T) = 1$. Accordingly, L^p-norms will refer to $L^p(\sigma)$. In particular,

$$\|f_r\|_p = \left\{ \int_T |f_r|^p\, d\sigma \right\}^{1/p} \qquad (0 < p < \infty), \tag{2}$$

$$\|f_r\|_\infty = \sup_\theta |f(re^{i\theta})|, \tag{3}$$

and we also introduce

$$\|f_r\|_0 = \exp \int_T \log^+ |f_r|\, d\sigma. \tag{4}$$

17.7 Definition If $f \in H(U)$ and $0 \le p \le \infty$, we put

$$\|f\|_p = \sup \{\|f_r\|_p : 0 \le r < 1\}. \tag{1}$$

If $0 < p \leq \infty$, H^p is defined to be the class of all $f \in H(U)$ for which $\|f\|_p < \infty$. (Note that this coincides with our previously introduced terminology in the case $p = \infty$.)

The class N consists of all $f \in H(U)$ for which $\|f\|_0 < \infty$.

It is clear that $H^\infty \subset H^p \subset H^s \subset N$ if $0 < s < p < \infty$.

17.8 Remarks (a) When $p < \infty$, Theorems 17.3 and 17.5 show that $\|f_r\|_p$ is a nondecreasing function of r, for every $f \in H(U)$; when $p = \infty$, the same follows from the maximum modulus theorem. Hence

$$\|f\|_p = \lim_{r \to 1} \|f_r\|_p. \tag{1}$$

(b) For $1 \leq p \leq \infty$, $\|f\|_p$ satisfies the triangle inequality, so that H^p is a normed linear space. To see this, note that the Minkowski inequality gives

$$\|(f+g)_r\|_p = \|f_r + g_r\|_p \leq \|f_r\|_p + \|g_r\|_p \tag{2}$$

if $0 < r < 1$. As $r \to 1$, we obtain

$$\|f + g\|_p \leq \|f\|_p + \|g\|_p. \tag{3}$$

(c) Actually, H^p is a *Banach space*, if $1 \leq p \leq \infty$: To prove completeness, suppose $\{f_n\}$ is a Cauchy sequence in H^p, $|z| \leq r < R < 1$, and apply the Cauchy formula to $f_n - f_m$, integrating around the circle of radius R, center 0. This leads to the inequalities

$$(R - r) |f_n(z) - f_m(z)| \leq \|(f_n - f_m)_R\|_1 \leq \|(f_n - f_m)_R\|_p \leq \|f_n - f_m\|_p$$

from which we conclude that $\{f_n\}$ converges uniformly on compact subsets of U to a function $f \in H(U)$. Given $\epsilon > 0$, there is an m such that $\|f_n - f_m\|_p < \epsilon$ for all $n > m$, and then, for every $r < 1$,

$$\|(f - f_m)_r\|_p = \lim_{n \to \infty} \|(f_n - f_m)_r\|_p \leq \epsilon. \tag{4}$$

This gives $\|f - f_m\|_p \to 0$ as $m \to \infty$.

(d) For $p < 1$, H^p is still a vector space, but the triangle inequality is no longer satisfied by $\|f\|_p$.

We saw in Theorem 15.23 that the zeros of any $f \in N$ satisfy the Blaschke condition $\Sigma(1 - |\alpha_n|) < \infty$. Hence the same is true in every H^p. It is interesting that the zeros of any $f \in H^p$ can be divided out without increasing the norm:

17.9 Theorem *Suppose $f \in N$, $f \not\equiv 0$, and B is the Blaschke product formed with the zeros of f. Put $g = f/B$. Then $g \in N$ and $\|g\|_0 = \|f\|_0$.*

Moreover, if $f \in H^p$, then $g \in H^p$ and $\|g\|_p = \|f\|_p$ $(0 < p \leq \infty)$.

PROOF Note first that

$$|g(z)| \geq |f(z)| \qquad (z \in U). \qquad (1)$$

In fact, strict inequality holds for every $z \in U$, unless f has no zeros in U, in which case $B = 1$ and $g = f$.

If s and t are nonnegative real numbers, the inequality

$$\log^+ (st) \leq \log^+ s + \log^+ t \qquad (2)$$

holds since the left side is 0 if $st < 1$ and is $\log s + \log t$ if $st \geq 1$. Since $|g| = |f|/|B|$, (2) gives

$$\log^+ |g| \leq \log^+ |f| + \log \frac{1}{|B|}. \qquad (3)$$

By Theorem 15.24, (3) implies that $\|g\|_0 \leq \|f\|_0$, and since (1) holds, we actually have $\|g\|_0 = \|f\|_0$.

Now suppose $f \in H^p$ for some $p > 0$. Let B_n be the finite Blaschke product formed with the first n zeros of f (we arrange these zeros in some sequence, taking multiplicities into account). Put $g_n = f/B_n$. For each n, $|B_n(re^{i\theta})| \to 1$ uniformly, as $r \to 1$. Hence $\|g_n\|_p = \|f\|_p$. As $n \to \infty$, $|g_n|$ increases to $|g|$, so that

$$\|g_r\|_p = \lim_{n \to \infty} \|(g_n)_r\|_p \qquad (0 < r < 1), \qquad (4)$$

by the monotone convergence theorem. The right side of (4) is at most $\|f\|_p$, for all $r < 1$. If we let $r \to 1$, we obtain $\|g\|_p \leq \|f\|_p$. Equality follows now from (1), as before. ////

17.10 Theorem *Suppose $0 < p < \infty$, $f \in H^p$, $f \not\equiv 0$, and B is the Blaschke product formed with the zeros of f. Then there is a zero-free function $h \in H^2$ such that*

$$f = B \cdot h^{2/p}. \qquad (1)$$

In particular, every $f \in H^1$ is a product

$$f = gh \qquad (2)$$

in which both factors are in H^2.

PROOF By Theorem 17.9, $f/B \in H^p$; in fact, $\|f/B\|_p = \|f\|_p$. Since f/B has no zero in U and U is simply connected, there exists $\varphi \in H(U)$ so that $\exp(\varphi) = f/B$ (Theorem 13.11). Put $h = \exp(p\varphi/2)$. Then $h \in H(U)$ and $|h|^2 = |f/B|^p$, hence $h \in H^2$, and (1) holds.

In fact, $\|h\|_2^2 = \|f\|_p^p$.

To obtain (2), write (1) in the form $f = (Bh) \cdot h$. ////

We can now easily prove some of the most important properties of the H^p-spaces.

17.11 Theorem *If $0 < p < \infty$ and $f \in H^p$, then*

(a) *the nontangential maximal functions $N_\alpha f$ are in $L^p(T)$, for all $\alpha < 1$;*
(b) *the nontangential limits $f^*(e^{i\theta})$ exist a.e. on T, and $f^* \in L^p(T)$;*
(c) $\lim_{r \to 1} \|f^* - f_r\|_p = 0$, *and*
(d) $\|f^*\|_p = \|f\|_p$.

If $f \in H^1$ then f is the Cauchy integral as well as the Poisson integral of f^*.

PROOF We begin by proving (a) and (b) for the case $p > 1$. Since holomorphic functions are harmonic, Theorem 11.30(b) shows that every $f \in H^p$ is then the Poisson integral of a function (call it f^*) in $L^p(T)$. Hence $N_\alpha f \in L^p(T)$, by Theorem 11.25(b), and $f^*(e^{i\theta})$ is the nontangential limit of f at almost every $e^{i\theta} \in T$, by Theorem 11.23.

If $0 < p \leq 1$ and $f \in H^p$, use the factorization

$$f = Bh^{2/p} \tag{1}$$

given by Theorem 17.10, where B is a Blaschke product, $h \in H^2$, and h has no zero in U. Since $|f| \leq |h|^{2/p}$ in U, it follows that

$$(N_\alpha f)^p \leq (N_\alpha h)^2, \tag{2}$$

so that $N_\alpha f \in L^p(T)$, because $N_\alpha h \in L^2(T)$.

Similarly, the existence of B^* and h^* a.e. on T implies that the nontangential limits of f (call them f^*) exist a.e. Obviously, $|f^*| \leq N_\alpha f$ wherever f^* exists. Hence $f^* \in L^p(T)$.

This proves (a) and (b), for $0 < p < \infty$.

Since $f_r \to f^*$ a.e. and $|f_r| < N_\alpha f$, the dominated convergence theorem gives (c).

If $p \geq 1$, (d) follows from (c), by the triangle inequality. If $p < 1$, use Exercise 24, Chap. 3, to deduce (d) from (c).

Finally, if $f \in H^1$, $r < 1$, and $f_r(z) = f(rz)$, then $f_r \in H(D(0, 1/r))$, and therefore f_r can be represented in U by the Cauchy formula

$$f_r(z) = \frac{1}{2\pi} \int_{-\pi}^{\pi} \frac{f_r(e^{it})}{1 - e^{-it}z} \, dt \tag{3}$$

and by the Poisson formula

$$f_r(z) = \frac{1}{2\pi} \int_{-\pi}^{\pi} P(z, e^{it}) f_r(e^{it}) \, dt. \tag{4}$$

For each $z \in U$, $|1 - e^{-it}z|$ and $P(z, e^{it})$ are bounded functions on T. The case $p = 1$ of (c) leads therefore from (3) and (4) to

$$f(z) = \frac{1}{2\pi} \int_{-\pi}^{\pi} \frac{f^*(e^{it})}{1 - e^{-it}z} \, dt \tag{5}$$

and
$$f(z) = \frac{1}{2\pi} \int_{-\pi}^{\pi} P(z, e^{it}) f^*(e^{it}) \, dt. \tag{6}$$

////

The space H^2 has a particularly simple characterization in terms of power series coefficients:

17.12 Theorem *Suppose $f \in H(U)$ and*

$$f(z) = \sum_0^\infty a_n z^n.$$

Then $f \in H^2$ if and only if $\sum_0^\infty |a_n|^2 < \infty$.

PROOF By Parseval's theorem, applied to f_r with $r < 1$,

$$\sum_0^\infty |a_n|^2 = \lim_{r \to 1} \sum_0^\infty |a_n|^2 r^{2n} = \lim_{r \to 1} \int_T |f_r|^2 \, d\sigma = \|f\|_2^2.$$

////

The Theorem of F. and M. Riesz

17.13 Theorem *If μ is a complex Borel measure on the unit circle T and*

$$\int_T e^{-int} \, d\mu(t) = 0 \tag{1}$$

for $n = -1, -2, -3, \ldots$, then μ is absolutely continuous with respect to Lebesgue measure.

PROOF Put $f = P[d\mu]$. Then f satisfies

$$\|f_r\|_1 \le \|\mu\| \qquad (0 \le r < 1). \tag{2}$$

(See Sec. 11.17.) Since, setting $z = re^{i\theta}$,

$$P(z, e^{it}) = \sum_{-\infty}^\infty r^{|n|} e^{in\theta} e^{-int}, \tag{3}$$

as in Sec. 11.5, the assumption (1), which amounts to saying that the Fourier coefficients $\hat{\mu}(n)$ are 0 for all $n < 0$, leads to the power series

$$f(z) = \sum_0^\infty \hat{\mu}(n) z^n \qquad (z \in U). \tag{4}$$

By (4) and (2), $f \in H^1$. Hence $f = P[f^*]$, by Theorem 17.11, where $f^* \in L^1(T)$. The uniqueness of the Poisson integral representation (Theorem 11.30) shows now that $d\mu = f^* \, d\sigma$. ////

The remarkable feature of this theorem is that it derives the absolute continuity of a measure from an apparently unrelated condition, namely, the vanishing of one-half of its Fourier coefficients. In recent years the theorem has been extended to various other situations.

Factorization Theorems

We already know from Theorem 17.9 that every $f \in H^p$ (except $f = 0$) can be factored into a Blaschke product and a function $g \in H^p$ which has no zeros in U. There is also a factorization of g which is of a more subtle nature. It concerns, roughly speaking, the rapidity with which g tends to 0 along certain radii.

17.14 Definition An *inner function* is a function $M \in H^\infty$ for which $|M^*| = 1$ a.e. on T. (As usual, M^* denotes the radial limits of M.)

If φ is a positive measurable function on T such that $\log \varphi \in L^1(T)$, and if

$$Q(z) = c \, \exp \left\{ \frac{1}{2\pi} \int_{-\pi}^{\pi} \frac{e^{it} + z}{e^{it} - z} \log \varphi(e^{it}) \, dt \right\} \tag{1}$$

for $z \in U$, then Q is called an *outer function*. Here c is a constant, $|c| = 1$.

Theorem 15.24 shows that every Blaschke product is an inner function, but there are others. They can be described as follows.

17.15 Theorem *Suppose c is a constant, $|c| = 1$, B is a Blaschke product, μ is a finite positive Borel measure on T which is singular with respect to Lebesgue measure, and*

$$M(z) = cB(z) \exp \left\{ -\int_{-\pi}^{\pi} \frac{e^{it} + z}{e^{it} - z} \, d\mu(t) \right\} \qquad (z \in U). \tag{1}$$

Then M is an inner function, and every inner function is of this form.

PROOF If (1) holds and $g = M/B$, then $\log |g|$ is the Poisson integral of $-d\mu$, hence $\log |g| \leq 0$, so that $g \in H^\infty$, and the same is true of M. Also $D\mu = 0$ a.e., since μ is singular (Theorem 7.13), and therefore the radial limits of log $|g|$ are 0 a.e. (Theorem 11.22). Since $|B^*| = 1$ a.e., we see that M is an inner function.

Conversely, let B be the Blaschke product formed with the zeros of a given inner function M and put $g = M/B$. Then $\log|g|$ is harmonic in U. Theorems 15.24 and 17.9 show that $|g| \le 1$ in U and that $|g^*| = 1$ a.e. on T. Thus $\log|g| \le 0$. We conclude from Theorem 11.30 that $\log|g|$ is the Poisson integral of $-d\mu$, for some positive measure μ on T. Since $\log|g^*| = 0$ a.e. on T, we have $D\mu = 0$ a.e. on T, so μ is singular. Finally, $\log|g|$ is the real part of

$$h(z) = -\int_{-\pi}^{\pi} \frac{e^{it} + z}{e^{it} - z}\, d\mu(t),$$

and this implies that $g = c \exp(h)$ for some constant c with $|c| = 1$. Thus M is of the form (1).

This completes the proof. ////

The simplest example of an inner function which is not a Blaschke product is the following: Take $c = 1$ and $B = 1$, and let μ be the unit mass at $t = 0$. Then

$$M(z) = \exp\left\{\frac{z+1}{z-1}\right\},$$

which tends to 0 very rapidly along the radius which ends at $z = 1$.

17.16 Theorem *Suppose Q is the outer function related to φ as in Definition 17.14. Then*

(a) $\log|Q|$ is the Poisson integral of $\log \varphi$.
(b) $\lim_{r \to 1} |Q(re^{i\theta})| = \varphi(e^{i\theta})$ a.e. on T.
(c) $Q \in H^p$ if and only if $\varphi \in L^p(T)$. In this case, $\|Q\|_p = \|\varphi\|_p$.

PROOF (a) is clear by inspection and (a) implies that the radial limits of $\log|Q|$ are equal to $\log \varphi$ a.e. on T, which proves (b). If $Q \in H^p$, Fatou's lemma implies that $\|Q^*\|_p \le \|Q\|_p$, so $\|\varphi\|_p \le \|Q\|_p$, by (b). Conversely, if $\varphi \in L^p(T)$, then

$$|Q(re^{i\theta})|^p = \exp\left\{\frac{1}{2\pi}\int_{-\pi}^{\pi} P_r(\theta - t) \log \varphi^p(e^{it})\, dt\right\}$$

$$\le \frac{1}{2\pi}\int_{-\pi}^{\pi} P_r(\theta - t) \varphi^p(e^{it})\, dt,$$

by the inequality between the geometric and arithmetic means (Theorem 3.3), and if we integrate the last inequality with respect to θ we find that $\|Q\|_p \le \|\varphi\|_p$ if $p < \infty$. The case $p = \infty$ is trivial. ////

17.17 Theorem *Suppose $0 < p \leq \infty$, $f \in H^p$, and f is not identically 0. Then $\log|f^*| \in L^1(T)$, the outer function*

$$Q_f(z) = \exp\left\{\frac{1}{2\pi}\int_{-\pi}^{\pi}\frac{e^{it}+z}{e^{it}-z}\log|f^*(e^{it})|\,dt\right\} \tag{1}$$

is in H^p, and there is an inner function M_f such that

$$f = M_f Q_f. \tag{2}$$

Furthermore,

$$\log|f(0)| \leq \frac{1}{2\pi}\int_{-\pi}^{\pi}\log|f^*(e^{it})|\,dt. \tag{3}$$

Equality holds in (3) if and only if M_f is constant.

The functions M_f and Q_f are called the *inner* and *outer factors* of f, respectively; Q_f depends only on the boundary values of $|f|$.

PROOF We assume first that $f \in H^1$. If B is the Blaschke product formed with the zeros of f and if $g = f/B$, Theorem 17.9 shows that $g \in H^1$; and since $|g^*| = |f^*|$ a.e. on T, it suffices to prove the theorem with g in place of f.

So let us assume that f has no zero in U and that $f(0) = 1$. Then $\log|f|$ is harmonic in U, $\log|f(0)| = 0$, and since $\log = \log^+ - \log^-$, the mean value property of harmonic functions implies that

$$\frac{1}{2\pi}\int_{-\pi}^{\pi}\log^-|f(re^{i\theta})|\,d\theta = \frac{1}{2\pi}\int_{-\pi}^{\pi}\log^+|f(re^{i\theta})|\,d\theta \leq \|f\|_0 \leq \|f\|_1 \tag{4}$$

for $0 < r < 1$. It now follows from Fatou's lemma that both $\log^+|f^*|$ and $\log^-|f^*|$ are in $L^1(T)$, hence so is $\log|f^*|$.

This shows that the definition (1) makes sense. By Theorem 17.16, $Q_f \in H^1$. Also, $|Q_f^*| = |f^*| \neq 0$ a.e., since $\log|f^*| \in L^1(T)$. If we can prove that

$$|f(z)| \leq |Q_f(z)| \qquad (z \in U), \tag{5}$$

then f/Q_f will be an inner function, and we obtain the factorization (2).

Since $\log|Q_f|$ is the Poisson integral of $\log|f^*|$, (5) is equivalent to the inequality

$$\log|f| \leq P[\log|f^*|], \tag{6}$$

which we shall now prove. Our notation is as in Chap. 11: $P[h]$ is the Poisson integral of the function $h \in L^1(T)$.

For $|z| \leq 1$ and $0 < R < 1$, put $f_R(z) = f(Rz)$. Fix $z \in U$. Then

$$\log|f_R(z)| = P[\log^+|f_R|](z) - P[\log^-|f_R|](z). \tag{7}$$

Since $|\log^+ u - \log^+ v| \le |u - v|$ for all real numbers u and v, and since $\|f_R - f^*\|_1 \to 0$ as $R \to 1$ (Theorem 17.11), the first Poisson integral in (7) converges to $P[\log^+ |f^*|]$, as $R \to 1$. Hence Fatou's lemma gives

$$P[\log^- |f^*|] \le \liminf_{R \to 1} P[\log^- |f_R|] = P[\log^+ |f^*|] - \log|f|, \quad (8)$$

which is the same as (6).

We have now established the factorization (2). If we put $z = 0$ in (5) we obtain (3); equality holds in (3) if and only if $|f(0)| = |Q_f(0)|$, i.e., if and only if $|M_f(0)| = 1$; and since $\|M_f\|_\infty = 1$, this happens only when M_f is a constant.

This completes the proof for the case $p = 1$.

If $1 < p \le \infty$, then $H^p \subset H^1$, hence all that remains to be proved is that $Q_f \in H^p$. But if $f \in H^p$, then $|f^*| \in L^p(T)$, by Fatou's lemma; hence $Q_f \in H^p$, by Theorem 17.16(c).

Theorem 17.10 reduces the case $p < 1$ to the case $p = 2$. ////

The fact that $\log |f^*| \in L^1(T)$ has a consequence which we have already used in the proof but which is important enough to be stated separately:

17.18 Theorem *If $0 < p \le \infty$, $f \in H^p$, and f is not identically 0, then at almost all points of T we have $f^*(e^{i\theta}) \ne 0$.*

PROOF If $f^* = 0$ then $\log|f^*| = -\infty$, and if this happens on a set of positive measure, then

$$\int_{-\pi}^{\pi} \log|f^*(e^{it})|\, dt = -\infty. \quad ////$$

Observe that Theorem 17.18 imposes a quantitative restriction on the location of the zeros of the radial limits of an $f \in H^p$. Inside U the zeros are also quantitatively restricted, by the Blaschke condition.

As usual, we can rephrase the above result about zeros as a uniqueness theorem:

If $f \in H^p$, $g \in H^p$, and $f^*(e^{i\theta}) = g^*(e^{i\theta})$ on some subset of T whose Lebesgue measure is positive, then $f(z) = g(z)$ for all $z \in U$.

17.19 Let us take a quick look at the class N, with the purpose of determining how much of Theorems 17.17 and 17.18 is true here. If $f \in N$ and $f \not\equiv 0$, we can divide by a Blaschke product and get a quotient g which has no zero in U and which is in N (Theorem 17.9). Then $\log|g|$ is harmonic, and since

$$|\log|g|| = 2\log^+|g| - \log|g| \quad (1)$$

and

$$\frac{1}{2\pi}\int_{-\pi}^{\pi} \log|g(re^{i\theta})|\, d\theta = \log|g(0)|, \qquad (2)$$

we see that $\log|g|$ satisfies the hypotheses of Theorem 11.30 and is therefore the Poisson integral of a real measure μ. Thus

$$f(z) = cB(z)\exp\left\{\int_T \frac{e^{it}+z}{e^{it}-z}\, d\mu(t)\right\}, \qquad (3)$$

where c is a constant, $|c|=1$, and B is a Blaschke product.

Observe how the assumption that the integrals of $\log^+|g|$ are bounded (which is a quantitative formulation of the statement that $|g|$ does not get too close to ∞) implies the boundedness of the integrals of $\log^-|g|$ (which says that $|g|$ does not get too close to 0 at too many places).

If μ is a negative measure, the exponential factor in (3) is in H^∞. Apply the Jordan decomposition to μ. This shows:

To every $f \in N$ there correspond two functions b_1 and $b_2 \in H^\infty$ such that b_2 has no zero in U and $f = b_1/b_2$.

Since $b_2^* \neq 0$ a.e., it follows that f has finite radial limits a.e. Also, $f^* \neq 0$ a.e.

Is $\log|f^*| \in L^1(T)$? Yes, and the proof is identical to the one given in Theorem 17.17.

However, the inequality (3) of Theorem 17.17 need no longer hold. For example, if

$$f(z) = \exp\left\{\frac{1+z}{1-z}\right\}, \qquad (4)$$

then $\|f\|_0 = e$, $|f^*| = 1$ a.e., and

$$\log|f(0)| = 1 > 0 = \frac{1}{2\pi}\int_{-\pi}^{\pi}\log|f^*(e^{it})|\, dt. \qquad (5)$$

The Shift Operator

17.20 Invariant Subspaces Consider a bounded linear operator S on a Banach space X; that is to say, S is a bounded linear transformation of X into X. If a closed subspace Y of X has the property that $S(Y) \subset Y$, we call Y an *S-invariant subspace*. Thus the S-invariant subspaces of X are exactly those which are mapped into themselves by S.

The knowledge of the invariant subspaces of an operator S helps us to visualize its action. (This is a very general—and hence rather vague—principle: In studying any transformation of any kind, it helps to know what the transformation leaves fixed.) For instance, if S is a linear operator on an n-dimensional vector space X and if S has n linearly independent characteristic vectors $x_1, \ldots,$

x_n, the one-dimensional spaces spanned by any of these x_i are S-invariant, and we obtain a very simple description of S if we take $\{x_1, \ldots, x_n\}$ as a basis of X.

We shall describe the invariant subspaces of the so-called "shift operator" S on ℓ^2. Here ℓ^2 is the space of all complex sequences

$$x = \{\xi_0, \xi_1, \xi_2, \xi_3, \ldots\} \tag{1}$$

for which

$$\|x\| = \left\{\sum_{n=0}^{\infty} |\xi_n|^2\right\}^{1/2} < \infty, \tag{2}$$

and S takes the element $x \in \ell^2$ given by (1) to

$$Sx = \{0, \xi_0, \xi_1, \xi_2, \ldots\}. \tag{3}$$

It is clear that S is a bounded linear operator on ℓ^2 and that $\|S\| = 1$.

A few S-invariant subspaces are immediately apparent: If Y_k is the set of all $x \in \ell^2$ whose first k coordinates are 0, then Y_k is S-invariant.

To find others we make use of a Hilbert space isomorphism between ℓ^2 and H^2 which converts the shift operator S to a multiplication operator on H^2. The point is that this multiplication operator is easier to analyze (because of the richer structure of H^2 as a space of holomorphic functions) than is the case in the original setting of the sequence space ℓ^2.

We associate with each $x \in \ell^2$, given by (1), the function

$$f(z) = \sum_{n=0}^{\infty} \xi_n z^n \qquad (z \in U). \tag{4}$$

By Theorem 17.12, this defines a linear one-to-one mapping of ℓ^2 onto H^2. If

$$y = \{\eta_n\}, \qquad g(z) = \sum_{n=0}^{\infty} \eta_n z^n \tag{5}$$

and if the inner product in H^2 is defined by

$$(f, g) = \frac{1}{2\pi} \int_{-\pi}^{\pi} f^*(e^{i\theta}) \overline{g^*(e^{i\theta})} \, d\theta, \tag{6}$$

the Parseval theorem shows that $(f, g) = (x, y)$. Thus we have a Hilbert space isomorphism of ℓ^2 onto H^2, and the shift operator S has turned into a multiplication operator (which we still denote by S) on H^2:

$$(Sf)(z) = zf(z) \qquad (f \in H^2, z \in U). \tag{7}$$

The previously mentioned invariant subspaces Y_k are now seen to consist of all $f \in H^2$ which have a zero of order at least k at the origin. This gives a clue: For any finite set $\{\alpha_1, \ldots, \alpha_k\} \subset U$, the space Y of all $f \in H^2$ such that $f(\alpha_1) = \cdots = f(\alpha_k) = 0$ is S-invariant. If B is the finite Blaschke product with zeros at $\alpha_1, \ldots, \alpha_k$, then $f \in Y$ if and only if $f/B \in H^2$. Thus $Y = BH^2$.

348 REAL AND COMPLEX ANALYSIS

This suggests that *infinite* Blaschke products may also give rise to S-invariant subspaces and, more generally, that Blaschke products might be replaced by arbitrary inner functions φ. It is not hard to see that each φH^2 is a closed S-invariant subspace of H^2, but that *every* closed S-invariant subspace of H^2 is of this form is a deeper result.

17.21 Beurling's Theorem

(a) *For each inner function φ the space*

$$\varphi H^2 = \{\varphi f : f \in H^2\} \tag{1}$$

is a closed S-invariant subspace of H^2.
(b) *If φ_1 and φ_2 are inner functions and if $\varphi_1 H^2 = \varphi_2 H^2$, then φ_1/φ_2 is constant.*
(c) *Every closed S-invariant subspace Y of H^2, other than $\{0\}$, contains an inner function φ such that $Y = \varphi H^2$.*

PROOF H^2 is a Hilbert space, relative to the norm

$$\|f\|_2 = \left\{ \frac{1}{2\pi} \int_{-\pi}^{\pi} |f^*(e^{i\theta})|^2 \, d\theta \right\}^{1/2}. \tag{2}$$

If φ is an inner function, then $|\varphi^*| = 1$ a.e. The mapping $f \to \varphi f$ is therefore an isometry of H^2 into H^2; being an isometry, its range φH^2 is a closed subspace of H^2. [*Proof*: If $\varphi f_n \to g$ in H^2, then $\{\varphi f_n\}$ is a Cauchy sequence, hence so is $\{f_n\}$, hence $f_n \to f \in H^2$, so $g = \varphi f \in \varphi H^2$.] The S-invariance of φH^2 is also trivial, since $z \cdot \varphi f = \varphi \cdot zf$. Hence (a) holds.

If $\varphi_1 H^2 = \varphi_2 H^2$, then $\varphi_1 = \varphi_2 f$ for some $f \in H^2$, hence $\varphi_1/\varphi_2 \in H^2$. Similarly, $\varphi_2/\varphi_1 \in H^2$. Put $\varphi = \varphi_1/\varphi_2$ and $h = \varphi + (1/\varphi)$. Then $h \in H^2$, and since $|\varphi^*| = 1$ a.e. on T, h^* is real a.e. on T. Since h is the Poisson integral of h^*, it follows that h is real in U, hence h is constant. Then φ must be constant, and (b) is proved.

The proof of (c) will use a method originated by Helson and Lowdenslager. Suppose Y is a closed S-invariant subspace of H^2 which does not consist of 0 alone. Then there is a smallest integer k such that Y contains a function f of the form

$$f(z) = \sum_{n=k}^{\infty} c_n z^n, \qquad c_k = 1. \tag{3}$$

Then $f \notin zY$, where we write zY for the set of all g of the form $g(z) = zf(z), f \in Y$. It follows that zY is a *proper* closed subspace of Y [closed by the argument used in the proof of (a)], so Y contains a nonzero vector which is orthogonal to zY (Theorem 4.11).

So there exists a $\varphi \in Y$ such that $\|\varphi\|_2 = 1$ and $\varphi \perp zY$. Then $\varphi \perp z^n\varphi$, for $n = 1, 2, 3, \ldots$. By the definition of the inner product in H^2 [see 17.20(6)] this means that

$$\frac{1}{2\pi}\int_{-\pi}^{\pi} |\varphi^*(e^{i\theta})|^2 e^{-in\theta}\, d\theta = 0 \qquad (n = 1, 2, 3, \ldots). \tag{4}$$

These equations are preserved if we replace the left sides by their complex conjugates, i.e., if we replace n by $-n$. Thus all Fourier coefficients of the function $|\varphi^*|^2 \in L^1(T)$ are 0, except the one corresponding to $n = 0$, which is 1. Since L^1-functions are determined by their Fourier coefficients (Theorem 5.15), it follows that $|\varphi^*| = 1$ a.e. on T. But $\varphi \in H^2$, so φ is the Poisson integral of φ^*, and hence $|\varphi| \leq 1$. We conclude that φ is an inner function.

Since $\varphi \in Y$ and Y is S-invariant, we have $\varphi z^n \in Y$ for all $n \geq 0$, hence $\varphi P \in Y$ for every polynomial P. The polynomials are dense in H^2 (the partial sums of the power series of any $f \in H^2$ converge to f in the H^2-norm, by Parseval's theorem), and since Y is closed and $|\varphi| \leq 1$, it follows that $\varphi H^2 \subset Y$. We have to prove that this inclusion is *not* proper. Since φH^2 is closed, it is enough to show that the assumptions $h \in Y$ and $h \perp \varphi H^2$ imply $h = 0$.

If $h \perp \varphi H^2$, then $h \perp \varphi z^n$ for $n = 0, 1, 2, \ldots$, or

$$\frac{1}{2\pi}\int_{-\pi}^{\pi} h^*(e^{i\theta})\overline{\varphi^*(e^{i\theta})} e^{-in\theta}\, d\theta = 0 \qquad (n = 0, 1, 2, \ldots). \tag{5}$$

If $h \in Y$, then $z^n h \in zY$ if $n = 1, 2, 3, \ldots$, and our choice of φ shows that $z^n h \perp \varphi$, or

$$\frac{1}{2\pi}\int_{-\pi}^{\pi} h^*(e^{i\theta})\overline{\varphi^*(e^{i\theta})} e^{-in\theta}\, d\theta = 0 \qquad (n = -1, -2, -3, \ldots). \tag{6}$$

Thus all Fourier coefficients of $h^*\overline{\varphi^*}$ are 0, hence $h^*\overline{\varphi^*} = 0$ a.e. on T; and since $|\varphi^*| = 1$ a.e., we have $h^* = 0$ a.e. Therefore $h = 0$, and the proof is complete. ////

17.22 Remark If we combine Theorems 17.15 and 17.21, we see that the S-invariant subspaces of H^2 are characterized by the following data: a sequence of complex numbers $\{\alpha_n\}$ (possibly finite, or even empty) such that $|\alpha_n| < 1$ and $\Sigma(1 - |\alpha_n|) < \infty$, and a positive Borel measure μ on T, singular with respect to Lebesgue measure (so $D\mu = 0$ a.e.). It is easy (we leave this as an exercise) to find conditions, in terms of $\{\alpha_n\}$ and μ, which ensure that one S-invariant subspace of H^2 contains another. The partially ordered set of all S-invariant subspaces is thus seen to have an extremely complicated structure, much more complicated than one might have expected from the simple definition of the shift operator on ℓ^2.

We conclude the section with an easy consequence of Theorem 17.21 which depends on the factorization described in Theorem 17.17.

17.23 Theorem *Suppose M_f is the inner factor of a function $f \in H^2$, and Y is the smallest closed S-invariant subspace of H^2 which contains f. Then*

$$Y = M_f H^2. \tag{1}$$

In particular, $Y = H^2$ if and only if f is an outer function.

PROOF Let $f = M_f Q_f$ be the factorization of f into its inner and outer factors. It is clear that $f \in M_f H^2$; and since $M_f H^2$ is closed and S-invariant, we have $Y \subset M_f H^2$.

On the other hand, Theorem 17.21 shows that there is an inner function φ such that $Y = \varphi H^2$. Since $f \in Y$, there exists an $h = M_h Q_h \in H^2$ such that

$$M_f Q_f = \varphi M_h Q_h. \tag{2}$$

Since inner functions have absolute value 1 a.e. on T, (2) implies that $Q_f = Q_h$, hence $M_f = \varphi M_h \in Y$, and therefore Y must contain the smallest S-invariant closed subspace which contains M_f. Thus $M_f H^2 \subset Y$, and the proof is complete. ////

It may be of interest to summarize these results in terms of two questions to which they furnish answers.

If $f \in H^2$, which functions $g \in H^2$ can be approximated in the H^2-norm by functions of the form fP, where P runs through the polynomials? *Answer*: Precisely those g for which $g/M_f \in H^2$.

For which $f \in H^2$ is it true that the set $\{fP\}$ is dense in H^2? *Answer*: Precisely for those f for which

$$\log |f(0)| = \frac{1}{2\pi} \int_{-\pi}^{\pi} \log |f^*(e^{it})|\, dt.$$

Conjugate Functions

17.24 Formulation of the Problem Every real harmonic function u in the unit disc U is the real part of one and only one $f \in H(U)$ such that $f(0) = u(0)$. If $f = u + iv$, the last requirement can also be stated in the form $v(0) = 0$. The function v is called the *harmonic conjugate* of u, or the *conjugate function* of u.

Suppose now that u satisfies

$$\sup_{r<1} \|u_r\|_p < \infty \tag{1}$$

for some p. Does it follow that (1) holds then with v in place of u?

Equivalently, does it follow that $f \in H^p$?

The answer (given by M. Riesz) is affirmative if $1 < p < \infty$. (For $p = 1$ and $p = \infty$ it is negative; see Exercise 24.) The precise statement is given by Theorem 17.26.

Let us recall that every harmonic u that satisfies (1) is the Poisson integral of a function $u^* \in L^p(T)$ (Theorem 11.30) if $1 < p < \infty$. Theorem 11.11 suggests therefore another restatement of the problem:

If $1 < p < \infty$, and if we associate to each $h \in L^p(T)$ the holomorphic function

$$(\psi h)(z) = \frac{1}{2\pi} \int_{-\pi}^{\pi} \frac{e^{it} + z}{e^{it} - z} h(e^{it})\, dt \qquad (z \in U), \tag{2}$$

do all of these functions ψh lie in H^p?

Exercise 25 deals with some other aspects of this problem.

17.25 Lemma *If* $1 < p \leq 2$, $\delta = \pi/(1+p)$, $\alpha = (\cos \delta)^{-1}$, *and* $\beta = \alpha^p(1 + \alpha)$, *then*

$$1 \leq \beta(\cos \varphi)^p - \alpha \cos p\varphi \qquad \left(-\frac{\pi}{2} \leq \varphi \leq \frac{\pi}{2}\right). \tag{1}$$

PROOF If $\delta \leq |\varphi| \leq \pi/2$, then the right side of (1) is not less than

$$-\alpha \cos p\varphi \geq -\alpha \cos p\delta = \alpha \cos \delta = 1,$$

and it exceeds $\beta(\cos \delta)^p - \alpha = 1$ if $|\varphi| \leq \delta$. ////

17.26 Theorem *If* $1 < p < \infty$, *then there is a constant* $A_p < \infty$ *such that the inequality*

$$\|\psi h\|_p \leq A_p \|h\|_p \tag{1}$$

holds for every $h \in L^p(T)$.

More explicitly, the conclusion is that ψh (defined in Sec. 17.24) is in H^p, and that

$$\int_T |(\psi h)_r|^p\, d\sigma \leq A_p^p \int_T |h|^p\, d\sigma \qquad (0 \leq r < 1) \tag{2}$$

where $d\sigma = d\theta/2\pi$ is the normalized Lebesgue measure on T.

Note that h is not required to be a real function in this theorem, which asserts that $\psi \colon L^p \to H^p$ is a bounded linear operator.

PROOF Assume first that $1 < p \leq 2$, that $h \in L^p(T)$, $h \geq 0$, $h \not\equiv 0$, and let u be the real part of $f = \psi h$. Formula 11.5(2) shows that $u = P[h]$, hence $u > 0$ in U. Since U is simply connected and f has no zero in U, there is a $g \in H(U)$ such that $g = f^p$, $g(0) > 0$. Also, $u = |f| \cos \varphi$, where φ is a real function with domain U that satisfies $|\varphi| < \pi/2$.

If $\alpha = \alpha_p$ and $\beta = \beta_p$ are chosen as in Lemma 17.25, it follows that

$$\int_T |f_r|^p\, d\sigma \leq \beta \int_T (u_r)^p\, d\sigma - \alpha \int_T |f_r|^p \cos(p\varphi_r)\, d\sigma \tag{3}$$

for $0 \leq r < 1$.

Note that $|f|^p \cos p\varphi = \operatorname{Re} g$. The mean value property of harmonic functions shows therefore that the last integral in (3) is equal to $\operatorname{Re} g(0) > 0$. Hence

$$\int_T |f_r|^p \, d\sigma \le \beta \int_T h^p \, d\sigma \qquad (0 \le r < 1) \tag{4}$$

because $u = P[h]$ implies $\|u_r\|_p \le \|h\|_p$. Thus

$$\|\psi h\|_p \le \beta^{1/p} \|h\|_p \tag{5}$$

if $h \in L^p(T)$, $h \ge 0$.

If h is an arbitrary (complex) function in $L^p(T)$, the preceding result applies to the positive and negative parts of the real and imaginary parts of h. This proves (2), for $1 < p \le 2$, with $A_p = 4\beta^{1/p}$.

To complete the proof, consider the case $2 < p < \infty$. Let $w \in L^q(T)$, where q is the exponent conjugate to p. Put $\tilde{w}(e^{i\theta}) = w(e^{-i\theta})$. A simple computation, using Fubini's theorem, shows for any $h \in L^p(T)$ that

$$\int_T (\psi h)_r \tilde{w} \, d\sigma = \int_T (\psi w)_r \tilde{h} \, d\sigma \qquad (0 \le r < 1). \tag{6}$$

Since $q < 2$, (2) holds with w and q in place of h and p, so that (6) leads to

$$\left| \int_T (\psi h)_r \tilde{w} \, d\sigma \right| \le A_q \|w\|_q \|h\|_p. \tag{7}$$

Now let w range over the unit ball of $L^q(T)$ and take the supremum on the left side of (7). The result is

$$\left\{ \int_T |(\psi h)_r|^p \, d\sigma \right\}^{1/p} \le A_q \left\{ \int_T |h|^p \, d\sigma \right\}^{1/p} \qquad (0 \le r < 1). \tag{8}$$

Hence (2) holds again, with $A_p \le A_q$. ////

(If we take the smallest admissible values for A_p and A_q, the last calculation can be reversed, and shows that $A_p = A_q$.)

Exercises

1 Prove Theorems 17.4 and 17.5 for upper semicontinuous subharmonic functions.

2 Assume $f \in H(\Omega)$ and prove that $\log(1 + |f|)$ is subharmonic in Ω.

3 Suppose $0 < p \le \infty$ and $f \in H(U)$. Prove that $f \in H^p$ if and only if there is a harmonic function u in U such that $|f(z)|^p \le u(z)$ for all $z \in U$. Prove that if there is one such *harmonic majorant* u of $|f|^p$, then there is a least one, say u_f. (Explicitly, $|f|^p \le u_f$ and u_f is harmonic; and if $|f|^p \le u$ and u is harmonic, then $u_f \le u$.) Prove that $\|f\|_p = u_f(0)^{1/p}$. *Hint*: Consider the harmonic functions in $D(0; R)$, $R < 1$, with boundary values $|f|^p$, and let $R \to 1$.

4 Prove likewise that $f \in N$ if and only if $\log^+ |f|$ has a harmonic majorant in U.

5 Suppose $f \in H^p$, $\varphi \in H(U)$, and $\varphi(U) \subset U$. Does it follow that $f \circ \varphi \in H^p$? Answer the same question with N in place of H^p.

6 If $0 < r < s \leq \infty$, show that H^s is a *proper* subclass of H^r.

7 Show that H^∞ is a *proper* subclass of the intersection of all H^p with $p < \infty$.

8 If $f \in H^1$ and $f^* \in L^p(T)$, prove that $f \in H^p$.

9 Suppose $f \in H(U)$ and $f(U)$ is not dense in the plane. Prove that f has finite radial limits at almost all points of T.

10 Fix $\alpha \in U$. Prove that the mapping $f \to f(\alpha)$ is a bounded linear functional on H^2. Since H^2 is a Hilbert space, this functional can be represented as an inner product with some $g \in H^2$. Find this g.

11 Fix $\alpha \in U$. How large can $|f'(\alpha)|$ be if $\|f\|_2 \leq 1$? Find the extremal functions. Do the same for $f^{(n)}(\alpha)$.

12 Suppose $p \geq 1$, $f \in H^p$, and f^* is real a.e. on T. Prove that f is then constant. Show that this result is false for every $p < 1$.

13 Suppose $f \in H(U)$, and suppose there exists an $M < \infty$ such that f maps every circle of radius $r < 1$ and center 0 onto a curve γ_r whose length is at most M. Prove that f has a continuous extension to $\bar{U}$ and that the restriction of f to T is absolutely continuous.

14 Suppose μ is a complex Borel measure on T such that

$$\int_T e^{int}\, d\mu(t) = 0 \quad (n = 1, 2, 3, \ldots).$$

Prove that then either $\mu = 0$ or the support of μ is all of T.

15 Suppose K is a *proper* compact subset of the unit circle T. Prove that every continuous function on K can be uniformly approximated on K by polynomials. *Hint:* Use Exercise 14.

16 Complete the proof of Theorem 17.17 for the case $0 < p < 1$.

17 Let φ be a nonconstant inner function with no zero in U.
  (a) Prove that $1/\varphi \notin H^p$ if $p > 0$.
  (b) Prove that there is at least one $e^{i\theta} \in T$ such that $\lim_{r \to 1} \varphi(re^{i\theta}) = 0$.
  Hint: $\log |\varphi|$ is a negative harmonic function.

18 Suppose φ is a nonconstant inner function, $|\alpha| < 1$, and $\alpha \notin \varphi(U)$. Prove that $\lim_{r \to 1} \varphi(re^{i\theta}) = \alpha$ for at least one $e^{i\theta} \in T$.

19 Suppose $f \in H^1$ and $1/f \in H^1$. Prove that f is then an outer function.

20 Suppose $f \in H^1$ and Re $[f(z)] > 0$ for all $z \in U$. Prove that f is an outer function.

21 Prove that $f \in N$ if and only if $f = g/h$, where g and $h \in H^\infty$ and h has no zero in U.

22 Prove the following converse of Theorem 15.24:
  If $f \in H(U)$ and if

$$\lim_{r \to 1} \int_{-\pi}^{\pi} |\log|f(re^{i\theta})||\, d\theta = 0, \tag{*}$$

then f is a Blaschke product. *Hint:* (*) implies

$$\lim_{r \to 1} \int_{-\pi}^{\pi} \log^+ |f(re^{i\theta})|\, d\theta = 0.$$

Since $\log^+ |f| \geq 0$, it follows from Theorems 17.3 and 17.5 that $\log^+ |f| = 0$, so $|f| \leq 1$. Now $f = Bg$, g has no zeros, $|g| \leq 1$, and (*) holds with $1/g$ in place of f. By the first argument, $|1/g| \leq 1$. Hence $|g| = 1$.

23 Find the conditions mentioned in Sec. 17.22.

24 The conformal mapping of U onto a vertical strip shows that M. Riesz's theorem on conjugate functions cannot be extended to $p = \infty$. Deduce that it cannot be extended to $p = 1$ either.

25 Suppose $1 < p < \infty$, and associate with each $f \in L^p(T)$ its Fourier coefficients

$$\hat{f}(n) = \frac{1}{2\pi} \int_{-\pi}^{\pi} f(e^{it}) e^{-int} \, dt \qquad (n = 0, \pm 1, \pm 2, \ldots).$$

Deduce the following statements from Theorem 17.26:

(a) To each $f \in L^p(T)$ there corresponds a function $g \in L^p(T)$ such that $\hat{g}(n) = \hat{f}(n)$ for $n \geq 0$ but $\hat{g}(n) = 0$ for all $n < 0$. In fact, there is a constant C, depending only on p, such that

$$\|g\|_p \leq C \|f\|_p.$$

The mapping $f \to g$ is thus a bounded linear projection of $L^p(T)$ into $L^p(T)$. The Fourier series of g is obtained from that of f by deleting the terms with $n < 0$.

(b) Show that the same is true if we delete the terms with $n < k$, where k is any given integer.

(c) Deduce from (b) that the partial sums s_n of the Fourier series of any $f \in L^p(T)$ form a bounded sequence in $L^p(T)$. Conclude further that we actually have

$$\lim_{n \to \infty} \|f - s_n\|_p = 0.$$

(d) If $f \in L^p(T)$ and if

$$F(z) = \sum_{n=0}^{\infty} \hat{f}(n) z^n,$$

then $F \in H^p$, and every $F \in H^p$ is so obtained. Thus the projection mentioned in (a) may be regarded as a mapping of $L^p(T)$ onto H^p.

26 Show that there is a much simpler proof of Theorem 17.26 if $p = 2$, and find the best value of A_2.

27 Suppose $f(z) = \sum_0^{\infty} a_n z^n$ in U and $\sum |a_n| < \infty$. Prove that

$$\int_0^1 |f'(re^{i\theta})| \, dr < \infty$$

for all θ.

28 Prove that the following statements are correct if $\{n_k\}$ is a sequence of positive integers which tends to ∞ sufficiently rapidly. If

$$f(z) = \sum_{k=1}^{\infty} \frac{z^{n_k}}{k}$$

then $|f'(z)| > n_k/(10k)$ for all z such that

$$1 - \frac{1}{n_k} < |z| < 1 - \frac{1}{2n_k}.$$

Hence

$$\int_0^1 |f'(re^{i\theta})| \, dr = \infty$$

for every θ, although

$$\lim_{R \to 1} \int_0^R f'(re^{i\theta}) \, dr$$

exists (and is finite) for almost all θ. Interpret this geometrically, in terms of the lengths of the images under f of the radii in U.

29 Use Theorem 17.11 to obtain the following characterization of the boundary values of H^p-functions, for $1 \leq p \leq \infty$:

A function $g \in L^p(T)$ is f^* (a.e.) for some $f \in H^p$ if and only if

$$\frac{1}{2\pi} \int_{-\pi}^{\pi} g(e^{it}) e^{-int} \, dt = 0$$

for all negative integers n.

CHAPTER
EIGHTEEN

ELEMENTARY THEORY OF BANACH ALGEBRAS

Introduction

18.1 Definitions A *complex algebra* is a vector space A over the complex field in which an associative and distributive multiplication is defined, i.e.,

$$x(yz) = (xy)z, \quad (x + y)z = xz + yz, \quad x(y + z) = xy + xz \tag{1}$$

for x, y, and $z \in A$, and which is related to scalar multiplication so that

$$\alpha(xy) = x(\alpha y) = (\alpha x)y \tag{2}$$

for x and $y \in A$, α a scalar.

If there is a norm defined in A which makes A into a normed linear space and which satisfies the multiplicative inequality

$$\|xy\| \le \|x\|\|y\| \quad (x \text{ and } y \in A), \tag{3}$$

then A is a *normed complex algebra*. If, in addition, A is a complete metric space relative to this norm, i.e., if A is a Banach space, then we call A a *Banach algebra*.

The inequality (3) makes multiplication a continuous operation. This means that if $x_n \to x$ and $y_n \to y$, then $x_n y_n \to xy$, which follows from (3) and the identity

$$x_n y_n - xy = (x_n - x)y_n + x(y_n - y). \tag{4}$$

Note that we have not required that A be commutative, i.e., that $xy = yx$ for all x and $y \in A$, and we shall not do so except when explicitly stated.

However, we *shall* assume that A has a *unit*. This is an element e such that

$$xe = ex = x \quad (x \in A). \tag{5}$$

It is easily seen that there is at most one such e ($e' = e'e = e$) and that $\|e\| \geq 1$, by (3). We shall make the additional assumption that

$$\|e\| = 1. \tag{6}$$

An element $x \in A$ will be called *invertible* if x has an *inverse* in A, i.e., if there exists an element $x^{-1} \in A$ such that

$$x^{-1}x = xx^{-1} = e. \tag{7}$$

Again, it is easily seen that no $x \in A$ has more than one inverse.

If x and y are invertible in A, so are x^{-1} and xy, since $(xy)^{-1} = y^{-1}x^{-1}$. The invertible elements therefore form a group with respect to multiplication.

The *spectrum* of an element $x \in A$ is the set of all complex numbers λ such that $x - \lambda e$ is *not* invertible. We shall denote the spectrum of x by $\sigma(x)$.

18.2 The theory of Banach algebras contains a great deal of interplay between algebraic properties on the one hand and topological ones on the other. We already saw an example of this in Theorem 9.21, and shall see others. There are also close relations between Banach algebras and holomorphic functions: The easiest proof of the fundamental fact that $\sigma(x)$ is never empty depends on Liouville's theorem concerning entire functions, and the spectral radius formula follows naturally from theorems about power series. This is one reason for restricting our attention to *complex* Banach algebras. The theory of real Banach algebras (we omit the definition, which should be obvious) is not so satisfactory.

The Invertible Elements

In this section, A will be a complex Banach algebra with unit e, and G will be the set of all invertible elements of A.

18.3 Theorem *If $x \in A$ and $\|x\| < 1$, then $e + x \in G$,*

$$(e + x)^{-1} = \sum_{n=0}^{\infty} (-1)^n x^n, \tag{1}$$

and

$$\|(e + x)^{-1} - e + x\| \leq \frac{\|x\|^2}{1 - \|x\|}. \tag{2}$$

PROOF The multiplicative inequality 18.1(3) shows that $\|x^n\| \leq \|x\|^n$. If

$$s_N = e - x + x^2 - \cdots + (-1)^N x^N, \tag{3}$$

it follows that $\{s_N\}$ is a Cauchy sequence in A, hence the series in (1) converges (with respect to the norm of A) to an element $y \in A$. Since multiplication is continuous and

$$(e + x)s_N = e + (-1)^N x^{N+1} = s_N(e + x), \tag{4}$$

we see that $(e + x)y = e = y(e + x)$. This gives (1), and (2) follows from

$$\left\| \sum_{n=2}^{\infty} (-1)^n x^n \right\| \le \sum_{n=2}^{\infty} \|x^n\| \le \sum_{n=2}^{\infty} \|x\|^n = \frac{\|x\|^2}{1 - \|x\|}. \tag{5}$$

////

18.4 Theorem *Suppose* $x \in G$, $\|x^{-1}\| = 1/\alpha$, $h \in A$, *and* $\|h\| = \beta < \alpha$. *Then* $x + h \in G$, *and*

$$\|(x + h)^{-1} - x^{-1} + x^{-1}hx^{-1}\| \le \frac{\beta^2}{\alpha^2(\alpha - \beta)}. \tag{1}$$

PROOF $\|x^{-1}h\| \le \beta/\alpha < 1$, hence $e + x^{-1}h \in G$, by Theorem 18.3; and since $x + h = x(e + x^{-1}h)$, we have $x + h \in G$ and

$$(x + h)^{-1} = (e + x^{-1}h)^{-1} x^{-1}. \tag{2}$$

Thus

$$(x + h)^{-1} - x^{-1} + x^{-1}hx^{-1} = [(e + x^{-1}h)^{-1} - e + x^{-1}h]x^{-1}, \tag{3}$$

and the inequality (1) follows from Theorem 18.3, with $x^{-1}h$ in place of x. ////

Corollary 1 *G is an open set, and the mapping $x \to x^{-1}$ is a homeomorphism of G onto G.*

For if $x \in G$ and $\|h\| \to 0$, (1) implies that $\|(x + h)^{-1} - x^{-1}\| \to 0$. Thus $x \to x^{-1}$ is continuous; it clearly maps G onto G, and since it is its own inverse, it is a homeomorphism.

Corollary 2 *The mapping $x \to x^{-1}$ is differentiable. Its differential at any $x \in G$ is the linear operator which takes $h \in A$ to $-x^{-1}hx^{-1}$.*

This can also be read off from (1). Note that the notion of the differential of a transformation makes sense in any normed linear space, not just in R^k, as in Definition 7.22. If A is commutative, the above differential takes h to $-x^{-2}h$, which agrees with the fact that the derivative of the holomorphic function z^{-1} is $-z^{-2}$.

Corollary 3 *For every $x \in A$, $\sigma(x)$ is compact, and $|\lambda| \le \|x\|$ if $\lambda \in \sigma(x)$.*

ELEMENTARY THEORY OF BANACH ALGEBRAS 359

For if $|\lambda| > \|x\|$, then $e - \lambda^{-1}x \in G$, by Theorem 18.3, and the same is true of $x - \lambda e = -\lambda(e - \lambda^{-1}x)$; hence $\lambda \notin \sigma(x)$. To prove that $\sigma(x)$ is closed, observe (a) $\lambda \in \sigma(x)$ if and only if $x - \lambda e \notin G$; (b) the complement of G is a closed subset of A, by Corollary 1; and (c) the mapping $\lambda \to x - \lambda e$ is a continuous mapping of the complex plane into A.

18.5 Theorem *Let Φ be a bounded linear functional on A, fix $x \in A$, and define*

$$f(\lambda) = \Phi[(x - \lambda e)^{-1}] \qquad (\lambda \notin \sigma(x)). \tag{1}$$

Then f is holomorphic in the complement of $\sigma(x)$, and $f(\lambda) \to 0$ as $\lambda \to \infty$.

PROOF Fix $\lambda \notin \sigma(x)$ and apply Theorem 18.4 with $x - \lambda e$ in place of x and with $(\lambda - \mu)e$ in place of h. We see that there is a constant C, depending on x and λ, such that

$$\|(x - \mu e)^{-1} - (x - \lambda e)^{-1} + (\lambda - \mu)(x - \lambda e)^{-2}\| \le C|\mu - \lambda|^2 \tag{2}$$

for all μ which are close enough to λ. Thus

$$\frac{(x - \mu e)^{-1} - (x - \lambda e)^{-1}}{\mu - \lambda} \to (x - \lambda e)^{-2} \tag{3}$$

as $\mu \to \lambda$, and if we apply Φ to both sides of (3), the continuity and linearity of Φ show that

$$\frac{f(\mu) - f(\lambda)}{\mu - \lambda} \to \Phi[(x - \lambda e)^{-2}]. \tag{4}$$

So f is differentiable and hence holomorphic outside $\sigma(x)$. Finally, as $\lambda \to \infty$ we have

$$\lambda f(\lambda) = \Phi[\lambda(x - \lambda e)^{-1}] = \Phi\left[\left(\frac{x}{\lambda} - e\right)^{-1}\right] \to \Phi(-e), \tag{5}$$

by the continuity of the inversion mapping in G. ////

18.6 Theorem *For every $x \in A$, $\sigma(x)$ is compact and not empty.*

PROOF We already know that $\sigma(x)$ is compact. Fix $x \in A$, and fix $\lambda_0 \notin \sigma(x)$. Then $(x - \lambda_0 e)^{-1} \ne 0$, and the Hahn-Banach theorem implies the existence of a bounded linear functional Φ on A such that $f(\lambda_0) \ne 0$, where f is defined as in Theorem 18.5. If $\sigma(x)$ were empty, Theorem 18.5 would imply that f is an entire function which tends to 0 at ∞, hence $f(\lambda) = 0$ for every λ, by Liouville's theorem, and this contradicts $f(\lambda_0) \ne 0$. So $\sigma(x)$ is not empty. ////

18.7 Theorem (Gelfand-Mazur) *If A is a complex Banach algebra with unit in which each nonzero element is invertible, then A is (isometrically isomorphic to) the complex field.*

An algebra in which each nonzero element is invertible is called a *division algebra*. Note that the commutativity of A is not part of the hypothesis; it is part of the conclusion.

PROOF If $x \in A$ and $\lambda_1 \neq \lambda_2$, at least one of the elements $x - \lambda_1 e$ and $x - \lambda_2 e$ must be invertible, since they cannot both be 0. It now follows from Theorem 18.6 that $\sigma(x)$ consists of exactly one point, say $\lambda(x)$, for each $x \in A$. Since $x - \lambda(x)e$ is not invertible, it must be 0, hence $x = \lambda(x)e$. The mapping $x \to \lambda(x)$ is therefore an isomorphism of A onto the complex field, which is also an isometry, since $|\lambda(x)| = \|\lambda(x)e\| = \|x\|$ for all $x \in A$. ////

18.8 Definition For any $x \in A$, the *spectral radius* $\rho(x)$ of x is the radius of the smallest closed disc with center at the origin which contains $\sigma(x)$ (sometimes this is also called the *spectral norm* of x; see Exercise 14):

$$\rho(x) = \sup \{|\lambda| : \lambda \in \sigma(x)\}.$$

18.9 Theorem (Spectral Radius Formula) *For every* $x \in A$,

$$\lim_{n \to \infty} \|x^n\|^{1/n} = \rho(x). \tag{1}$$

(*This existence of the limit is part of the conclusion.*)

PROOF Fix $x \in A$, let n be a positive integer, λ a complex number, and assume $\lambda^n \notin \sigma(x^n)$. We have

$$(x^n - \lambda^n e) = (x - \lambda e)(x^{n-1} + \lambda x^{n-2} + \cdots + \lambda^{n-1} e). \tag{2}$$

Multiply both sides of (2) by $(x^n - \lambda^n e)^{-1}$. This shows that $x - \lambda e$ is invertible, hence $\lambda \notin \sigma(x)$.

So if $\lambda \in \sigma(x)$, then $\lambda^n \in \sigma(x^n)$ for $n = 1, 2, 3, \ldots$. Corollary 3 to Theorem 18.4 shows that $|\lambda^n| \leq \|x^n\|$, and therefore $|\lambda| \leq \|x^n\|^{1/n}$. This gives

$$\rho(x) \leq \liminf_{n \to \infty} \|x^n\|^{1/n}. \tag{3}$$

Now if $|\lambda| > \|x\|$, it is easy to verify that

$$(\lambda e - x) \sum_{n=0}^{\infty} \lambda^{-n-1} x^n = e. \tag{4}$$

The above series is therefore $-(x - \lambda e)^{-1}$. Let Φ be a bounded linear functional on A and define f as in Theorem 18.5. By (4), the expansion

$$f(\lambda) = -\sum_{n=0}^{\infty} \Phi(x^n)\lambda^{-n-1} \tag{5}$$

is valid for all λ such that $|\lambda| > \|x\|$. By Theorem 18.5, f is holomorphic outside $\sigma(x)$, hence in the set $\{\lambda: |\lambda| > \rho(x)\}$. It follows that the power series (5) converges if $|\lambda| > \rho(x)$. In particular,

$$\sup_n |\Phi(\lambda^{-n} x^n)| < \infty \qquad (|\lambda| > \rho(x)) \tag{6}$$

for every bounded linear functional Φ on A.

It is a consequence of the Hahn-Banach theorem (Sec. 5.21) that the norm of any element of A is the same as its norm as a linear functional on the dual space of A. Since (6) holds for every Φ, we can now apply the Banach-Steinhaus theorem and conclude that to each λ with $|\lambda| > \rho(x)$ there corresponds a real number $C(\lambda)$ such that

$$\|\lambda^{-n} x^n\| \le C(\lambda) \qquad (n = 1, 2, 3, \ldots). \tag{7}$$

Multiply (7) by $|\lambda|^n$ and take nth roots. This gives

$$\|x^n\|^{1/n} \le |\lambda| [C(\lambda)]^{1/n} \qquad (n = 1, 2, 3, \ldots) \tag{8}$$

if $|\lambda| > \rho(x)$, and hence

$$\limsup_{n \to \infty} \|x^n\|^{1/n} \le \rho(x). \tag{9}$$

The theorem follows from (3) and (9). ////

18.10 Remarks

(a) Whether an element of A is or is not invertible in A is a purely algebraic property. Thus the spectrum of x, and likewise the spectral radius $\rho(x)$, are defined in terms of the algebraic structure of A, regardless of any metric (or topological) considerations. The limit in the statement of Theorem 18.9, on the other hand, depends on metric properties of A. This is one of the remarkable features of the theorem: It asserts the equality of two quantities which arise in entirely different ways.

(b) Our algebra may be a subalgebra of a larger Banach algebra B (an example follows), and then it may very well happen that some $x \in A$ is not invertible in A but is invertible in B. The spectrum of x therefore depends on the algebra; using the obvious notation, we have $\sigma_A(x) \supset \sigma_B(x)$, and the inclusion may be proper. The *spectral radius* of x, however, is unaffected by this, since Theorem 18.9 shows that it can be expressed in terms of metric properties of powers of x, and these are independent of anything that happens outside A.

18.11 Example Let $C(T)$ be the algebra of all continuous complex functions on the unit circle T (with pointwise addition and multiplication and the supremum norm), and let A be the set of all $f \in C(T)$ which can be extended to a continuous function F on the closure of the unit disc U, such that F is holomorphic in U. It is easily seen that A is a subalgebra of $C(T)$. If $f_n \in A$

and $\{f_n\}$ converges uniformly on T, the maximum modulus theorem forces the associated sequence $\{F_n\}$ to converge uniformly on the closure of U. This shows that A is a *closed* subalgebra of $C(T)$, and so A is itself a Banach algebra.

Define the function f_0 by $f_0(e^{i\theta}) = e^{i\theta}$. Then $F_0(z) = z$. The spectrum of f_0 as an element of A consists of the closed unit disc; with respect to $C(T)$, the spectrum of f_0 consists only of the unit circle. In accordance with Theorem 18.9, the two spectral radii coincide.

Ideals and Homomorphisms

From now on we shall deal only with *commutative* algebras.

18.12 Definition A subset I of a commutative complex algebra A is said to be an *ideal* if (a) I is a subspace of A (in the vector space sense) and (b) $xy \in I$ whenever $x \in A$ and $y \in I$. If $I \neq A$, I is a *proper* ideal. *Maximal ideals* are proper ideals which are not contained in any larger proper ideals. Note that no proper ideal contains an invertible element.

If B is another complex algebra, a mapping φ of A into B is called a *homomorphism* if φ is a linear mapping which also preserves multiplication: $\varphi(x)\varphi(y) = \varphi(xy)$ for all x and $y \in A$. The *kernel* (or null space) of φ is the set of all $x \in A$ such that $\varphi(x) = 0$. It is trivial to verify that the kernel of a homomorphism is an ideal. For the converse, see Sec. 18.14.

18.13 Theorem *If A is a commutative complex algebra with unit, every proper ideal of A is contained in a maximal ideal. If, in addition, A is a Banach algebra, every maximal ideal of A is closed.*

PROOF The first part is an almost immediate consequence of the Hausdorff maximality principle (and holds in any commutative ring with unit). Let I be a proper ideal of A. Partially order the collection $\mathscr{P}$ of all proper ideals of A which contain I (by set inclusion), and let M be the union of the ideals in some maximal linearly ordered subcollection $\mathscr{Q}$ of $\mathscr{P}$. Then M is an ideal (being the union of a *linearly* ordered collection of ideals), $I \subset M$, and $M \neq A$, since no member of $\mathscr{P}$ contains the unit of A. The maximality of $\mathscr{Q}$ implies that M is a maximal ideal of A.

If A is a Banach algebra, the closure $\overline{M}$ of M is also an ideal (we leave the details of the proof of this statement to the reader). Since M contains no invertible element of A and since the set of all invertible elements is open, we have $\overline{M} \neq A$, and the maximality of M therefore shows that $\overline{M} = M$. ////

18.14 Quotient Spaces and Quotient Algebras Suppose J is a subspace of a vector space A, and associate with each $x \in A$ the coset

$$\varphi(x) = x + J = \{x + y : y \in J\}. \tag{1}$$

If $x_1 - x_2 \in J$, then $\varphi(x_1) = \varphi(x_2)$. If $x_1 - x_2 \notin J$, $\varphi(x_1) \cap \varphi(x_2) = \varnothing$. The set of all cosets of J is denoted by A/J; it is a vector space if we define

$$\varphi(x) + \varphi(y) = \varphi(x + y), \qquad \lambda\varphi(x) = \varphi(\lambda x) \tag{2}$$

for x and $y \in A$ and scalars λ. Since J is a vector space, the operations (2) are well defined; this means that if $\varphi(x) = \varphi(x')$ and $\varphi(y) = \varphi(y')$, then

$$\varphi(x) + \varphi(y) = \varphi(x') + \varphi(y'), \qquad \lambda\varphi(x) = \lambda\varphi(x'). \tag{3}$$

Also, φ is clearly a linear mapping of A onto A/J; the zero element of A/J is $\varphi(0) = J$.

Suppose next that A is not merely a vector space but a commutative algebra and that J is a proper ideal of A. If $x' - x \in J$ and $y' - y \in J$, the identity

$$x'y' - xy = (x' - x)y' + x(y' - y) \tag{4}$$

shows that $x'y' - xy \in J$. Therefore multiplication can be defined in A/J in a consistent manner:

$$\varphi(x)\varphi(y) = \varphi(xy) \qquad (x \text{ and } y \in A). \tag{5}$$

It is then easily verified that A/J is an algebra, and φ is a homomorphism of A onto A/J whose kernel is J.

If A has a unit element e, then $\varphi(e)$ is the unit of A/J, and *A/J is a field if and only if J is a maximal ideal*.

To see this, suppose $x \in A$ and $x \notin J$, and put

$$I = \{ax + y : a \in A, y \in J\}. \tag{6}$$

Then I is an ideal in A which contains J properly, since $x \in I$. If J is maximal, $I = A$, hence $ax + y = e$ for some $a \in A$ and $y \in J$, hence $\varphi(a)\varphi(x) = \varphi(e)$; and this says that every nonzero element of A/J is invertible, so that A/J is a field. If J is not maximal, we can choose x as above so that $I \neq A$, hence $e \notin I$, and then $\varphi(x)$ is not invertible in A/J.

18.15 Quotient Norms Suppose A is a normed linear space, J is a *closed* subspace of A, and $\varphi(x) = x + J$, as above. Define

$$\|\varphi(x)\| = \inf\{\|x + y\| : y \in J\}. \tag{1}$$

Note that $\|\varphi(x)\|$ is the greatest lower bound of the norms of those elements which lie in the coset $\varphi(x)$; this is the same as the distance from x to J. We call the norm defined in A/J by (1) the *quotient norm* of A/J. It has the following properties:

(a) A/J is a normed linear space.
(b) If A is a Banach space, so is A/J.
(c) If A is a commutative Banach algebra and J is a proper closed ideal, then A/J is a commutative Banach algebra.

These are easily verified:

If $x \in J$, $\|\varphi(x)\| = 0$. If $x \notin J$, the fact that J is closed implies that $\|\varphi(x)\| > 0$. It is clear that $\|\lambda\varphi(x)\| = |\lambda|\|\varphi(x)\|$. If x_1 and $x_2 \in A$ and $\epsilon > 0$, there exist y_1 and $y_2 \in J$ so that

$$\|x_i + y_i\| < \|\varphi(x_i)\| + \epsilon \qquad (i = 1, 2). \tag{2}$$

Hence

$$\|\varphi(x_1 + x_2)\| \le \|x_1 + x_2 + y_1 + y_2\| < \|\varphi(x_1)\| + \|\varphi(x_2)\| + 2\epsilon, \tag{3}$$

which gives the triangle inequality and proves (a).

Suppose A is complete and $\{\varphi(x_n)\}$ is a Cauchy sequence in A/J. There is a subsequence for which

$$\|\varphi(x_{n_i}) - \varphi(x_{n_{i+1}})\| < 2^{-i} \qquad (i = 1, 2, 3, \ldots), \tag{4}$$

and there exist elements z_i so that $z_i - x_{n_i} \in J$ and $\|z_i - z_{i+1}\| < 2^{-i}$. Thus $\{z_i\}$ is a Cauchy sequence in A; and since A is complete, there exists $z \in A$ such that $\|z_i - z\| \to 0$. It follows that $\varphi(x_{n_i})$ converges to $\varphi(z)$ in A/J. But if a Cauchy sequence has a convergent subsequence, then the full sequence converges. Thus A/J is complete, and we have proved (b).

To prove (c), choose x_1 and $x_2 \in A$ and $\epsilon > 0$, and choose y_1 and $y_2 \in J$ so that (2) holds. Note that $(x_1 + y_1)(x_2 + y_2) \in x_1 x_2 + J$, so that

$$\|\varphi(x_1 x_2)\| \le \|(x_1 + y_1)(x_2 + y_2)\| \le \|x_1 + y_1\|\|x_2 + y_2\|. \tag{5}$$

Now (2) implies

$$\|\varphi(x_1 x_2)\| \le \|\varphi(x_1)\|\|\varphi(x_2)\|. \tag{6}$$

Finally, if e is the unit element of A, take $x_1 \notin J$ and $x_2 = e$ in (6); this gives $\|\varphi(e)\| \ge 1$. But $e \in \varphi(e)$, and the definition of the quotient norm shows that $\|\varphi(e)\| \le \|e\| = 1$. So $\|\varphi(e)\| = 1$, and the proof is complete.

18.16 Having dealt with these preliminaries, we are now in a position to derive some of the key facts concerning commutative Banach algebras.

Suppose, as before, that A is a commutative complex Banach algebra with unit element e. We associate with A the set Δ of all complex homomorphisms of A; these are the homomorphisms of A onto the complex field, or, in different terminology, the *multiplicative linear functionals* on A which are not identically 0. As before, $\sigma(x)$ denotes the spectrum of the element $x \in A$, and $\rho(x)$ is the spectral radius of x.

Then the following relations hold:

18.17 Theorem

(a) *Every maximal ideal M of A is the kernel of some $h \in \Delta$.*
(b) *$\lambda \in \sigma(x)$ if and only if $h(x) = \lambda$ for some $h \in \Delta$.*
(c) *x is invertible in A if and only if $h(x) \ne 0$ for every $h \in \Delta$.*
(d) *$h(x) \in \sigma(x)$ for every $x \in A$ and $h \in \Delta$.*
(e) *$|h(x)| \le \rho(x) \le \|x\|$ for every $x \in A$ and $h \in \Delta$.*

PROOF If M is a maximal ideal of A, then A/M is a field; and since M is closed (Theorem 18.13), A/M is a Banach algebra. By Theorem 18.7 there is an isomorphism j of A/M onto the complex field. If $h = j \circ \varphi$, where φ is the homomorphism of A onto A/M whose kernel is M, then $h \in \Delta$ and the kernel of h is M. This proves (a).

If $\lambda \in \sigma(x)$, then $x - \lambda e$ is not invertible; hence the set of all elements $(x - \lambda e)y$, where $y \in A$, is a proper ideal of A, which lies in a maximal ideal (by Theorem 18.13), and (a) shows that there exists an $h \in \Delta$ such that $h(x - \lambda e) = 0$. Since $h(e) = 1$, this gives $h(x) = \lambda$.

On the other hand, if $\lambda \notin \sigma(x)$, then $(x - \lambda e)y = e$ for some $y \in A$. It follows that $h(x - \lambda e)h(y) = 1$ for every $h \in \Delta$, so that $h(x - \lambda e) \neq 0$, or $h(x) \neq \lambda$. This proves (b).

Since x is invertible if and only if $0 \notin \sigma(x)$, (c) follows from (b).

Finally, (d) and (e) are immediate consequences of (b). ////

Note that (e) implies that the norm of h, as a linear functional, is at most 1. In particular, each $h \in \Delta$ is continuous. This was already proved earlier (Theorem 9.21).

Applications

We now give some examples of theorems whose statements involve no algebraic concepts but which can be proved by Banach algebra techniques.

18.18 Theorem *Let $A(U)$ be the set of all continuous functions on the closure $\bar{U}$ of the open unit disc U whose restrictions to U are holomorphic. Suppose $f_1, \ldots, f_n$ are members of $A(U)$, such that*

$$|f_1(z)| + \cdots + |f_n(z)| > 0 \tag{1}$$

for every $z \in \bar{U}$. Then there exist $g_1, \ldots, g_n \in A(U)$ such that

$$\sum_{i=1}^{n} f_i(z)g_i(z) = 1 \qquad (z \in \bar{U}). \tag{2}$$

PROOF Since sums, products, and uniform limits of holomorphic functions are holomorphic, $A(U)$ is a Banach algebra, with the supremum norm. The set J of all functions $\Sigma f_i g_i$, where the g_i are arbitrary members of $A(U)$, is an ideal of $A(U)$. We have to prove that J contains the unit element 1 of $A(U)$. By Theorem 18.13 this happens if and only if J lies in no maximal ideal of $A(U)$. By Theorem 18.17(a) it is therefore enough to prove that there is no homomorphism h of $A(U)$ onto the complex field such that $h(f_i) = 0$ for every i ($1 \leq i \leq n$).

Before we determine these homomorphisms, let us note that the polynomials form a dense subset of $A(U)$. To see this, suppose $f \in A(U)$ and $\epsilon > 0$; since f is uniformly continuous on $\bar{U}$, there exists an $r < 1$ such that $|f(z) - f(rz)| < \epsilon$ for all $z \in \bar{U}$; the expansion of $f(rz)$ in powers of z converges if $|rz| < 1$, hence converges to $f(rz)$ uniformly for $z \in \bar{U}$, and this gives the desired approximation.

Now let h be a complex homomorphism of $A(U)$. Put $f_0(z) = z$. Then $f_0 \in A(U)$. It is obvious that $\sigma(f_0) = \bar{U}$. By Theorem 18.17(d) there exists an $\alpha \in \bar{U}$ such that $h(f_0) = \alpha$. Hence $h(f_0^n) = \alpha^n = f_0^n(\alpha)$, for $n = 1, 2, 3, \ldots$, so $h(P) = P(\alpha)$ for every polynomial P. Since h is continuous and since the polynomials are dense in $A(U)$, it follows that $h(f) = f(\alpha)$ for every $f \in A(U)$.

Our hypothesis (1) implies that $|f_i(\alpha)| > 0$ for at least one index i, $1 \leq i \leq n$. Thus $h(f_i) \neq 0$.

We have proved that to each $h \in \Delta$ there corresponds at least one of the given functions f_i such that $h(f_i) \neq 0$, and this, as we noted above, is enough to prove the theorem. ////

Note: We have also determined all maximal ideals of $A(U)$, in the course of the preceding proof, since each is the kernel of some $h \in \Delta$: *If $\alpha \in \bar{U}$ and if M_α is the set of all $f \in A(U)$ such that $f(\alpha) = 0$, then M_α is a maximal ideal of $A(U)$, and all maximal ideals of $A(U)$ are obtained in this way.*

$A(U)$ is often called the *disc algebra*.

18.19 The restrictions of the members of $A(U)$ to the unit circle T form a closed subalgebra of $C(T)$. This is the algebra A discussed in Example 18.11. In fact, A is a *maximal* subalgebra of $C(T)$. More explicitly, *if $A \subset B \subset C(T)$ and B is a closed (relative to the supremum norm) subalgebra of $C(T)$, then either $B = A$ or $B = C(T)$.*

It is easy to see (compare with Exercise 29, Chap. 17) that A consists precisely of those $f \in C(T)$ for which

$$\hat{f}(n) = \frac{1}{2\pi} \int_{-\pi}^{\pi} f(e^{i\theta}) e^{-in\theta} \, d\theta = 0 \qquad (n = -1, -2, -3, \ldots). \tag{1}$$

Hence the above-mentioned maximality theorem can be stated as an approximation theorem:

18.20 Theorem *Suppose $g \in C(T)$ and $\hat{g}(n) \neq 0$ for some $n < 0$. Then to every $f \in C(T)$ and to every $\epsilon > 0$ there correspond polynomials*

$$P_n(e^{i\theta}) = \sum_{k=0}^{m(n)} a_{n,k} e^{ik\theta} \qquad (n = 0, \ldots, N) \tag{1}$$

such that

$$\left| f(e^{i\theta}) - \sum_{n=0}^{N} P_n(e^{i\theta}) g^n(e^{i\theta}) \right| < \epsilon \qquad (e^{i\theta} \in T). \tag{2}$$

PROOF Let B be the closure in $C = C(T)$ of the set of all functions of the form

$$\sum_{n=0}^{N} P_n g^n. \tag{3}$$

The theorem asserts that $B = C$. Let us assume $B \neq C$.

The set of all functions (3) (note that N is not fixed) is a complex algebra. Its closure B is a Banach algebra which contains the function f_0, where $f_0(e^{i\theta}) = e^{i\theta}$. Our assumption that $B \neq C$ implies that $1/f_0 \notin B$, for otherwise B would contain f_0^n for all integers n, hence all trigonometric polynomials would be in B; and since the trigonometric polynomials are dense in C (Theorem 4.25) we should have $B = C$.

So f_0 is not invertible in B. By Theorem 18.17 there is a complex homomorphism h of B such that $h(f_0) = 0$. Every homomorphism onto the complex field satisfies $h(1) = 1$; and since $h(f_0) = 0$, we also have

$$h(f_0^n) = [h(f_0)]^n = 0 \qquad (n = 1, 2, 3, \ldots). \tag{4}$$

We know that h is a linear functional on B, of norm at most 1. The Hahn-Banach theorem extends h to a linear functional on C (still denoted by h) of the same norm. Since $h(1) = 1$ and $\|h\| \leq 1$, the argument used in Sec. 5.22 shows that h is a *positive* linear functional on C. In particular, $h(f)$ is real for real f; hence $h(\bar{f}) = \overline{h(f)}$. Since f_0^{-n} is the complex conjugate of f_0^n, it follows that (4) also holds for $n = -1, -2, -3, \ldots$. Thus

$$h(f_0^n) = \begin{cases} 1 & \text{if } n = 0, \\ 0 & \text{if } n \neq 0. \end{cases} \tag{5}$$

Since the trigonometric polynomials are dense in C, there is only one bounded linear functional on C which satisfies (5). Hence h is given by the formula

$$h(f) = \frac{1}{2\pi} \int_{-\pi}^{\pi} f(e^{i\theta}) \, d\theta \qquad (f \in C). \tag{6}$$

Now if n is a positive integer, $gf_0^n \in B$; and since h is multiplicative on B, (6) gives

$$\hat{g}(-n) = \frac{1}{2\pi} \int_{-\pi}^{\pi} g(e^{i\theta}) e^{in\theta} \, d\theta = h(gf_0^n) = h(g)h(f_0^n) = 0, \tag{7}$$

by (5). This contradicts the hypothesis of the theorem. ////

We conclude with a theorem due to Wiener.

18.21 Theorem *Suppose*

$$f(e^{i\theta}) = \sum_{-\infty}^{\infty} c_n e^{in\theta}, \qquad \sum_{-\infty}^{\infty} |c_n| < \infty, \tag{1}$$

and $f(e^{i\theta}) \neq 0$ for every real θ. Then

$$\frac{1}{f(e^{i\theta})} = \sum_{-\infty}^{\infty} \gamma_n e^{in\theta} \quad \text{with} \quad \sum_{-\infty}^{\infty} |\gamma_n| < \infty. \tag{2}$$

PROOF We let A be the space of all complex functions f on the unit circle which satisfy (1), with the norm

$$\|f\| = \sum_{-\infty}^{\infty} |c_n|. \tag{3}$$

It is clear that A is a Banach space. In fact, A is isometrically isomorphic to ℓ^1, the space of all complex functions on the integers which are integrable with respect to the counting measure. But A is also a commutative Banach algebra, under pointwise multiplication. For if $g \in A$ and $g(e^{i\theta}) = \Sigma b_n e^{in\theta}$, then

$$f(e^{i\theta})g(e^{i\theta}) = \sum_n \left(\sum_k c_{n-k} b_k \right) e^{in\theta} \tag{4}$$

and hence

$$\|fg\| = \sum_n \left| \sum_k c_{n-k} b_k \right| \leq \sum_k |b_k| \sum_n |c_{n-k}| = \|f\| \cdot \|g\|. \tag{5}$$

Also, the function 1 is the unit of A, and $\|1\| = 1$.

Put $f_0(e^{i\theta}) = e^{i\theta}$, as before. Then $f_0 \in A$, $1/f_0 \in A$, and $\|f_0^n\| = 1$ for $n = 0, \pm 1, \pm 2, \ldots$. If h is any complex homomorphism of A and $h(f_0) = \lambda$, the fact that $\|h\| \leq 1$ implies that

$$|\lambda^n| = |h(f_0^n)| \leq \|f_0^n\| = 1 \quad (n = 0, \pm 1, \pm 2, \ldots). \tag{6}$$

Hence $|\lambda| = 1$. In other words, to each h corresponds a point $e^{i\alpha} \in T$ such that $h(f_0) = e^{i\alpha}$, so

$$h(f_0^n) = e^{in\alpha} = f_0^n(e^{i\alpha}) \quad (n = 0, \pm 1, \pm 2, \ldots). \tag{7}$$

If f is given by (1), then $f = \Sigma c_n f_0^n$. This series converges in A; and since h is a continuous linear functional on A, we conclude from (7) that

$$h(f) = f(e^{i\alpha}) \quad (f \in A). \tag{8}$$

Our hypothesis that f vanishes at no point of T says therefore that f is not in the kernel of any complex homomorphism of A, and now Theorem 18.17 implies that f is invertible in A. But this is precisely what the theorem asserts.
////

Exercises

1 Suppose $B(X)$ is the algebra of all bounded linear operators on the Banach space X, with

$$(A_1 + A_2)(x) = A_1 x + A_2 x, \quad (A_1 A_2)(x) = A_1(A_2 x), \quad \|A\| = \sup \frac{\|Ax\|}{\|x\|},$$

if A, A_1, and $A_2 \in B(X)$. Prove that $B(X)$ is a Banach algebra.

2 Let n be a positive integer, let X be the space of all complex n-tuples (normed in any way, as long as the axioms for a normed linear space are satisfied), and let $B(X)$ be as in Exercise 1. Prove that the spectrum of each member of $B(X)$ consists of at most n complex numbers. What are they?

3 Take $X = L^2(-\infty, \infty)$, suppose $\varphi \in L^\infty(-\infty, \infty)$, and let M be the multiplication operator which takes $f \in L^2$ to φf. Show that M is a bounded linear operator on L^2 and that the spectrum of M is equal to the essential range of φ (Chap. 3, Exercise 19).

4 What is the spectrum of the shift operator on ℓ^2? (See Sec. 17.20 for the definition.)

5 Prove that the closure of an ideal in a Banach algebra is an ideal.

6 If X is a compact Hausdorff space, find all maximal ideals in $C(X)$.

7 Suppose A is a commutative Banach algebra with unit, which is generated by a single element x. This means that the polynomials in x are dense in A. Prove that the complement of $\sigma(x)$ is a connected subset of the plane. *Hint*: If $\lambda \notin \sigma(x)$, there are polynomials P_n such that $P_n(x) \to (x - \lambda e)^{-1}$ in A. Prove that $P_n(z) \to (z - \lambda)^{-1}$ uniformly for $z \in \sigma(x)$.

8 Suppose $\sum_0^\infty |c_n| < \infty$, $f(z) = \sum_0^\infty c_n z^n$, $|f(z)| > 0$ for every $z \in \bar{U}$, and $1/f(z) = \sum_0^\infty a_n z^n$. Prove that $\sum_0^\infty |a_n| < \infty$.

9 Prove that a closed linear subspace of the Banach algebra $L^1(R^1)$ (see Sec. 9.19) is translation invariant if and only if it is an ideal.

10 Show that $L^1(T)$ is a commutative Banach algebra (without unit) if multiplication is defined by

$$(f * g)(t) = \frac{1}{2\pi} \int_{-\pi}^{\pi} f(t - s) g(s) \, ds.$$

Find all complex homomorphisms of $L^1(T)$, as in Theorem 9.23. If E is a set of integers and if I_E is the set of all $f \in L^1(T)$ such that $\hat{f}(n) = 0$ for all $n \in E$, prove that I_E is a closed ideal in $L^1(T)$, and prove that every closed ideal in $L^1(T)$ is obtained in this manner.

11 The *resolvent* $R(\lambda, x)$ of an element x in a Banach algebra with unit is defined as

$$R(\lambda, x) = (\lambda e - x)^{-1}$$

for all complex λ for which this inverse exists. Prove the identity

$$R(\lambda, x) - R(\mu, x) = (\mu - \lambda) R(\lambda) R(\mu)$$

and use it to give an alternative proof of Theorem 18.5.

12 Let A be a commutative Banach algebra with unit. The *radical* of A is defined to be the intersection of all maximal ideals of A. Prove that the following three statements about an element $x \in A$ are equivalent:

(a) x is in the radical of A.

(b) $\lim\limits_{n \to \infty} \|x^n\|^{1/n} = 0$.

(c) $h(x) = 0$ for every complex homomorphism of A.

13 Find an element x in a Banach algebra A (for instance, a bounded linear operator on a Hilbert space) such that $x^n \neq 0$ for all $n > 0$, but $\lim_{n \to \infty} \|x^n\|^{1/n} = 0$.

14 Suppose A is a commutative Banach algebra with unit, and let Δ be the set of all complex homomorphisms of A, as in Sec. 18.16. Associate with each $x \in A$ a function $\hat{x}$ on Δ by the formula

$$\hat{x}(h) = h(x) \qquad (h \in \Delta).$$

$\hat{x}$ is called the Gelfand transform of x.

Prove that the mapping $x \to \hat{x}$ is a homomorphism of A onto an algebra $\hat{A}$ of complex functions on Δ, with pointwise multiplication. Under what condition on A is this homomorphism an isomorphism? (See Exercise 12.)

Prove that the spectral radius $\rho(x)$ is equal to

$$\|\hat{x}\|_\infty = \sup\{|\hat{x}(h)| : h \in \Delta\}.$$

Prove that the range of the function $\hat{x}$ is exactly the spectrum $\sigma(x)$.

15 If A is a commutative Banach algebra *without* unit, let A_1 be the algebra of all ordered pairs (x, λ), with $x \in A$ and λ a complex number; addition and multiplication are defined in the "obvious" way, and $\|(x, \lambda)\| = \|x\| + |\lambda|$. Prove that A_1 is a commutative Banach algebra with unit and that the mapping $x \to (x, 0)$ is an isometric isomorphism of A onto a maximal ideal of A_1. This is a standard embedding of an algebra without unit in one with unit.

16 Show that H^∞ is a commutative Banach algebra with unit, relative to the supremum norm and pointwise addition and multiplication. The mapping $f \to f(\alpha)$ is a complex homomorphism of H^∞, whenever $|\alpha| < 1$. Prove that there must be others.

17 Show that the set of all functions $(z - 1)^2 f$, where $f \in H^\infty$, is an ideal in H^∞ which is not closed. *Hint*:

$$|(1 - z)^2(1 + \epsilon - z)^{-1} - (1 - z)| < \epsilon \qquad \text{if } |z| < 1, \epsilon > 0.$$

18 Suppose φ is an inner function. Prove that $\{\varphi f : f \in H^\infty\}$ is a closed ideal in H^∞. In other words, prove that if $\{f_n\}$ is a sequence in H^∞ such that $\varphi f_n \to g$ uniformly in U, then $g/\varphi \in H^\infty$.

CHAPTER
NINETEEN

HOLOMORPHIC FOURIER TRANSFORMS

Introduction

19.1 In Chap. 9 the Fourier transform of a function f on R^1 was defined to be a function $\hat{f}$ on R^1. Frequently $\hat{f}$ can be extended to a function which is holomorphic in a certain region of the plane. For instance, if $f(t) = e^{-|t|}$, then $\hat{f}(x) = (1 + x^2)^{-1}$, a rational function. This should not be too surprising. For each real t, the kernel e^{itz} is an entire function of z, so one should expect that there are conditions on f under which $\hat{f}$ will be holomorphic in certain regions.

We shall describe two classes of holomorphic functions which arise in this manner.

For the first one, let F be any function in $L^2(-\infty, \infty)$ which vanishes on $(-\infty, 0)$ [i.e., take $F \in L^2(0, \infty)$] and define

$$f(z) = \int_0^\infty F(t)e^{itz}\, dt \qquad (z \in \Pi^+), \tag{1}$$

where Π^+ is the set of all $z = x + iy$ with $y > 0$. If $z \in \Pi^+$, then $|e^{itz}| = e^{-ty}$, which shows that the integral in (1) exists as a Lebesgue integral.

If $\operatorname{Im} z > \delta > 0$, $\operatorname{Im} z_n > \delta$, and $z_n \to z$, the dominated convergence theorem shows that

$$\lim_{n \to \infty} \int_0^\infty |\exp(itz_n) - \exp(itz)|^2\, dt = 0$$

because the integrand is bounded by the L^1-function $4\exp(-2\delta t)$ and tends to 0 for every $t > 0$. The Schwarz inequality implies therefore that f is continuous in Π^+. The theorems of Fubini and Cauchy show that $\int_\gamma f(z)\, dz = 0$ for every closed path γ in Π^+. By Morera's theorem, $f \in H(\Pi^+)$.

Let us rewrite (1) in the form

$$f(x+iy) = \int_0^\infty F(t)e^{-ty}e^{itx}\,dt, \qquad (2)$$

regard y as fixed, and apply Plancherel's theorem. We obtain

$$\frac{1}{2\pi}\int_{-\infty}^\infty |f(x+iy)|^2\,dx = \int_0^\infty |F(t)|^2 e^{-2ty}\,dt \le \int_0^\infty |F(t)|^2\,dt \qquad (3)$$

for every $y > 0$. [Note that our notation now differs from that in Chap. 9. There the underlying measure was Lebesgue measure divided by $\sqrt{2\pi}$. Here we just use Lebesgue measure. This accounts for the factor $1/(2\pi)$ in (3).] This shows:

(a) *If f is of the form (1), then f is holomorphic in Π^+ and its restrictions to horizontal lines in Π^+ form a bounded set in $L^2(-\infty, \infty)$.*

Our second class consists of all f of the form

$$f(z) = \int_{-A}^{A} F(t)e^{itz}\,dt \qquad (4)$$

where $0 < A < \infty$ and $F \in L^2(-A, A)$. These functions f are entire (the proof is the same as above), and they satisfy a growth condition:

$$|f(z)| \le \int_{-A}^{A} |F(t)|e^{-ty}\,dt \le e^{A|y|}\int_{-A}^{A} |F(t)|\,dt. \qquad (5)$$

If C is this last integral, then $C < \infty$, and (5) implies that

$$|f(z)| \le Ce^{A|z|}. \qquad (6)$$

[Entire functions which satisfy (6) are said to be of *exponential type*.] Thus:

(b) *Every f of the form (4) is an entire function which satisfies (6) and whose restriction to the real axis lies in L^2 (by the Plancherel theorem).*

It is a remarkable fact that the converses of (a) and (b) are true. This is the content of Theorems 19.2 and 19.3.

Two Theorems of Paley and Wiener

19.2 Theorem *Suppose $f \in H(\Pi^+)$ and*

$$\sup_{0<y<\infty} \frac{1}{2\pi}\int_{-\infty}^\infty |f(x+iy)|^2\,dx = C < \infty. \qquad (1)$$

Then there exists an $F \in L^2(0, \infty)$ such that

$$f(z) = \int_0^\infty F(t)e^{itz}\,dt \qquad (z \in \Pi^+) \qquad (2)$$

and

$$\int_0^\infty |F(t)|^2 \, dt = C. \tag{3}$$

Note: The function F we are looking for is to have the property that $f(x + iy)$ is the Fourier transform of $F(t)e^{-yt}$ (we regard y as a positive constant). Let us apply the inversion formula (whether or not this is correct does not matter; we are trying to motivate the proof that follows): The desired F should be of the form

$$F(t) = e^{ty} \cdot \frac{1}{2\pi} \int_{-\infty}^\infty f(x + iy)e^{-itx} \, dx = \frac{1}{2\pi} \int f(z)e^{-itz} \, dz. \tag{4}$$

The last integral is over a horizontal line in Π^+, and if this argument is correct at all, the integral will not depend on the particular line we happen to choose. This suggests that the Cauchy theorem should be invoked.

PROOF Fix y, $0 < y < \infty$. For each $\alpha > 0$ let Γ_α be the rectangular path with vertices at $\pm\alpha + i$ and $\pm\alpha + iy$. By Cauchy's theorem

$$\int_{\Gamma_\alpha} f(z)e^{-itz} \, dz = 0. \tag{5}$$

We consider only real values of t. Let $\Phi(\beta)$ be the integral of $f(z)e^{-itz}$ over the straight line interval from $\beta + i$ to $\beta + iy$ (β real). Put $I = [y, 1]$ if $y < 1$, $I = [1, y]$ if $1 < y$. Then

$$|\Phi(\beta)|^2 = \left| \int_I f(\beta + iu)e^{-it(\beta + iu)} \, du \right|^2 \le \int_I |f(\beta + iu)|^2 \, du \int_I e^{2tu} \, du. \tag{6}$$

Put

$$\Lambda(\beta) = \int_I |f(\beta + iu)|^2 \, du. \tag{7}$$

Then (1) shows, by Fubini's theorem, that

$$\frac{1}{2\pi} \int_{-\infty}^\infty \Lambda(\beta) \, d\beta \le Cm(I). \tag{8}$$

Hence there is a sequence $\{\alpha_j\}$ such that $\alpha_j \to \infty$ and

$$\Lambda(\alpha_j) + \Lambda(-\alpha_j) \to 0 \quad (j \to \infty). \tag{9}$$

By (6), this implies that

$$\Phi(\alpha_j) \to 0, \quad \Phi(-\alpha_j) \to 0 \quad \text{as } j \to \infty. \tag{10}$$

Note that this holds for every t and that the sequence $\{\alpha_j\}$ does not depend on t.

Let us define

$$g_j(y, t) = \frac{1}{2\pi} \int_{-\alpha_j}^{\alpha_j} f(x + iy)e^{-itx}\, dx. \tag{11}$$

Then we deduce from (5) and (10) that

$$\lim_{j \to \infty} [e^{ty}g_j(y, t) - e^t g_j(1, t)] = 0 \quad (-\infty < t < \infty). \tag{12}$$

Write $f_y(x)$ for $f(x + iy)$. Then $f_y \in L^2(-\infty, \infty)$, by hypothesis, and the Plancherel theorem asserts that

$$\lim_{j \to \infty} \int_{-\infty}^{\infty} |\hat{f}_y(t) - g_j(y, t)|^2\, dt = 0, \tag{13}$$

where $\hat{f}_y$ is the Fourier transform of f_y. A subsequence of $\{g_j(y, t)\}$ converges therefore pointwise to $\hat{f}_y(t)$, for almost all t (Theorem 3.12). If we define

$$F(t) = e^t \hat{f}_1(t), \tag{14}$$

it now follows from (12) that

$$F(t) = e^{ty}\hat{f}_y(t). \tag{15}$$

Note that (14) does not involve y and that (15) holds for every $y \in (0, \infty)$. Plancherel's theorem can be applied to (15):

$$\int_{-\infty}^{\infty} e^{-2ty}|F(t)|^2\, dt = \int_{-\infty}^{\infty} |\hat{f}_y(t)|^2\, dt = \frac{1}{2\pi} \int_{-\infty}^{\infty} |f_y(x)|^2\, dx \leq C. \tag{16}$$

If we let $y \to \infty$, (16) shows that $F(t) = 0$ a.e. in $(-\infty, 0)$.
If we let $y \to 0$, (16) shows that

$$\int_0^{\infty} |F(t)|^2\, dt \leq C. \tag{17}$$

It now follows from (15) that $\hat{f}_y \in L^1$ if $y > 0$. Hence Theorem 9.14 gives

$$f_y(x) = \int_{-\infty}^{\infty} \hat{f}_y(t)e^{itx}\, dt \tag{18}$$

or

$$f(z) = \int_0^{\infty} F(t)e^{-yt}e^{itx}\, dt = \int_0^{\infty} F(t)e^{itz}\, dt \quad (z \in \Pi^+). \tag{19}$$

This is (2), and now (3) follows from (17) and formula 19.1(3). ////

19.3 Theorem *Suppose A and C are positive constants and f is an entire function such that*

$$|f(z)| \leq Ce^{A|z|} \tag{1}$$

for all z, and

$$\int_{-\infty}^{\infty} |f(x)|^2 \, dx < \infty. \tag{2}$$

Then there exists an $F \in L^2(-A, A)$ such that

$$f(z) = \int_{-A}^{A} F(t)e^{itz} \, dt \tag{3}$$

for all z.

PROOF Put $f_\epsilon(x) = f(x)e^{-\epsilon|x|}$, for $\epsilon > 0$ and x real. We shall show that

$$\lim_{\epsilon \to 0} \int_{-\infty}^{\infty} f_\epsilon(x)e^{-itx} \, dx = 0 \qquad (t \text{ real}, |t| > A). \tag{4}$$

Since $\|f_\epsilon - f\|_2 \to 0$ as $\epsilon \to 0$, the Plancherel theorem implies that the Fourier transforms of f_ϵ converge in L^2 to the Fourier transform F of f (more precisely, of the restriction of f to the real axis). Hence (4) will imply that F vanishes outside $[-A, A]$, and then Theorem 9.14 shows that (3) holds for almost every real z. Since each side of (3) is an entire function, it follows that (3) holds for every complex z.

Thus (4) implies the theorem.

For each real α, let Γ_α be the path defined by

$$\Gamma_\alpha(s) = se^{i\alpha} \qquad (0 \leq s < \infty), \tag{5}$$

put

$$\Pi_\alpha = \{w : \text{Re}(we^{i\alpha}) > A\}, \tag{6}$$

and if $w \in \Pi_\alpha$, define

$$\Phi_\alpha(w) = \int_{\Gamma_\alpha} f(z)e^{-wz} \, dz = e^{i\alpha} \int_0^\infty f(se^{i\alpha}) \exp(-wse^{i\alpha}) \, ds. \tag{7}$$

By (1) and (5), the absolute value of the integrand is at most

$$C \exp\{-[\text{Re}(we^{i\alpha}) - A]s\},$$

and it follows (as in Sec. 19.1) that Φ_α is holomorphic in the half plane Π_α.

However, more is true if $\alpha = 0$ and if $\alpha = \pi$: We have

$$\Phi_0(w) = \int_0^\infty f(x)e^{-wx} \, dx \qquad (\text{Re } w > 0), \tag{8}$$

$$\Phi_\pi(w) = -\int_{-\infty}^{0} f(x)e^{-wx}\, dx \qquad (\text{Re } w < 0). \tag{9}$$

Φ_0 and Φ_π are holomorphic in the indicated half planes because of (2).

The significance of the functions Φ_α to (4) lies in the easily verified relation

$$\int_{-\infty}^{\infty} f_\epsilon(x) e^{-itx}\, dx = \Phi_0(\epsilon + it) - \Phi_\pi(-\epsilon + it) \qquad (t \text{ real}). \tag{10}$$

Hence we have to prove that the right side of (10) tends to 0 as $\epsilon \to 0$, if $t > A$ and if $t < -A$.

We shall do this by showing that any two of our functions Φ_α agree in the intersection of their domains of definition, i.e., that they are analytic continuations of each other. Once this is done, we can replace Φ_0 and Φ_π by $\Phi_{\pi/2}$ in (10) if $t < -A$, and by $\Phi_{-\pi/2}$ if $t > A$, and it is then obvious that the difference tends to 0 as $\epsilon \to 0$.

So suppose $0 < \beta - \alpha < \pi$. Put

$$\gamma = \frac{\alpha + \beta}{2}, \qquad \eta = \cos\frac{\beta - \alpha}{2} > 0. \tag{11}$$

If $w = |w| e^{-i\gamma}$, then

$$\text{Re } (we^{i\alpha}) = \eta |w| = \text{Re } (we^{i\beta}) \tag{12}$$

so that $w \in \Pi_\alpha \cap \Pi_\beta$ as soon as $|w| > A/\eta$. Consider the integral

$$\int_\Gamma f(z) e^{-wz}\, dz \tag{13}$$

over the circular arc Γ given by $\Gamma(t) = re^{it}$, $\alpha \leq t \leq \beta$. Since

$$\text{Re } (-wz) = -|w|r \cos(t - \gamma) \leq -|w|r\eta, \tag{14}$$

the absolute value of the integrand in (13) does not exceed

$$C \exp\{(A - |w|\eta)r\}.$$

If $|w| > A/\eta$ it follows that (13) tends to 0 as $r \to \infty$.

We now apply the Cauchy theorem. The integral of $f(z)e^{-wz}$ over the interval $[0, re^{i\beta}]$ is equal to the sum of (13) and the integral over $[0, re^{i\alpha}]$. Since (13) tends to 0 as $r \to \infty$, we conclude that $\Phi_\alpha(w) = \Phi_\beta(w)$ if $w = |w|e^{-i\gamma}$ and $|w| > A/\eta$, and then Theorem 10.18 shows that Φ_α and Φ_β coincide in the intersection of the half planes in which they were originally defined.

This completes the proof. ////

19.4 Remarks Each of the two preceding proofs depended on a typical application of Cauchy's theorem. In Theorem 19.2 we replaced integration over one horizontal line by integration over another to show that 19.2(15) was independent of y. In Theorem 19.3, replacement of one ray by another was used to construct analytic continuations; the result actually was that the functions Φ_α are restrictions of one function Φ which is holomorphic in the complement of the interval $[-Ai, Ai]$.

The class of functions described in Theorem 19.2 is the half plane analogue of the class H^2 discussed in Chap. 17. Theorem 19.3 will be used in the proof of the Denjoy-Carleman theorem (Theorem 19.11).

Quasi-analytic Classes

19.5 If Ω is a region and if $z_0 \in \Omega$, every $f \in H(\Omega)$ is uniquely determined by the numbers $f(z_0), f'(z_0), f''(z_0), \ldots$. On the other hand, there exist infinitely differentiable functions on R^1 which are not identically 0 but which vanish on some interval. Thus we have here a uniqueness property which holomorphic functions possess but which does not hold in C^∞ (the class of all infinitely differentiable complex functions on R^1).

If $f \in H(\Omega)$, the growth of the sequence $\{|f^{(n)}(z_0)|\}$ is restricted by Theorem 10.26. It is therefore reasonable to ask whether the above uniqueness property holds in suitable subclasses of C^∞ in which the growth of the derivatives is subject to some restrictions. This motivates the following definitions; the answer to our question is given by Theorem 19.11.

19.6 The Classes $C\{M_n\}$ If $M_0, M_1, M_2, \ldots$ are positive numbers, we let $C\{M_n\}$ be the class of all $f \in C^\infty$ which satisfy inequalities of the form

$$\|D^n f\|_\infty \leq \beta_f B_f^n M_n \qquad (n = 0, 1, 2, \ldots). \tag{1}$$

Here $D^0 f = f$, $D^n f$ is the nth derivative of f if $n \geq 1$, the norm is the supremum norm over R^1, and β_f and B_f are positive constants (depending on f, but not on n).

If f satisfies (1), then

$$\limsup_{n \to \infty} \left\{ \frac{\|D^n f\|_\infty}{M_n} \right\}^{1/n} \leq B_f. \tag{2}$$

This shows that B_f is a more significant quantity than β_f. However, if β_f were omitted in (1), the case $n = 0$ would imply $\|f\|_\infty \leq M_0$, an undesirable restriction. The inclusion of β_f makes $C\{M_n\}$ into a vector space.

Each $C\{M_n\}$ is invariant under affine transformations. More explicitly, suppose $f \in C\{M_n\}$ and $g(x) = f(ax + b)$. Then g satisfies (1), with $\beta_g = \beta_f$ and $B_g = aB_f$.

We shall make two standing assumptions on the sequences $\{M_n\}$ under consideration:

$$M_0 = 1. \tag{3}$$

$$M_n^2 \leq M_{n-1} M_{n+1} \qquad (n = 1, 2, 3, \ldots). \tag{4}$$

Assumption (4) can be expressed in the form: $\{\log M_n\}$ *is a convex sequence.*

These assumptions will simplify some of our work, and they involve no loss of generality. [One can prove, although we shall not do so, that every class $C\{M_n\}$ is equal to a class $C\{\overline{M}_n\}$, where $\{\overline{M}_n\}$ satisfies (3) and (4).]

The following result illustrates the utility of (3) and (4):

19.7 Theorem *Each $C\{M_n\}$ is an algebra, with respect to pointwise multiplication.*

PROOF Suppose f and $g \in C\{M_n\}$, and β_f, B_f, β_g, and B_g are the corresponding constants. The product rule for differentiation shows that

$$D^n(fg) = \sum_{j=0}^{n} \binom{n}{j}(D^j f) \cdot (D^{n-j}g). \tag{1}$$

Hence

$$|D^n(fg)| \le \beta_f \beta_g \sum_{j=0}^{n} \binom{n}{j} B_f^j B_g^{n-j} M_j M_{n-j}. \tag{2}$$

The convexity of $\{\log M_n\}$, combined with $M_0 = 1$, shows that $M_j M_{n-j} \le M_n$ for $0 \le j \le n$. Hence the binomial theorem leads from (2) to

$$\|D^n(fg)\|_\infty \le \beta_f \beta_g (B_f + B_g)^n M_n \qquad (n = 0, 1, 2, \ldots), \tag{3}$$

so that $fg \in C\{M_n\}$. ////

19.8 Definition A class $C\{M_n\}$ is said to be *quasi-analytic* if the conditions

$$f \in C\{M_n\}, \qquad (D^n f)(0) = 0 \qquad (\text{for } n = 0, 1, 2, \ldots) \tag{1}$$

imply that $f(x) = 0$ for all $x \in R^1$.

The content of the definition is of course unchanged if $(D^n f)(0)$ is replaced by $(D^n f)(x_0)$, where x_0 is any given point.

The quasi-analytic classes are thus the ones which have the uniqueness property we mentioned in Sec. 19.6. One of these classes is very intimately related to holomorphic functions:

19.9 Theorem *The class $C\{n!\}$ consists of all f to which there corresponds a $\delta > 0$ such that f can be extended to a bounded holomorphic function in the strip defined by $|\operatorname{Im}(z)| < \delta$.*

Consequently $C\{n!\}$ is a quasi-analytic class.

PROOF Suppose $f \in H(\Omega)$ and $|f(z)| < \beta$ for all $z \in \Omega$, where Ω consists of all $z = x + iy$ with $|y| < \delta$. It follows from Theorem 10.26 that

$$|(D^n f)(x)| \le \beta \delta^{-n} n! \qquad (n = 0, 1, 2, \ldots) \tag{1}$$

for all real x. The restriction of f to the real axis therefore belongs to $C\{n!\}$.

Conversely, suppose f is defined on the real axis and $f \in C\{n!\}$. In other words,

$$\|D^n f\|_\infty \leq \beta B^n n! \qquad (n = 0, 1, 2, \ldots). \tag{2}$$

We claim that the representation

$$f(x) = \sum_{n=0}^{\infty} \frac{(D^n f)(a)}{n!} (x - a)^n \tag{3}$$

is valid for all $a \in R^1$ if $a - B^{-1} < x < a + B^{-1}$. This follows from Taylor's formula

$$f(x) = \sum_{j=0}^{n-1} \frac{(D^j f)(a)}{j!} (x - a)^j + \frac{1}{(n-1)!} \int_a^x (x - t)^{n-1} (D^n f)(t)\, dt, \tag{4}$$

which one obtains by repeated integrations by part. By (2) the last term in (4) (the "remainder") is dominated by

$$n\beta B^n \left| \int_a^x (x - t)^{n-1}\, dt \right| = \beta |B(x - a)|^n. \tag{5}$$

If $|B(x - a)| < 1$, this tends to 0 as $n \to \infty$, and (3) follows.

We can now replace x in (3) by any complex number z such that $|z - a| < 1/B$. This defines a holomorphic function F_a in the disc with center at a and radius $1/B$, and $F_a(x) = f(x)$ if x is real and $|x - a| < 1/B$. The various functions F_a are therefore analytic continuations of each other; they form a holomorphic extension F of f in the strip $|y| < 1/B$.

If $0 < \delta < 1/B$ and $z = a + iy$, $|y| < \delta$, then

$$|F(z)| = |F_a(z)| = \left| \sum_{n=0}^{\infty} \frac{(D^n f)(a)}{n!} (iy)^n \right| \leq \beta \sum_{n=0}^{\infty} (B\delta)^n = \frac{\beta}{1 - B\delta}.$$

This shows that F is bounded in the strip $|y| < \delta$, and the proof is complete. ////

19.10 Theorem *The class $C\{M_n\}$ is quasi-analytic if and only if $C\{M_n\}$ contains no nontrivial function with compact support.*

PROOF If $C\{M_n\}$ is quasi-analytic, if $f \in C\{M_n\}$, and if f has compact support, then evidently f and all its derivatives vanish at some point, hence $f(x) = 0$ for all x.

Suppose $C\{M_n\}$ is not quasi-analytic. Then there exists an $f \in C\{M_n\}$ such that $(D^n f)(0) = 0$ for $n = 0, 1, 2, \ldots$, but $f(x_0) \neq 0$ for some x_0. We may assume $x_0 > 0$. If $g(x) = f(x)$ for $x \geq 0$ and $g(x) = 0$ for $x < 0$, then $g \in C\{M_n\}$. Put $h(x) = g(x)g(2x_0 - x)$. By Theorem 19.7, $h \in C\{M_n\}$. Also, $h(x) = 0$ if $x < 0$ and if $x > 2x_0$. But $h(x_0) = f^2(x_0) \neq 0$. Thus h is a nontrivial member of $C\{M_n\}$ with compact support. ////

We are now ready for the fundamental theorem about quasi-analytic clases.

The Denjoy-Carleman Theorem

19.11 Theorem *Suppose* $M_0 = 1$, $M_n^2 \leq M_{n-1} M_{n+1}$ *for* $n = 1, 2, 3, \ldots$, *and*

$$Q(x) = \sum_{n=0}^{\infty} \frac{x^n}{M_n}, \qquad q(x) = \sup_{n \geq 0} \frac{x^n}{M_n},$$

for $x > 0$. *Then each of the following five conditions implies the other four*:

(a) $C\{M_n\}$ is not quasi-analytic.

(b) $\displaystyle\int_0^{\infty} \log Q(x) \frac{dx}{1 + x^2} < \infty.$

(c) $\displaystyle\int_0^{\infty} \log q(x) \frac{dx}{1 + x^2} < \infty.$

(d) $\displaystyle\sum_{n=1}^{\infty} \left(\frac{1}{M_n}\right)^{1/n} < \infty.$

(e) $\displaystyle\sum_{n=1}^{\infty} \frac{M_{n-1}}{M_n} < \infty.$

Note: If $M_n \to \infty$ very rapidly as $n \to \infty$, then $Q(x)$ tends to infinity slowly as $x \to \infty$. Thus each of the five conditions says, in its own way, that $M_n \to \infty$ rapidly. Note also that $Q(x) \geq 1$ and $q(x) \geq 1$. The integrals in (b) and (c) are thus always defined. It may happen that $Q(x) = \infty$ for some $x < \infty$. In that case, the integral (b) is $+\infty$, and the theorem asserts that $C\{M_n\}$ is quasi-analytic.

If $M_n = n!$, then $M_{n-1}/M_n = 1/n$, hence (e) is violated, and the theorem asserts that $C\{n!\}$ is quasi-analytic, in accordance with Theorem 19.9.

PROOF THAT (a) IMPLIES (b) Assume that $C\{M_n\}$ is not quasi-analytic. Then $C\{M_n\}$ contains a nontrivial function with compact support (Theorem 19.10). An affine change of variable gives a function $F \in C\{M_n\}$, with support in some interval $[0, A]$, such that

$$\|D^n F\|_{\infty} \leq 2^{-n} M_n \qquad (n = 0, 1, 2, \ldots) \tag{1}$$

and such that F is not identically zero. Define

$$f(z) = \int_0^A F(t) e^{itz} \, dt \tag{2}$$

and

$$g(w) = f\left(\frac{i - iw}{1 + w}\right). \tag{3}$$

Then f is entire. If Im $z > 0$, the absolute value of the integrand in (2) is at most $|F(t)|$. Hence f is bounded in the upper half plane; therefore g is bounded in U. Also, g is continuous on $\bar{U}$, except at the point $w = -1$. Since f is not identically 0 (by the uniqueness theorem for Fourier transforms) the same is true of g, and now Theorem 15.19 shows that

$$\frac{1}{2\pi} \int_{-\pi}^{\pi} \log |g(e^{i\theta})| \, d\theta > -\infty. \tag{4}$$

If $x = i(1 - e^{i\theta})/(1 + e^{i\theta}) = \tan(\theta/2)$, then $d\theta = 2(1 + x^2)^{-1} \, dx$, so (4) is the same as

$$\frac{1}{\pi} \int_{-\infty}^{\infty} \log |f(x)| \frac{dx}{1 + x^2} > -\infty. \tag{5}$$

On the other hand, partial integration of (2) gives

$$f(z) = (iz)^{-n} \int_0^A (D^n F)(t) e^{itz} \, dt \qquad (z \neq 0) \tag{6}$$

since F and all its derivatives vanish at 0 and at A. It now follows from (1) and (6) that

$$|x^n f(x)| \leq 2^{-n} A M_n \qquad (x \text{ real}, n = 0, 1, 2, \ldots). \tag{7}$$

Hence

$$Q(x)|f(x)| = \sum_{n=0}^{\infty} \frac{x^n |f(x)|}{M_n} \leq 2A \qquad (x \geq 0), \tag{8}$$

and (5) and (8) imply that (b) holds.   ////

PROOF THAT (b) IMPLIES (c) $q(x) \leq Q(x)$.   ////

PROOF THAT (c) IMPLIES (d) Put $a_n = M_n^{1/n}$. Since $M_0 = 1$ and since $M_n^2 \leq M_{n-1} M_{n+1}$, it is easily verified that $a_n \leq a_{n+1}$, for $n > 0$. If $x \geq ea_n$, then $x^n/M_n \geq e^n$, so

$$\log q(x) \geq \log \frac{x^n}{M_n} \geq \log e^n = n. \tag{9}$$

Hence

$$e \int_{ea_1}^{\infty} \log q(x) \cdot \frac{dx}{x^2} \geq e \sum_{n=1}^{N} n \int_{ea_n}^{ea_{n+1}} x^{-2} \, dx + e \int_{ea_{N+1}}^{\infty} (N+1) x^{-2} \, dx$$

$$= \sum_{n=1}^{N} n \left(\frac{1}{a_n} - \frac{1}{a_{n+1}} \right) + \frac{N+1}{a_{N+1}} = \sum_{n=1}^{N+1} \frac{1}{a_n} \tag{10}$$

for every N. This shows that (c) implies (d).   ////

PROOF THAT (d) IMPLIES (e) Put

$$\lambda_n = \frac{M_{n-1}}{M_n} \qquad (n = 1, 2, 3, \ldots). \tag{11}$$

Then $\lambda_1 \geq \lambda_2 \geq \lambda_3 \geq \cdots$, and if $a_n = M_n^{1/n}$, as above, we have

$$(a_n \lambda_n)^n \leq M_n \cdot \lambda_1 \lambda_2 \cdots \lambda_n = 1. \tag{12}$$

Thus $\lambda_n \leq 1/a_n$, and the convergence of $\Sigma(1/a_n)$ implies that of $\Sigma \lambda_n$. ////

PROOF THAT (e) IMPLIES (a) The assumption now is that $\Sigma \lambda_n < \infty$, where λ_n is given by (11). We claim that the function

$$f(z) = \left(\frac{\sin z}{z}\right)^2 \prod_{n=1}^{\infty} \frac{\sin \lambda_n z}{\lambda_n z} \tag{13}$$

is an entire function of exponential type, not identically zero, which satisfies the inequalities

$$|x^k f(x)| \leq M_k \left(\frac{\sin x}{x}\right)^2 \qquad (x \text{ real}, k = 0, 1, 2, \ldots). \tag{14}$$

Note first that $1 - z^{-1} \sin z$ has a zero at the origin. Hence there is a constant B such that

$$\left|1 - \frac{\sin z}{z}\right| \leq B|z| \qquad (|z| \leq 1). \tag{15}$$

It follows that

$$\left|1 - \frac{\sin \lambda_n z}{\lambda_n z}\right| \leq B\lambda_n |z| \qquad \left(|z| \leq \frac{1}{\lambda_n}\right), \tag{16}$$

so that the series

$$\sum_{n=1}^{\infty} \left|1 - \frac{\sin \lambda_n z}{\lambda_n z}\right| \tag{17}$$

converges uniformly on compact sets. (Note that $1/\lambda_n \to \infty$ as $n \to \infty$, since $\Sigma \lambda_n < \infty$.) The infinite product (13) therefore defines an entire function f which is not identically zero.

Next, the identity

$$\frac{\sin z}{z} = \frac{1}{2} \int_{-1}^{1} e^{itz} \, dt \tag{18}$$

shows that $|z^{-1} \sin z| \leq e^{|y|}$ if $z = x + iy$. Hence

$$|f(z)| \leq e^{A|z|}, \qquad \text{with } A = 2 + \sum_{n=1}^{\infty} \lambda_n. \tag{19}$$

For real x, we have $|\sin x| \leq |x|$ and $|\sin x| \leq 1$. Hence

$$|x^k f(x)| \leq |x^k| \left(\frac{\sin x}{x}\right)^2 \prod_{n=1}^{k} \left|\frac{\sin \lambda_n x}{\lambda_n x}\right|$$

$$\leq \left(\frac{\sin x}{x}\right)^2 (\lambda_1 \cdots \lambda_k)^{-1} = M_k \left(\frac{\sin x}{x}\right)^2. \tag{20}$$

This gives (14), and if we integrate (14) we obtain

$$\frac{1}{\pi} \int_{-\infty}^{\infty} |x^k f(x)| \, dx \leq M_k \qquad (k = 0, 1, 2, \ldots). \tag{21}$$

We have proved that f satisfies the hypotheses of Theorem 19.3. The Fourier transform of f,

$$F(t) = \frac{1}{2\pi} \int_{-\infty}^{\infty} f(x) e^{-itx} \, dx \qquad (t \text{ real}) \tag{22}$$

is therefore a function with compact support, not identically zero, and (21) shows that $F \in C^\infty$ and that

$$(D^k F)(t) = \frac{1}{2\pi} \int_{-\infty}^{\infty} (-ix)^k f(x) e^{-itx} \, dx, \tag{23}$$

by repeated application of Theorem 9.2(f). Hence $\|D^k F\|_\infty \leq M_k$, by (21), which shows that $F \in C\{M_n\}$.

Hence $C\{M_n\}$ is not quasi-analytic, and the proof is complete. ////

Exercises

1 Suppose f is an entire function of exponential type and

$$\varphi(y) = \int_{-\infty}^{\infty} |f(x + iy)|^2 \, dx.$$

Prove that either $\varphi(y) = \infty$ for all real y or $\varphi(y) < \infty$ for all real y. Prove that $f = 0$ if φ is a bounded function.

2 Suppose f is an entire function of exponential type such that the restriction of f to two nonparallel lines belongs to L^2. Prove that $f = 0$.

3 Suppose f is an entire function of exponential type whose restriction to two nonparallel lines is bounded. Prove that f is constant. (Apply Exercise 9 of Chap. 12.)

4 Suppose f is entire, $|f(z)| < C \exp(A|z|)$, and $f(z) = \Sigma a_n z^n$. Put

$$\Phi(w) = \sum_{n=0}^{\infty} \frac{n! \, a_n}{w^{n+1}}.$$

Prove that the series converges if $|w| > A$, that

$$f(z) = \frac{1}{2\pi i} \int_\Gamma \Phi(w) e^{wz} \, dw$$

if $\Gamma(t) = (A + \epsilon)e^{it}$, $0 \le t \le 2\pi$, and that Φ is the function which occurred in the proof of Theorem 19.3. (See also Sec. 19.4.)

5 Suppose f satisfies the hypothesis of Theorem 19.2. Prove that the Cauchy formula

$$f(z) = \frac{1}{2\pi i} \int_{-\infty}^{\infty} \frac{f(\xi + i\epsilon)}{\xi + i\epsilon - z} \, d\xi \qquad (0 < \epsilon < y) \qquad (*)$$

holds; here $z = x + iy$. Prove that

$$f^*(x) = \lim_{y \to 0} f(x + iy)$$

exists for almost all x. What is the relation between f^* and the function F which occurs in Theorem 19.2? Is (*) true with $\epsilon = 0$ and with f^* in place of f in the integrand?

6 Suppose $\varphi \in L^2(-\infty, \infty)$ and $\varphi > 0$. Prove that there exists an f with $|f| = \varphi$ such that the Fourier transform of f vanishes on a half line if and only if

$$\int_{-\infty}^{\infty} \log \varphi(x) \frac{dx}{1 + x^2} > -\infty.$$

Suggestion: Consider f^*, as in Exercise 5, where $f = \exp(u + iv)$ and

$$u(z) = \frac{1}{\pi} \int_{-\infty}^{\infty} \frac{y}{(x - t)^2 + y^2} \log \varphi(t) \, dt.$$

7 Let f be a complex function on a closed set E in the plane. Prove that the following two conditions on f are equivalent:
  (a) There is an open set $\Omega \supset E$ and a function $F \in H(\Omega)$ such that $F(z) = f(z)$ for $z \in E$.
  (b) To each $\alpha \in E$ there corresponds a neighborhood V_α of α and a function $F_\alpha \in H(V_\alpha)$ such that $F_\alpha(z) = f(z)$ in $V_\alpha \cap E$.
(A special case of this was proved in Theorem 19.9.)

8 Prove that $C\{n!\} = C\{n^n\}$.

9 Prove that there are quasi-analytic classes which are larger than $C\{n!\}$.

10 Put $\lambda_n = M_{n-1}/M_n$, as in the proof of Theorem 19.11. Pick $g_0 \in C_c(R^1)$, and define

$$g_n(x) = (2\lambda_n)^{-1} \int_{-\lambda_n}^{\lambda_n} g_{n-1}(x - t) \, dt \qquad (n = 1, 2, 3, \ldots).$$

Prove directly (without using Fourier transforms or holomorphic functions) that $g = \lim g_n$ is a function which demonstrates that (e) implies (a) in Theorem 19.11. (You may choose any g_0 that is convenient.)

11 Find an explicit formula for a function $\varphi \in C^\infty$, with support in $[-2, 2]$, such that $\varphi(x) = 1$ if $-1 \le x \le 1$.

12 Prove that to every sequence $\{\alpha_n\}$ of complex numbers there corresponds a function $f \in C^\infty$ such that $(D^n f)(0) = \alpha_n$ for $n = 0, 1, 2, \ldots$. *Suggestion*: If φ is as in Exercise 11, if $\beta_n = \alpha_n/n!$, if $g_n(x) = \beta_n x^n \varphi(x)$, and if

$$f_n(x) = \lambda_n^{-n} g_n(\lambda_n x) = \beta_n x^n \varphi(\lambda_n x),$$

then $\|D^k f_n\|_\infty < 2^{-n}$ for $k = 0, \ldots, n - 1$, provided that λ_n is large enough. Take $f = \Sigma f_n$.

13 Construct a function $f \in C^\infty$ such that the power series

$$\sum_{n=0}^{\infty} \frac{(D^n f)(a)}{n!} (x - a)^n$$

has radius of convergence 0 for every $a \in R^1$. *Suggestion*: Put

$$f(z) = \sum_{k=1}^{\infty} c_k e^{i\lambda_k x},$$

where $\{c_k\}$ and $\{\lambda_k\}$ are sequences of positive numbers, chosen so that $\Sigma c_k \lambda_k^n < \infty$ for $n = 0, 1, 2, \ldots$ and so that $c_n \lambda_n^n$ increases very rapidly and is much larger than the sum of all the other terms in the series $\Sigma c_k \lambda_k^n$.

For instance, put $c_k = \lambda_k^{1-k}$, and choose $\{\lambda_k\}$ so that

$$\lambda_k > 2 \sum_{j=1}^{k-1} c_j \lambda_j^k \quad \text{and} \quad \lambda_k > k^{2k}.$$

14 Suppose $C\{M_n\}$ is quasi-analytic, $f \in C\{M_n\}$, and $f(x) = 0$ for infinitely many $x \in [0, 1]$. What follows?

15 Let X be the vector space of all entire functions f that satisfy $|f(z)| \le Ce^{\pi|z|}$ for some $C < \infty$, and whose restriction to the real axis is in L^2. Associate with each $f \in X$ its restriction to the integers. Prove that $f \to \{f(n)\}$ is a linear one-to-one mapping of X onto ℓ^2.

16 Assume f is a measurable function on $(-\infty, \infty)$ such that $|f(x)| < e^{-|x|}$ for all x. Prove that its Fourier transform $\hat{f}$ cannot have compact support, unless $f(x) = 0$ a.e.

CHAPTER
TWENTY

UNIFORM APROXIMATION BY POLYNOMIALS

Introduction

20.1 Let K^0 be the interior of a compact set K in the complex plane. (By definition, K^0 is the union of all open discs which are subsets of K; of course, K^0 may be empty even if K is not.) Let $P(K)$ denote the set of all functions on K which are uniform limits of polynomials in z.

Which functions belong to $P(K)$?

Two necessary conditions come to mind immediately: If $f \in P(K)$, then $f \in C(K)$ and $f \in H(K^0)$.

The question arises whether these necessary conditions are also sufficient. The answer is negative whenever K separates the plane (i.e., when the complement of K is not connected). We saw this in Sec. 13.8. On the other hand, if K is an interval on the real axis (in which case $K^0 = \emptyset$), the Weierstrass approximation theorem asserts that

$$P(K) = C(K).$$

So the answer is positive if K is an interval. Runge's theorem also points in this direction, since it states, for compact sets K which do not separate the plane, that $P(K)$ contains at least all those $f \in C(K)$ which have holomorphic extensions to some open set $\Omega \supset K$.

In this chapter we shall prove the theorem of Mergelyan which states, without any superfluous hypotheses, that the above-mentioned necessary conditions are also sufficient if K does not separate the plane.

The principal ingredients of the proof are: Tietze's extension theorem, a smoothing process invoving convolutions, Runge's theorem, and Lemma 20.2, whose proof depends on properties of the class $\mathscr{S}$ which was introduced in Chap. 14.

Some Lemmas

20.2 Lemma *Suppose D is an open disc of radius $r > 0$, $E \subset D$, E is compact and connected, $\Omega = S^2 - E$ is connected, and the diameter of E is at least r. Then there is a function $g \in H(\Omega)$ and a constant b, with the following property: If*

$$Q(\zeta, z) = g(z) + (\zeta - b)g^2(z), \tag{1}$$

the inequalities

$$|Q(\zeta, z)| < \frac{100}{r} \tag{2}$$

$$\left| Q(\zeta, z) - \frac{1}{z - \zeta} \right| < \frac{1{,}000 r^2}{|z - \zeta|^3} \tag{3}$$

hold for all $z \in \Omega$ and for all $\zeta \in D$.

We recall that S^2 is the Riemann sphere and that the diameter of E is the supremum of the numbers $|z_1 - z_2|$, where $z_1 \in E$ and $z_2 \in E$.

PROOF We assume, without loss of generality, that the center of D is at the origin. So $D = D(0; r)$.

The implication $(d) \to (b)$ of Theorem 13.11 shows that Ω is simply connected. (Note that $\infty \in \Omega$.) By the Riemann mapping theorem there is therefore a conformal mapping F of U onto Ω such that $F(0) = \infty$. F has an expansion of the form

$$F(w) = \frac{a}{w} + \sum_{n=0}^{\infty} c_n w^n \qquad (w \in U). \tag{4}$$

We define

$$g(z) = \frac{1}{a} F^{-1}(z) \qquad (z \in \Omega), \tag{5}$$

where F^{-1} is the mapping of Ω onto U which inverts F, and we put

$$b = \frac{1}{2\pi i} \int_\Gamma z g(z) \, dz, \tag{6}$$

where Γ is the positively oriented circle with center 0 and radius r.

By (4), Theorem 14.15 can be applied to F/a. It asserts that the diameter of the complement of $(F/a)(U)$ is at most 4. Therefore $\operatorname{diam} E \leq 4|a|$. Since $\operatorname{diam} E \geq r$, it follows that

$$|a| \geq \frac{r}{4}. \tag{7}$$

Since g is a conformal mapping of Ω onto $D(0; 1/|a|)$, (7) shows that

$$|g(z)| < \frac{4}{r} \qquad (z \in \Omega) \tag{8}$$

and since Γ is a path in Ω, of length $2\pi r$, (6) gives

$$|b| < 4r. \tag{9}$$

If $\zeta \in D$, then $|\zeta| < r$, so (1), (8), and (9) imply

$$|Q| \le \frac{4}{r} + 5r\left(\frac{16}{r^2}\right) < \frac{100}{r}. \tag{10}$$

This proves (2).

Fix $\zeta \in D$.

If $z = F(w)$, then $zg(z) = wF(w)/a$; and since $wF(w) \to a$ as $w \to 0$, we have $zg(z) \to 1$ as $z \to \infty$. Hence g has an expansion of the form

$$g(z) = \frac{1}{z - \zeta} + \frac{\lambda_2(\zeta)}{(z - \zeta)^2} + \frac{\lambda_3(\zeta)}{(z - \zeta)^3} + \cdots \qquad (|z - \zeta| > 2r). \tag{11}$$

Let Γ_0 be a large circle with center at 0; (11) gives (by Cauchy's theorem) that

$$\lambda_2(\zeta) = \frac{1}{2\pi i} \int_{\Gamma_0} (z - \zeta) g(z) \, dz = b - \zeta. \tag{12}$$

Substitute this value of $\lambda_2(\zeta)$ into (11). Then (1) shows that the function

$$\varphi(z) = \left[Q(\zeta, z) - \frac{1}{z - \zeta} \right](z - \zeta)^3 \tag{13}$$

is bounded as $z \to \infty$. Hence φ has a removable singularity at ∞. If $z \in \Omega \cap D$, then $|z - \zeta| < 2r$, so (2) and (13) give

$$|\varphi(z)| < 8r^3 |Q(\zeta, z)| + 4r^2 < 1{,}000 r^2. \tag{14}$$

By the maximum modulus theorem, (14) holds for all $z \in \Omega$. This proves (3).
////

20.3 Lemma *Suppose $f \in C'_c(R^2)$, the space of all continuously differentiable functions in the plane, with compact support. Put*

$$\bar{\partial} = \frac{1}{2}\left(\frac{\partial}{\partial x} + i \frac{\partial}{\partial y}\right). \tag{1}$$

Then the following "Cauchy formula" holds:

$$f(z) = -\frac{1}{\pi} \iint_{R^2} \frac{(\bar{\partial} f)(\zeta)}{\zeta - z} \, d\xi \, d\eta \qquad (\zeta = \xi + i\eta). \tag{2}$$

PROOF This may be deduced from Green's theorem. However, here is a simple direct proof:

Put $\varphi(r, \theta) = f(z + re^{i\theta})$, $r > 0$, θ real. If $\zeta = z + re^{i\theta}$, the chain rule gives

$$(\bar{\partial}f)(\zeta) = \frac{1}{2} e^{i\theta} \left[\frac{\partial}{\partial r} + \frac{i}{r} \frac{\partial}{\partial \theta} \right] \varphi(r, \theta). \tag{3}$$

The right side of (2) is therefore equal to the limit, as $\epsilon \to 0$, of

$$-\frac{1}{2\pi} \int_\epsilon^\infty \int_0^{2\pi} \left(\frac{\partial \varphi}{\partial r} + \frac{i}{r} \frac{\partial \varphi}{\partial \theta} \right) d\theta \, dr. \tag{4}$$

For each $r > 0$, φ is periodic in θ, with period 2π. The integral of $\partial \varphi / \partial \theta$ is therefore 0, and (4) becomes

$$-\frac{1}{2\pi} \int_0^{2\pi} d\theta \int_\epsilon^\infty \frac{\partial \varphi}{\partial r} dr = \frac{1}{2\pi} \int_0^{2\pi} \varphi(\epsilon, \theta) \, d\theta. \tag{5}$$

As $\epsilon \to 0$, $\varphi(\epsilon, \theta) \to f(z)$ uniformly. This gives (2).   ////

We shall establish Tietze's extension theorem in the same setting in which we proved Urysohn's lemma, since it is a fairly direct consequence of that lemma.

20.4 Tietze's Extension Theorem *Suppose K is a compact subset of a locally compact Hausdorff space X, and $f \in C(K)$. Then there exists an $F \in C_c(X)$ such that $F(x) = f(x)$ for all $x \in K$.*

(As in Lusin's theorem, we can also arrange it so that $\|F\|_X = \|f\|_K$.)

PROOF Assume f is real, $-1 \leq f \leq 1$. Let W be an open set with compact closure so that $K \subset W$. Put

$$K^+ = \{x \in K : f(x) \geq \tfrac{1}{3}\}, \qquad K^- = \{x \in K : f(x) \leq -\tfrac{1}{3}\}. \tag{1}$$

Then K^+ and K^- are disjoint compact subsets of W. As a consequence of Urysohn's lemma there is a function $f_1 \in C_c(X)$ such that $f_1(x) = \tfrac{1}{3}$ on K^+, $f_1(x) = -\tfrac{1}{3}$ on K^-, $-\tfrac{1}{3} \leq f_1(x) \leq \tfrac{1}{3}$ for all $x \in X$, and the support of f_1 lies in W. Thus

$$|f - f_1| \leq \tfrac{2}{3} \text{ on } K, \qquad |f_1| \leq \tfrac{1}{3} \text{ on } X. \tag{2}$$

Repeat this construction with $f - f_1$ in place of f: There exists an $f_2 \in C_c(X)$, with support in W, so that

$$|f - f_1 - f_2| \leq (\tfrac{2}{3})^2 \text{ on } K, \qquad |f_2| \leq \tfrac{1}{3} \cdot \tfrac{2}{3} \text{ on } X. \tag{3}$$

In this way we obtain functions $f_n \in C_c(X)$, with supports in W, such that

$$|f - f_1 - \cdots - f_n| \leq (\tfrac{2}{3})^n \text{ on } K, \qquad |f_n| \leq \tfrac{1}{3} \cdot (\tfrac{2}{3})^{n-1} \text{ on } X. \tag{4}$$

Put $F = f_1 + f_2 + f_3 + \cdots$. By (4), the series converges to f on K, and it converges uniformly on X. Hence F is continuous. Also, the support of F lies in $\overline{W}$. ////

Mergelyan's Theorem

20.5 Theorem *If K is a compact set in the plane whose complement is connected, if f is a continuous complex function on K which is holomorphic in the interior of K, and if $\epsilon > 0$, then there exists a polynomial P such that $|f(z) - P(z)| < \epsilon$ for all $z \in K$.*

If the interior of K is empty, then part of the hypothesis is vacuously satisfied, and the conclusion holds for every $f \in C(K)$. Note that K need not be connected.

PROOF By Tietze's theorem, f can be extended to a continuous function in the plane, with compact support. We fix one such extension, and denote it again by f.

For any $\delta > 0$, let $\omega(\delta)$ be the supremum of the numbers

$$|f(z_2) - f(z_1)|$$

where z_1 and z_2 are subject to the condition $|z_2 - z_1| \leq \delta$. Since f is uniformly continuous, we have

$$\lim_{\delta \to 0} \omega(\delta) = 0. \tag{1}$$

From now on, δ will be fixed. We shall prove that there is a polynomial P such that

$$|f(z) - P(z)| < 10{,}000\,\omega(\delta) \qquad (z \in K). \tag{2}$$

By (1), this proves the theorem.

Our first objective is the construction of a function $\Phi \in C_c'(R^2)$, such that for all z

$$|f(z) - \Phi(z)| \leq \omega(\delta), \tag{3}$$

$$|(\bar{\partial}\Phi)(z)| < \frac{2\omega(\delta)}{\delta}, \tag{4}$$

and

$$\Phi(z) = -\frac{1}{\pi} \iint_X \frac{(\bar{\partial}\Phi)(\zeta)}{\zeta - z} \, d\xi \, d\eta \qquad (\zeta = \xi + i\eta), \tag{5}$$

where X is the set of all points in the support of Φ whose distance from the complement of K does not exceed δ. (Thus X contains no point which is "far within" K.)

We construct Φ as the convolution of f with a smoothing function A. Put $a(r) = 0$ if $r > \delta$, put

$$a(r) = \frac{3}{\pi\delta^2}\left(1 - \frac{r^2}{\delta^2}\right)^2 \quad (0 \le r \le \delta), \tag{6}$$

and define

$$A(z) = a(|z|) \tag{7}$$

for all complex z. It is clear that $A \in C'_c(R^2)$. We claim that

$$\iint_{R^2} A = 1, \tag{8}$$

$$\iint_{R^2} \bar{\partial}A = 0, \tag{9}$$

$$\iint_{R^2} |\bar{\partial}A| = \frac{24}{15\delta} < \frac{2}{\delta}. \tag{10}$$

The constants are so adjusted in (6) that (8) holds. (Compute the integral in polar coordinates.) (9) holds simply because A has compact support. To compute (10), express $\bar{\partial}A$ in polar coordinates, as in the proof of Lemma 20.3, and note that $\partial A/\partial \theta = 0$, $|\partial A/\partial r| = -a'(r)$.

Now define

$$\Phi(z) = \iint_{R^2} f(z - \zeta)A(\zeta) \, d\xi \, d\eta = \iint_{R^2} A(z - \zeta)f(\zeta) \, d\xi \, d\eta. \tag{11}$$

Since f and A have compact support, so does Φ. Since

$$\Phi(z) - f(z) = \iint_{R^2} [f(z - \zeta) - f(z)]A(\zeta) \, d\xi \, d\eta \tag{12}$$

and $A(\zeta) = 0$ if $|\zeta| > \delta$, (3) follows from (8). The difference quotients of A converge boundedly to the corresponding partial derivatives of A, since

$A \in C'_c(R^2)$. Hence the last expression in (11) may be differentiated under the integral sign, and we obtain

$$(\bar\partial \Phi)(z) = \iint_{R^2} (\bar\partial A)(z - \zeta) f(\zeta) \, d\xi \, d\eta$$

$$= \iint_{R^2} f(z - \zeta)(\bar\partial A)(\zeta) \, d\xi \, d\eta$$

$$= \iint_{R^2} [f(z - \zeta) - f(z)](\bar\partial A)(\zeta) \, d\xi \, d\eta. \qquad (13)$$

The last equality depends on (9). Now (10) and (13) give (4). If we write (13) with Φ_x and Φ_y in place of $\bar\partial \Phi$, we see that Φ has continuous partial derivatives. Hence Lemma 20.3 applies to Φ, and (5) will follow if we can show that $\bar\partial \Phi = 0$ in G, where G is the set of all $z \in K$ whose distance from the complement of K exceeds δ. We shall do this by showing that

$$\Phi(z) = f(z) \qquad (z \in G); \qquad (14)$$

note that $\bar\partial f = 0$ in G, since f is holomorphic there. (We recall that $\bar\partial$ is the Cauchy-Riemann operator defined in Sec. 11.1.) Now if $z \in G$, then $z - \zeta$ is in the interior of K for all ζ with $|\zeta| < \delta$. The mean value property for harmonic functions therefore gives, by the first equation in (11),

$$\Phi(z) = \int_0^\delta a(r) r \, dr \int_0^{2\pi} f(z - re^{i\theta}) \, d\theta$$

$$= 2\pi f(z) \int_0^\delta a(r) r \, dr = f(z) \iint_{R^2} A = f(z) \qquad (15)$$

for all $z \in G$.

We have now proved (3), (4), and (5).

The definition of X shows that X is compact and that X can be covered by finitely many open discs $D_1, \ldots, D_n$, of radius 2δ, whose centers are *not* in K. Since $S^2 - K$ is connected, the center of each D_j can be joined to ∞ by a polygonal path in $S^2 - K$. It follows that each D_j contains a compact connected set E_j, of diameter at least 2δ, so that $S^2 - E_j$ is connected and so that $K \cap E_j = \varnothing$.

We now apply Lemma 20.2, with $r = 2\delta$. There exist functions

$g_j \in H(S^2 - E_j)$ and constants b_j so that the inequalities

$$|Q_j(\zeta, z)| < \frac{50}{\delta}, \tag{16}$$

$$\left| Q_j(\zeta, z) - \frac{1}{z - \zeta} \right| < \frac{4{,}000\delta^2}{|z - \zeta|^3} \tag{17}$$

hold for $z \notin E_j$ and $\zeta \in D_j$, if

$$Q_j(\zeta, z) = g_j(z) + (\zeta - b_j)g_j^2(z). \tag{18}$$

Let Ω be the complement of $E_1 \cup \cdots \cup E_n$. Then Ω is an open set which contains K.

Put $X_1 = X \cap D_1$ and $X_j = (X \cap D_j) - (X_1 \cup \cdots \cup X_{j-1})$, for $2 \le j \le n$. Define

$$R(\zeta, z) = Q_j(\zeta, z) \qquad (\zeta \in X_j, z \in \Omega) \tag{19}$$

and

$$F(z) = \frac{1}{\pi} \iint_X (\bar{\partial}\Phi)(\zeta)R(\zeta, z)\, d\xi\, d\eta \qquad (z \in \Omega). \tag{20}$$

Since

$$F(z) = \sum_{j=1}^{n} \frac{1}{\pi} \iint_{X_j} (\bar{\partial}\Phi)(\zeta)Q_j(\zeta, z)\, d\xi\, d\eta, \tag{21}$$

(18) shows that F is a finite linear combination of the functions g_j and g_j^2. Hence $F \in H(\Omega)$.

By (20), (4), and (5) we have

$$|F(z) - \Phi(z)| < \frac{2\omega(\delta)}{\pi\delta} \iint_X \left| R(\zeta, z) - \frac{1}{z - \zeta} \right| d\xi\, d\eta \qquad (z \in \Omega). \tag{22}$$

Observe that the inequalities (16) and (17) are valid with R in place of Q_j if $\zeta \in X$ and $z \in \Omega$. For if $\zeta \in X$ then $\zeta \in X_j$ for some j, and then $R(\zeta, z) = Q_j(\zeta, z)$ for all $z \in \Omega$.

Now fix $z \in \Omega$, put $\zeta = z + \rho e^{i\theta}$, and estimate the integrand in (22) by (16) if $\rho < 4\delta$, by (17) if $4\delta \le \rho$. The integral in (22) is then seen to be less than the sum of

$$2\pi \int_0^{4\delta} \left(\frac{50}{\delta} + \frac{1}{\rho} \right) \rho\, d\rho = 808\pi\delta \tag{23}$$

and

$$2\pi \int_{4\delta}^{\infty} \frac{4{,}000\delta^2}{\rho^3} \rho \, d\rho = 2{,}000\pi\delta. \tag{24}$$

Hence (22) yields

$$|F(z) - \Phi(z)| < 6{,}000\omega(\delta) \qquad (z \in \Omega). \tag{25}$$

Since $F \in H(\Omega)$, $K \subset \Omega$, and $S^2 - K$ is connected, Runge's theorem shows that F can be uniformly approximated on K by polynomials. Hence (3) and (25) show that (2) can be satisfied.

This completes the proof. ////

One unusual feature of this proof should be pointed out. We had to prove that the given function f is in the closed subspace $P(K)$ of $C(K)$. (We use the terminology of Sec. 20.1.) Our first step consisted in approximating f by Φ. But this step took us outside $P(K)$, since Φ was so constructed that in general Φ will not be holomorphic in the whole interior of K. Hence Φ is at some positive distance from $P(K)$. However, (25) shows that this distance is less than a constant multiple of $\omega(\delta)$. [In fact, having proved the theorem, we know that this distance is at most $\omega(\delta)$, by (3), rather than 6,000 $\omega(\delta)$.] The proof of (25) depends on the inequality (4) and on the fact that $\bar{\partial}\Phi = 0$ in G. Since holomorphic functions φ are characterized by $\bar{\partial}\varphi = 0$, (4) may be regarded as saying that Φ is not too far from being holomorphic, and this interpretation is confirmed by (25).

Exercises

1 Extend Mergelyan's theorem to the case in which $S^2 - K$ has finitely many components: Prove that then every $f \in C(K)$ which is holomorphic in the interior of K can be uniformly approximated on K by rational functions.

2 Show that the result of Exercise 1 does not extend to arbitrary compact sets K in the plane, by verifying the details of the following example. For $n = 1, 2, 3, \ldots$, let $D_n = D(\alpha_n; r_n)$ be disjoint open discs in U whose union V is dense in U, such that $\Sigma r_n < \infty$. Put $K = \bar{U} - V$. Let Γ and γ_n be the paths

$$\Gamma(t) = e^{it}, \qquad \gamma_n(t) = \alpha_n + r_n e^{it}, \qquad 0 \le t \le 2\pi,$$

and define

$$L(f) = \int_{\Gamma} f(z) \, dz - \sum_{n=1}^{\infty} \int_{\gamma_n} f(z) \, dz \qquad (f \in C(K)).$$

Prove that L is a bounded linear functional on $C(K)$, prove that $L(R) = 0$ for every rational function R whose poles are outside K, and prove that there exists an $f \in C(K)$ for which $L(f) \ne 0$.

3 Show that the function g constructed in the proof of Lemma 20.2 has the smallest supremum norm among all $f \in H(\Omega)$ such that $zf(z) \to 1$ as $z \to \infty$. (This motivates the proof of the lemma.)

Show also that $b = c_0$ in that proof and that the inequality $|b| < 4r$ can therefore be replaced by $|b| < r$. In fact, b lies in the convex hull of the set E.

APPENDIX
HAUSDORFF'S MAXIMALITY THEOREM

We shall first prove a lemma which, when combined with the axiom of choice, leads to an almost instantaneous proof of Theorem 4.21.

If $\mathscr{F}$ is a collection of sets and $\Phi \subset \mathscr{F}$, we call Φ a *subchain* of $\mathscr{F}$ provided that Φ is totally ordered by set inclusion. Explicitly, this means that if $A \in \Phi$ and $B \in \Phi$, then either $A \subset B$ or $B \subset A$. The union of all members of Φ will simply be referred to as the *union of* Φ.

Lemma *Suppose $\mathscr{F}$ is a nonempty collection of subsets of a set X such that the union of every subchain of $\mathscr{F}$ belongs to $\mathscr{F}$. Suppose g is a function which associates to each $A \in \mathscr{F}$ a set $g(A) \in \mathscr{F}$ such that $A \subset g(A)$ and $g(A) - A$ consists of at most one element. Then there exists an $A \in \mathscr{F}$ for which $g(A) = A$.*

PROOF Fix $A_0 \in \mathscr{F}$. Call a subcollection $\mathscr{F}'$ of $\mathscr{F}$ a *tower* if $\mathscr{F}'$ has the following three properties:

(a) $A_0 \in \mathscr{F}'$.
(b) The union of every subchain of $\mathscr{F}'$ belongs to $\mathscr{F}'$.
(c) If $A \in \mathscr{F}'$, then also $g(A) \in \mathscr{F}'$.

The family of all towers is nonempty. For if $\mathscr{F}_1$ is the collection of all $A \in \mathscr{F}$ such that $A_0 \subset A$, then $\mathscr{F}_1$ is a tower. Let $\mathscr{F}_0$ be the intersection of all towers. Then $\mathscr{F}_0$ is a tower (the verification is trivial), but no proper subcollection of $\mathscr{F}_0$ is a tower. Also, $A_0 \subset A$ if $A \in \mathscr{F}_0$. The idea of the proof is to show that $\mathscr{F}_0$ is a subchain of $\mathscr{F}$.

Let Γ be the collection of all $C \in \mathscr{F}_0$ such that every $A \in \mathscr{F}_0$ satisfies either $A \subset C$ or $C \subset A$.

For each $C \in \Gamma$, let $\Phi(C)$ be the collection of all $A \in \mathscr{F}_0$ such that either $A \subset C$ or $g(C) \subset A$.

Properties (a) and (b) are clearly satisfied by Γ and by each $\Phi(C)$. Fix $C \in \Gamma$, and suppose $A \in \Phi(C)$. We want to prove that $g(A) \in \Phi(C)$. If $A \in \Phi(C)$, there are three possibilities: Either $A \subset C$ and $A \neq C$, or $A = C$, or $g(C) \subset A$. If A is a proper subset of C, then C cannot be a proper subset of $g(A)$, otherwise $g(A) - A$ would contain at least two elements; since $C \in \Gamma$, it follows that $g(A) \subset C$. If $A = C$, then $g(A) = g(C)$. If $g(C) \subset A$, then also $g(C) \subset g(A)$, since $A \subset g(A)$. Thus $g(A) \in \Phi(C)$, and we have proved that $\Phi(C)$ is a tower. The minimality of $\mathscr{F}_0$ implies now that $\Phi(C) = \mathscr{F}_0$, for every $C \in \Gamma$.

In other words, if $A \in \mathscr{F}_0$ and $C \in \Gamma$, then either $A \subset C$ or $g(C) \subset A$. But this says that $g(C) \in \Gamma$. Hence Γ is a tower, and the minimality of $\mathscr{F}_0$ shows that $\Gamma = \mathscr{F}_0$. It follows now from the definition of Γ that $\mathscr{F}_0$ is totally ordered.

Let A be the union of $\mathscr{F}_0$. Since $\mathscr{F}_0$ satisfies (b), $A \in \mathscr{F}_0$. By (c), $g(A) \in \mathscr{F}_0$. Since A is the largest member of $\mathscr{F}_0$ and since $A \subset g(A)$, it follows that $A = g(A)$. ////

Definition A *choice function* for a set X is a function f which associates to each nonempty subset E of X an element of E: $f(E) \in E$.

In more informal terminology, f "chooses" an element out of each nonempty subset of X.

The Axiom of Choice *For every set there is a choice function.*

Hausdorff's Maximality Theorem *Every nonempty partially ordered set P contains a maximal totally ordered subset.*

PROOF Let $\mathscr{F}$ be the collection of all totally ordered subsets of P. Since every subset of P which consists of a single element is totally ordered, $\mathscr{F}$ is not empty. Note that the union of any chain of totally ordered sets is totally ordered.

Let f be a choice function for P. If $A \in \mathscr{F}$, let A^* be the set of all x in the complement of A such that $A \cup \{x\} \in \mathscr{F}$. If $A^* \neq \varnothing$, put

$$g(A) = A \cup \{f(A^*)\}.$$

If $A^* = \varnothing$, put $g(A) = A$.

By the lemma, $A^* = \varnothing$ for at least one $A \in \mathscr{F}$, and any such A is a maximal element of $\mathscr{F}$. ////

NOTES AND COMMENTS

Chapter 1

The first book on the modern theory of integration and differentiation is Lebesgue's "Leçons sur l'intégration," published in 1904.

For an illuminating history of the earlier attempts that were made toward the construction of a satisfactory theory of integration, see [42]†, which contains interesting discussions of the difficulties that even first-rate mathematicians had with simple set-theoretic concepts before these were properly defined and well understood.

The approach to abstract integration presented in the text is inspired by Saks [28]. Greater generality can be attained if σ-algebras are replaced by σ-rings (*Axioms*: $\bigcup A_i \in \mathscr{R}$ and $A_1 - A_2 \in \mathscr{R}$ if $A_i \in \mathscr{R}$ for $i = 1, 2, 3, \ldots$; it is not required that $X \in \mathscr{R}$), but at the expense of a necessarily fussier definition of measurability. See Sec. 18 of [7]. In all classical applications the measurability of X is more or less automatic. This is the reason for our choice of the somewhat simpler theory based on σ-algebras.

Sec. 1.11. This definition of $\mathscr{B}$ is as in [12]. In [7] the Borel sets are defined as the σ-ring generated by the compact sets. In spaces which are not σ-compact, this is a smaller family than ours.

Chapter 2

Sec. 2.12. The usual statement of Urysohn's lemma is: If K_0 and K_1 are disjoint closed sets in a *normal* Hausdorff space X, then there is a continuous function on X which is 0 on K_0 and 1 on K_1. The proof is exactly as in the text.

Sec. 2.14. The original form of this theorem, with $X = [0, 1]$, is due to F. Riesz (1909). See [5], pp. 373, 380–381, and [12], pp. 134–135 for its further history. The theorem is here presented in the same generality as in [12]. The set function μ which is defined for all subsets of X in the proof of Theorem 2.14 is called an *outer measure* because of its countable subadditivity (Step I). For systematic exploitations (originated by Carathéodory) of this notion, see [25] and [28].

Sec. 2.20. For direct constructions of Lebesgue measure, along more classical lines, see [31], [35], [26] and [53].

† Numbers in brackets refer to the Bibliography.

Sec. 2.21. A proof that the cardinality of every countably generated σ-algebra is $\leq c$ may be found on pp. 133–134 of [44]. That this cardinality is either finite or $\geq c$ should be clear after doing Exercise 1 of Chap. 1.

Sec. 2.22. A very instructive paper on the subject of nonmeasurable sets in relation to measures invariant under a group is: J. von Neumann, Zur allgemeinen Theorie des Masses, *Fundamenta Math.*, vol. 13, pp. 73–116, 1929. See also Halmos's article in the special (May, 1958) issue of *Bull. Am. Math. Soc.* devoted to von Neumann's work.

Sec. 2.24. [28], p. 72.

Sec. 2.25. [28], p. 75. There is another approach to the Lebesgue theory of integration, due to Daniell (*Ann. Math.*, vol. 19, pp. 279–294, 1917–1918) based on extensions of positive linear functionals. When applied to $C_c(X)$ it leads to a construction which virtually turns the Vitali-Carathéodory theorem into the definition of measurability. See [17] and, for the full treatment, [18].

Exercise 8. Two interesting extensions of this phenomenon have appeared in *Amer. Math. Monthly*: see vol. 79, pp. 884–886, 1972 (R. B. Kirk) and vol. 91, pp. 564–566, 1984 (F. S. Cater).

Exercise 17. This example appears in A Theory of Radon Measures on Locally Compact Spaces, by R. E. Edwards, *Acta Math.*, vol. 89, p. 160, 1953. Its existence was unfortunately overlooked in [27].

Exercise 18. [7], p. 231; originally due to Dieudonné.

Chapter 3

The best general reference is [9]. See also Chap. 1 of [36].

Exercise 3. Vol. 1 (1920) of *Fundamenta Math.* contains three papers relevant to the parenthetical remark.

Exercise 7. A very complete answer to this question was found by A. Villani, in *Am. Math. Monthly*, vol. 92, pp. 485–487, 1985.

Exercise 16. [28], p. 18. When t ranges over all positive real numbers, a measurability problem arises in the suggested proof. This is the reason for including (ii) as a hypothesis. See W. Walter, *Am. Math. Monthly*, vol. 84, pp. 118–119, 1977.

Exercise 17. The second suggested proof of part (*b*) was published by W. P. Novinger, in *Proc. Am. Math. Soc.*, vol. 34, pp. 627–628, 1972. It was also discovered by David Hall.

Exercise 18. Convergence in measure is a natural concept in probability theory. See [7], Chap. IX.

Chapter 4

There are many books dealing with Hilbert space theory. We cite [6] and [24] as main references. See also [5], [17], and [19].

The standard work on Fourier series is [36]. For simpler introductions, see [10], [31], [43], and [45].

Exercise 2. This is the so-called Gram-Schmidt orthogonalization process.

Exercise 18. The functions that are uniform limits (on R^1) of members of X are called *almost periodic*.

Chapter 5

The classical work here is [2]. More recent treatises are [5], [14], and [24]. See also [17] and [49].

Sec. 5.7. The relations between measure theory on the one hand and Baire's theorem on the other are discussed in great detail in [48].

Sec. 5.11. Although there are continuous functions whose Fourier series diverge on a dense G_δ, the set of divergence must have measure 0. This was proved by L. Carleson (*Acta Math.*, vol. 116, pp. 135–157, 1966) for all f in $L^2(T)$; the proof was extended to $L^p(T)$, $p > 1$, by R. A. Hunt. See also [45], especially Chap. II.

Sec. 5.22. For a deeper discussion of representing measures see Arens and Singer, *Proc. Am. Math. Soc.*, vol. 5, pp. 735–745, 1954. Also [39], [52].

Chapter 6

Sec. 6.3. The constant $1/\pi$ is best possible. See R. P. Kaufman and N. W. Rickert, *Bull. Am. Math. Soc.*, vol. 72, pp. 672–676, 1966, and (for a simpler treatment) W. W. Bledsoe, *Am. Math. Monthly*, vol. 77, pp. 180–182, 1970.

Sec. 6.10. von Neumann's proof is in a section on measure theory in his paper: On Rings of Operators, III, *Ann. Math.*, vol. 41, pp. 94–161, 1940. See pp. 124–130.

Sec. 6.15. The phenomenon $L^\infty \neq (L^1)^*$ is discussed by J. T. Schwartz in *Proc. Am. Math. Soc.*, vol. 2, pp. 270–275, 1951, and by H. W. Ellis and D. O. Snow in *Can. Math. Bull.*, vol. 6, pp. 211–229, 1963. See also [7], p. 131, and [28], p. 36.

Sec. 6.19. The references to Theorem 2.14 apply here as well.

Exercise 6. See [17], p. 43.

Exercise 10(g). See [36], vol. I, p. 167.

Chapter 7

Sec. 7.3. This simple covering lemma seems to appear for the first time in a paper by Wiener on the ergodic theorem (*Duke Math. J.*, vol. 5, pp. 1-18, 1939). Covering lemmas play a central role in the theory of differentiation. See [50], [53], and, for a very detailed treatment, [41].

Sec. 7.4. Maximal functions were first introduced by Hardy and Littlewood, in *Acta Math.*, vol. 54, pp. 81–116, 1930. That paper contains proofs of Theorems 8.18, 11.25(b), and 17.11.

Sec. 7.21. The same conclusion can be obtained under somewhat weaker hypotheses; see [16], Theorems 260–264. Note that the proof of Theorem 7.21 uses the existence and integrability of only the *right-hand* derivative of f, plus the continuity of f. For a further refinement, see P. L. Walker, *Amer. Math. Monthly*, vol. 84, pp. 287–288, 1977.

Secs. 7.25, 7.26. This treatment of the change of variables formula is quite similar to D. Varberg's in *Amer. Math. Monthly*, vol. 78, pp. 42–45, 1971.

Exercise 5. A very simple proof, due to K. Stromberg, is in *Proc. Amer. Math. Soc.*, vol. 36, p. 308, 1972.

Exercise 12. For an elementary proof that every monotone function (hence every function of bounded variation) is differentiable a.e., see [24], pp. 5–9. In that work, this theorem is made the starting point of the Lebesgue theory. Another, even simpler, proof by D. Austin is in *Proc. Amer. Math. Soc.*, vol. 16, pp. 220–221, 1965.

Exercise 18. These functions φ_n are the so-called Rademacher functions. Chap. V of [36] contains further theorems about them.

Chapter 8

Fubini's theorem is developed here as in [7] and [28]. For a different approach, see [25]. Sec. 8.9(c) is in *Fundamenta Math.*, vol. 1, p. 145, 1920.

Sec. 8.18. This proof of the Hardy-Littlewood theorem (see the reference to Sec. 7.4) is essentially that of a very special case of the Marcinkiewicz interpolation theorem. A full discussion of the latter may be found in Chap. XII of [36]. See also [50].

Exercise 2. Corresponding to the idea that an integral is an area under a curve, the theory of the Lebesgue integral can be developed in terms of measures of ordinate sets. This is done in [16].

Exercise 8. Part (b), in even more precise form, was proved by Lebesgue in *J. Mathématiques*, ser. 6, vol. 1, p. 201, 1905, and seems to have been forgotten. It is quite remarkable in view of another example of Sierpinski (*Fundamenta Math.*, vol. 1, p. 114, 1920): There is a plane set E which is not Lebesgue measurable and which has at most two points on each straight line. If $f = \chi_E$, then f is not Lebesgue measurable, although all of the sections f_x and f^y are upper semicontinuous; in fact, each has at most two points of discontinuity. (This example depends on the axiom of choice, but not on the continuum hypothesis.)

Chapter 9

For another brief introduction, see [36], chap. XVI. A different proof of Plancherel's theorem is in [33]. Group-theoretic aspects and connections with Banach algebras are discussed in [17], [19], and [27]. For more on invariant subspaces (Sec. 9.16) see [11]; the corresponding problem in L^1 is described in [27], Chap. 7.

Chapter 10

General references: [1], [4], [13], [20], [29], [31] and [37].

Sec. 10.8. Integration can also be defined over arbitrary rectifiable curves. See [13], vol. I, Appendix C.

Sec. 10.10. The topological concept of index is applied in [29] and is fully utilized in [1]. The computational proof of Theorem 10.10 is as in [1], p. 93.

Sec. 10.13. Cauchy published his theorem in 1825, under the additional assumption that f' is continuous. Goursat showed that this assumption is redundant, and stated the theorem in its present form. See [13], p. 163, for further historical remarks.

Sec. 10.16. The standard proofs of the power series representation and of the fact that $f \in H(\Omega)$ implies $f' \in H(\Omega)$ proceed via the Cauchy integral formula, as here. Recently proofs have been constructed which use the winding number but make no appeal to integration. For details see [34].

Sec. 10.25. A very elementary proof of the algebraic completeness of the complex field is in [26], p. 170.

Sec. 10.30. The proof of part (b) is as in [47].

Sec. 10.32. The open mapping theorem and the discreteness of $Z(f)$ are topological properties of the class of all nonconstant holomorphic functions which characterize this class up to homeomorphisms. This is Stoilov's theorem. See [34].

Sec. 10.35. This strikingly simple and elementary proof of the global version of Cauchy's theorem was discovered by John D. Dixon, *Proc. Am. Math. Soc.*, vol. 29, pp. 625–626, 1971. In [1] the proof is based on the theory of exact differentials. In the first edition of the present book it was deduced from Runge's theorem. That approach was used earlier in [29], p. 177. There, however, it was applied in simply connected regions only.

Chapter 11

General references: [1], Chap. 5; [20], Chap. 1.

Sec. 11.14. The reflection principle was used by H. A. Schwarz to solve problems concerning conformal mappings of polygonal regions. See Sec. 17.6 of [13]. Further results along these lines were obtained by Carathéodory; see [4], vol. II, pp. 88–92, and *Commentarii Mathematici Helvetici*, vol. 19, pp. 263–278, 1946–1947.

Secs. 11.20, 11.25. This is the principal result of the Hardy-Littlewood paper mentioned in the reference to Sec. 7.4. The proof of the second inequality in Theorem 11.20 is as in [40], p. 23.

Sec. 11.23. The first theorems of this type are in Fatou's thesis, Séries trigonométriques et séries de Taylor, *Acta Math.*, vol. 30, pp. 335–400, 1906. This is the first major work in which Lebesgue's theory of integration is applied to the study of holomorphic functions.

Sec. 11.30. Part (c) is due to Herglotz, *Leipziger Berichte*, vol. 63, pp. 501–511, 1911.

Exercise 14. This was suggested by W. Ramey and D. Ullrich.

Chapter 12

Sec. 12.7. For further examples, see [31], pp. 176–186.

Sec. 12.12. This theorem was first proved for trigonometric series by W. H. Young (1912; $q = 2$, 4, 6, ...) and F. Hausdorff (1923; $2 \leq q \leq \infty$). F. Riesz (1923) extended it to uniformly bounded orthonormal sets, M. Riesz (1926) derived this extension from a general interpolation theorem, and G. O. Thorin (1939) discovered the complex-variable proof of M. Riesz's theorem. The proof of the text is the Calderón-Zygmund adaptation (1950) of Thorin's idea. Full references and discussions of other interpolation theorems are in Chap. XII of [36].

Sec. 12.13. In slightly different form, this is in *Duke Math. J.*, vol. 20, pp. 449–458, 1953.

Sec. 12.14. This proof is essentially that of R. Kaufman (*Math. Ann.*, vol. 169, p. 282, 1967). E. L. Stout (*Math. Ann.*, vol. 177, pp. 339–340, 1968) obtained a stronger result.

Chapter 13

Sec. 13.9. Runge's theorem was published in *Acta Math.*, vol. 6, 1885. (Incidentally, this is the same year in which the Weierstrass theorem on uniform approximation by polynomials on an interval was published; see *Mathematische Werke*, vol. 3, pp. 1–37.) See [29], pp. 171–177, for a proof which is close to the original one. The functional analysis proof of the text is known to many analysts and has probably been independently discovered several times in recent years. It was communicated to me by L. A. Rubel. In [13], vol. II, pp. 299–308, attention is paid to the closeness of the approximation if the degree of the polynomial is fixed.

Exercises 5, 6. For yet another method, see D. G. Cantor, *Proc. Am. Math. Soc.*, vol. 15, pp. 335–336, 1964.

Chapter 14

General reference: [20]. Many special mapping functions are described there in great detail.

Sec. 14.3. More details on linear fractional transformations may be found in [1], pp. 22–35; in [13], pp. 46–57; in [4]; and especially in Chap. 1 of L. R. Ford's book "Automorphic Functions," McGraw-Hill Book Company, New York, 1929.

Sec. 14.5. Normal families were introduced by Montel. See Chap. 15 of [13].

Sec. 14.8. The history of Riemann's theorem is discussed in [13], pp. 320–321, and in [29], p. 230. Koebe's proof (Exercise 26) is in *J. für reine und angew. Math.*, vol. 145, pp. 177–223, 1915; doubly connected regions are also considered there.

Sec. 14.14. Much more is true than just $|a_2| \leq 2$: In fact, $|a_n| \leq n$ for all $n \geq 2$. This was conjectured by Bieberbach in 1916, and proved by L. de Branges in 1984 [*Acta Math.*, vol. 154, pp. 137–152, (1985)]. Moreover, if $|a_n| = n$ for just one $n \geq 2$, then f is one of the Koebe functions of Example 14.11.

Sec. 14.19. The boundary behavior of conformal mappings was investigated by Carathéodory in *Math. Ann.*, vol. 73, pp. 323–370, 1913. Theorem 14.19 was proved there for regions bounded by Jordan curves, and the notion of prime ends was introduced. See also [4], vol. II, pp. 88–107.

Exercise 24. This proof is due to Y. N. Moschovakis.

Chapter 15

Sec. 15.9. The relation between canonical products and entire functions of finite order is discussed in Chap. 2 of [3], Chap. VII of [29], and Chap. VIII of [31].

Sec. 15.25. See Szasz, *Math. Ann.*, vol. 77, pp. 482–496, 1916, for further results in this direction. Also Chap. II of [21].

Chapter 16

The classical work on Riemann surfaces is [32]. (The first edition appeared in 1913.) Other references: Chap. VI of [1], Chap. 10 of [13], Chap. VI of [29], and [30].

Sec. 16.5. Ostrowski's theorem is in *J. London Math. Soc.*, vol. 1, pp. 251–263, 1926. See J.-P. Kahane, Lacunary Taylor and Fourier Series, *Bull. Am. Math. Soc.*, vol. 70, pp. 199–213, 1964, for a more recent account of gap series.

Sec. 16.15. The approach to the monodromy theorem was a little simpler in the first edition of this book. It used the fact that every simply connected plane region is homeomorphic to a convex one, namely U. The present proof is so arranged that it applies without change to holomorphic functions of several complex variables. (Note that when $k > 2$ there exist simply connected open sets in R^k which are not homeomorphic to any convex set; spherical shells furnish examples of this.)

Sec. 16.17. Chap. 13 of [13], Chap. VIII of [29], and part 7 of [4].

Sec. 16.21. Picard's big theorem is proved with the aid of modular functions in part 7 of [4]. "Elementary" proofs may be found in [31], pp. 277–284, and in Chap. VII of [29].

Exercise 10. Various classes of removable sets are discussed by Ahlfors and Beurling in Conformal Invariants and Function-theoretic Null-Sets, *Acta Math.*, vol. 83, pp. 101–129, 1950.

Chapter 17

The classical reference here is [15]. See also [36], Chap. VII. Although [15] deals mainly with the unit disc, most proofs are so constructed that they apply to other situations which are described there. Some of these generalizations are in Chap. 8 of [27]. Other recent books on these topics are [38], [40], and [46].

Sec. 17.1. See [22] for a thorough treatment of subharmonic functions.

Sec. 17.13. For a different proof, see [15], or the paper by Helson and Lowdenslager in *Acta Math.*, vol. 99, pp. 165–202, 1958. An extremely simple proof was found by B. K. Øksendal, in *Proc. Amer. Math. Soc.*, vol. 30, p. 204, 1971.

Sec. 17.14. The terms "inner function" and "outer function" were coined by Beurling in the paper in which Theorem 17.21 was proved: On Two Problems Concerning Linear Transformations in Hilbert Space, *Acta Math.*, vol. 81, pp. 239–255, 1949. For further developments, see [11].

Secs. 17.25, 17.26. This proof of M. Riesz's theorem is due to A. P. Calderón. See *Proc. Am. Math. Soc.*, vol. 1, pp. 533–535, 1950. See also [36], vol. I, pp. 252–262.

Exercise 3. This forms the basis of a definition of H^p-spaces in other regions. See *Trans. Am. Math. Soc.*, vol. 78, pp. 46–66, 1955.

Chapter 18

General references: [17], [19], and [23]; also [14]. The theory was originated by Gelfand in 1941.

Sec. 18.18. This was proved in elementary fashion by P. J. Cohen in *Proc. Am. Math. Soc.*, vol. 12, pp. 159–163, 1961.

Sec. 18.20. This theorem is Wermer's, *Proc. Am. Math. Soc.*, vol. 4, pp. 866–869, 1953. The proof of the text is due to Hoffman and Singer. See [15], pp. 93–94, where an extremely short proof by P. J. Cohen is also given. (See the reference to Sec. 18.18.)

Sec. 18.21. This was one of the major steps in Wiener's original proof of his Tauberian theorem. See [33], p. 91. The painless proof given in the text was the first spectacular success of the Gelfand theory.

Exercise 14. The set Δ can be given a compact Hausdorff topology with respect to which the function $\hat{x}$ is continuous. Thus $x \to \hat{x}$ is a homomorphism of A into $C(\Delta)$. This representation of A as an algebra of continuous functions is a most important tool in the study of commutative Banach algebras.

Chapter 19

Secs. 19.2, 19.3: [21], pp. 1–13. See also [3], where functions of exponential type are the main subject.

Sec. 19.5. For a more detailed introduction to the classes $C\{M_n\}$, see S. Mandelbrojt, "Séries de Fourier et classes quasi-analytiques," Gauthier-Villars, Paris, 1935.

Sec. 19.11. In [21], the proof of this theorem is based on Theorem 19.2 rather than on 19.3.

Exercise 4. The function Φ is called the Borel transform of f. See [3], Chap. 5.

Exercise 12. The suggested proof is due to H. Mirkil, *Proc. Am. Math. Soc.*, vol. 7, pp. 650–652, 1956. The theorem was proved by Borel in 1895.

Chapter 20

See S. N. Mergelyan, Uniform Approximations to Functions of a Complex Variable, *Uspehi Mat. Nauk* (N.S.) 7, no. 2 (48), 31–122, 1952; Amer. Math. Soc. Translation No. 101, 1954. Our Theorem 20.5 is Theorem 1.4 in Mergelyan's paper.

A functional analysis proof, based on measure-theoretic considerations, has recently been published by L. Carleson in *Math. Scandinavica*, vol. 15, pp. 167–175, 1964.

Appendix

The maximality theorem was first stated by Hausdorff on p. 140 of his book "Grundzüge der Mengenlehre," 1914. The proof of the text is patterned after Sec. 16 of Halmos's book [8]. The idea to choose g so that $g(A) - A$ has at most one element appears there, as does the term "tower." The proof is similar to one of Zermelo's proofs of the well-ordering theorem; see *Math. Ann.*, vol. 65, pp. 107–128, 1908.

BIBLIOGRAPHY

1. L. V. Ahlfors: "Complex Analysis," 3d ed., McGraw-Hill Book Company, New York, 1978.
2. S. Banach: Théorie des Opérations linéaires, "Monografie Matematyczne," vol. 1, Warsaw, 1932.
3. R. P. Boas: "Entire Functions," Academic Press Inc., New York, 1954.
4. C. Carathéodory: "Theory of Functions of a Complex Variable," Chelsea Publishing Company, New York, 1954.
5. N. Dunford and J. T. Schwartz: "Linear Operators," Interscience Publishers, Inc., New York, 1958.
6. P. R. Halmos: "Introduction to Hilbert Space and the Theory of Spectral Multiplicity," Chelsea Publishing Company, New York, 1951.
7. P. R. Halmos: "Measure Theory," D. Van Nostrand Company Inc., Princeton, N. J., 1950.
8. P. R. Halmos: "Naive Set Theory," D. Van Nostrand Company, Inc., Princeton, N. J., 1960.
9. G. H. Hardy, J. E. Littlewood, and G. Pólya: "Inequalities," Cambridge University Press, New York, 1934.
10. G. H. Hardy and W. W. Rogosinski: "Fourier Series," Cambridge Tracts no. 38, Cambridge, London, and New York, 1950.
11. H. Helson: "Lectures on Invariant Subspaces," Academic Press Inc., New York, 1964.
12. E. Hewitt and K. A. Ross: "Abstract Harmonic Analysis," Springer-Verlag, Berlin, vol. I, 1963; vol. II, 1970.
13. E. Hille: "Analytic Function Theory," Ginn and Company, Boston, vol. I, 1959; vol. II, 1962.
14. E. Hille and R. S. Phillips: "Functional Analysis and Semigroups," Amer. Math. Soc. Colloquium Publ. 31, Providence, 1957.
15. K. Hoffman: "Banach Spaces of Analytic Functions," Prentice-Hall, Inc., Englewood Cliffs, N. J., 1962.
16. H. Kestelman: "Modern Theories of Integration," Oxford University Press, New York, 1937.
17. L. H. Loomis: "An Introduction to Abstract Harmonic Analysis," D. Van Nostrand Company, Inc., Princeton, N. J., 1953.
18. E. J. McShane: "Integration," Princeton University Press, Princeton, N. J. 1944.
19. M. A. Naimark: "Normed Rings," Erven P. Noordhoff, NV, Groningen Netherlands, 1959.
20. Z. Nehari: "Conformal Mapping," McGraw-Hill Book Company, New York 1952.
21. R. E. A. C. Paley and N. Wiener: "Fourier Transforms in the Complex Domain," Amer. Math. Soc. Colloquium Publ. 19, New York, 1934.
22. T. Rado: Subharmonic Functions, *Ergeb. Math.*, vol. 5, no. 1, Berlin, 1937.
23. C. E. Rickart: "General Theory of Banach Algebras," D. Van Nostrand Company, Inc., Princeton, N. J., 1960.

24. F. Riesz and B. Sz.-Nagy: "Leçons d'Analyse Fonctionnelle," Akadémiai Kiadó, Budapest, 1952.
25. H. L. Royden: "Real Analysis," The Macmillan Company, New York, 1963.
26. W. Rudin: "Principles of Mathematical Analysis," 3d ed., McGraw-Hill Book Company, New York, 1976.
27. W. Rudin: "Fourier Analysis on Groups," Interscience Publishers, Inc., New York, 1962.
28. S. Saks: "Theory of the Integral," 2d ed., "Monografie Matematyczne," vol. 7, Warsaw, 1937. Reprinted by Hafner Publishing Company, Inc., New York.
29. S. Saks and A. Zygmund: "Analytic Functions," "Monografie Matematyzcne," vol. 28, Warsaw, 1952.
30. G. Springer: "Introduction to Riemann Surfaces," Addison-Wesley Publishing Company, Inc., Reading, Mass., 1957.
31. E. C. Titchmarsh: "The Theory of Functions," 2d ed., Oxford University Press, Fair Lawn, N. J., 1939.
32. H. Weyl: "The Concept of a Riemann Surface," 3d ed., Addison-Wesley Publishing Company, Inc., Reading, Mass., 1964.
33. N. Wiener: "The Fourier Integral and Certain of Its Applications," Cambridge University Press, New York, 1933. Reprinted by Dover Publications, Inc., New York.
34. G. T. Whyburn: "Topological Analysis," 2d ed., Princeton University Press, Princeton, N. J., 1964.
35. J. H. Williamson: "Lebesgue Integration," Holt, Rinehart and Winston, Inc., New York, 1962.
36. A. Zygmund: "Trigonometric Series," 2d ed., Cambridge University Press, New York, 1959.

Supplementary References

37. R. B. Burckel: "An Introduction to Classical Complex Analysis," Birkhäuser Verlag, Basel, 1979.
38. P. L. Duren: "Theory of H^p Spaces," Academic Press, New York, 1970.
39. T. W. Gamelin: "Uniform Algebras," Prentice-Hall, Englewood Cliffs, N. J., 1969.
40. J. B. Garnett: "Bounded Analytic Functions," Academic Press, New York, 1981.
41. M. de Guzman: "Differentiation of Integrals in R^n," Lecture Notes in Mathematics 481, Springer-Verlag, Berlin, 1975.
42. T. Hawkins: "Lebesgue's Theory of Integration," University of Wisconsin Press, Madison, 1970.
43. H. Helson: "Harmonic Analysis," Addison-Wesley Publishing Company, Inc., Reading, Mass., 1983.
44. E. Hewitt and K. Stromberg: "Real and Abstract Analysis," Springer-Verlag, New York, 1965.
45. Y. Katznelson: "An Introduction to Harmonic Analysis," John Wiley and Sons, Inc., New York, 1968.
46. P. Koosis: "Lectures on H_p Spaces," London Math. Soc. Lecture Notes 40, Cambridge University Press, London, 1980.
47. R. Narasimhan: "Several Complex Variables," University of Chicago Press, Chicago, 1971.
48. J. C. Oxtoby: "Measure and Category," Springer-Verlag, New York, 1971.
49. W. Rudin: "Functional Analysis," McGraw-Hill Book Company, New York, 1973.
50. E. M. Stein: "Singular Integrals and Differentiability Properties of Functions," Princeton University Press, Princeton, N. J., 1970.
51. E. M. Stein and G. Weiss: "Introduction to Fourier Analysis on Euclidean Spaces," Princeton University Press, Princeton, N. J., 1971.
52. E. L. Stout: "The Theory of Uniform Algebras," Bogden and Quigley, Tarrytown-on-Hudson, 1971.
53. R. L. Wheeden and A. Zygmund: "Measure and Integral," Marcel Dekker Inc., New York, 1977.

LIST OF SPECIAL SYMBOLS AND ABBREVIATIONS†

$\exp(z)$	1	$\hat{x}(\alpha)$	82
τ	8	T	88
$\mathfrak{M}$	8	$L^p(T), C(T)$	88
χ_E	11	Z	89
lim sup	14	$\hat{f}(n)$	91
lim inf	14	c_0	104
f^+, f^-	15	$\|f\|_E$	108
$L^1(\mu)$	24	U	110
a.e.	27	$P_r(\theta - t)$	111
$\bar{E}$	35	Lip α	113
$C_c(X)$	38	$|\mu|(E)$	116
$K \prec f \prec V$	39	μ^+, μ^-	119
$\mathfrak{M}_F$	42	$\lambda \ll \mu$	120
m, m_k	51	$\lambda_1 \perp \lambda_2$	120
$\Delta(T)$	51, 150	$B(x, r)$	136
$L^1(R^k), L^1(E)$	53	$Q_r \mu$	136
$\|f\|_p, \|f\|_\infty$	65, 66	$D\mu$	136, 241
$L^p(\mu), L^p(R^k), \ell^p$	65	$M\mu$	136, 241
$L^\infty(\mu), L^\infty(R^k), \ell^\infty$	66	Mf	138
$C_0(X), C(X)$	70	AC	145
(x, y)	76	$T'(x)$	150
$\|x\|$	76	J_T	150
$x \perp y, M^\perp$	79	BV	157

† The standard set-theoretic symbols are described on pages 6 to 8 and are not listed here.

LIST OF SPECIAL SYMBOLS AND ABBREVIATIONS

E_x, E^y	161	$P[d\mu]$	240
f_x, f^y	162	Ω_α	240
$\mu \times \lambda$	164	N_α	241
$f * g$	171	M_{rad}	241
$\mu * \lambda$	175	σ	241
$\hat{f}(t)$	178	H^∞	248
C^∞	194	$f^*(e^{it})$	249
C_c^∞	194	$\varphi_\alpha(z)$	254
$D(a;r), D'(a;r), \bar{D}(a;r)$	196	S^2	266
Ω	197	$\mathscr{S}$	285
$H(\Omega)$	197	$E_p(z)$	301
γ, γ^*	200	$\log^+ t$	311
$\partial\Delta$	202	N	311
$\text{Ind}_\gamma(z)$	203	$(f_0, D_0) \sim (f_1, D_1)$	323
$Z(f)$	208	H^p	338
$\dotplus$	217	M_f, Q_f	344
$I^2 = I \times I$	222	$\sigma(x)$	357
$\partial, \bar{\partial}$	231	$\rho(x)$	360
Δ	232	$C\{M_n\}$	377
$P[f]$	233	$P(K)$	386
Π^+, Π^-	237	$C'_c(R^2)$	388

INDEX

Absolute continuity, 120
   of functions, 145
Absolute convergence, 116
Absolutely convergent Fourier series, 367
Addition formula, 1
Affine transformation, 377
Ahlfors, L. V., 402
Algebra, 356
   of measures, 175
   of sets, 10
Algebraically closed field, 213
Almost everywhere, 27
Almost periodic function, 94
Almost uniform convergence, 214
Analytic continuation, 323, 377, 379
Analytic function, 197
Annulus, 229, 264, 292
Approach region, 240
Area, 250
Area theorem, 286
Arens, R., 398
Argument, 204
Arithmetic mean, 63
Arzela-Ascoli theorem, 245
Associative law, 18
Asymptotic value, 265
Austin, D., 399
Average, 30
Axiom of choice, 396

Baire's theorem, 97
Balanced collection, 268

Ball, 9
Banach, S., 105
Banach algebra, 190, 356
Banach space, 95
Banach-Steinhaus theorem, 98
Bessel's inequality, 85, 260
Beurling, A., 335, 402
Beurling's theorem, 348
Bieberbach conjecture, 401
Blaschke product, 310, 317, 318, 338, 353
Bledsoe, W. W., 399
Borel, E., 5, 402
Borel function, 12
Borel measure, 47
Borel set, 12
Borel transform, 402
Boundary, 108, 202, 289
Bounded function, 66
Bounded linear functional, 96, 113, 127, 130
Bounded linear transformation, 96
Bounded variation, 117, 148, 157
Box, 50
Brouwer's fixed point theorem, 151

Calderón, A. P., 401, 402
Cancellation law, 19
Canonical product, 302
Cantor, D. G., 401
Cantor set, 58, 145
Carathéodory, C., 397, 400, 401
Carleson, L., 398
Carrier, 58

Cartesian product, 7, 160
Category theorem, 98
Cater, F. S., 398
Cauchy, A., 400
Cauchy formula, 207, 219, 229, 268, 341, 384, 388
Cauchy-Riemann equations, 232
Cauchy sequence, 67
Cauchy theorem, 205, 206, 207, 218
Cauchy's estimates, 213
Cell, 50
Chain, 218
Chain rule, 197
Change of variables, 153, 156
Character, 179
Characteristic function, 11
Choice function, 396
Class, 6
Closed curve, 200
Closed graph theorem, 114
Closed path, 201
Closed set, 13, 35
Closed subspace, 78
Closure, 35
Cohen, P. J., 402
Collection, 6
Commutative algebra, 356
Commutative law, 18
Compact set, 35
Complement, 7
Complete measure, 28
Complete metric space, 67
Completion:
  of measure space, 28, 168
  of metric space, 70, 74
Complex algebra, 356
Complex field, 213, 359
Complex homomorphism, 191, 362
Complex-linear functional, 105
Complex measure, 16, 116
Complex vector space, 33
Component, 197
Composite function, 7
Concentrated, 119
Conformal equivalence, 282
Conjugate exponents, 63
Conjugate function, 350
Connected set, 196
Continuity, 8, 9
Continuous function, 8
Continuous linear functional, 81, 96, 113, 127, 130
Continuous measure, 175
Continuum hypothesis, 167

Convergence:
  dominated, 26
  in L^p, 67
  in measure, 74
  monotone, 21
  almost uniform, 214
  uniform on compact subsets, 214
  weak, 245
Convex function, 61
Convex sequence, 410
Convex set, 79
Convexity theorem, 257
Convolution, 170, 175, 178
Coset, 362
Cosine, 2, 265
Countable additivity, 6, 16
Counting measure, 17
Cover, 35, 324
Covering lemma, 137, 399
Curve, 200
  with orthogonal increments, 94
Cycle, 218

Daniell, P. J., 398
de Branges, L., 401
Denjoy, A., 144
Denjoy-Carleman theorem, 380
Dense set, 58
Derivative, 135, 197
  of Fourier transform, 179
  of function of bounded variation, 157
  of integral, 141
  of measure, 136, 142, 143, 241
  symmetric, 136
  of transformation, 150
Determinant, 54
Diagonal process, 246
Dieudonné, J., 398
Differentiable transformation, 150
Differential, 150
Direct continuation, 323
Dirichlet kernel, 101
Dirichlet problem, 235
Disc, 9, 196
Discrete measure, 175
Disjoint sets, 7
Distance function, 9
Distribution function, 172
Distributive law, 18
Division algebra, 360
Dixon, J. D., 400
Domain, 7
Dominated convergence theorem, 26, 29, 180

Double integral, 165
Dual space, 108, 112, 127, 130

Eberlein, W. F., 58
Edwards, R. E., 398
Egoroff's theorem, 73
Elementary factor, 301
Elementary set, 161
Ellipse, 287
Ellis, H. W., 399
Empty set, 6
End point, 200
Entire function, 198
Equicontinuous family, 245
Equivalence classes, 67
Equivalent paths, 201
Essential range, 74
Essential singularity, 211, 267
Essential supremum, 66
Euclidean space, 34, 49
Euclid's inequality, 77
Euler's identity, 2
Exponential function, 1, 198
Exponential type, 372, 382, 383
Extended real line, 7, 9
Extension, 105
Extension theorem, 105
Extremal function, 255
Extreme point, 251

F_σ-set, 12
Factorization, 303, 338, 344
Family, 6
Fatou, P., 400
Fatou's lemma, 23, 68, 309, 344
Fatou's theorem, 249
Fejér kernel, 252
Fejér's theorem, 91
Field, 394
Finite additivity, 17
First category, 98
Fixed point, 151, 229, 247, 293, 297, 314, 318, 395
Ford, L. R., 401
Fourier coefficients, 82, 91
 of measure, 133
Fourier series, 83, 91
Fourier transform, 178
Fubini's theorem, 164, 168
Function, 7
 absolutely continuous, 145
 analytic, 197
 Borel, 13
 bounded, 66
 of bounded variation, 148

 complex, 8
 continuous, 8
 convex, 61
 entire, 198
 essentially bounded, 66
 exponential, 1
 of exponential type, 372, 383
 harmonic, 232
 holomorphic, 197
 Lebesgue integrable, 24
 left-continuous, 157
 locally integrable, 194
 lower semicontinuous, 37
 measurable, 8, 29
 meromorphic, 224
 modular, 328
 nowhere differentiable, 114
 rational, 267
 real, 8
 simple, 15
 subharmonic, 335
 summable, 24
 upper semicontinuous, 37
Function element, 323
Functional:
 bounded, 96
 on C_0, 130
 complex-linear, 105
 continuous, 96
 on Hilbert space, 81
 on L^p, 127
 linear, 33
 multiplicative, 191, 364
 positive, 34
 real-linear, 105
Functional analysis, 108
Fundamental domain, 329
Fundamental theorem of calculus, 144, 148

G_δ-set, 12
Gap series, 276, 321, 334, 354, 385
Gelfand, I., 402
Gelfand-Mazur theorem, 359
Gelfand transform, 370
Geometric mean, 63
Goursat, E., 400
Graph, 114, 174
Greatest lower bound, 7
Green's theorem, 389

Hadamard, J., 264, 321
Hahn-Banach theorem, 104, 107, 270, 313, 359
Hahn decomposition, 126
Hall, D., 398

Halmos, P. R., 398, 403
Hardy, G. H., 173, 335, 399
Hardy's inequality, 72, 177
Harmonic conjugate, 350
Harmonic function, 232
Harmonic majorant, 352
Harnack's theorem, 236, 250
Hausdorff, F., 12, 400
Hausdorff maximality theorem, 87, 107, 195, 362, 396
Hausdorff separation axiom, 36
Hausdorff space, 36
Hausdorff-Young theorem, 261
Heine-Borel theorem, 36
Helson, H., 348
Herglotz, G., 400
Hilbert cube, 92
Hilbert space, 77
   isomorphism, 86
Hoffman, K., 402
Hölder's inequality, 63, 66
Holomorphic function, 197
Homomorphism, 179, 191, 364, 402
Homotopy, 222
Hunt, R. A., 398

Ideal, 175, 305, 362
Image, 7
Independent set, 82
Index:
   of curve, 223, 230
   of cycle, 218
   of path, 203
Infimum, 7
Infinite product, 298
Initial point, 200
Inner factor, 344
Inner function, 342
Inner product, 76, 89
Inner regular set, 47
Integral, 19, 24
Integration, 19
   over cycle, 218
   of derivative, 146, 148, 149
   over measurable set, 20
   by parts, 157
   over path, 200
   with respect to complex measure, 129
Interior, 267
Interpolation, 173, 260, 304
Intersection, 6
Interval, 7
Invariant subspace, 188, 346
Inverse image, 7

Inverse mapping, 7, 215
Inversion, 280
Inversion formula, 181
Inversion theorem, 185, 186
Invertible element, 357
Isolated singularity, 210, 266
Isometry, 84
Isomorphism, 86
Iterated integral, 165

Jacobian, 150, 250
Jensen's formula, 307
Jensen's inequality, 62
Jordan, C., 5
Jordan curve, 291
Jordan decomposition, 119

Kahane, J. P., 401
Kaufman, R. P., 399, 437
Kirk, R. B., 398
Koebe function, 285, 401

Laplace equation, 232
Laplacian, 195
Laurent series, 230
Least upper bound, 7
Lebesgue, H. J., 5, 21, 397, 399
Lebesgue decomposition, 121
Lebesgue integrable function, 24
Lebesgue integral, 19
Lebesgue measurable set, 51
Lebesgue measure, 51
Lebesgue point, 138, 159, 241
Left-continuous function, 157
Left-hand limit, 157
Length, 159, 202
Limit, pointwise, 14
   in mean, 67
   of measurable functions, 14
   in measure, 74
Lindelöf's theorem, 259
Linear combination, 82
Linear fractional transformation, 279, 296
Linear independence, 82
Linear transformation, 33
Linearly ordered set, 87
Liouville's theorem, 212, 220, 359
Lipschitz condition, 113
Littlewood, J. E., 173, 399
Locally compact space, 36
Locally integrable function, 194
Logarithm, 227, 274
Lowdenslager, D., 348
Lower half plane, 237

Lower limit, 14
Lower semicontinuous function, 37
Lusin's theorem, 55

Mandelbrojt, S., 402
Mapping, 7
  continuous, 8
  one-to-one, 7
  open, 99, 214
  (*See also* Function)
Marcinkiewicz, J., 173
Maximal function, 136, 138, 241
Maximal ideal, 362, 364
Maximal orthonormal set, 85
Maximal subalgebra, 366
Maximality theorem, 87, 396
Maximum modulus theorem, 110, 212
Mean value property, 237
Measurable function, 8, 29
Measurable set, 8
Measurable space, 8
Measure, 17
  absolutely continuous, 120
  Borel, 47
  complete, 28
  complex, 16
  continuous, 175
  counting, 18
  discrete, 175
  Lebesgue, 51
  positive, 16
  real, 16
  regular, 47
  representing, 109
  σ-finite, 47
  signed, 119
  singular, 120
  translation-invariant, 51
Measure space, 16
Mergelyan's theorem, 390, 394
Meromorphic function, 224, 304
Metric, 9
Metric density, 141
Metric space, 9
Minkowski's inequality, 63, 177
Mirkil, H., 402
Mittag-Leffler theorem, 273
Modular function, 328
Modular group, 328
Monodromy theorem, 327
Monotone class, 160
Monotone convergence theorem, 21
Monotonicity, 17, 42
Morera's theorem, 208

Moschovakis, Y. N., 401
Multiplication operator, 114, 347
Multiplicative inequality, 356
Multiplicative linear functional, 364
Multiplicity of a zero, 209
Müntz-Szasz theorem, 313, 318

Natural boundary, 320, 330
Negative part, 15
Negative variation, 119
Neighborhood, 9, 35
Neumann, J. von, 122, 398, 399
Nevanlinna, R., 311
Nicely shrinking sequence, 140
Nonmeasurable set, 53, 157, 167
Nontangential approach region, 240
Nontangential limit, 243, 340
Nontangential maximal function, 241, 340
Norm, 65, 76, 95, 96
Norm-preserving extension, 106
Normal family, 281
Normed algebra, 356
Normed linear space, 95
Novinger, W. P., 398
Nowhere dense, 98
Nowhere differentiable function, 114
Null-homotopic curve, 222
Null space, 362

One-to-one mapping, 7
One-parameter family, 222, 326
Onto, 7
Open ball, 9
Open cover, 35
Open mapping theorem, 99, 214, 216
Open set, 8
Opposite path, 201
Orbit, 317
Order:
  of entire function, 315
  of pole, 210
  of zero, 209
Ordinal, 59
Ordinate set, 174, 399
Oriented interval, 202
Orthogonal projection, 80
Orthogonality, 79
Orthogonality relations, 82
Orthonormal basis, 85
Orthonormal set, 82
Ostrowski, A., 321
Outer factor, 344
Outer function, 342
Outer measure, 397

Outer regular set, 47
Overconvergence, 321

Paley-Wiener theorems, 372, 375
Parallelogram law, 80
Parameter interval, 200
Parseval's identity, 85, 91, 187, 212
Partial derivative, 231
Partial fractions, 267
Partial product, 298
Partial sum of Fourier series, 83, 91, 101, 354
Partially ordered set, 86
Partition:
   of set, 116
   of unity, 40
Path, 200
Periodic function, 2, 88, 93, 156
Perron, O., 144
Phragmen-Lindelöf method, 256
π, 3
Picard theorem, 332
Plancherel theorem, 186
Plancherel transform, 186
Pointwise limit, 14
Poisson integral, 112, 233, 240, 252
Poisson kernel, 111, 233
Poisson summation formula, 195
Polar coordinates, 175
Polar representation of measure, 125
Pole, 210, 267
Polynomial, 110
Positive linear functional, 34, 40, 109
Positive measure, 16
Positive part, 15
Positive variation, 119
Positively oriented circle, 202
Power series, 198
Pre-image, 7
Preservation of angles, 278
Prime end, 401
Principal ideal, 305
Principal part, 211
Product measure, 164
Projection, 80
Proper subset, 6
Punctured disc, 196

Quasi-analytic class, 378
Quotient algebra, 363
Quotient norm, 363
Quotient space, 363

Rademacher functions, 158
Radial limit, 239

Radial maximal function, 241
Radical, 369
Radius of convergence, 198
Radon-Nikodym derivative, 122, 140
Radon-Nikodym theorem, 121
Rado's theorem, 263
Ramey, W., 400
Range, 7
Rational function, 267, 276
Real line, 7
Real-linear functional, 105
Real measure, 16
Rectangle, 160
Reflection principle, 237
Region, 197
Regular Borel measure, 47
Regular point, 319
Removable set, 333
Removable singularity, 210
Representable by power series, 198
Representation theorems, 40, 81, 130
Representing measure, 109, 398
Residue, 224
Residue theorem, 224
Resolvent, 369
Restriction, 109
Rickert, N. W., 399
Riemann integral, 5, 34
Riemann-Lebesgue lemma, 103
Riemann mapping theorem, 283, 295
Riemann sphere, 266
Riesz, F., 34, 341, 397, 400
Riesz, M., 341, 350, 401
Riesz-Fischer theorem, 85, 91
Riesz representation theorem, 34, 40, 130
Right-hand derivative, 399
Root test, 198
Rotation, 279
Rotation-invariance, 51, 195
Rouché's theorem, 225, 229
Rubel, L. A., 401
Runge's theorem, 270, 272, 387

Saks, S., 397
Scalar, 33
Scalar product, 76
Schwartz, J. T., 399
Schwarz, H. A., 400
Schwarz inequality, 49, 63, 77
Schwarz lemma, 254
Schwarz reflection principle, 237
Second category, 98
Section, 161
Segment, 7

Separable space, 92, 247
Set, 6
    Borel, 13
    closed, 13, 35
    compact, 36
    connected, 196
    convex, 79
    dense, 58
    elementary, 161
    empty, 6
    F_σ, 12
    G_δ, 12
    inner regular, 47
    measurable, 8, 51
    nonmeasurable, 53, 157, 167
    open, 7
    outer regular, 47
    partially ordered, 86
    strictly convex, 112
    totally disconnected, 58
    totally ordered, 87
Shift operator, 347
Sierpinski, W., 167, 399
σ-algebra, 8
σ-compact set, 47
σ-finite measure, 47
σ-ring, 397
Signed measure, 119
Simple boundary point, 289
Simple function, 15
Simply connected, 222, 274
Sine, 2, 265, 316
Singer, I. M., 398, 402
Singular measure, 120
Singular point, 319
Snow, D. O., 399
Space:
    Banach, 95
    compact, 35
    complete metric, 67
    dual, 108, 112, 127, 130
    Hausdorff, 36
    Hilbert, 77
    inner product, 76
    locally compact, 36
    measurable, 8
    metric, 9
    normed linear, 95
    separable, 92
    topological, 8
    unitary, 76
    vector, 33
Span, 82
Spectral norm, 360

Spectral radius, 360
Spectrum, 357
Square root, 274
Stoilov's theorem, 400
Stout, E. L., 401
Strictly convex set, 112
Stromberg, K., 399
Subadditivity, 397
Subchain, 395
Subharmonic function, 335
Subset, 6
Subspace, 78
Sum of paths, 217
Summability method, 114
Summable function, 24
Supremum, 7
Supremum norm, 70
Support, 38, 58
Symmetric derivative, 136
Szasz, O., 401

Tauberian theorem, 402
Taylor's formula, 379
Thorin, G. O., 401
Three-circle theorem, 264
Tietze's extension theorem, 389
Topological space, 8
Topology, 8
Total variation, 117, 148
Totally disconnected set, 58
Totally ordered set, 87
Tower, 395
Transcendental number, 170
Transformation:
    affine, 377
    bounded linear, 96
    differentiable, 150
    linear, 33
    linear fractional, 279, 296
    (*See also* Function)
Transitivity, 324
Translate:
    of function, 182
    of set, 50
Translation-invariance, 51
Translation-invariant measure, 51
Translation-invariant subspace, 188
Triangle, 202
Triangle inequality, 9, 49, 77
Trigonometric polynomial, 88
Trigonometric system, 89

Ullrich, D., 400
Uniform boundedness principle, 98

Uniform continuity, 51
Uniform integrability, 133
Union, 6
Unit, 357
Unit ball, 96
Unit circle, 2
Unit disc, 110
Unit mass, 17
Unit vector, 96
Unitary space, 76
Upper half plane, 237
Upper limit, 14
Upper semicontinuous function, 37
Urysohn's lemma, 39

Vanish at infinity, 70
Varberg, D. E., 399
Vector space, 33
Villani, A., 398
Vitali-Carathéodory theorem, 56
Vitali's theorem, 133
Volume, 50
von Neumann, J., 122

Walker, P. L., 399
Weak convergence, 245, 246, 251
Weak L^1, 138
Weierstrass, K., 301, 332
Weierstrass approximation theorem, 312, 387
Weierstrass factorization theorem, 303
Wermer, J., 402
Wiener, N., 367
Winding number, 204

Young, W. H., 5, 400

Zermelo, E., 403
Zero set, 209
Zorn's lemma, 87
Zygmund, A., 401